Introduction to Bioanalytical Sensors

TECHNIQUES IN ANALYTICAL CHEMISTRY SERIES

BAKER · CAPILLARY ELECTROPHORESIS

CUNNINGHAM · INTRODUCTION TO BIOANALYTICAL SENSORS

LACOURSE · PULSED ELECTROCHEMICAL DETECTORS IN HPLC

MCNAIR AND MILLER · BASIC GAS CHROMATOGRAPHY

MIRABELLA · MODERN TECHNIQUES IN APPLIED MOLECULAR SPECTROSCOPY

STEWART and EBEL · CHEMICAL MEASUREMENTS IN BIOLOGICAL SYSTEMS

TAYLOR · SUPERCRITICAL FLUID EXTRACTION

Introduction to Bioanalytical Sensors

ALICE J. CUNNINGHAM
Professor Emerita
Agnes Scott College
Decatur, Georgia

A Wiley-Interscience Publication
JOHN WILEY & SONS, INC
New York • Chichester • Weinheim • Brisbane • Singapore • Toronto

This book is printed on acid-free paper. ♾

Published simultaneously in Canada.

Library of Congress Cataloging-in-Publication Data:

Cunningham, Alice J., 1937–

Introduction to bioanalytical sensors / Alice J. Cunningham.

p. cm. — (Techniques in analytical chemistry series)

"A Wiley-Interscience publication."

Includes bibliographical references and index.

ISBN 0–471–11861–3 (cloth : alk. paper)

1. Biosensors. I. Title. II. Series.

R857.B54C86 1998 97–38100

610′.28–dc21 CIP

Printed in the United States of America.

10 9 8 7 6 5 4 3 2 1

Contents

Preface

Modern research problems are frequently so multidimensional that it is difficult to delineate each of the contributing scientific subdisciplines. One of the best examples of this multidimensionality lies in research dealing with bioanalytical sensors. Potential applications cut across the analytical landscape from the environment to the brain. Various biosensing devices have been designed utilizing almost every type of basic instrumental technique for measurement. Each analytical problem requiring a specific type of biosensor is unique, yet there are integrating principles of design and operation that bring some degree of cohesion to the field as a whole.

While broadly based research problems present endless opportunities and exciting scientific challenges, they also impose an extraordinary need for maintaining currency across a wide spectrum of developments in the field. The dynamic tension between deciding which experiments to perform and knowing what others are doing is, however, the very basis of the collaborative nature of science. Cross-disciplinary problems simply magnify the challenge of the collaboration while enriching the rewards of accomplishment. For someone just embarking on research dealing with bioanalytical sensors, this dual driving force should make the choice of being a scientist seem like just the right thing at a time when that confirmation is needed most!

Borrowing from the vernacular of research supervisors, "to be up and running fairly quickly" is the goal when starting a new project. That is no small task when embarking on a cross-disciplinary research problem. Obviously, someone who has used the formal educational experience primarily for *learning to learn* will be at great advantage. By combining this introductory monograph with selections from the copious references that person should be able to be up and running reasonably well. At least, the learner should attain fairly quickly the status of a conversant, contributing member of a research team.

It is always a challenge for an author to produce an effective "printed teacher" for the self-learner. So much is lost when the one-on-one interactions of a classroom are omitted from the learning process. In the absence of customized adjustments for individual readers, it is essential that there be a well-defined starting point in content. Keeping this in mind, I have prepared this book for those who have, at least, an undergraduate degree in chemistry or a chemistry-related field that requires some advanced chemistry courses. While some may argue about the clarity of that boundary condition, I have assumed that the reader knows fundamentals of college chemistry and biochemistry, as well as undergraduate physics, math and electronics. In the modern context of undergraduate degree requirements the assumed background for chemistry includes syntheses and mechanisms, chemical measurements and physical chemistry, both theoretical and experimental.

For those who need review of fundamental principles there are references for further background reading. In some cases this will involve little more than revisiting the textbooks of undergraduate courses. Many of those texts, as well as topical monographs, are referred to in this book. The expansiveness of the scientific literature in the burgeoning field of biosensing strategies is reflected in the extensive references provided for further study and scientific exploration. Those references range from multi-volume treatises to hundreds of individual reports in the original literature. The incredible redundancy in the literature of biosensors attests not only to the intense activity in the field, but also to the rapidity with which new developments are occurring. The maturation and self-realization of a cross-disciplinary field seems to require a certain amount of thrashing around in each of the contributing disciplines before settling into a comfortable niche of its own. Perhaps electronic research reports will accelerate that process.

In a cross-disciplinary field such as bioanalytical sensors it is tempting to produce one more treatise to add to the tomes which already exist. In fact, the very existence of those other references precludes the necessity for following that path. A newcomer to a particular research area needs to understand how the basic principles and techniques of chemistry pertain to that particular area of investigation. Further, there must be an operative comprehension of current developments, presented within a framework of contextual history on which current and future research might be based. Accordingly, I have attempted to integrate topics related to current developments with the basic science and fundamental knowledge on which those developments depend. I have not tried to write a history of the development of bioanalytical sensors, nor have I tried to produce an all encompassing treatment of every type of biosensor for every conceivable application.

In keeping with the purpose of this series in analytical chemistry, this book should provide a succinct, foundational introduction to the topic of bioanalytical sensors. It is my hope that after perusal of this volume a novice in the area will be comfortable with the basic principles and terminologies, understand how we got to our present point in the evolution of the field and feel confident about embarking upon a new intellectual adventure as a potential contributor to the field.

We feel ourselves hurtling toward greater understanding of increasingly complex processes and diminishing scales of matter, time and energy. The dynamics of progress, *per se*, make writing an introduction such as this very difficult. Just as you think you have reached a time when new findings have coalesced to enable an accurate overview of the field, some new breakthrough expands the boundaries of our understandings and accomplishments, thus suggesting the next new experiments. I will have succeeded if some of my readers are the ones who design those next new experiments. Enjoy!

ACKNOWLEDGMENTS

During preparation of this book more patience than I have deserved has been exhibited by family and friends. That patience is greatly appreciated. The scheduling conflicts and aborted activities in order to finish "the book" have seemed unending. Through it all, usually with good humor, Mary Alice Clower has been particularly helpful by proof-reading, organizing data and references and providing much needed support and encouragement.

Neta Counts of Computer & Information Technology Consultants, Stone Mountain, GA, has been invaluable in producing the illustrations. She may have even learned a little chemistry along the way as we worked through the concepts to be depicted.

Finally, since the beginning of this project Betty Sun of John Wiley & Sons provided continued guidance and encouragement. She also exhibited extreme patience with incredible civility as deadline after deadline was missed.

Even with the able assistance of Mac, Patty and Roz, I have undoubtedly missed some errors in the manuscript, references or data compilations. In anticipation of that likelihood, I offer apologies in advance. I have decided that omissions and oversights are unavoidable pitfalls of using extensive references and trying to cover multiple aspects of complex problems.

Stone Mountain GA ALICE J. CUNNINGHAM

Selected Acronyms and Abbreviations for Biosensing Devices and Methodologies

ANN	artificial neural network
APM	acoustic plate mode
AT	(XYl) –35° cut of quartz
ATR	attenuated total reflectance
AW	acoustic wave
BOD	biological oxygen demand
CCD	charge-coupled device
CE	capillary electrophoresis
CHEMFET	chemically modified field-effect transistor
CL	chemiluminescence
CPE	carbon paste electrode
CWE	coated-wire electrode
D	diffusion coefficient
DR	dynamic range
E	potential of an electrode versus a reference
E°	standard potential of a redox couple versus normal hydrogen electrode at standard conditions

$E^{\circ\prime}$	formal potential, usually at pH = 7, versus a reference
ECIA	electrochemical immunoassay
ECL	electrogenerated chemiluminescence
ENFET	enzyme-modified field-effect transistor
ET	enzyme thermistor
FET	field-effect transistor
FIA	flow-injection analysis
FOCS (or FOB)	fiber-optic chemical sensor (or biosensor)
FOM	figure of merit
GCE	glassy carbon electrode
GSAM	generalized standard addition method
HPLC	high-performance liquid chromatography
IDE	interdigitated electrode
IDT	interdigitated transducer
IGFET	insulated-gate field-effect transistor
IMER	immobilized enzyme reactor
IOW	integrated optical waveguide
ISE	ion-selective electrode
ISFET	ion-selective field-effect transistor
K_m	Michaelis constant for an enzyme-substrate combination
K_{SV}	Stern–Volmer fluorescence quenching constant
LAPS	light-addressable potentiometric sensor
L-B film	Langmuir-Blodgett film
LED	light-emitting diode
LIF	laser-induced fluorescence
LOD	limit of determination
MD	microdialysis
MIP	molecularly imprinted polymer
NAL	nucleic acid ligand
NAS	net analyte signal
N–E equation	Nicolsky–Eisenman equation
NEE	nanoelectrode ensemble
NIR	near-infrared
NFO	near-field optics
NHE	normal hydrogen electrode
NSOM	near-field scanning optical microscopy
NSOS	near-field scanning optical spectroscopy
OPEE	organic-phase enzyme electrode
OTF	organized thin film

PCR	1) principal components regression or 2) polymerase chain reaction
pI	pH at which a protein has no net charge
PLS	partial least squares
QCM	quartz crystal microbalance
REFET	reference field-effect transistor
RIfS	reflectometric interference spectroscopy
RVC	reticulated vitreous carbon
SAM	1) standard addition method or 2) self-assembled monolayer
SAW	surface acoustic wave
SCE	saturated calomel electrode
SECM	scanning electrochemical microscopy
SEL	selectivity
SEN	sensitivity
SH-APM or SH-SAW	shear-horizontal acoustic-wave device(s)
SPR	surface plasmon resonance
STW	surface transverse wave (refers to acoustic-wave devices)
μ-TAS	micro-total analytical system
TEP	thermal enzyme probe
TIRF	total internal reflection fluorescence
TSM	thickness shear mode (refers to acoustic-wave devices, QCMs)
UME	ultramicroelectrode
V	volts, stated versus a reference
V_g	voltage applied to the gate-metal of an insulated-gate field-effect transistor, IGFET
V_{max}	maximum reaction rate for an enzymatic reaction

Introduction to Bioanalytical Sensors

1. Biosensors and Bioanalytical Challenges

Accurate measurement of biologically related substances has been a major goal in analytical sciences throughout the twentieth century. Bioanalytical problems are among the most challenging due to the variety of substances in biological samples, the complex molecular structures and time-dependent concentrations. Most of the original biochemical protocols, many of which are still in use, require isolation of components from natural environments and separations of complex mixtures. In some cases model compounds and artificially prepared solutions are used as an alternative to coping with complex structures and mixtures. Oftentimes a method, such as microbiological assay, designed for acquisition of information about a physiological process, is not easily translated into a quantitative method for determination of individual analytes.

It was the need for more detailed information about analytes in naturally occurring matrices that led to the conception and development of bioanalytical sensors, referred to simply as biosensors. Biosensors were envisioned originally as discrete analytical devices that would measure analytes selectively, often in a natural matrix, without prior separation of multicomponent samples. The high degree of selectivity required for such measurements was to be based on biological molecular recognition processes occurring via immobilized biocomponents on ordinary analytical detectors. For example, an enzymatic preparation attached to an electrode should enable selective monitoring of a substrate of that enzyme. Antibodies immobilized on an optical surface should provide a signal indicative of interactions with complementary antigens of the antibody.

From a purely physicochemical point of view biosensors were to do even more than provide selective detection. They were to:

- eliminate or simplify sample separation steps
- provide new options for high selectivity in direct measurements
- achieve better accessibility for challenging measurements within the natural location or environment
- respond in a continuous, reversible manner
- accomplish a measurement with minimum perturbation of the sample
- be compatible with the environment in which the measurement is made

Added to this list were other goals of convenience and economics such as miniaturization, portability, low cost of mass production, and ease of use.

There are many types of effective biosensors in use today. Rarely, however, does one find an individual device that meets all the goals above. On the other hand, as operational principles and performance characteristics are understood more thoroughly, experimentally sound options for designs and sensing strategies have expanded significantly beyond the originally envisioned bounds. Actually, as will be discussed later in this chapter, recent technological advances in analytical instrumentation, surface modifications and newer methods for extracting information from data have prompted a re-examination of the analytical role for biosensors in the future. The multiple options for sensor designs, plus additional sampling and measurement capabilities made possible by miniaturization and integration of components, have changed perspectives with regard to applications ranging from inexpensive, disposable devices to sophisticated coupling of biosensors with microseparation methods.

1.1. A HISTORICAL PERSPECTIVE

Since the 1950's there have been extensive developments in analytical measurement techniques, new approaches to difficult separations and elucidation of mechanisms of biochemical reactions. These advances have improved enormously bioanalytical capabilities and have led to astounding new knowledge of natural biochemical sequences. One of the major bioanalytical outcomes of research in the mid-twentieth century was the introduction of bioanalytical probes for measuring biochemical species in natural matrices. Clark's group (Clark *et al.*, 1958; Clark and Lyons, 1962) designed electroanalytical devices with expanded capabilities for bioanalytical measurements, specifically oxygen and glucose in tissues and glucose in blood samples.

The first of Clark's devices consisted of a platinum electrode protected by a membrane which excluded proteins but was permeable to oxygen. An amperometric signal resulting from the electrochemical reduction of oxygen at the electrode was monitored:

$$O_2 + 2H_2O + 4e^- \rightarrow 4OH^-$$

The magnitude of the cathodic current served as an indicator of the amount of oxygen. Later the same general approach was modified for glucose determinations by interposing between the sample and electrode a catalytic layer containing the soluble enzyme, glucose oxidase (GOx). Glucose oxidase catalyzed the reaction between glucose and oxygen:

$$\begin{array}{l} \text{Glucose} + O_2 \xrightarrow[\text{Oxidase}]{\text{Glucose}} H_2O_2 + \text{Gluconolactone} \\ \qquad\qquad\qquad\qquad\qquad\qquad\qquad \Downarrow\!\Uparrow \\ \qquad\qquad\qquad\qquad\qquad\qquad\quad \text{Gluconate}^- + H^+ \end{array}$$

This device, depicted in Figure 1-1, provided indirect measurement of glucose. An O_2 electrode measured depletion of O_2 , thus indicating the extent of the enzyme-catalyzed reaction, which depends on the amount of glucose in the sample. Alternatively, the anodic process for oxidation of H_2O_2 could be monitored:

$$H_2O_2 \rightarrow O_2 + 2H^+ + 2e^-$$

Those early amperometric probes are still viewed as prototypes for many modern devices. Commercially available versions of Clark-type oxygen electrodes are widely used. Clark (1993) recently provided a historical recounting of the development of his oxygen and glucose sensing devices, and included sketches prepared for the patent applications. His scientific tale offers an interesting perspective on the whole matter of bioanalytical sensors from then to now.

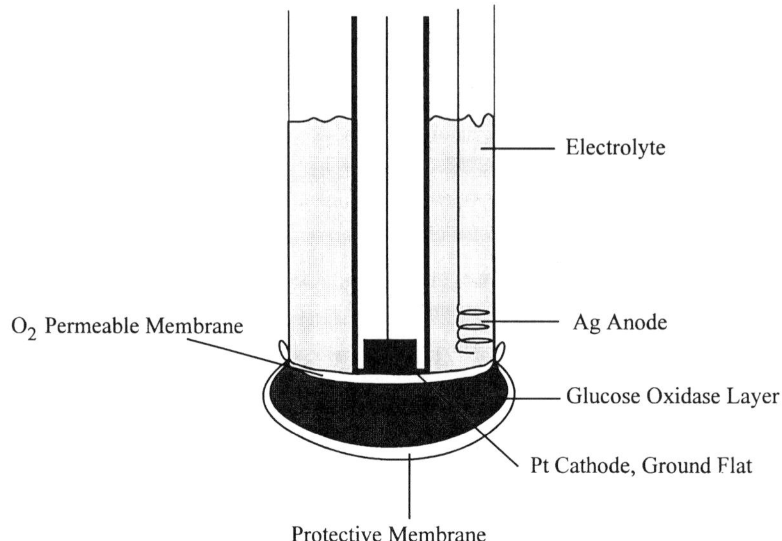

FIG. 1-1. Clark-type oxygen electrode. Sensing is based on the reduction of oxygen which is a cosubstrate for the enzyme, glucose oxidase, contained in solution behind the protective membrane.

In 1967 Updike and Hicks introduced their version of a glucose electrode with the GOx immobilized in a polyacrylamide gel on the surface of the O_2 electrode. This design for using an immobilized biological agent rather than a soluble species necessitated re-examination of the mechanism by which the catalyzed reactions occur—i.e., immobilized agents behave differently than soluble species. The amperometric detection mode was the same as Clark's, but the use of an affixed matrix on the transducer surface opened up many possibilities for a more durable reaction layer with more effective coupling to the energy transducer.

In the 1970s the general approach of incorporating enzymatically catalyzed reactions into other detection schemes was expanded to include thermal and optical devices. These approaches expanded chemical options to enzymatically-catalyzed reactions other than oxidation–reduction processes—e.g., enzymatically catalyzed hydrolyses, group transfers, etc. Danielsson and Mosbach (1989) and Seitz (1988) have provided reviews and references for much of the early work in thermometric and optical sensors, respectively.

The thermal enzyme probe (TEP) was simply a glass-encased thermistor overlaid with an enzymatic layer to provide selective detection based on heats of reaction (Carr and Bowers, 1980; Danielsson and Mosbach, 1989). The molar heats of reaction for some of the common enzymatic reactions are on the order of 20–250 kJ mol^{-1}.

			Approximate $-\Delta H$(kJ mol^{-1})
Glucose + O_2	$\xrightarrow[\text{Oxidase}]{\text{Glucose}}$	Gluconolactone + H_2O_2	80
H_2O_2	$\xrightarrow{\text{Catalase}}$	$\frac{1}{2}O_2 + H_2O$	100
Urea + H_2O	$\xrightarrow{\text{Urease}}$	CO_2 + $2NH_3$	60
Uric acid	$\xrightarrow[\text{Oxidase}]{\text{Urate}}$	Allantoin + CO_2 + H_2O_2	49

As the heat of the enzymatic reaction evolved, the thermistor's temperature-dependent resistance change could be monitored through use of a Wheatstone bridge arrangement. Temperature changes in millidegree ranges could be observed. The TEP was moderately successful, but heat loss problems, unacceptable for an adiabatic measurement, were common.

The enzyme thermistor (ET) was found to circumvent some of those problems (Mosbach and Danielsson, 1974; Mosbach *et al.*, 1975). The ET was designed as a flow system with small reaction columns containing immobilized enzyme(s) in the reactor. A sample with analyte, which is a substrate for the enzyme(s), flows through the reactor bed to effect the catalyzed reaction. A measuring thermistor was inserted, usually at the outlet of the column. Temperature measurements in this mode were more efficient than with the TEP, thus improving sensitivity. In this thermal mode it was also possible to enhance sensitivity via signal amplification by coupling enzymatic reactions:

			$-\Delta H$
Glucose + O_2	$\xrightarrow{\text{GOx}}$	Gluconolactone + H_2O_2	80
H_2O_2	$\xrightarrow{\text{Catalase}}$	$\frac{1}{2}O_2 + H_2O$	100
Glucose + $\frac{1}{2}O_2$	$\longrightarrow$	Gluconolactone + H_2O	180

Though the ET design did not conform exactly to the format of the other biosensing devices—i.e., no immobilized component on the detector—the use of immobilized enzyme reactors (IMER) with outlet detection has continued to be highly applicable. Much of the technology for that approach has been developed in parallel with affinity chromatography, which also involves immobilization of components on a variety of stationary supports.

The ET with its associated flow control through the reaction columns was one of the first examples of a biosensing system, which will be distinguished from discrete biosensors in later discussions. In addition to numerous updated thermometric sensing systems, that experimental design has been adopted very effectively for other detection modes, as will be illustrated in subsequent discussions.

Also in the 1970's Lübbers and Opitz (1975) developed an optical analog of the electrochemical and thermal devices. Their fiber-optic device was designated as an *optode,* based on the Greek word for optical way. The optical device was designed with immobilized indicators for monitoring CO_2 via pH changes within the reaction layer and O_2 by fluorescence quenching. Although the immobilized components were not of biological origin, thus would not qualify as a biosensor according to some definitions, the device was used in biological fluids and is considered the forerunner of today's fiber-optic biosensors.

The evolution of biosensing devices has been an interesting process. The basic sensor designs mentioned above can be recognized easily in some present-day devices, though newer devices are considered to be more sophisticated as a result of newer optical materials and technologies for immobilizing components on detectors. What may seem to be minor variations and progress since mid-century actually represent years of effort in some cases. A good example lies in the comparison of the original Clark glucose electrode with more recent miniaturized versions.

To say that many glucose sensors have been based on Clark's original design would be an understatement. One such device is Jung and Wilson's (1996) miniaturized amperometric glucose sensor for subcutaneous monitoring of glucose. Hydrogen peroxide, produced in the GOx-catalyzed reaction of glucose and oxygen, is measured amperometrically just as Clark did. In the Jung and Wilson version, however, Clark's Pt disk has been replaced by a Pt–Ir wire with a sensing area of only $\sim 1\ mm^2$. Membranes for control of response characteristics and fouling have been updated to newer polymeric materials, a 3-mercaptopropylsilane inner layer and a polyurethane outer membrane. Performance, of course, has been upgraded by orders of magnitude through these seemingly minor changes.

In the thermal realm there has been a transition to resistive films, ultramicrobead thermistors or thermopiles for miniaturized versions of the TEP and ET analogs (Bataillard *et al.*, 1993; Xie *et al.*, 1994a,b, 1995; Shimohigoshi *et al.*, 1995; Shimohigoshi and Karube, 1996). Most of these sensors are still used in flow system configurations, but miniaturization and new thermally sensitive materials have made possible thermal arrays which amplify signals and circumvent earlier problems of signal degradation due to heat losses.

The updating to new materials and technologies is also reflected in more modern optical approaches to biosensing. Optical fibers of various types and sizes have been developed for a vast assortment of analytes and applications. Using a very clever photopolymerization method for immobilizations, Walt's group pioneered the use of optical fiber bundles for combining sensing and imaging (Pantano and Walt, 1995) and creating optical sensing arrays (White *et al.*, 1996; Healy and Walt, 1997). While the fundamental approach of immobilizing pH and oxygen-sensitive indicators on optical fibers still pertains in some of their applications, the sophisticated photopolymerization techniques are utilized for site-specific thin layers of optically sensitive reagents on single fibers or individual fibers of imaging bundles. Kopelman's group has translated the basic immobilization technique to the production of nanoptodes for near-field spectroscopic measurements, some of which are used for intracellular measurements (Tan and Kopelman, 1996).

Currently the dynamic revolution occurring in instrumentation and analytical systems is having a great impact on the whole bioanalytical scene. New materials and technologies are providing a new generation of very sophisticated analytical devices which should be invaluable in biochemical analysis. Integration of components and miniaturization are significant characteristics of this new generation of instruments (Bard, 1994). As a result, there are numerous reports of reliable portable or implantable devices that have been designed and commercialized in many cases. Biosensors, as discrete units or as integral components of miniaturized analytical systems, will continue to be important devices for applying newer strategies in instrumental analysis.

As a result of the miniaturization–integration trend, capabilities with regard to portability, modes of sampling and temporal and spatial resolution have expanded considerably. Methods for extracting more information from data and measuring extremely low concentrations or masses, often on short time-scales or in minute volumes, have effected analytical limits few envisioned 10–15 years ago. Measurements in single cells, inside heart muscles, and in spatially resolved areas of the brain or on active surfaces are now being made. Pollutants that threaten the ecological balance of a biologically diverse population are being measured on-site at diminishingly low concentrations, thus facilitating immediate remediation. Intricate biological processes fundamental to biotechnology industries are being monitored and controlled with high precision.

It would be difficult to find an area of modern science and technology more exemplary of multidisciplinary efforts than that of developing biosensors. Research teams and collaborative groups are usually comprised of individual scientists whose educational specialities encompass a variety of traditional areas in physical and biological sciences, as well as engineering. Analytical chemists, biochemists, bioengineers, physicists, environmental scientists, medical clinicians, material scientists, chemometricians and electronics specialists are among those who have special expertise to contribute in meeting the evolving goals for bioanalytical methods.

Research and development related to biosensors provides an excellent example of applied science driven in part by the marketplace. As analytical sciences and technologies have matured, industries and governments have become more acutely aware of potential uses of biosensors in healthcare, environmental monitoring and industrial process control. Active markets emerge faster than researchers can meet the challenges of commercial devices for human needs. Reducing healthcare costs and implementing environmental regulations are only two of the factors to be considered. McCann (1989), Hall (1991), Sethi (1994), Rogers (1995) and Weetall (1996) have discussed economic factors and market forces. A global biosensor market in the billions of US dollars has been projected by some. Weetall (1996) provides an updated perspective based on considerations of limited niches for unique applications and newer options among alternative analytical methodologies. While market forces constitute only a part of driving forces for developments in biosensors, the magnitude of the potential market for these devices reflects the scope of needs and applications.

Some additional comment is in order about the economic driving forces for developing biosensors. In the early literature there was considerable emphasis on eventual commercialization of essentially all biosensing devices. Design criteria almost always included the goals for all such devices to be self-contained, reagentless, inexpensive and usable by individuals with relatively little or no prior training—e.g., at-home or in-the-field use. Certainly, at-home and in-the-office devices, such as glucose monitors and other rapid diagnostic tests, have been invaluable, as have some of the field-tested environmental monitoring devices.

Particularly with the advent of miniaturized, multianalyte devices, one can only assume that the market for these mass-produced, reagentless, easy-to-use devices will continue to swell. Alvarez-Icaza and Bilitewski (1993) have summarized some of the techological considerations involved in mass production. If a biosensor is to be mass produced for commercialization, the economic and usage criteria are of paramount importance. If the device is designed for use by specially trained personnel doing multi-sample measurements or by researchers for in-the-laboratory studies, however, priorities among the criteria may be different. It is not uncommon to find a research report of a biosensing device that functions quite well for its intended purpose, yet is not amenable to commercialization and mass production. That distinction should be borne in mind as the various types of biosensors are discussed.

1.2. BIOANALYTICAL SENSORS AND SYSTEMS

What is a biosensor? What is the difference between a biosensor and a biosensing system? What are some current variations on the original concept of a biosensor? These are pertinent questions, though not always easy to answer any way other than by example and experience. There have been continuing arguments within the scientific community about usual definitions and distinctions. Lines become even more blurred when advances such as integrated arrays and miniaturized flow and separation systems are considered. In many circles there has been total abandonment of efforts toward definition. In the most general terms a *bioanalytical sensor* is an analytical device modified specifically for chemical selectivity in measurements related to solving bioanalytical problems. *Biosensor* may be thought of as an acceptable shortened form. The terms will be used interchangeably in this book. An examination of some definitions provided in the literature reveals the difficulties encountered in reaching semantic agreement.

1.2.1. Distinctions and Dilemmas

An acceptable definition for biosensors should encompass all types of sensors and sensing configurations. That is not an easy task. Definitions should distinguish sensors from ordinary instrumental detectors, which do not necessarily exhibit biochemical selectivity, and from threshhold monitors, which function solely as alarm devices. Further, there has been an effort to set apart *bio*sensors as a particular type of chemical sensor, distinguishable from those used for non-bioanalytical applications such as monitoring automobile exhausts or manufactured chemicals unrelated to biochemical processes.

In order to appreciate some the distinctions prompting diversity of nomenclature, one must examine some of the dilemmas arising from real problems. Is a pH-sensing device a biosensor if it is used in urine but not if it is used in river water? When is an electroanalytical mercury-sensing device a biosensor as opposed to being merely an ordinary mercury-selective electrode? Is a sensor for neurotransmitters always considered a biosensor, even if the device does not include as part of its structure a molecular recognition component of biological origin? If additional components, such as flow controllers or micro-separation devices, are coupled to a bioselective detector, does that disqualify the whole assembly as a biosensor? All of these questions emphasize the need for clarification and operational standardization of terminologies when discussing biosensors.

1.2.2. Attempts To Define

There have been almost as many operational definitions of the basic term biosensors as there have been authors on the subject:

> ". . . small-sized devices comprising a recognition element, a transduction element, and a signal processor capable of continuously and reversibly reporting a chemical concentration" (Wolfbeis, 1990)
>
> ". . . device [that] incorporates a biological sensing element in close proximity or integrated with the signal transducer. . . " (Hall, 1991)
>
> ". . . biosensors comprise an analyte selective interface in close proximity or integrated with a transducer, whose function it is to relay the interaction between the surface and analyte either directly or through a chemical mediator. . ." (Hall, 1992)
>
> ". . . A biosensor can be described as an analytical detector (typically an electrode or fiber optic transducer) whose selectivity is enhanced by immobilizing a sensitive and selective biological element (typically an enzyme) within close proximity of the sensor" (Pantano and Kuhr,1995)

In 1991 the International Union of Pure and Applied Chemistry (IUPAC) published definitions and classifications of chemical sensors, the broad genus under which they include biosensors as one type:

> "A chemical sensor is a device that transforms chemical information, ranging from the concentration of a specific sample component to total composition analysis, into an analytically useful signal. The chemical information, mentioned above, may originate from a chemical reaction of the analyte or from a physical property of the system" (IUPAC, 1991)

The IUPAC report goes on to say that chemical sensors contain two essential "functional units: a receptor part and a transducer part. . . The receptor part of chemical sensors may be based upon various principles . . . [including] biochemical, in which a biochemical process is the source of the analytical signal. . ."

In 1996 the Physical Chemistry and Analytical Chemistry Divisions of IUPAC proposed their own definition applicable to electrochemical biosensors:

". . . A biosensor is a self-contained integrated device that is capable of providing specific quantitative or semi-quantitative analytical information using a biological recognition element (biochemical receptor) which is in direct spatial contact with a transduction element" (IUPAC, 1996)

That group pointed out the desirability of distinguishing a biosensor from (1) a bioanalytical system, which "requires additional processing steps," [meaning flow control,reagent injection, etc.] and (2) a bioprobe, which they described as a single-use device incapable of monitoring on a continuous basis.

1.2.3. Toward More Integrated Systems

The various definitions of biosensors reflect the actual evolution of modern devices. There is common agreement that there are two essential components, an energy transducer and a molecularly selective chemical component. There is also agreement that this physicochemical tandem design is for the purpose of imparting to the energy transducer an enhanced chemical selectivity toward analytes.

This basic design of a generic biosensor is illustrated in Figure 1-2. A generalized energy transducer, modified with a chemically selective reaction layer applied to the surface and an exclusionary or protective membrane, is depicted. Analytes and interferences approach the assembly by any one of several mass transport processes—e.g., stirring, flowing stream, diffusion. Diffusion only is illustrated in Figure 1-2 for the purposes of contrasting diffusion properties in solution and in the membrane and for emphasizing the various diffusional pathways that must be considered. For this particular sensor selective permeation of the membrane by analyte only is implied. Membrane-related reactions according to chemically selective mechansims, typically those shown at the bottom of the figure, are also depicted.

Operational essentials for the biosensing process are included in the illustrated sequence of Figure 1-2. Chemical transduction occurs on or within the film or membrane. Selectivity of the device arises primarily from differing permeabilities and reaction chemistries of the analyte and interferences. Coupled to the chemical transduction step is signal generation which is often based on creation of a detectable product of the reaction sequence(s). This indirect mode of detection in which a species other than the target analyte is actually measured is common to many types of sensors.

According to earlier definitions the chemically selectively component of a biosensor was to be of biological origin—an enzyme, an immunochemical species, cells, or tissues. The biological molecular recognition of enzymes for substrates, antibodies for antigens and cell receptors for triggering chemicals was invoked as the epitome of chemical selectivity. But not all molecular recognition agents in newer devices are of biological origin. Other forms of

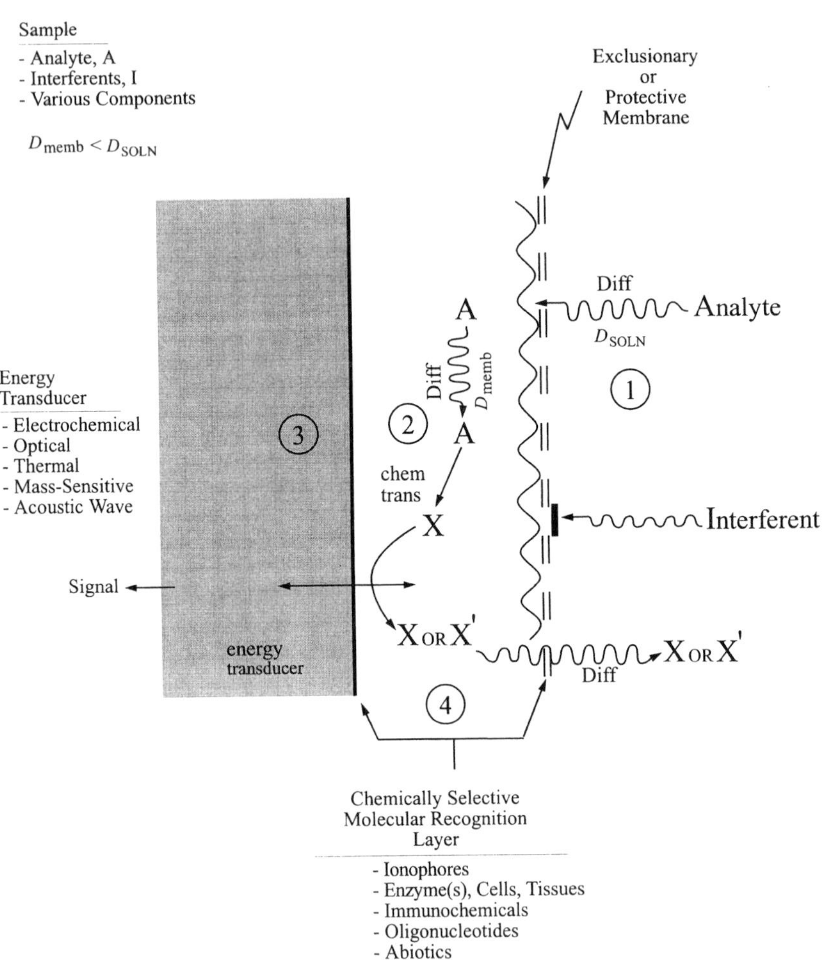

FIG. 1-2. General scheme of biosensor configurations. *D*'s denote diffusion constants in the solution and chemically selective membrane.

molecular recognition have been found to convey equally important selective mechanisms. For example, synthetically produced host–guest molecules such as crown ethers, cryptands, calixarenes, and molecularly imprinted polymers are abiotic—i.e., not of biological origin. Further, the future use of molecularly engineered species is bound to increase, thus providing even more options for altered biotics or abiotics that are effective selective agents. The 1996 IUPAC recommendation broadened the possibilities by their wording: "The biological recognition element may be based on a chemical reaction catalyzed by, or on an equilibrium reaction with, macromolecules that have been isolated, engineered or are present in their original biological environment."

There is potential argument about the use of the term *receptor* in definitions of sensors. Often the term has been used by chemists to refer to the chemically selective components of all chemical sensors, including biosensors. The established usage and importance of the term in its usual meaning within cell biology leaves room for possible confusion. The advisability of using interchangeably the terms *molecular recognition* element and *biochemical receptor* is questionable. Cellular receptors are quite specialized entities of Nature. Cellular receptors are molecular recognition components, but not all molecular recognition processes involve biologically defined cell receptors.

Zare's use of single cells and arrays of cells *as* biosensors, via the natural cells' receptors, points out quite clearly the potential for confusion in this respect (Shear *et al.*, 1995; Fishman, *et al.* 1996). Reiken *et al.* (1996) have also reconstituted actual nicotinic acetylcholine receptors in bilayer liquid membranes to study effects of bispecific antibodies on activation/inactivation processes involving neuronal receptors. With increasing use of biological cellular receptors for sensing and studying biochemical processes, it is probably wise to avoid the use of the term receptor when discussing generally the chemical selectivity components of biosensors unless actual cellular receptors are being utilized. Chemically (or biochemically) selective agents, biological recognition components, molecular selectivity components or molecular recognition agents are just a few of the options that would preserve the use of the term receptor in its original biological connotation.

The distinction between a biosensor and a bioanalytical system, mentioned in the previous section, has remained a problem over the years. The Enzyme Thermistor discussed earlier was actually designed in a flow-cell configuration with enzymatically catalyzed reactions occurring in the column preceding a thermal detector. Consequently, the ET was often excluded as a "pure" biosensor because it did not conform to the strictest definition of a surface-modified energy transducer. Yet, that sensing configuration is increasingly adopted with a variety of detection modes, and has some definite analytical advantages for selectivity and multicomponent analyses.

In spite of some very clever molecular recognition schemes over the years, adequate selectivity of discrete sensors has been difficult to attain in many cases. Modern microseparation devices have enabled integrated biosensing systems which can embody both functions, separation and bioselective detection. The chip-sized μ-TAS (micro-*t*otal *a*nalytical *s*ystem) represents a major development in this regard (Effenhauser *et al.*, 1993, 1994; Fan and

Harrison, 1994; Jacobson *et al.*, 1994a–e; Fluri *et al.*, 1996; Jacobson and Ramsey, 1996; Roberts *et al.*, 1997; Woolley *et al.*, 1997). Additionally, biosensors may be incorporated as the detection unit for flow-injection analyses (FIA), microdialysis (MD) and capillary electrophoresis (CE). For example, Niwa *et al.* (1996) have introduced a glutamate sensing system using a microdialysis probe for sampling from cells, an enzymatic reactor column for oxidizing glutamate on-line, and an enzyme-modified electrode for electrochemical detection of the hydrogen peroxide produced in the enzymatic reaction. This design incorporates elements for intracellular sampling and microseparation, the IMER approach to chemical transduction, and a biosensing detector that conforms to the more conventional definition of a biosensor. Obviously, this biosensing strategy requires an expanded perspective for defining bioanalytical sensors.

Cosford and Kuhr (1996) have also reported a fluorescence-based glutamate sensor utilizing glutamate dehydrogenase immobilized inside a capillary. Reduced nicotinamide cofactor (NADH), produced by the enzymatic reaction inside the capillary, is measured by fluorescence emitted through the capillary wall. The Cosford and Kuhr biosensor was designed for eventual coupling as a bioselective detector in microdialysis or capillary electrophoresis.

The preceding examples, together with others discussed in subsequent sections, illustrate how newer technologies for integrating separation with bioselective detection have enhanced analytical capabilities beyond those possible with basic, discrete biosensors. This approach can be expected to increase significantly in the future as various automated and micromachined separation devices become more commonplace (Robinson and Justice, 1991; Effenhauser *et al.*, 1993, 1994; Fan and Harrison, 1994; Jacobson,1994 a–e, 1996; Fluri, 1996; Liang *et al.*, 1996; Schneiderheinze and Hogan, 1996; Swerdlow *et al.*, 1997).

Rather than asking the usual question about where to draw the line between sensors and systems, the more pertinent question at this time may be: when should a given approach be adopted? This is particularly true if the analytical problem at hand can be solved more efficiently and accurately by a more complex integrated device. For all practical purposes future options will range from using a biosensor, discrete or in array format, without prior separation to a compound device that includes separation capability coupled to a biosensor as the selective detector. Consternation over exactly whose definition is appropriate pales in the light of what can be accomplished by a proper analytical choice.

The point to be emphasized by the previous examples is that after almost forty years of research and development there are now optional *sensing strategies*, which probably should be the focal point of future designations. That is essentially the approach reflected in the 1996 IUPAC recommendations. Bioprobes, biosensors and bioanalytical systems are distinguished according to the sensing strategy—(a) single use, disposable; (b) stand-alone device which might be used for continuous monitoring; and (c) a systems approach in which the sensor's function complements other processing steps, such as time-resolved sampling, on-line separation, flow control, etc.

Considering newer sensing strategies and technologies, descriptive terminologies rather than steadfast definitions are useful, perhaps even necessary. For experienced researchers in the field the intention is usually clarified by the bioanalytical problem and designs for particular applications. For the novice an understanding of the working terminologies and distinctions should prove useful.

1.3. A WORKING VOCABULARY

In addition to the descriptions and distinctions of the previous section, there are other terminologies that form a working vocabulary of sensing strategies. Some of the terms are familiar to those in any scientific discipline and need only to be put into the context of bioanalytical problems. In other cases there are commonly encountered descriptors and variations that arise from cross-disciplinary differences or non-universal language origins. This section summarizes some of the additional terminologies that will used freely in later sections.

1.3.1. Transduction Modes and Classifications

A novice confronting the literature of biosensors for the first time finds that much of the research is presented according to various classification schemes. Some writers refer to biosensors in terms of the energy transduction mode—e.g., potentiometric, chemiluminescent, thermal or piezoelectric. Others might speak of their biosensor as an enzymatically based device or an immunosensor. Sometimes the sensors are referred to simply in terms of the analyte, such as a glucose sensor or a pesticide biosensor. Compound descriptors, such as amperometric dopamine sensor or a fiber-optic sensor for penicillin, may also be encountered. This diversity of nomenclature creates a nightmarish situation with regard to literature searching strategies and lengths of titles. The various "monikers" assigned are, however, usually fairly descriptive of the device being reported.

The word, transducer, comes from the Latin word *transducere* which means to lead across. The analytical function of a transducer is to lead across chemical information about analyte(s) so it can be processed as an electrical signal for data acquisition and interpretation. For example, a phototube transduces light into an electrical signal which can be captured by a recorder, meter or computer. This would represent an **energy transduction** process typical of any instrumental method. For an optically based biosensor a chemically selective molecular recognition layer might produce a luminescent species from a multicomponent sample, then transmit the emitted light via an optical fiber to a phototube for conversion to an electrical signal. The reaction layer has merely converted by chemical means (**chemical transduction**) an ordinary detector into a more selective detector.

The total transducing function of a biosensor is accomplished by a designed combination of **chemical transduction** and **energy transduction**. Those distinguishing elements of a whole process are frequently delineated when discussing theory and principles of biosensors, but work seamlessly in tandem for generation of a biosensor's response.

1.3.1.1. Energy Transduction Modes

Various energy transducing principles, essentially the same ones operative in conventional instrumentation, are utilized in biosensors. These major modes are shown in Figure 1-3. This abbreviated outline obscures the fact that

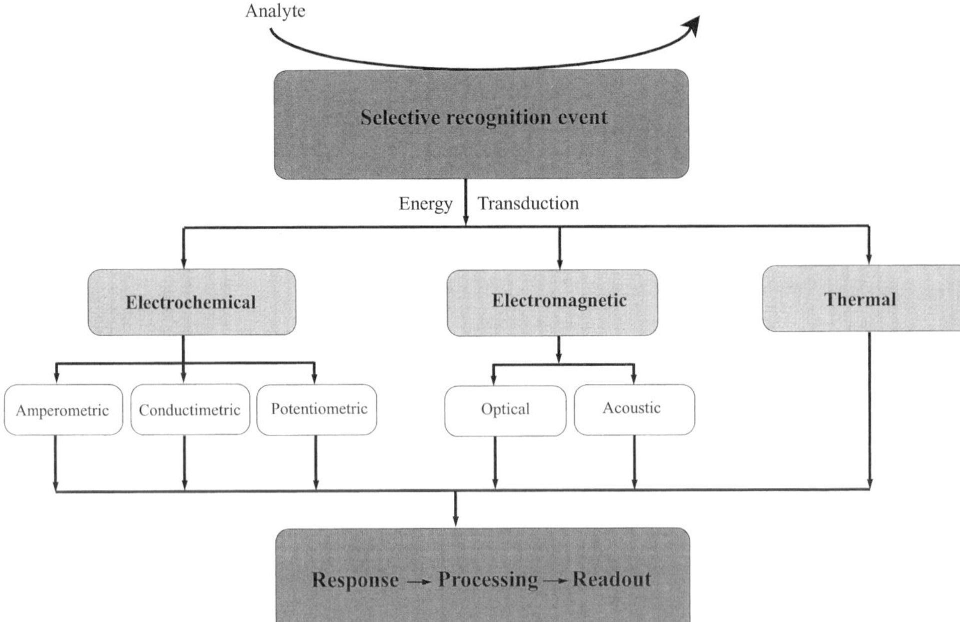

FIG. 1-3. Sequence of process(es) for biosensing operation. Chemical transduction occurs in the shaded porti labeled Selective Recognition Event. Various modes of energy transduction are indicated. In the lower porti signal capture for output is depicted.

there are multiple options within each group. For example, only three of the more common modes for electrochemical sensors are shown in Figure 1-3, yet there are many more that depend primarily on interfacial properties affected by chemical modification layers. Optical modes include the familiar absorbance and fluorescence approaches, but there is also a whole series of optical detection schemes based on internal reflections. Finally, while many practicing chemists are quite familiar with the quartz crystal microbalance as a useful mass-sensitive device, there are several other acoustic-wave devices that provide better sensitivity in some situations and potentially more information about the sensing layers. Table 1-1 extends the categories to illustrate more of the sensing modes within each major category.

1.3.1.2. Chemical Transduction Options

Four of the most useful chemical transduction schemes involve ion-selective sensing, enzymatically catalyzed reactions, immunochemical binding, and the binding complementarity of oligonucleotides. Some argue that ion-selective sensing does not belong in categories for biosensors, but when coupled to enzymatically catalyzed processes or used for determining ion ratios in living cells, ion-selective detection becomes very important as a biosensing strategy.

TABLE 1-1 Extended Categorization of Energy Transduction Modes

Electrochemical	Acoustic waves
Amperometric	Acoustic plate mode
Conductimetric	Flexural plate mode
Impedimetric	Surface acoustic wave
Potentiometric	Surface transverse wave
Potentiometric stripping analysis	Thickness shear mode (quartz crystal microbalance)
Optical	Thermal
Absorbance	Adiabatic
Chemiluminescence	Heat conduction
Electrogenerated chemiluminescence	
Fluorescence	Other
Fluorescence lifetime	Biomagnetic
Fiber optic waveguides	Light addressable potentiometric
Near-field microscopy and spectroscopy	Scanning electrochemical microscopy
Near-infrared	Separation/detection systems
Planar waveguides	
Surface enhanced raman	
Surface plasmon resonance	

As indicated in Section 1.2.3 it is often not the original analyte of interest that is actually measured by the detector, but a reaction layer product, co-reactant or molecular label. In describing enzymatically catalyzed sensors, Clark has used the term "transformate" for a product that is actually measured at the detector (Clark, 1989). Others use "measurand" as the generic designation of the detected species, which may or may not be the target analyte. In Chapter 2 several examples of repeatedly encountered measurands will be given. For example, hydrogen peroxide occurs as a product in many of the oxidation–reduction reactions catalyzed by a certain class of enzymes, the oxidoreductases. The flavin ($FAD/FADH_2$) and nicotinamide ($NAD^+/NADH$) enzymatic cofactors may also be the detectable species in a great many enzymatically catalyzed oxidation–reduction reactions. Protons and ammonia are commonly detected as products of enzymatically catalyzed hydrolysis reactions used for chemical transduction. Oxygen, either as the dissolved gas as analyte or as a co-reactant, is often measured by amperometric detection or fluorescence quenching of metal complexes immobilized in a sensor's reaction layer(s). Thus, there is a need to develop substance-specific sensors such as those for H_2O_2, NADH, H^+, NH_3 or O_2 even when they are not the target analytes. Variations of these general sensors can then be customized for specific applications in different sensing environments.

Arnold and Meyerhoff (1988) discussed the pros and cons of various classification schemes for biosensors. They adopted a general categorization based on the nature of the chemical transduction mode. According to their classification, based on chemical transduction, biosensors are either **biocatalytic** or **bioaffinity** types, regardless of the energy transduction mode. The former designation refers to chemical transduction based on enzymatically catalyzed processes utilizing isolated enzymes, cells or tissues. Chemical transduction of the second type, bioaffinity, involves ligand binding phenomena classically illustrated by antibody–antigen interactions, but also applicable to other biological or chemical processes that depend on ligand binding. The Arnold–Meyerhoff designation has proven to be one of the most useful classifications and has persisted in the literature as a popular framework for discussing biosensors. Figure 1-4 outlines usual molecular recognition components, including categorizations according to the Arnold–Meyerhoff scheme.

1.3.1.3. Approaches to Immobilization

In order to couple appropriately the chemical and energy transduction steps leading to a biosensing response, chemically selective components must be associated intimately with an energy transducer. A survey of the most common methods of immobilization of chemically selective agents is the topic of

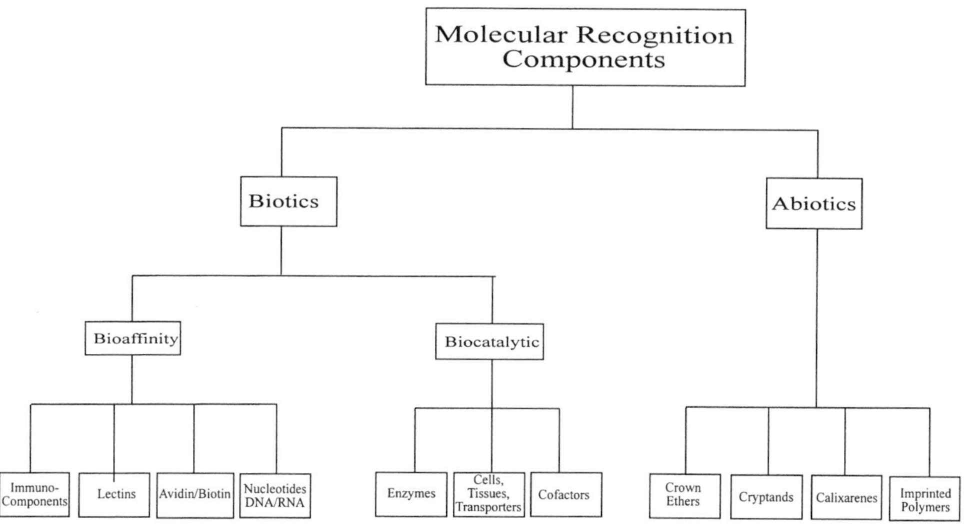

FIG. 1-4. Major categories of molecular recognition components used for selectivity in biosensors.

Chapter 3. The general discussion of this section merely provides a minimal vocabulary for discussing immobilized species and the basic approaches to immobilization.

In the literature of biosensors there are established methods of immobilization that have been utilized for almost thirty years. They are usually discussed in terms of adsorption, entrapment, cross-linking, and covalent bonding. During the past decade more controlled molecular design has been effected with newer immobilization and coupling chemistries, though most fit into the four major general strategies. Specifics of all modification processes depend on the exact nature of the energy transducers' surfaces as well as the immobilized species' properties. Immobilization strategies are often broadly applicable, however, with little variation in the basic chemistry of the method.

It is instructive to consider the general direction and trends in development of newer methods for immobilization. Since the early days of designing biosensors, biochemical components for chemical selectivity have often been isolated from the natural sources, purified and affixed by relatively simple methods for containment or attachment to a transducer surface. A variety of membranes for permeability, protection or additional selectivity have also been used to build the selection and reaction layers. The membranes might be available in pre-cast forms, but those applied to the transducer surface by various dip, drop or spin-casting techniques have been more common. The sensors produced early on using this type of design sequence were usually manually produced in limited quantities for investigating a specific bioanalytical problem, elucidating the operational mechanism and nuances of prototypical devices, or testing a new concept in bioselective sensing. Control of thicknesses and reproducibility of the relatively thick reaction layers sometimes proved to be an experimental problem.

More efficient, controlled methods for designing bioanalytical sensors have resulted from development of new technologies for miniaturization, better knowledge of fundamental behavior of chemically modified transducers and new methods for creating thin layers and highly structured molecular architectures at surfaces. Oriented binding and patterned surfaces have resulted from many of the advances. While there are still needs for manually produced, one-of-a-kind sensors for certain applications, there is a notable trend and driving force toward applying the accumulated knowledge of modified surfaces to designs for miniaturized devices which can be mass produced with carefully defined reaction layers of controlled dimensions.

An examination of the evolution of immobilization methods illustrates some of the trends mentioned above. Isolated, soluble enzymes, cells and tissues may be held to a surface by simple **containment** behind a permeable, protective membrane such as dialysis tubing or microporous polymers. This is an arrangment reminiscent of Clark's original electrodes, and may be as simple as using an O-ring to attach the containment membrane to the sensor body. Soluble enzymes used in this manner should exhibit the same enzyme kinetics widely reported from conventional enzymatic assays.

Cells and tissues with their native enzymes intact may also be immobilized by a containment approach. It is usually much more difficult to know or characterize the kinetic behavior of intact enzymes of cells and tissues. The intact enzymes often exhibit a greater efficiency of chemical transduction, presumably because they have undergone less alteration during acquisition and preparation. Yet, selectivity problems may arise from the presence of competing reactions due to multiple enzymes in the preparations. Arnold and Rechnitz (1989) and Karube (1989a) have provided numerous examples to demonstrate these differences between soluble and intact enzymes of cells and tissues.

Adsorption of enzymes and antibodies has also been used for a variety of sensor designs. Typical of proteins in general, these macromolecules adsorb on many of the metal and metal oxide surfaces used for transducers, as well as on carbon and glasses. In fact, before biosensors were designed specifically using adsorbed materials for chemical transduction, the macromolecular adsorption phenomenon was often considered a nuisance to be avoided in bioanalytical procedures. Spurious electrochemical behavior and residual optical signals due to adsorption of proteins complicated experimentation considerably until it was realized that the whole phenomenon could be used to good advantage! Not only is purposeful adsorption now used for creating some chemical transduction layers, but the behavior of surface-adsorbed species is understood far more thoroughly, particularly in the realm of electrochemical and optical methods. Thus, controlled, monitored adsorption can be an effective immobilization technique.

As will be discussed in Chapter 3, the simplicity of direct surface adsorption of macromolecules for immobilization must be weighed against disadvantages arising from problems related to ease of subsequent desorption and lack of definition of the layer's structure. The problem of non-specific adsorption of unwanted species is also still a problem, just as it was in earlier days. It is also important to note that immobilization by simple adsorption does not normally promote molecular orientation of the macromolecules for optimal efficiency in chemical transduction.

With increasing use of flowing samples through capillaries for modern separation methods, it has been necessary to re-visit the whole problem of preventing unwanted protein adsorptions. Several approaches for preventing non-specific adsorption have been introduced. Hydrophobic areas may be created on a sensor's surface (e.g., see Bhatia *et al.*, 1993). Chiem and Harrison (1997) have summarized the major approaches to adsorption prevention and have chosen to use zwitterionic buffers containing surface-active agents. Córdova *et al.* (1997) studied the effectiveness of polycationic polymeric coatings on the inside of capillaries to inhibit adsorption of positively charged proteins. Their relatively thorough study emphasized the importance of electrostatic interactions in protein adsorption, and demonstrated that adsorption is a function of both the isoelectric point and molecular weight of the protein. Their findings, while related directly to CE studies, have wide-ranging applicability in all areas of designing and optimizing biosensors of various types.

In spite of what has been said above about problems arising from direct macromolecular adsorption with little molecular-level control, the increasing use of spontaneously adsorbed species to produce **organized thin films (OTF)** has fostered a revolution in abilities to control and design interfaces (Zhong and Porter, 1995; Finklea, 1996; Sackmann,1996). These reports provide good overviews of advances and possibilities that arise particularly from the use of self-assembled monolayers (SAMs) of organic thiols on gold (Au–S–R–X) and Langmuir–Blodgett films. The Finklea chapter is quite thorough and comprehensive in discusssing SAMs with emphasis on electrochemical applications.

In the case of the Au–S–R–X monolayers, all of the above authors discuss how the interfacial properties and selectivity can be manipulated by the nature of the end group, X. By altering that portion of the film, fine tuning of chemical selectivity by hydrophobic/hydrophilic, electrostatic or complexation interactions may be achieved. The ramifications of this for designing selective biosensing surfaces are far reaching. Fortunately, much of the same chemistry that has been developed for coupling biomolecules to surfaces, as discussed in Chapter 3, works for modifying the SAMs' end groups. Thus, intricately designed molecular architectures for thin films are achievable. Even oriented binding of macromolecules can be effected by manipulation of the end-group chemistry (Sigal *et al.*, 1996).

Entrapment and **cross-linking** of biochemically selective components and auxiliary reactants in polymeric networks are significant immobilization strategies. The polymeric matrices range from some of the gels used traditionally as biological media to a variety of organic polymers produced on the transducer surface by electrochemical, photochemical or plasma polymerization. Polymeric networks may serve strictly as a structural framework in which the biocomponents are enmeshed, thereby reducing leakage and desorption of reactants. In other cases the polymeric materials may serve additional functions, such as selective ion permeability, enhanced electrochemical conductivity, or mediation of electron transfer processes. As in the case of immobilization by adsorption, orientation of the macromolecules is not one of the usual outcomes of immobilization by polymeric entrapment.

Sol–gel technology has enabled extension of the entrapment principle to silicate networks that have some advantageous characteristics for immobilizing biocomponents (Dave *et al.*, 1994; Klein, 1994; Tsionsky *et al.*, 1994; Lev *et al.*, 1995). The matrix preparation is conducted under experimental conditions that are milder than some of the polymerizations mentioned previously. Thus, the integrity of the biocomponents may be preserved to a higher degree. Also, the sol–gel matrices seem to exhibit good mass transport and molecular-access properties, particularly for electrochemical and optical transduction modes (Wang, R. *et al.*,1993, 1995; Dave *et al.*, 1994; Jordan *et al.*, 1995; Petit-Dominguez *et al.*, 1997).

In addition to the above methods for immobilization, **covalent attachment** of reaction layer components is a well-developed approach, particularly as applied to proteins. Much of the technology for covalent immobilizations was derived from or developed in parallel to affinity chromatography applications. Chapter 3 includes numerous references to reviews and monographs dealing with the subject of covalent binding of macromolecules to surfaces and to tethering groups extending outward from surfaces. Because enzymes, antibodies, carbohydrates and oligonucleotides have class-specific molecular characteristics, general approaches for each class have been devised through years of progress in surface modifications. As an example, proteins are normally bound through involvement of the amino, carboxyl, sulfhydryl or aromatic side chains of the amino acids in the macromolecule. Coupling reagents targeted for each of these groups have been developed and well tested. Several of the most commonly used covalent methods have been summarized by Weetall (1993). For molecules such as smaller organics and metal complexes, surface modification methods must be customized sometimes to individual species and functional groups.

On a practical basis covalent binding is often more laborious and tedious than some of the immobilization methods discussed earlier. In light of that fact, it is reasonable to ask the question, "why is covalent binding of chemical transduction agents sometimes preferable to other methods?" The answer is twofold. Compared to adsorption and some entrapment techniques covalent binding often produces a better designed surface structure that enhances the amount of reagent on the surface. Second, covalent binding can produce a more durable surface modification. Covalently attached components do not leach as badly from a reaction layer as some entrapped species do, and the layer tends to exhibit a longer-term stability under more strenuous experimental conditions, such as exposure to flowing samples, stirring, washing, etc.

Having examined the basic concepts of containment, adsorption, polymeric entrapment, and covalent attachment, several experimental challenges in immobilization techniques become more obvious. Because enzymes and antibodies are so widely used for biocatalytic and bioaffinity sensors, the major points for considerations can be framed in terms of those species. The overall chemical transduction efficiency will be dependent on several factors which are controlled to a large degree by the immobilization process. Newer, more effective immobilization methods have addressed those factors.

1. Not only is the amount of the biocomponent important, as indicated usually by a loading factor or fractional surface coverage, but also the energy transducer must remain accessible for signal generation and transmission. This means that an optimal coverage should have the highest loading of the selective biomolecule without inducing molecular "crowding" or blocking the energy transducer.
2. In cases where mass transport must be controlled or manipulated for optimized signal generation, it is often desirable to produce the thinnest layers possible which exhibit the necessary molecular architecture for the intended biomolecular interactions of the sensing mechanism.
3. Assuming that optimal loading for chemical and energy transduction can be achieved, the next consideration is that of the detectable species' access to the molecular recognition agent. For example, an enzyme's reactive site may be obscured either by immobilization using a side chain near that site or by the crowding phenomenon mentioned above. An antibody also has specific regions for binding antigens. Immobilization methods that randomize the attachment of antibody molecules will decrease the efficiency of the antigen–antibody interaction being detected by the sensor. Thus, not only is the amount of

reactive component important on the surface, but also the orientation necessary for the intended biomolecular interactions.

4. With the advent of microscopic methods for spatial resolution and the increasing use of analytical sensing arrays, patterning of surfaces and deposition of reactants at microsites have taken on new importance. Being able to create microdomains for specific biomolecular interactions opens up amazing new possibilities in biosensing. Nowhere is this more apparent than in the use of high density arrays of oligonucleotides that have revolutionized genetic analyses (Hunkapiller *et al.*, 1991; Pease *et al.*, 1994; Beattie *et al.*, 1995; Blanchard *et al.*, 1996; Gette and Kreine, 1997). Other applications of patterning are equally important, however. In the realm of sensing arrays, individual units may be only a few μm^2 or nm^2 in area. The ability to deposit chemically selective agents within the confined areas is an essential factor in the fabrication of such devices.

Many of the newer surface modification techniques have been developed with the four major factors above as the challenges: optimal loading of selective components, properly structured thin films, proper orientation for efficient biomolecular binding, and patterned surfaces for well-designed miniaturization. Solving these problems has been the goal of much of the work in self-assembled monolayers (SAMs), light-directed synthesis, photolithography, micromachining, ink-jet deposition and microscopic patterning of surfaces. Examples of all of these approaches are included where appropriate in the discussions of subsequent chapters, as well as the last section of this chapter.

1.3.1.4. Basic Designs of Discrete Sensors

Having introduced the basic terminologies of transduction processes and immobilization methods for chemical transduction, general designs of sensors can be examined. Some selected types of useful sensors are illustrated in Figure 1-5. In order to illustrate pertinent features of each type, there has been no attempt to scale either the internal parts of the sensors or the relative sizes among the types shown.

At the top of the figure two versions of electrochemical sensors are illustrated. In (a) a typical carbon-fiber tip is shown with the usual glass sheathing that is created by automated pipet-pulling devices. The shaded material represents a conductive epoxy material; electrical connection would be made from the top of the assembly through the open end of the pipet. Typical diameters for such ultramicroelectrodes might range from 5μm to 10 μm with a glass sheathing thickness of ~1-2 μm. Wightman and Wipf (1989), Heinze (1993) and Zoski (1996) have discussed these types of electrodes quite thoroughly. The electrode in Figure 1-5(b) is a different type

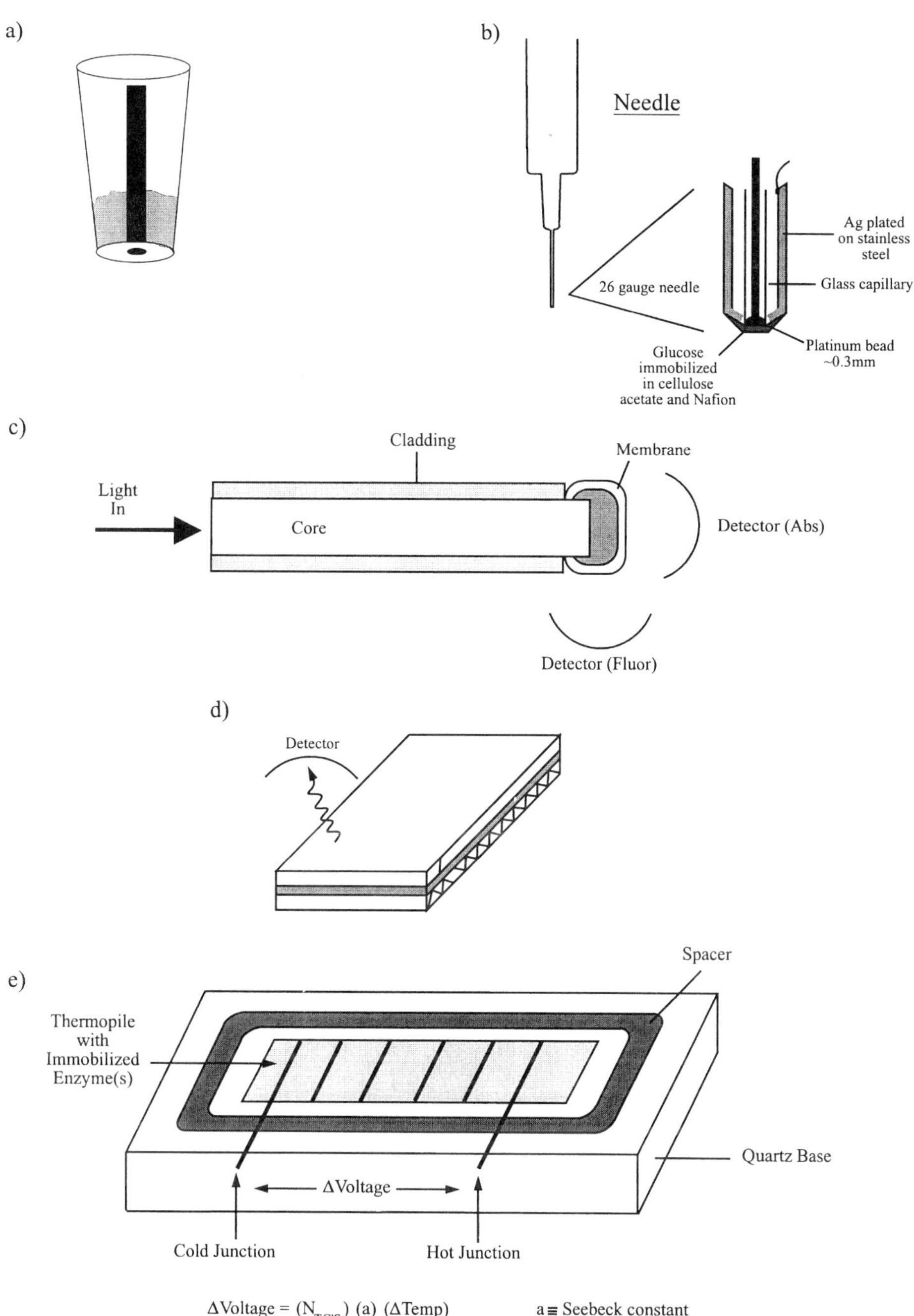

FIG. 1-5. Examples of biosensors: (a) carbon-fiber electrode, (b) needle type which might be used for implantation in tissues, (c) fiber-optic sensor with a distal-end placement of bioselective agent, (d) planar waveguide, and (e) thermopile.

designed for tissue implantation for monitoring glucose. The resemblance to the original Clark-type electrode is obvious, even in this miniaturized form.

A generic fiber-optic chemical sensor (FOCS) is illustrated in Figure 1-5(c). The design shown is referred to as a distal-end type, indicating that the immobilized reagents are deposited on the tip of the fiber. The sensing film might include pH-sensitive chromophores that would be detected as a function of proton production in the chemical transduction step, or the sensing layer could be immobilized antibodies that would bind an antigen tagged with a fluorescent label. Light is directed in and either an absorbance or fluorescence signal may be captured, as illustrated.

There are many other formats for designing optical sensors, as will be discussed in Chapter 5. Several variations of optical sensors utilize the interaction of an immobilized species with the evanescent wave resulting from internally reflected light in a fiber or waveguide. For this format using the fiber of Figure 1-5(c), cladding would be removed from a portion of the fiber to accommodate an immobilized reaction layer along the wall of the fiber. The principles of internal reflections and evanescent waves are included in Chapter 5. Suffice it to say at this point that the evanescent wave is an extension of radiative intensity emanating from the points of internal reflection. The device depicted in Figure 1-5(d) is a planar waveguide also based on internally reflected light, as shown in the lower optical flat of the drawing. The immobilized species is represented by the shaded region. Fluorescence, induced by interaction of the recognition layer with the evanescent wave, is depicted by the detection of emitted light.

The device shown in Figure 1-5(e) is a thermal unit based on a thermopile, which is essentially a series of thermocouple pairs. The component illustrated would probably be the sensing unit of a flow cell through which analyte would flow. If the analyte is a substrate for the immobilized enzyme, the heat of reaction would be detected by heat conduction to the thermopile. The quantity of heat would then relate to the analyte concentration.

1.3.1.5. Translating to Arrays

Energy transduction, chemical transduction, immobilizations, and basic designs have been introduced in terms of discrete sensors, i.e., individual units that may be of varying sizes and geometries. A higher level of integration using the same sensing principles, but more complex devices, is represented by **sensing arrays**. The translation of the basic principles to these more powerful devices is a significant aspect of current research. The term array implies a regular order or arrangement of discrete units. Most undergraduates have probably had experience with arrays of diodes or charge-coupled devices as photodetectors. Electrochemical arrays are also used in some systems as chromatographic detectors. Numerous analytical advantages accrue from the conversion of discrete sensors to array formats. An especially important advantage relates to resolution of responses from partially selective sensors, thus facilitating multicomponent analysis.

In the biosensing world arrays are utilized for several different purposes. Three of these general applications will be described below. Any of the usual energy transduction modes—electrochemical, optical, thermal, or acoustic waves—can be fabricated in an array format (e.g., see Diamond, 1993; Menon and Martin, 1995; Xie *et al.*, 1995; Zellers *et al.*, 1995; Healy and Walt, 1997; Sutter and Jurs, 1997). Furthermore, combinations of sensing modes, such as potentiometric and amperometric (Goldberg *et al.*, 1994; Silber *et al.*, 1996), within an array can be accommodated. Microfabrication techniques have revolutionized the capabilities of producing arrays, thus accelerating their utility. Arrays of transducers can now be designed with overall dimensions in the mm^2 range or less. Individual elements of the array can have dimensions on the order of nanometers to micrometers, depending on the microfabrication method. The present discussion is directed to the general purposes for which arrays might be utilized.

As a first general example of a useful array, consider multiple, identical microtransducers—e.g., electrodes or thermistors—configured in a defined geometric arrangement. Because the elements of the array are identical, two possibilites exist for effects on the signal. The signal will be either enhanced (amplified) beyond that attainable with a single unit or the signal may be spatially resolved within the limits of the array's dimensions. These advantages can be of paramount importance when working with trace and ultratrace concentrations or in confined areas of cells and tissues.

For the second example, consider that each element in the array described above is now chemically modified by addition of a molecular recognition layer, but with different molecular recognition agents for different members of the array. A third of the elements may be made selective for an ion M^+. Another third of the array is modified to sense selectively O_2 in the sample, and the remaining third is coated with an enzyme which shows a high degree of specificity for X. The array has now become more than a device for signal enhancement or spatial resolution. It is now a multicomponent sensor that can measure the three analytes M^+, O_2 and X simultaneously, perhaps with spatial resolution. When this general design is coupled to the power of modern chemometric techniques for extraction of information the possibilities for solving some very difficult analytical problems are enormous.

Finally, an important variation of the previous example has vast implications for studies in immunochemistry and nucleic acid interactions. The ability to "build" chemical arrays of different macromolecules on a surface has resulted from new technologies for creating well-defined chemical patterns. Photolithographic techniques with or without patterned masks (Kovacs *et al.*, 1996; Dontha *et al.*, 1997), light-directed synthesis (Pease *et al.*, 1994; Chee *et al.*, 1996) and ink-jet technology (Blanchard *et al.*, 1996; Lemmo *et al.*, 1997) are widely used methods. Using micropatterned reaction sites it is possible to detect antigens to a variety of antibodies (Pritchard *et al.*, 1995a,b; Sundberg *et al.*, 1995) or to determine complementary nucleotide sequences (Fodor *et al.*, 1991; Pease *et al.*, 1994; Chee *et al.*, 1996; Gette and Kreiner, 1997).

1.3.2. Chemical Selectivity and Molecular Recognition

Molecular recognition is a term adopted from biological chemistry, but sometimes used too loosely by chemists describing chemical transduction processes on which selectivity is based. Selectivity is a much broader term than molecular recognition and may be achieved by chemical processes other than molecular recognition, as defined by the biochemists. In a biochemical sense, molecular recognition pertains to very specific interactions implying both bonding complementarity and biological function (see Alberts *et al.*, 1989 and references therein). Structural complementarity results in multiple weak bonds—e.g., hydrogen, ionic, and van der Waals—which create bonding amplification evidenced by overall strong binding. Thus, a substrate binds strongly to an active site of an enzyme. An antigen binds strongly to the appropriate portions of a complementary antibody. A cellular receptor derives its biological function from molecular recognition of a specific chemical trigger for the receptor's response, such as channel opening, exocytosis, or neuronal transmission.

For many biosensor designs, biological molecular recognition is utilized in the chemical transduction step. Immobilized biotics—e.g., enzymes, antibodies, cells, and tissues—in those devices actually exhibit their biological function derived from molecular recognition through binding specificity. In other sensors, particularly some of those with selectivity based on abiotic chemical transduction, molecular recognition may not necessarily form the actual basis of selectivity. Partially selective metal–ligand binding, partitioning between phases, and size or charge exclusion are examples of selectivity mechanisms not based on true molecular recognition. An interesting discussion with regard to some acoustic-wave results has arisen recently in the literature. Grate (1996) proposed that an acoustic-wave response mechanism for organic vapors may be based purely on partitioning into polymeric layers (vapor sorption) rather than on an actual molecular recognition event involving binding by immobilized cavitand molecules in the layers, as reported by Schierbaum *et al.* (1994). Dickert *et al.* (1996, 1997) have reported a substantial amount of data in support of the host–guest inclusion mechanism.

1.3.3. Dimensional Considerations

As a point of reference for relative sizes, consider the usual pH probe, an ordinary macroscopic glass electrode. A typical probe is approximately 0.5 inch (~13 mm) in diameter and ~5.5 inches (138 mm) in length. This might be the pH probe chosen for an undergraduate's measurements in a beaker in an introductory laboratory. If one must work with smaller volumes or in differently shaped containers, such as a volumetric flask, a smaller probe would be more appropriate. In this case long narrow probes are commercially available. One popular device is ~ 0.16 inch (< 5 mm) in diameter and of various lengths.

Even those reduced dimensions, which some vendors advertise as microdimensions, are considerably oversized for measurements which must be made in a blood vessel, the brain, tissue slices, or an individual cell. For these examples of Nature's small volumes truly miniaturized devices, perhaps ultramicroelectrodes or nanoptodes, must be used. Ultramicroelectrodes are < 20 μm in diameter while nanodes are < 1 μm. These are sizes comparable to hairs or even less. Relative dimensions pertinent to bioanalytical strategies are illustrated in Figure 1-6.

Throughout the literature there are references to the descriptors, macro-, micro-, submicro-, ultramicro-, and nano-. The prefixes , micro and nano, should create no problem. These have the usual meanings of SI units, base unit times 10^{-6} and 10^{-9} , respectively. The modifiers to the prefixes are sometimes not so clearly defined. For example, terminologies regarding electrodes have arisen from both the physiology and electrochemistry communities. For many years physiologists used 20–25 μm electrodes, which they referred to as microelectrodes. In the early days of electrochemistry, macroelectrodes much larger than that were utilized as electrochemical methods were being developed. As methods became available for making tips and disks of much smaller dimensions, it became necessary to create a descriptive subdivision for ultramicroelectrodes and nanodes (Wightman and Wipf, 1989; Abe *et al.*, 1992; Strein and Ewing, 1992; Heinze, 1993; Martin, 1994; Zhang *et al.*, 1996).

The expansion of resources has necessitated some standardization of nomenclature. A **nanode** is an electrode with a diameter of < 1 μm. Electrodes in the ~1–20 μm range are **ultramicroelectrodes**. Though the limits are not universally defined, a **microelectrode** is normally considered to be ~20–200 μm in diameter. Anything larger than that would be considered a **macroelectrode** , such as those routinely used in analytical procedures.

Other biosensing devices of comparable dimensions have been fabricated. Among the recent examples are the miniaturized optical fibers produced by Kopelman's group for near-field optical spectroscopy (Tan *et al.*, 1992a,b;

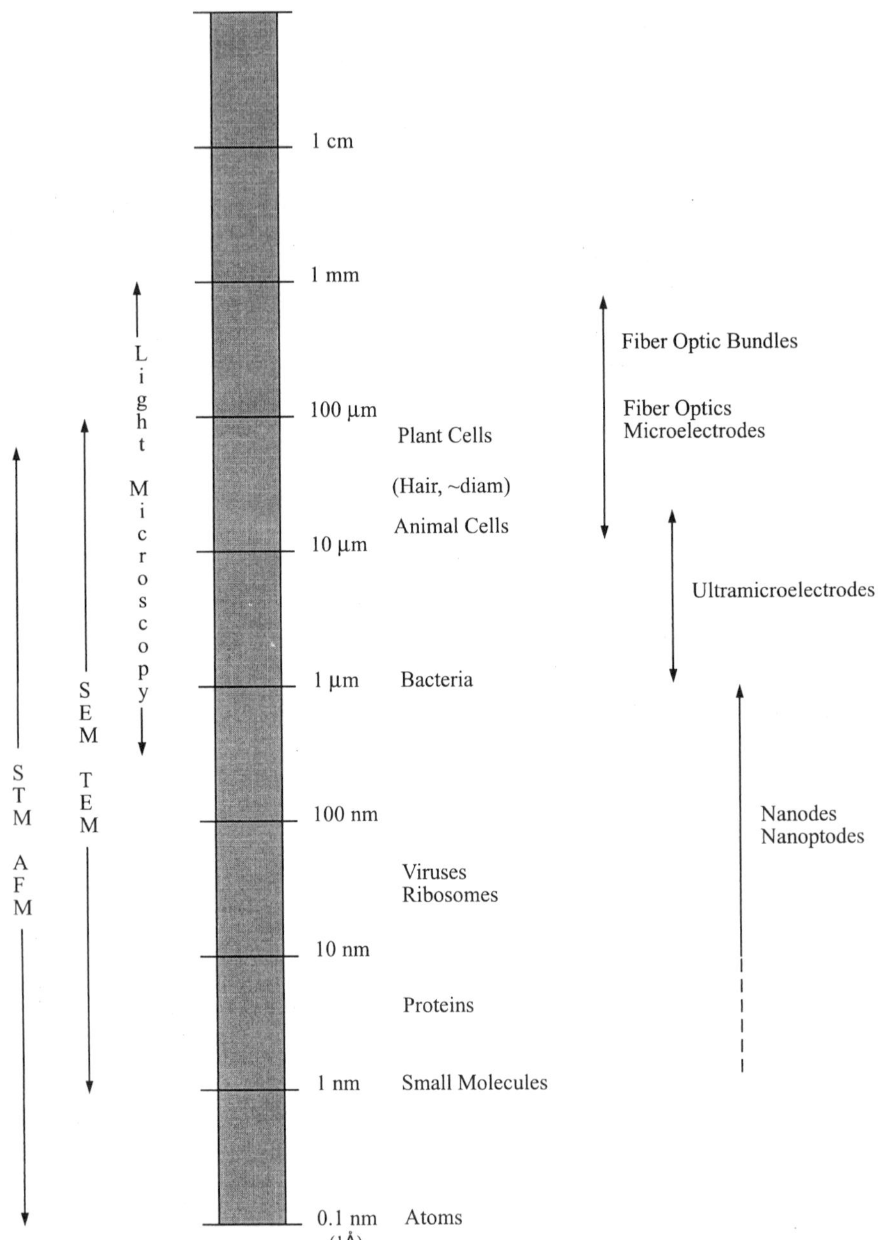

FIG. 1-6. Relative dimensions: center scale indicates a broad comparison of dimensional ranges encountered, the left side shows approximate ranges on that scale for three forms of microscopy, and the right side indicates the relative sizes of some types of sensors.

Kopelman and Tan, 1993; Kopelman and Tan, 1994; Tan and Kopelman, 1996). As is the case with some electrodes, a carefully controlled pipet-tapering assembly is utilized. Kopelman has reported optical fiber biosensors with tips less than 1 μm. In early reports on near-field spectroscopy, Kopelman referred to submicrometer optical sensors, but later that group adopted terminology parallel to that of the electrochemists—i.e., "nano-optodes."

Theories of expected sensor response are quite dependent on the relative values of sensor size and sample size. A theoretical treatment of a macrosensor is not appropriate for an ultramicro- or nanosensor. Diffusion characteristics, considerations of charge balances, and relative effects of coupled chemical reactions can all vary between macroscopic and ultramicroscopic or nanoscopic devices. As different types of miniaturized devices become more common, adequate attention must be given to scaling limits. Bioanalytical problems involving spatial dimensions of cells and tissues will be investigated at an increasing rate. Therefore, spatial resolution related to site-directed immobilization methods and mapping of cell and tissue reactions requires thorough understanding of effects due to relative sizes and geometries.

Finally, there is also a dimensional consideration with regard to the molecular recognition layers—the membranes or films—incorporated into biosensors. Chemically selective films, usually polymeric in nature, may be applied to an energy transducer by various means. Structural definition and characterization of the membranes involve dimensions of both area and depth. Films may be deposited by several techniques, which are discussed in more detail in Chapter 3. Early on in the development of biosensors, relatively primitive methods of applying films and membranes hindered reproducibility, controlled thickness and structural definition. Newer methods have improved the situation considerably.

Stable, reproducible layering is dependent on the surface material and the technique for applying reaction layers. Dip or drop coating, photolithography, photo-, electro-, and plasma polymerization, spin casting, screen printing, and ink-jet deposition are commonly used procedures. **Ultrathin films** are those with thicknesses < 100 nm. **Thin films** are considered to be in the 100 nm–1 μm range. Films with a depth > 1 μm above the deposition surface are **thick films**.

Whether or not the film exhibits properties of the bulk material is a function of the macromolecular characterisics and the thickness of the film. Thinner films may exhibit behavior different from that predicted from the polymer's bulk properties. In working with biosensors the molecular architectures necessary for defined and reproducible behavior often mandate the use of thin or ultrathin films. Influences of film thickness are discussed further for applications in subsequent chapters.

1.3.4. Calibration and Figures of Merit

1.3.4.1. Calibration

In accordance with terminology of general analytical chemistry there is a technical language related to operational characteristics and performance criteria of biosensing devices and methods. The quantifiable measures of performance of a device or a method are referred to as the *figures of merit* (FOM). The FOM for a biosensing device can be determined formally using chemometric analysis of data. Calibration through the use of samples with known concentrations allows creation of a response model from which prediction for future results can be ascertained and the FOMs can be derived. The ultimate goal in obtaining calibration information is to predict quantity of an unknown analyte, either concentration or amount, based on observed responses for a calibration set of samples.

Calibration may be univariate or multivariate, the latter meaning that the response is a function of more than one variable or component. For those who may not have extensive experience with statistical treatment of data, chemometric techniques and calibration procedures, the books by Sharaf *et al.* (1986), Haswell (1992), and Meier and Zund (1993) should prove useful. General articles on theory of analytical chemistry and multivariate calibration should also be consulted (Sanchez and Kowalski, 1988a,b; Brown and Bear, 1993; Booksh and Kowalski, 1994; Thomas, 1994). The net analyte signal, on which many of the FOM definitions depend, has been the topic of a recent report by Lorber *et al.* (1997). Selectivity for *n*-order data has recently been treated by Messick *et. al.* (1996). The Taylor and Schultz (1996) handbook includes Kowalski's updated summary of basic chemometric approaches, as presented in some of the above references.

In the Booksh and Kowalski (1994) article, concise definitions and descriptions of instrumental orders have been summarized from the references therein. The categorizations are according to the types of data provided by the instrument or device. These categorizations apply directly to biosensors. Discrete biosensors, usually calibrated in a univariate mode, are classified as zero order—i.e., a device that provides only a single datum per sample or at one time. For example, a device designed for measuring an absorbance at one wavelength for a sample is zero order. Further, a measurement involving a single electrode potential or current at a given applied potential for a sample provides zero-order data. The relationship between response and concentration may be known, *a priori*, as a result of a theoretical connection such as the Lambert–Beer–Bouger law. If such a theoretical basis is unknown, then regression analysis is ordinarily used to determine the relationship on an empirical basis. Both linear and nonlinear relationships between response and concentration can be encountered and analyzed (Haswell, 1992; Sekulic *et al.*, 1993; Booksh and Kowalski, 1994).

A first-order instrument, such as an array of zero-order biosensors or a flow system with a zero-order detector, provides a series of or multiple measurements per sample or at a particular time. There must be at least s sensors for a analytes ($s \geq a$). Absorbance at multiple wavelengths, potentiometric response of an array of ion-selective electrodes, or responses of a chromatographic detector at various times during the elution process are also examples of first-order data. The responses can be represented as a vector of data, which is a first-order tensor.

Second-order instruments are exemplified by two first-order devices operating in tandem. The configuration might involve a combination of chromatography/spectroscopy or a flow system with a sensing array. Second-order devices provide a matrix of data (a second-order tensor) for each sample. The typical example involves the acquisition of both temporal and physical characteristic data, as in spectral, electrochemical or acoustic-wave output. This multiple measurement approach provides two domains for discrimination among the species present. As described by Booksh and Kowalski (1994), "the instrument in the first order, such as the gas chromatograph in GC/MS, modulates the analyte concentration or environment such that the net signal from the instrument in the second order, the mass spectrometer in GC/MS, changes." Jointly published reports by Lin and Burgess (1994) and Lin *et al.* (1994) illustrate a transition from first order to second order. The authors reported an optical sensor for heavy metal ions using in the first instance a first-order configuration with filter photometry and in the second case a second-order design in which an array spectrometer was utilized.

As can be gleaned from the above descriptions, the three major biosensing strategies discussed earlier fit very nicely into this analytical organization. Discrete sensors used for single measurements are zero-order devices. Some continuous monitors and sensing arrays are first order. Sensing sytems may be configured to perform as second-order instruments, though not all are so configured. The obvious question is "what are the experimental advantages of each of these experimental designs?" The answer lies primarily in the power of the analytical information that emerges. Much of the following discussion relates primarily to multicomponent samples, which are typical of biosensing problems.

Data obtained from a zero-order device do not provide information about the presence or effects of interferences that may be present in the sample. Using a discrete sensor with no prior separation steps, a single response to a target analyte is recorded. That result may be affected by the presence of other species, but that effect cannot be discerned from the acquired data. This type of information must be acquired as empirical data for a chosen set of interferents known to exist commonly with the target analyte. For example, an amperometric sensor may be calibrated as a zero-order instrument by univariate calibration. The measured currents for standard amounts of target analyte are determined. Then, the effects of commonly occurring interferents must be determined as separate experiments, and are usually reported as % change in the signal in the presence of the interferent or as a pre-defined preference ratio for target analyte.

A first-order instrument or device will detect the presence of interferents through orthogonal vectors that emerge in the data analysis, *if the interferents were present in the calibration standards*. The bias, or effects on signal of the target analyte, will not be chemometrically extractable, however. It should be noted that the set of interferents to include in the calibration set is often unknown for complex biochemical or environmental samples.

Second-order instrumentation provides sufficient information not only for detecting interferents, whether present in the calibration set or not, but also for accurate determination of the target analyte in the presence of interferents. Data analysis through matrix methods provides the basis vectors for each linearly independent component of a sample. This is referred to as the "second-order advantage", which is of utmost importance when dealing with multicomponent samples, particularly when the exact nature of the interferents is often unknown or widely variable.

Calibration in the real world of biosensing, even if univariate, is sometimes a difficult task. There are many cases involving interferents that are difficult to profile in number or composition. This creates a fundamental problem in preparing calibration standards that mimic exactly the analytical situation with regard to interferents and matrix effects. A basic approach for this problem in univariate calibration is the standard additions method to compensate for matrix effects. On a practical basis this involves acquired samples to which standards may be added. This is not an option for many of the bioanalytical problems in which sample cannot be withdrawn from its natural matrix. A generalized standard additions method has also been applied for multivariate situations (Sharaf *et al.*, 1986). In some analytical situations recipes for "mock" samples—e.g., artificial cerebrospinal fluid (CSF)—have been described in the biochemical literature and can be utilized for a calibration process.

For *in situ* measurements—i.e., in place or in site—there are additional challenges. *In situ* calibration is not often an option, and thus there is the practical question of when to perform the calibration, before or after exposure of the device to the *in situ* environment. Using biosensing devices implanted in living tissues provides a good example of this dilemma. A usual practice in performing those experiments is to conduct both pre- and post-calibrations—i.e., before placement and after removal of the sensor from the tissue. Justice (1987) provides a succinct discussion of this type of experimental dilemma.

One-point calibration, assuming a valid (0,0) point, is often a desirable goal for *in situ* situations. This allows pre-*in situ* calibration with no further data necessary. Heller's group has reported an implantable glucose sensor that performed well with one-point calibration (Csöregi *et al.*, 1994b). That same group also performed statistical testing for valid clinical decisions based on one-point calibration for pairs of sensors (Schmidtke *et al.*, 1996).

1.3.4.2. Figures of Merit

In order to characterize a biosensor's performance in analysis the universal criteria of **sensitivity (SEN), limit of determination (LOD)** and **concentration range for linear response (dynamic range, DR)** are normally reported with additional data concerning **selectivity (SEL)** and the **response time (t_R)**. Each of these will be discussed in this section. The sketches of Figures 1-7 through 1-9 illustrate these terms.

Several fundamental points should be made about origin of the figures of merit, particularly as related to multivariate calibration. As indicated above, calibration relates an instrumental response to an amount or concentration of analyte. Using the calibration data, prediction of unknown concentrations is possible. This a three-step process: calibration, formulation of a regression model based on that calibration, then prediction based on the model. Using calibration data the model may be derived from various mathematical procedures such as linear regression for univariate cases, multiple linear regression (MLR), principal components regression (PCR), or partial least squares (PLS). The validity of the model, and the resulting prediction for unknowns, will be partly dependent on the way in which data are pre-treated—e.g., smoothing, centering, filtering, weighting, and choice of multiple variables for measurement.

Once the data set has been pre-processed, matrix methods may be used to reduce the set and create a regression model based on linear combination of orthogonal vectors for each analyte in a multicomponent sample. A method such as cross-validation (Sharaf *et al.*, 1986) may be used to determine the statistically significant number of vectors (factors) for optimal model size with minimal prediction error. Coefficients in the regression model are related to the sensitivities toward each analyte. Thus, they are related to the net analyte signal (NAS) necessary for deriving the FOMs.

Before embarking on operational definitions of the FOMs for biosensors, it should be mentioned that other criteria might be imposed by applications, though not included in the technically defined performance criteria. Reliability for repeated measurements under a variety of experimental conditions or by a diverse population of analysts is usually referred to as the **ruggedness** or **robustness** of a sensor or method. This characteristic is particularly important for devices or methods utilized for *in situ* measurements,

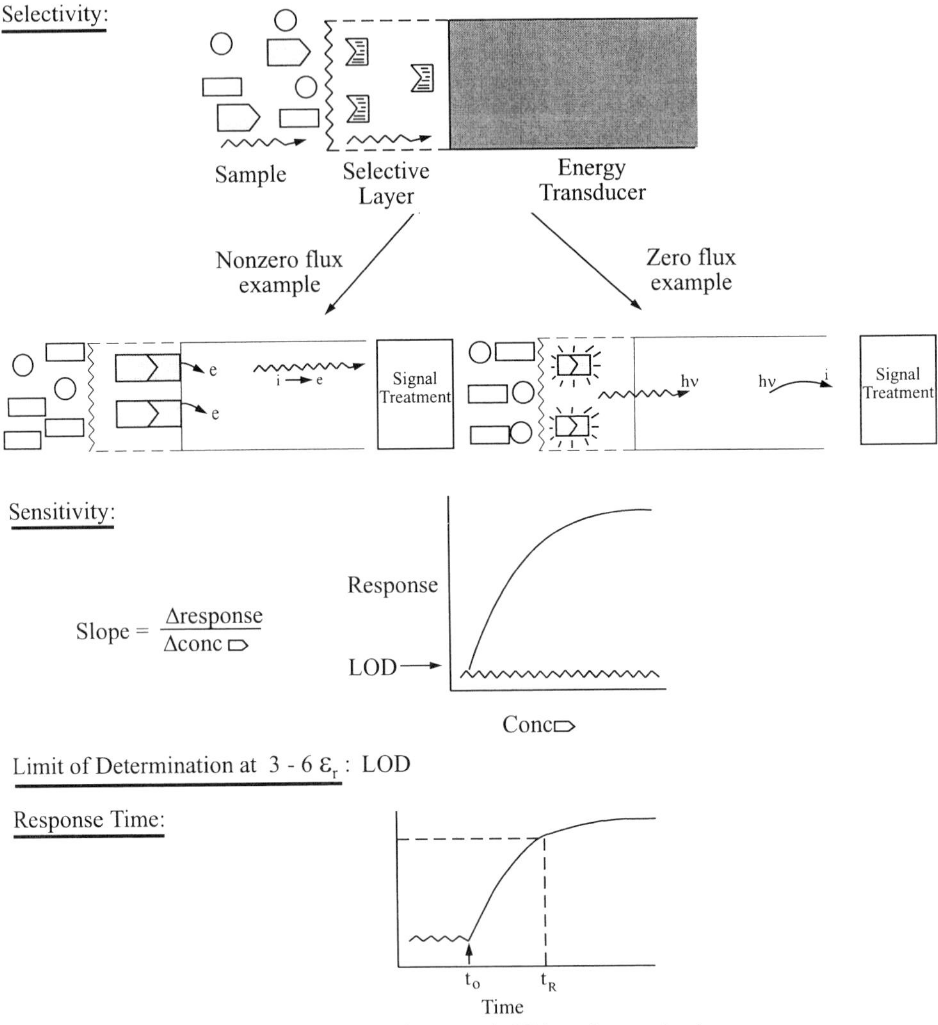

FIG. 1-7. Basic relationships for figures of merit for biosensors: selectivity, sensitivity, limit of determination and response time.

in-the-field measurements, or use by untrained individuals. Practical devices must be characterized not only by reliability or ruggedness, but also by ease of use, economical production and widespread availability. These are not criteria that are technically quantifiable as figures of merit, but may be very important for specifications affecting designs.

Size and packaging are also factors to be considered in terms of the specific application. Some of the most exacting reqirements for miniaturiza-

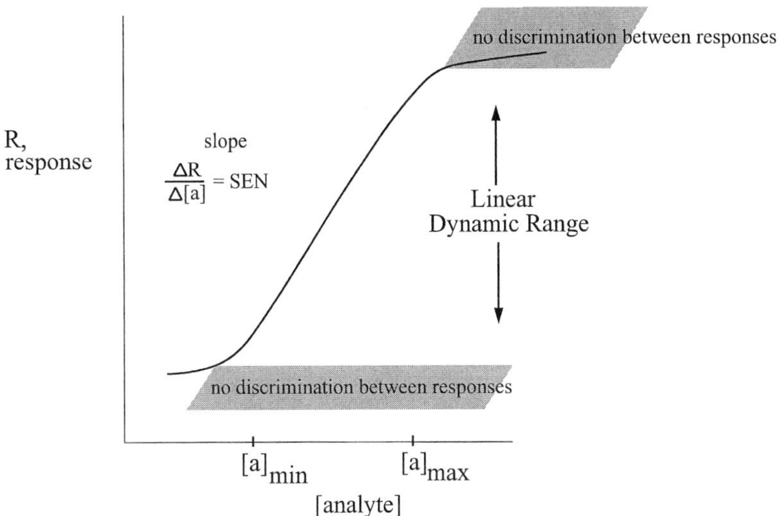

FIG. 1-8. Schematic of a calibration curve showing relationships for determining sensitivity (SEN) and linear dynamic range. Concentration of analyte is indicated as [a].

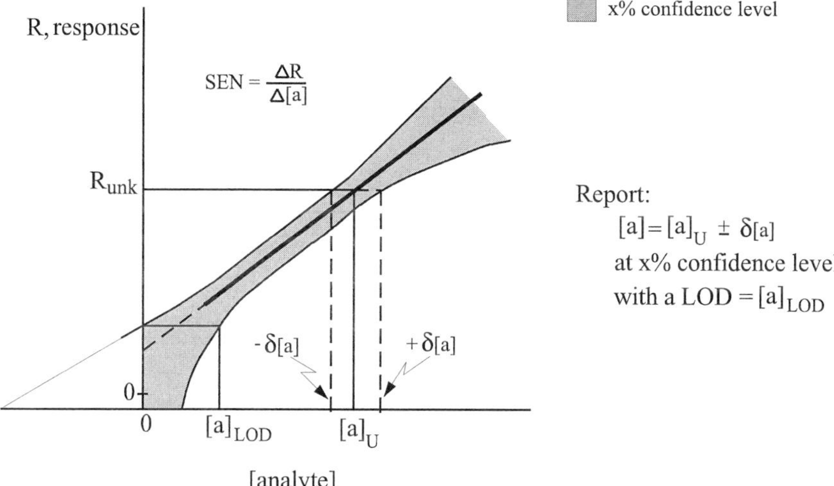

FIG. 1-9. Schematic of calibration curve showing confidence interval (shaded) and limit of determination (LOD).

tion and biocompatibility arise from biomedical applications, but challenges for packaging could be just as difficult, though different, for *in situ* probes of underground soil measurements or a fermentation vat. All of these characteristics may pertain to ultimate design as associated criteria in development, but they are not stated as figures of merit.

Although figures of merit are defined fundamentally for any particular analytical method, each criterion has its own relative importance for different types of biosensing applications. Adequate sensitivity and selectivity are, of course, essential in any of the applications. It is not, however, essential to have a "perfectly" selective discrete device if other strategies involving chemometric extraction of information can be devised. Because sensing arrays may be utilized primarily for partially or non-selective sensors, selectivity becomes an especially important criterion and determinant in design and assessment of methods based on sensing arrays. Response time is a more ambiguous FOM. For example, t_R for a glucose sensor used in diabetic control is dictated by the physiological time-frame for abnormal blood sugar excursions. Ideally this response time should be on the order of 30 seconds or less. Many of the proposed glucose sensors do not meet this criterion. The optical sensor reported by Li and Walt (1995) is an example of one that does. Depending on the thickness of the membrane on the optical fiber, response times of 8–29 seconds were observed. The Healy and Walt (1997) optical sensing arrays for pH and oxygen exhibit a t_R on the order of a few hundred milliseconds. The response time constraints might be far less stringent for a glucose sensor designed for use in industrial process control.

Sensitivity. Sensitivity (SEN) with regard to an individual analyte is the change in response that results from a unit change in concentration of analyte, dr/dc. For the simplest case of a univariate calibration scheme, the slope of the curve representing response versus quantity of analyte, in concentation or amount, is the sensitivity relative to that analyte. This is illustrated in Figures 1-7 and 1-9 for linear response whereas Figure 1-8 represents a case of nonlinear response. One of the simplest examples is the conventional Beer's law constant ($k = \varepsilon b$) relating absorbance (A) to concentration (c) at a given wavelength, $A = kc$. A higher extinction coefficient, ε, or a longer pathlength, b, increases the change in absorbance per unit of concentration of absorbing species. Thus, the sensitivity is higher.

The sensitivity relative to a particular analyte is affected by the presence of interferents. The effects of interferents vary. Some may increase sensitivity and some may decrease it. The importance of the presence of the real interferents during calibration is emphasized by this point.

For multicomponent situations there is a sensitivity assignable to each component contributing to a response. In the classic undergraduate experiment for determining photometrically two components using measurements at two wavelengths, the concentration for each component is obtainable by solving two equations in two unknowns (see Figure 1-10). Assuming a pathlength of 1 cm, the coefficients would correspond to the four extinction coefficients, one for each substance at each wavelength. Four sensitivities would be operable, one for each component at each wavelength.

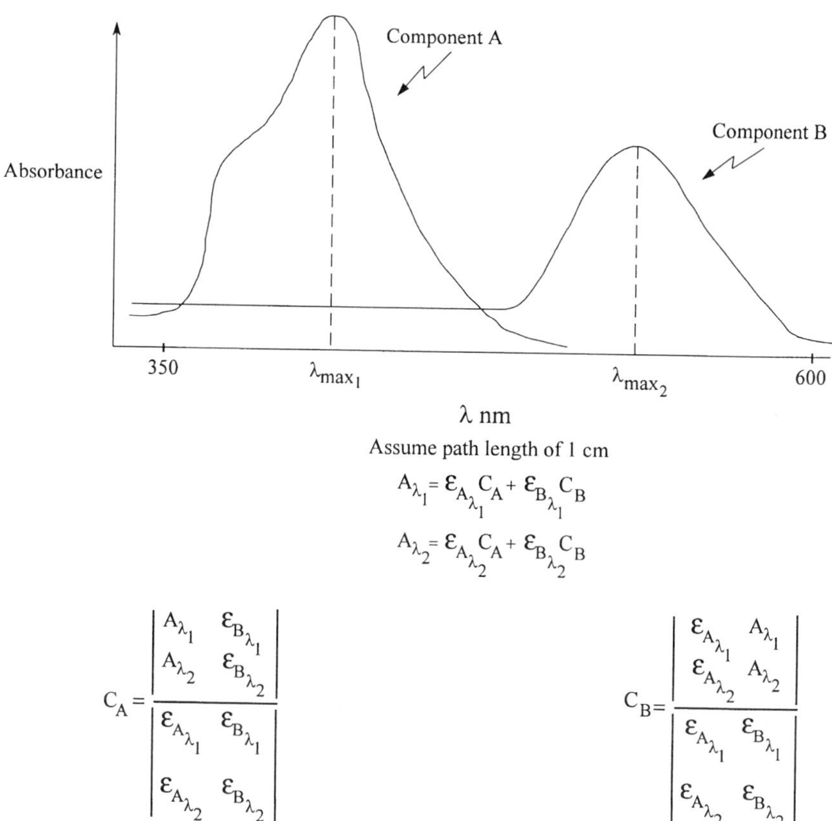

$$A_{\lambda_1} = \varepsilon_{A_{\lambda_1}} C_A + \varepsilon_{B_{\lambda_1}} C_B$$

$$A_{\lambda_2} = \varepsilon_{A_{\lambda_2}} C_A + \varepsilon_{B_{\lambda_2}} C_B$$

$$C_A = \frac{\begin{vmatrix} A_{\lambda_1} & \varepsilon_{B_{\lambda_1}} \\ A_{\lambda_2} & \varepsilon_{B_{\lambda_2}} \end{vmatrix}}{\begin{vmatrix} \varepsilon_{A_{\lambda_1}} & \varepsilon_{B_{\lambda_1}} \\ \varepsilon_{A_{\lambda_2}} & \varepsilon_{B_{\lambda_2}} \end{vmatrix}} \qquad C_B = \frac{\begin{vmatrix} \varepsilon_{A_{\lambda_1}} & A_{\lambda_1} \\ \varepsilon_{A_{\lambda_2}} & A_{\lambda_2} \end{vmatrix}}{\begin{vmatrix} \varepsilon_{A_{\lambda_1}} & \varepsilon_{B_{\lambda_1}} \\ \varepsilon_{A_{\lambda_2}} & \varepsilon_{B_{\lambda_2}} \end{vmatrix}}$$

FIG. 1-10. Example of spectrophotometric determination of two components in a mixture: ϵ, molar extinction coefficients. Pathlength of 1 cm is assumed.

Slama *et al.* (1996) have reported a two-component chemometric application involving a microbial sensor for mixture analysis. The method is based on differences between effects of biologically relevant organics on the microbial activity, as monitored with an oxygen electrode. The proof-of-concept study was conducted with two-component mixtures of gluconate + acetate and serine + threonine. The operational principle of the "dynamic microbial sensor" is that of an amperometric device which responds discriminately according to both concentration-dependent and time-dependent effects of each component of the mixture on the microbes. Thus, the chemometric treatment was formulated around sequential sampling of current at different times in an FIA system. Responses were taken every 5 s for standard solutions, thus generating the data for the concentration and time-dependent response vectors. Partial least squares (PLS) algorithms were used for the multivariate calibration. Prediction errors ranged from ~4% to 9%. The authors suggested that with more extensive pattern recognition methods and the use of microbial arrays the method should be adaptable to multicomponent analysis.

Beyond the case of two components or for measurements at multiple wavelengths multiple sensitivities can be extracted chemometrically. A series of reports from Arnold's group demonstrates the power of chemometric extraction of information for multicomponent analyses. That group has presented foundational work for non-invasive glucose determination in complex matrices via near-infrared spectroscopy with chemometric analysis of data acquired at multiple wavelengths (Small *et al.*, 1993; Bangalore *et al.*, 1996; Pan *et al.*, 1996; Shaffer *et al.*, 1996).

Limit of Determination. Whole chapters and books have been written on the matter of limits to obtaining reliable quantitative information. Haswell (1992), Booksh and Kowalski (1994), Boumans (1994) and Ferrus and Egea (1994), including the many references therein, cover the topic in detail. Over the years numerous terminologies based on different statistical criteria have been utilized: limit of detection, limit of determination, limit of quantitation, and limit of discrimination. For purposes of this discussion the definitions of Booksh and Kowalski (1994) will be adopted (r is the measured response):

> "... The precision of a measurement is traditionally defined as S/N, [r/ε_r], in which ε_r is the standard deviation of repeated measurements of r. . . The precision of analysis can be defined as the minimum difference between two concentrations that can be distinguished from the effect of random instrumental errors. . . This is termed the limit of determination, defined by the International Union of Pure and Applied Chemistry as [LOD $= k\varepsilon_r/$ SEN], in which k is an integer that defines the number of standard deviations of separation that constitutes 'different'. . . Usually k is equal to 3, which is nearly 99% probability that the two measurements differ."

There are three important points to be made about the preceding statements. (1) In the S/N expression, the response factor and standard deviation of repeated measurements do not necessarily change in the same manner over the whole concentration range. Accordingly, as Booksh and Kowalski (1994) state, "... S/N is defined at an arbitrary unit concentration." (2) Second, there is argument about the value of the constant k in the limit of determination expression. Borrowing from the language of statistics, if both Type I errors and Type II errors are to be avoided at the 99% confidence level, the number k should be ~6 rather than 3. Based on decisions related to the null hypothesis, a Type I error would occur if it were decided that something is there when it is not there (false positive). A Type II error would occur if the null hypothesis were accepted, leading to a decision that nothing is there when there really is something (false negative). This choice of k and the disagreements surrounding its value are the reasons Booksh and Kowalski used quotation marks on the word 'different'. (3) Finally, the sensitivity and limit of determination are two entirely different FOMs, though they are related. If a sensor exhibits low sensitivity to an analyte, then the ability to distinguish beween zero concentration and a nonzero value will be impaired. This implies a poor limit of determination. A minimal illustration of the relationships among terms defined in this section is shown in Figure 1-9.

Dynamic range. The linear range of concentration dependence is referred to as the dynamic range (DR) of a sensor. Linear responses are helpful, though not absolutely necessary. Chemometric treatment of data is more straightforward for linear responses, even in the case of second-order instrumentation where linear responses in each order are assumed in certain mathematical approaches to the data analysis (Booksh and Kowalski, 1994). Several of the energy transduction modes are inherently linear over large concentration ranges, though overall responses may be modulated by the coupled chemical transduction mode. This implies that the design and fabrication of the chemical transduction layer(s) can be manipulated to alter the dynamic range of some sensors. Behavior similar to that illustrated in Figure 1-8 is typical of a sensor exhibiting an inherently nonlinear response, such as a potentiometric device. There is a linear segment of the calibration curve which can be utilized, but there are concentration values for which no discrimination among concentrations can be discerned.

While the plateau regions of the Figure 1-8 curve might be characteristic of the energy transduction mode—e.g., a logarithmic response according to the Nernst equation—some response curves result in nonlinear behavior due solely to the chemical transduction mechanism affecting response. For example, an enzymatically controlled signal may exhibit a plateau for concentrations approaching substrate amounts necessary for maximum velocity of the enzymatic reaction. There are limited sites for substrate binding to the immobilized enzyme(s), and thus a limiting value for response of the sensor. A similar result of "saturation" behavior is observed for immunosensors in which site-limited bound antibody can accommodate only so many antigen molecules. In this case the binding curves are reminiscent of those observed for the classical Langmuir adsorption isotherms (Atkins, 1990).

Selectivity. Selectivity (SEL) refers to capability for discriminating among components in a mixture. If prior separation of a complex sample precedes detection of an analyte, the selectivity required of the detector is reduced because interferents have supposedly been removed by the separation step. This approach would be applicable to the sensing systems discussed earlier—e.g., MD, CE or a μTAS. As indicated in that discussion the design is that of a first- or second-order instrument.

Even in the absence of prior separation, an ideally or perfectly selective sensor will be sensitive to only one analyte. In the real world of biosensors and natural samples, however, perfectly selective sensors are essentially non-existent. Thus, appropriate methods must be devised for dealing with partially selective sensors. There are multiple origins or sources of selectivity. An energy transduction method may be partially selective. For example, there may be unique wavelengths at which only one component of a mixture absorbs. Also, as illustrated in Figure 1-11, characteristic redox potentials may be exhibited by various components of a mixture. Rarely are either of these phenomena sufficient to resolve directly the signals of complex samples. For example, in using amperometric sensors for glucose in blood, normally occurring interferents with overlapping redox responses are acetaminophen, ascorbic acid and uric acid. Using the same type of glucose sensor for food processing control, other sugars might be the interferents. Ascorbic acid is also a notorious interferent for amperometric measurement of neurotransmitters in the brain.

For biosensors the next level of selectivity, beyond that inherent in the energy transduction process, is designed through the molecularly selective reaction mechanisms of chemical transduction. Thus, by coupling an analyte's characteristic wavelengths or redox potentials with the molecular recognition of host–guest chemistry, enzyme–substrate specificities or antigen–antibody binding enhanced selectivity may be achieved. Many sensors are essentially non-selective with regard to energy transduction—e.g., conductimetric and acoustic-wave devices. In those cases, selectivity is conveyed entirely by the chosen chemical transduction mechanism. For multicomponent samples configuring sensors (even non-selective ones) in array formats allows the analyst to capitalize on not only the options for chemical transduction, but also the chemometric extraction of information for various components.

When using discrete, zero-order sensors with univariate calibration, empirical methods for stating selectivities must be utilized. The approach with which most chemists are most familiar is that of selectivity coefficients for ion-selective devices, both potentiometric and optical (see Umezawa, 1990; Bakker and Simon, 1992; Bakker *et al.*, 1994; Umezawa *et al.*, 1995). Frequently, experimentally required selectivity, $k^{\mathrm{pot}}_{\mathrm{A,B}}$ or $k^{\mathrm{opt}}_{\mathrm{A,B}}$, of one analyte relative to another component can be calculated based on prior knowledge of usual relative concentrations in a certain type of sample (Bakker *et al.*, 1994; Lerchi *et al.*, 1994; Bakker, 1997). Thus, informed by this required selectivity, the sensor may be designed and assessed for a prescribed level of selectivity. Wang has also proposed empirically determined selectivity coefficients for amperometric sensors and for class-selective enzyme electrodes (Wang, J., 1994, Wang, J. and Chen, 1996). In

(a)

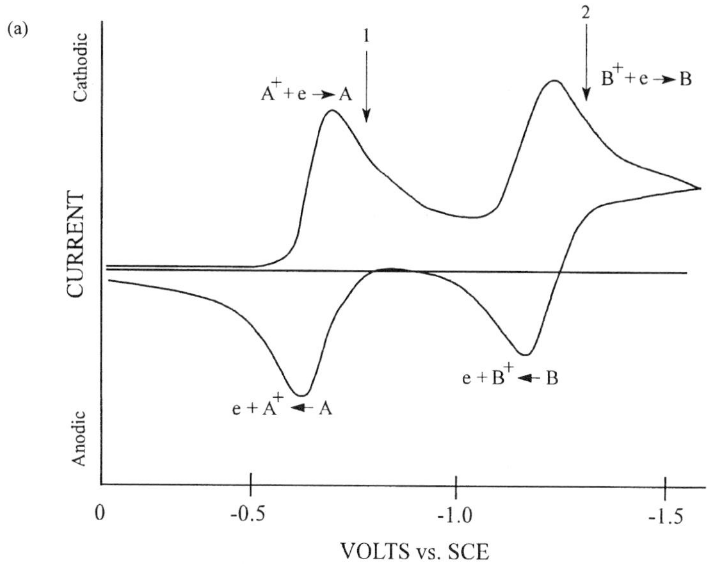

(b) At applied potential 1 only A will be reduced.

At applied potential 2 both A and B will be reduced.

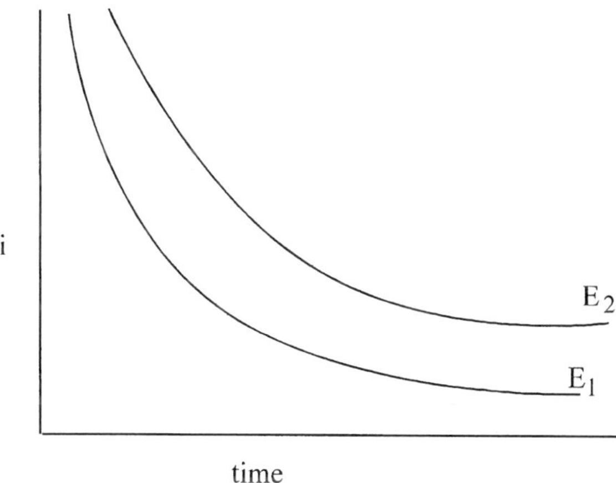

FIG. 1-11. (a) Example of the inherent selectivity conveyed by differing redox potentials for two electroactive species. Curve represents a cyclic voltammetric scan from 0.0 V versus SCE to ~ -1.6 V versus SCE, then reversal of scan to the initial potential (SCE is standard calomel electrode). On the forward scan A^+ is reduced, then at a more negative potential B^+ is also reduced, thus producing additional cathodic current. On the reverse scan B is first reoxidized, then A is reoxidized. At ~ -1.0 V versus SCE, B could be reoxidized without affecting A, but at potentials more positive than ~ -0.55 V versus SCE both A and B would be oxidized. (b) Sketch of the current–time curve expected for electrolysis at the potentials shown.

practice, various methods of reporting experimentally observed selectivities for univariate sensors are employed.

As an FOM for a device or a method, selectivity is quantifiable by chemometric extraction of information. Selectivity (SEL) has been defined as the fraction of response, r^*/r, attributable to the analyte of interest which is designated by the asterisk. There is an SEL value for each component relative to any other component in a mixture. SEL may also be thought of as the degree of overlap between response vectors for various components contributing to a signal. From chemometric definitions it may be discerned that selectivity ranges from a value of 0 to 1. A value of 1 for SEL indicates perfect selectivity and orthogonality of the response vectors. When SEL = 0 it indicates a totally non-selective response with no unique sensitivity for the target analyte. Selectivity may be calculated as the ratio of the net analyte signal for a given component, NAS_k, and the magnitude of the total response vector, $NAS_k/\| r \|$. Booksh and Kowalski (1994), Messick *et al.* (1996) and Lorber *et al.* (1997) provide details of the meaning and practical application of the chemometric derivation of selectivity.

Response time. Though not an FOM derived from calibration data, the response time t_R of a sensor is an important consideration and should be reported as a matter of characterization. Ordinary response times range from milliseconds to minutes, and must be tailored to the time-scale of chemical events producing the detectable species. When a sensor is to be used for multi-sample measurements both the inherent response time of the sensor and the sample handling or regeneration time must be considered with regard to throughput. The importance of the latter factors are evident in automatic systems for repeated measurements (Trojanowicz, 1996).

There have been various accepted procedures for reporting t_R's and relatively little standardization across types of sensors or applications. The manner in which sample is introduced to the sensor, the sensor design, as well as rates and equilibria of the chemical transduction step(s), all affect transient, steady-state and equilibrium signals. Analyte may be introduced to a sensor in three ways:

1. the sensing probe may be inserted into a sample, stirred or unstirred;
2. sample may be injected into the detection cell, with or without stirring; or
3. sample may be transported to the sensor by a flowing stream as in FIA, CE or a flow-cell sensing system.

The response time of the sensor, *per se*, will be a result of the controlling factors related to convection, diffusion within the reaction layer(s) or any chemical processes that occur in the chemical transduction step(s). Some generalizations are possible. With modern electronic components and circuitry, response times should be dependent only on the design of the sensor and the mode of sample introduction. In the absence of slow chemical processes within a sensor's selective membrane or rate-limited interfacial phenomena, thinner immobilization matrices exhibiting good permeability to the analyte normally provide faster responses. Investigation of the interplay between convection parameters—e.g., stirring rate—and response time can provide valuable information about the controlling factors of response.

Because most sensors produce eventually a stable plateau of signal, the approach to a stable signal has become the criterion for some applications with regard to t_R. In these cases the time from introduction of sample to the time for stable response would be considered 100% signal. Values are then frequently stated relative to that maximum signal—e.g., t_{95} for time to reach 95% of stable signal. In modeling the steady state and transient responses of amperometric sensors, Tatsuma *et al.* (1992b) determined that a t_{80} was an adequate value for good agreement between theory and experiment. In characterizing a potentiometric salicylate-selective sensor Bachas' group adopted the IUPAC-preferred convention of determining t_R as the time from analyte introduction to time for attainment of a stable signal with less than $< 0.1\ \mathrm{mV\,min^{-1}}$ variation (Hutchins *et al.*, 1997).

Immunosensors present a different situation with regard to response time. Binding between antigen and antibody often requires a relatively lengthy equilibration time. Due to strong antigen–antibody binding, regeneration of the sensing surface for repeated measurements is also a significant consideration. Consequently, the time factor for making a meaningful measurement for analyte challenge to a stablized sensor can be several minutes, or perhaps an hour or more.

Kirkbright *et al.* (1984), Seitz (1988) and Wolfbeis (1991) discussed some of the original, relatively simple optical pH-sensors based on distal-end immobilized pH-indicators. The approach to stable signal in these cases depends on both the composition of the immobilizing matrix and the characteristics of the IN^-/HIN equilibrium. This means that mass transport considerations as well as equilibration time will affect the t_R. In many cases the t_{90} or t_{63} were reported in the literature. The latter value $\left(1 - \frac{1}{e}\right)$ assumes an exponential response function. Although the chosen % value varies among researchers, the time to reach 90% and 95% steady signal has been widely used for reporting response times of various kinds of sensors.

1.3.5. Sampling Modes

For those who have not studied Latin, there may be some confusion about terms that are used to describe sampling modes. The term *in situ* is often used. That means that the measurement is made in place—i.e., analytes in their natural environment, which could be in a fermentation broth, in the brain, in a well, etc. The opposite situation would be *ex situ*, meaning out of its place. This refers to removal of the analytes from the original environment for out-of-place analysis. An *in vivo* measurement would be made in a living organism, cell or organ. If a sample is transferred to a test tube, beaker, reaction well, etc., the measurement is designated as *in vitro* (literally meaning in glass). A measurement might be made outside the living system, *ex vivo*, but be **in-line**. A measurement made in a heart–lung machine would be an example of the in-line approach. **On-line** systems in which the sample is not returned to the source are far more common. Microdialysis (Newton and Justice, 1994; Niwa *et al.*, 1996) and capillary electrophoresis (Shear *et al.*, 1995; Fishman *et al.*, 1996; Tao and Kennedy, 1996) are now well-established on-line techniques in biosensing. The *in vivo* and *ex vivo* terms, when used for animal studies, might be transformed into intracorporeal and extracorporeal, meaning in or out of the body, respectively. An example of an extracorporeal arrangement is shown in Figure 1-12.

1.4. AN OVERVIEW OF APPLICATIONS

An examination of some applications explains partially the investigative intensity concerning biosensors. The multidimensional character of the whole biosensor area is illustrated in Figure 1-13. While there are some commonalities across the applications spectrum, each measurement problem has its own set of challenges for the analyst. Some of the major challenges in developing biosensors have been related to calibration, stability for *in situ* measurements, miniaturization and interferences in multicomponent samples.

There are several aspects of biosensors that must be examined when considering variations across applications. Design and performance characteristics are dictated largely by the application. As indicated in Table 1-2, the usual analytes among the various applications are no different from those of analytical chemistry as a whole. However, if biosensors were used to quantify the same analyte—e.g., lead ions or carbon dioxide—in widely variant media, the concentration ranges, measurement conditions, interferents, and matrix effects would be distinctly different. For example, the constraints for measuring accurately the oxygen in blood during surgery are significantly different from those of monitoring oxygen in a bioprocessing fermentor. Another example pertains to the response time of a sensor. Requirements for response times of sensors may vary from fractions of seconds to several minutes or longer, depending on the application. The response time of a particular biosensor must be reasonable within the natural time-frame of the

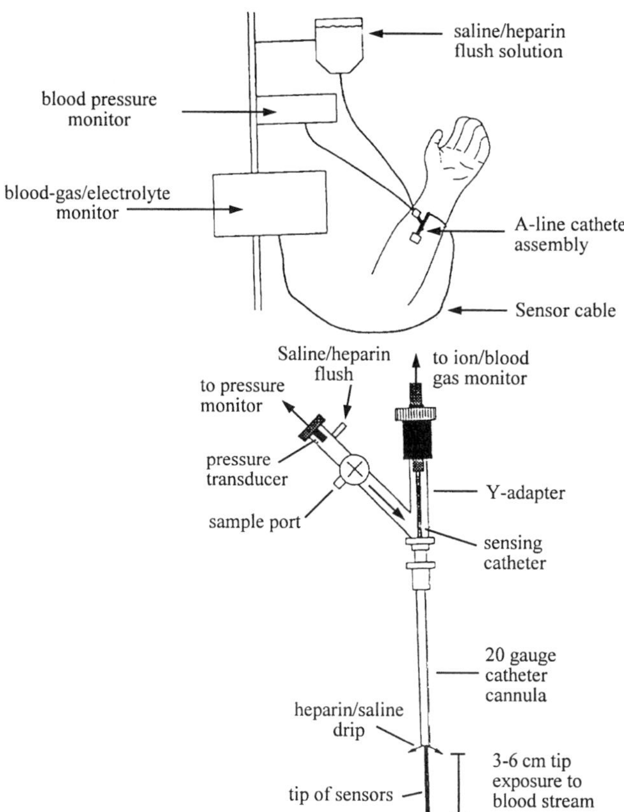

FIG. 1-12. Arterial line catheter for continuous monitoring of blood gases and electrolytes. Reprinted from *Trends Anal. Chem.*, **12**(6). Meyerhoff, M. E., In vivo blood-gas and electrolyte sensors: progress and challenges, p. 258. (1993) with kind permission of Elsevier Science B. V., Amsterdam, The Netherlands.

bioanalytical problem. Applications in critical-care medicine must meet some of the most stringent requirements in this regard.

1.4.1. Environmental Control

Potential applications for biosensors in environmental control are quite extensive (Damgaard *et al.*, 1995; Rogers, 1995). Consider the need for measuring toxic vapors in the atmosphere or polluting substances in river waters. Environmental control problems might involve the measurement of

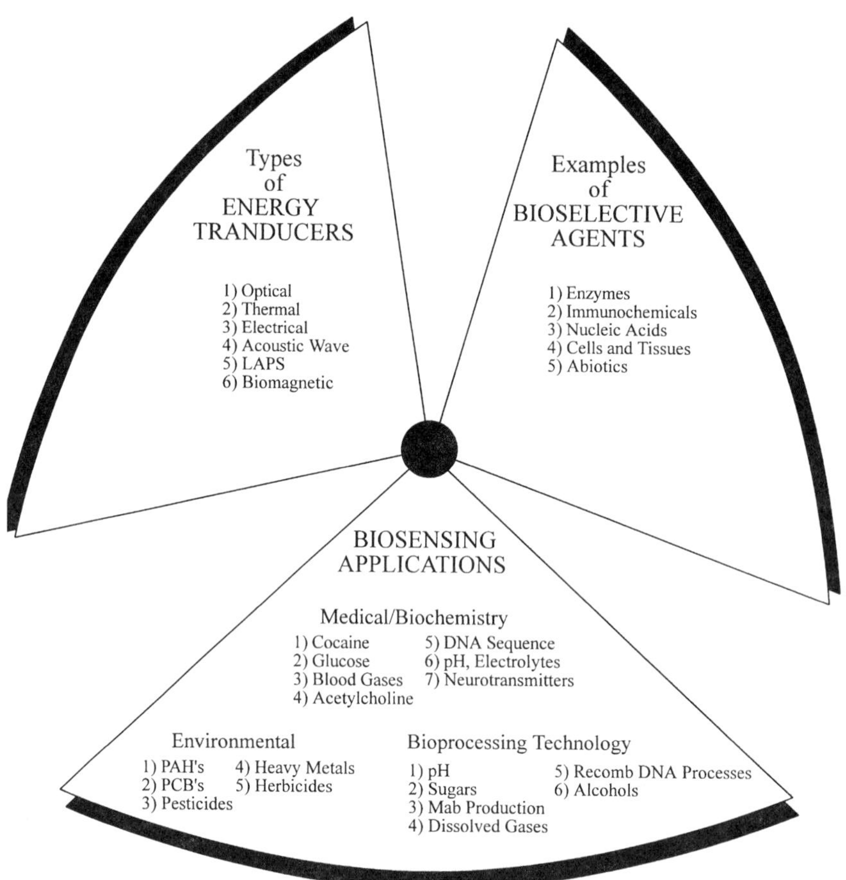

FIG. 1-13. Overview of the biosensor field: types of energy transducers, examples of bioselective agents, and biosensing applications.

heavy metals (Booksh *et al.*, 1994; Lin and Burgess, 1994; Lin *et al.*, 1994; Turyan and Mandler, 1994, 1997) or toxic hydrocarbons (Krska *et al.*, 1993; Henshaw *et al.*, 1994; Rosenberg *et al.*, 1994; Smilde *et al.*, 1994; Tauler and Kowalski, 1994; Roberts and Durst, 1995). Pesticides and herbicides have become primary analytical targets with increasing environmental control. In these latter cases pollutant-sensitive biocomponents have been used with a variety of detection modes. Brewster *et al.* (1995) and Preuss and Hall (1995) used photosystem II cells (PET cells) with amperometric detection. Sadik and Van Emon (1996) and Bauer, C. G.

TABLE 1-2 Ordinary Analytes

Applications	Analytes	
Clinical/Biochemical	Electrolytes	Blood gases
	Sodium ion	Oxygen
	Potassium ion	Carbon dioxide
	Chloride ion	
	Calcium ion	
	pH	
	Metabolites	Enzymes
	Ascorbic acid	Alanine aminotransferase
	Bilirubin	Alkaline phosphatase
	Cholesterol	alpha-amylase
	Creatinine	Aspartate aminotransferase
	L-Lactate	Creatinine kinase
	Urea	Lactate dehydrogenase
	Uric Acid	
	Co-factors:	Drugs
	Nicotinamides	Acetaminophen
	Flavins	Digoxin
		Penicillin
		Platinum anti-cancer
		Salicylate
	Immunocomponents	Neurotransmitters
	Hepatitis B antigen	Acetylcholine
	Herpes virus	Choline
	AIDS virus antigen	Dopamine and metabolites
	IgG	Epinephrine, Norepinephrine
		Serotonin
Bioprocess control	Alcohols	Lactose
	Amino acids	Maltose
	Ammonia-N	Monoclonal antibodies
	Cell concentrations	Oxygen
	Carbon dioxide	pH
	Glucose	Sucrose
	Glutamine	
Environmental	Heavy metals	Polychlorinated biphenyls
	Herbicides	Polyaromatic hydrocarbons
	Nitrate, Nitrite	Trichloroethylene
	Pesticides	TNT

et al. (1996) used different types of electrochemically based immunoassays. Mionetto *et al.* (1994) relied on a response effected by inhibition of an enzyme, acetycholinesterase.

Rugged, reliable sensors are required for on-site testing. At the same time these devices must have sufficient sensitivity and limits of determination to satisfy environmental control regulations. An interesting example of a chemical sensor design tailored to agency regulations involved development of lead- and silver-selective optodes (Lerchi *et al.*, 1992, 1996). The resulting sensing strategy is quite effective and the authors describe the development process explicitly.

Rogers (1995) raises an interesting point about the use of selective sensors in environmental problems. The discussion emphasizes that for many cases in environmental monitoring multianalyte sensors are far more in demand and are more economical than single-analyte sensors. This need is not unique to environmental applications. Multianalyte sensors are being designed at an increasingly rapid pace, but most have not yet reached the commercialization stage.

1.4.2. Process Control

Industries involved in pharmaceuticals, food processing and recombinant DNA technology need effective biosensors for monitoring various processes, including microbial and cell cultures. Maintaining conditions for optimal growth and productivity involves, for example, control of pH, temperature, alcohol and sugar levels, dissolved gases, or polynucleotide fragments. Some of the bioprocess control applications also provide good examples of biosensors in computerized feedback control. Nutrient feeds, product removal, etc., must be adjusted on a real-time basis during processing. The use of *in situ* sensors would seem to be the ideal measurement strategy. There are, however, severe practical limitations to this approach.

The system described by Hitzman *et al.* (1995) illustrates some important points about the use of biosensors in these applications. The multianalyte nature of biotechnology processes is illustrated very well by the multichannel flow-injection system shown in Figure 1-14. In this system three analytes are being subjected to FIA with detectors commonly used in biosensors. The fourth measurement involves a stopped-flow photometric determination of enzymatic activity. It should be noted that this on-line approach, rather than *in situ* biosensors, is utilized in a great many bioprocess control situations. Why? There are very few biosensors commercially available that can meet the requirements for these *in situ* measurements.

Three of the biggest problems of *in situ* applications are related to sterility, calibration and operational lifetime. Every component of the

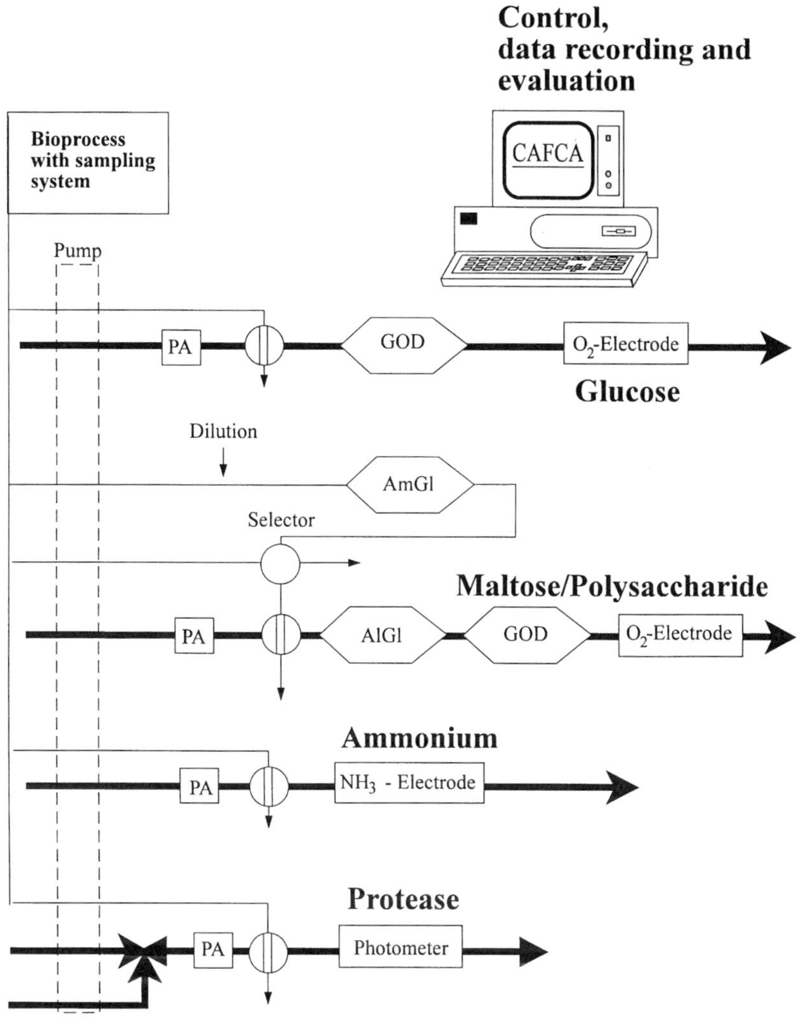

GOD glucose oxidase
AmGl amylo glucosidase
AlGl α-glucosidase
PA pulse absorber

FIG. 1-14. Multi-channel flow-injection system supervised by expert system. Reprinted, with permission, from Hitzmann *et al.*, In Rogers, K. R., Mulchandani, A. and Zhou, W., *Biosensor and Chemical Sensor Technology* (p. 134). ACS Symposium Series #613. Copyright 1995, American Chemical Society, Washington, DC.

system can be an added source of contamination. This includes joints, valves, gaskets, etc. Further, periodic calibrations might require removal and reinsertion of a sensor. This would increase the probability of contamination. Enfors (1989) has suggested one solution that involves an enclosed adjustable sensor housing for periodic partial withdrawal for calibration. Until more stable, accurate sensors for *in situ* operation are developed, off-line or non-invasive methods will remain the best options.

A second point illustrated by the Hitzman configuration is the high degree of automation. The entire system is presented as an example of a process controlled by an expert system. Real-time monitoring and feedback-control of the process can improve efficiency and productivity. This closed-loop function could be facilitated even further by biosensors for monitoring, if the problems mentioned in the preceding paragraph were solved. It may be, however, that newer non-invasive methods and the use of microsystems for separation/detection will prove to be the best solutions. Mulchandani and Bassi (1995) have reviewed the principles and applications for biosensors in bioprocess control. There is also available the report of a symposium on biosensors in process monitoring and control and environmental control (Rogers *et al.*, 1995).

1.4.3. Clinical/Biochemical

Although environmental measurements and process control offer excellent examples of the utility of bioanalytical sensors, there has been no greater influence on development than the universal need for better medical diagnostic and monitoring devices. Development of accurate, portable, relatively inexpensive and easy-to-use biosensors in healthcare has become a high priority all over the world. There is a general trend toward more decentralized and immediate diagnostics—e.g., in-the-office or at-home measurements, bedside assessments, and rapidly responsive critical-care monitors.

Table 1-3 lists some types of bioanalytical sensors utilized for a variety of healthcare challenges. In some cases the devices given as examples have been developed to a fairly mature stage of actual use in medical practice. In other cases original goals for effectiveness have not been met because of practical limitations. A sensor that functions adequately as a prototype during development with controlled samples may not perform according to specifications when used in the natural environment for which it was designed. Or, perhaps, the device works well on a large scale, but miniaturization problems are difficult to overcome. Stability is frequently a problem when a

TABLE 1-3 Typical Biosensors Used in Healthcare

Analyte	Normal Range (in blood)	Chemical Transduction	Examples of Energy Transduction
Sodium ion	136–143 mmol/L	Ionophore complexation	Potentiometric
Potassium ion	3.6–5.0 mmol/L	Ionophore complexation	Potentiometric
Calcium ion	1.15–1.31 mmol/L	Ionophore complexation	Potentiometric
Chloride ion	98–107 mmol/L	Ionophore complexation	Potentiometric
pH	7.31–7.45 blood 5–8 urine	Ionophore pH sensitive chromophore	Potentiometric Optical
Oxygen	80–104 Torr	Gas permeable membrane Perylene or Ru fluorophore	Amperometric Optical(quenching)
Carbon dioxide	33–48 Torr	Gas permeable membrane with bicarbonate solution Membrane with bicarbonate solution and pH chromophore	Potentiometric Optical
Glucose	65–105 mg/dL	Enzymatic catalysis	Amperometric Potentiometric Optical Thermal
L-Lactate	3–7 mg/dL	Enzymatic catalysis	Amperometric Potentiometric Optical
Urea	7–18 mg/dL (BUN)	Enzymatic catalysis (+ phenol)	Potentiometric Thermal Optical

device, encased in supposedly biocompatible packaging materials, is implanted into a living system. Buck *et al.* (1992), Meyerhoff (1993), Yim *et al.* (1993), Goldberg *et al.* (1994), Cosofret *et al.* (1995b), Nishida *et al.* (1995), Quinn *et al.* (1995a) and Yun *et al.* (1997) have addressed many of the pertinent issues related to biocompatibility of amperometric and potentiometric devices.

Among healthcare applications, there is no better example to illustrate the cumulative effect of biosensor development than the glucose sensor. Since Clark's original work with an amperometric device, the search for an optimal glucose sensor has been ongoing. Almost every generic type of sensor has been adapted for measurement of glucose, as evidenced by the entries in Appendix A-2. Good glucose sensors beget better glucose sensors. More importantly, good glucose sensors beget sensors for other analytes. In the process of perfecting sensors for one analyte, e.g., glucose, alternatives in chemical and energy transduction are developed and new technologies, translatable to other analytes, are applied.

The enormous global problem of diabetes is a major reason for continuing interest in glucose sensors. An ultimate goal is the artificial pancreas which can respond dynamically to glucose levels and control insulin supply based on the sensor's response. The problems in reaching the goal are not unlike those of the *in situ* sensors in bioprocess control. Biocompatibility, calibration, operational lifetime and rapid dynamic response head the list of major challenges. The article by Reach and Wilson (1992) provides a good summary of the most pressing considerations.

1.5. CONNECTING THEORY AND PRACTICE

1.5.1. Ion-selective Mechanisms

It is not unfair to say that full understanding of ion-selective mechanisms has been an arduous task. Practice definitely led theory for many years, relying on Nernstian behavior of potentiometric devices with selectivity described by the familiar Nicolsky–Eisenman formalism (Nicolsky *et al.*, 1967; Morf, 1981; Janata, 1989; Umezawa, 1990; Umezawa *et al.*, 1995).

$$E = \text{constant} + \frac{RT}{z_{\text{A}}F}\ln\left[a_{\text{A}} + \sum_{\text{B}\neq\text{A}}^{\text{B}} k_{\text{A,B}}^{\text{pot}}(a_{\text{B}})^{z_A/z_{\text{B}}}\right]$$

Much of the current work in ion-selective sensing is based on polymer-membrane based chemistries. Moody *et al.* (1970) developed a calcium-selective polymer-membrane electrode that is still the basis for most modern devices. It has taken almost three decades for the theory to lead practice and inform design, even that for optical ion-selective sensing. Initially primary emphasis was placed on cation-selective devices because they were more effective than corresponding anion-selective electrodes—i.e., selectivity was easier to control and optimize. As newer, more general formalisms for describing the ion-selective response have been developed factors controlling selectivity are now fairly clear. The theories apply equally well to anions and cations. Meyerhoff's group has also extended the theory to polyions (Fu *et al.*, 1994; Wang, E. *et al.*, 1995; Meyerhoff *et al.*, 1996). Chapter 2 includes more detail concerning ion-selective mechanisms and theoretical developments in recent years.

The most common type of ion-selective sensing, illustrated in Figure 1-15, is based on the inclusion of dissociated ion-exchangers (E^+E^-), neutral (L)

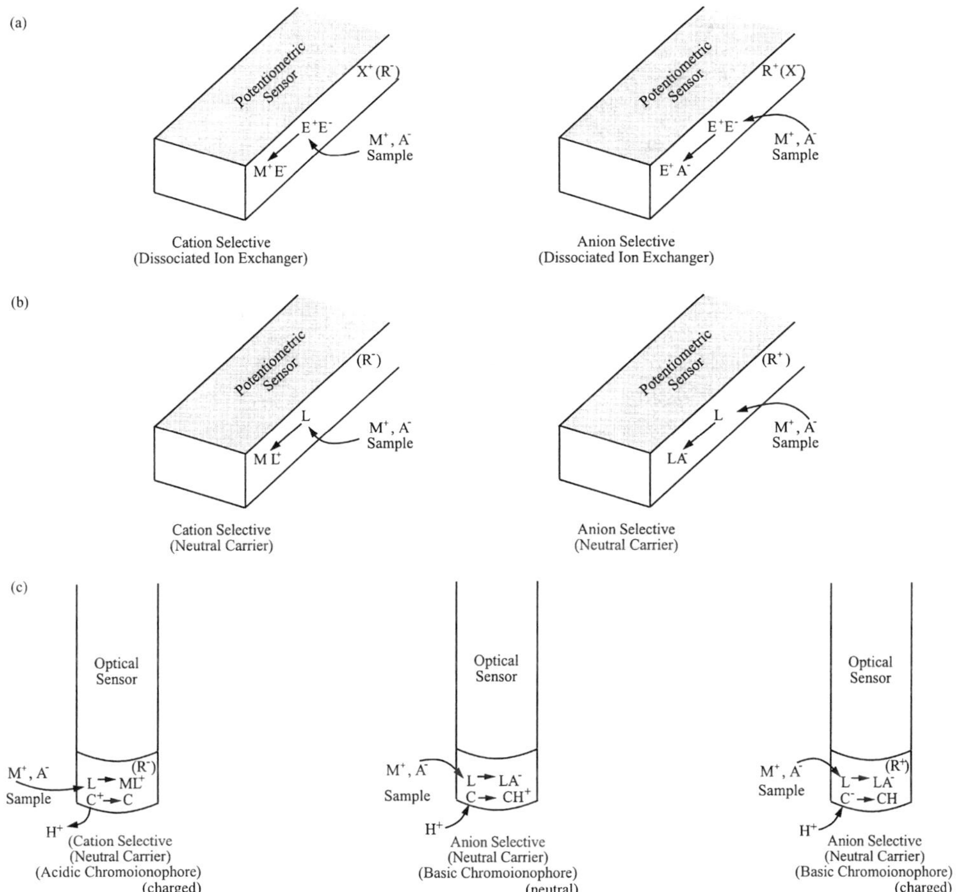

FIG. 1-15. Basic principles of polymer membrane-based ion-selective sensing: (a) and (b) illustrate potentiometric devices using dissociated ion exchanger and neutral carriers; and (c) shows the same type of ion-selective sensing with the addition of chromoionophores for optical detection. The cases for charged and reactive carriers have been omitted for simplicity. See Bakker *et al.* (1994) and Barker *et al.* (1997). M^+, A^- are the cations and anions of a sample. The carriers are denoted as L, the chromoionophores, protonated and unprotonated, are the C species, and the R^+ and R^- represent lipophilic additives (with their counterions, X) used to maintain electroneutrality and to lower the membrane resistance.

or charged (L^+ or L^-) carriers in a polymeric matrix that contains lipophilic (hydrophobic) additives (R^+ or R^-) to lower the membrane resistance and maintain electroneutrality. Reactive carriers, such as metal complexes for anion sensing, are also useful (Yim *et al.*, 1993; Barker *et al.*, 1997), but are not shown in Figure 1-15.

Anions partition into the membrane according to their lipophilic nature, which may be indicated by a lipophilic constant, k_{lp}, which is dependent on an ion's relative solvation energies in the two phases. Consequently, the more hydrophilic anions require a significant free energy for partitioning. This is a pivotal concept for understanding the difficulties that have arisen with anion selectivity using polymeric membrane devices. Anions tend to partition always according to the classical Hofmeister (1888) pattern (see Yim *et al.* ,1993; Xiao *et al.*, 1997):

$$\text{ClO}_4^- > \text{SCN}^- > \text{I}^- > \text{Sal}^- > \text{NO}_3^- > \text{Br}^- > \text{Cl}^- > \text{HCO}_3^- > \text{OAc}^- > \text{SO}_4^{2-} > \text{HPO}_4^{2-}$$

A change in the sequence is required for selective sensing of the more hydrophilic (lipophobic) ions, such as the oxoanions. Thus, creating anion sensors that will exhibit a different preferential sequence requires some selectivity mechanism to complement the partitioning effect. Significant progress has been made toward using complexation by various agents within the membrane (Barker *et al.*, 1997; Hutchins *et al.*, 1997; Xiao *et al.*, 1997). This approach is discussed more thoroughly in Chapters 2 and 4. For an interesting, short description of the Hofmeister effect, as it pertains to salting out proteins, Abeles *et al.* (1992) should be examined.

Detection may be either potentiometric or optical with the polymer-based membranes. For potentiometric sensors the response is governed by the phase boundary potential between sample and membrane (Bakker *et al.*, 1994):

$$E_{\text{A}} = E_{\text{A}}^0 + \frac{RT}{z_{\text{A}}F}\ln\frac{k_{\text{lp,A}}a_{\text{A}}}{[A^{z^+}]} \qquad \text{for a cation}$$

For ionophores that are dissociated ion-exchangers, the selectivity between two ions, primary and interferent, is solely dependent on the relative k_{lp}'s of the ions. For neutral and charged carriers, however, the selectivity depends on not only the partitioning phenomena, but also the ion–carrier complexation constants (Bakker *et al.* 1994).

For a neutral carrier, L, and a lipophilic additive, R^-, z's > O,

$$k_{\text{AB}}^{\text{SEL}} = \left(\frac{z_{\text{B}}\text{L} - \text{R}^-}{z_{\text{A}}\text{L} - \text{R}^-}\right)\left(\frac{k_{\text{lp,B}}\ \beta_{\text{BL}}}{k_{\text{lp,A}}\ \beta_{\text{AL}}}\right)\ \exp[(E - E^0)(z_{\text{A}} - z_{\text{B}})F/RT]$$

The β's are ionophore complexation constants. Thus, by varying the amounts and ratio of carrier to lipophilic additive, the response function can be chemically "tuned".

The basic chemical mechanisms for potentiometric and optical ion-selective sensing are the same, but in the case of optical detection the added relationships for absorbance or emission changes of an auxiliary chromophore or fluorophore must be considered. In the Bibliography there are numerous references which trace out the history of both potentiometric and optical ion-selective sensing. The reports (and references therein) by Schaller *et al.* (1994, 1995), Lerchi *et al.* (1992, 1994, 1996), Bakker *et al.* (1994), Meyerhoff *et al.* (1996), Fu *et al.* (1994), Wang, E. *et al.* (1995) and Barker *et al.* (1997) should be consulted.

1.5.2. Biocatalytic Sensors

The development of theory for enzymatic electrodes illustrates beautifully the power of coupling chemical and energy transduction. Theoretical results have led to a much greater understanding of the design parameters necessary for required performance. Variations of enzyme loads in membranes, thicknesses of membranes and alternatives among enzymes have all been refined toward optimization by modeling of biocatalytic electrochemical devices. The modeling of these electrodes has advanced perhaps more than any other area for connecting transduction theory and experiment in biosensing. Some of the earliest theoretical work on amperometric sensors was conducted by Mell and Maloy (1975,1976). Carr and Bowers (1980) presented a very thorough examination of both theory and experiment for potentiometric and amperometric sensors in their book on immobilized enzymes. That book is still an excellent reference for foundational material in potentiometric, thermal and amperometric biocatalytic devices, as well as immobilized enzyme reactors used in flow systems. Chapter 2 includes some of the most basic elements in the development of the diffusion-reaction mechanisms that are operative in these types of sensors.

Because of the widespread interest in amperometric enzyme electrodes, that emphasis is evident in the literature. Most models have dealt with either the simple one-substrate, one-product reaction or the ping-pong mechanism of enzymatic catalysis.

A:

$$\mathrm{S} + \mathrm{E} \underset{k_{-1}}{\overset{k_1}{\rightleftharpoons}} \mathrm{ES} \xrightarrow{k_2} \mathrm{E} + \mathrm{P}$$

B:

$$S_1 + E \rightarrow P_1 + E'$$
$$S_2 + E' \rightarrow P_2 + E$$

This latter mechanism is of particular importance since it applies to many of the oxidases used in enzymatic electrodes. Bartlett and Pratt (1993) have provided a thorough summary of modeling approaches for both potentiometric and amperometric biosensors. Scheller and Schubert (1992) also present good detail of the necessary interrelationships to be considered for a variety of enzymatic cases. Speiser's (1996) contribution updates much of the earlier work in digital simulations of the systems, including comparisons of methods and different programming packages. The studies of Tatsuma and Watanabe (1992b), Tatsuma *et al.* (1992b) and Bacha *et al.* (1995) dealing with response times and of Martens and Hall (1994) and Martens *et al.* (1995) on the subject of two substrate reactions illustrate the utility of the modeling approaches in solving very practical problems.

1.5.3. Acoustic-wave Devices

Initially the piezoelectric effect, included in Chapter 6, was utilized primarily as a mass-sensitive technique based on the Sauerbrey relationship between acoustic wave frequency changes and surface modified quartz crystals, the quartz crystal microbalance or QCM (Sauerbrey, 1959; Guilbault and Jordan, 1988; Guilbault and Suleiman, 1990). A schematic of the basic experimental setup is shown in Figure 1-16. Much of the earliest work dealt only with vapors and gas phase phenomena. With increasing use of these devices, including a transition to sensing in liquids, it was found that resonant frequency changes of the oscillator could not be explained solely on the basis of mass changes. Full consideration must be given to the mechanical and electrical elements of the equivalent circuit for the resonator. Lasky and Buttry's (1989, 1990) reports of a QCM glucose sensor exhibiting signals much higher than predicted serve as a good example of the realization that effects other than mass changes are operative. Muratsugu *et al.* (1993) reported similar effects in their study of human serum albumin (HSA) and HSA anitibodies with a QCM.

With accumulated experience with different types of acoustic-wave sensors and further examination of the nature of the response, the influence of changes in the surface-modifying layers was elucidated. For example, viscoelastic effects, conductivity and water sorption all affected signals. Based on these observations the theory of acoustic-wave devices, particularly in solutions, has been studied more thoroughly (Lu and Czanderna, 1984; Mandelis and Christofides, 1993; Martin *et al.*, 1997; Thompson and

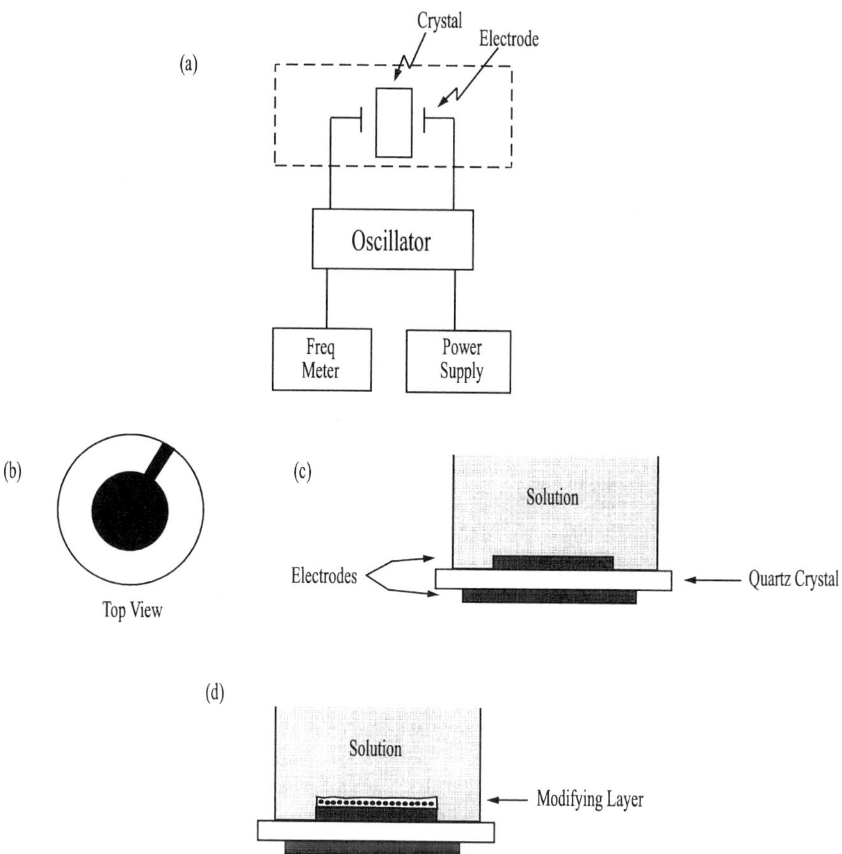

FIG. 1-16. Basic quartz crystal microbalance (QCM) or thickness shear mode (TSM) acoustic-wave device: (a) instrumental setup for simple resonant frequency monitoring is shown; (b) one side only of the crystal in (a) and another electrode of the "flag" type would be on the opposite side of the crystal; and (c) and (d) side-views of a QCM (TSM) cell with liquid sample, where (d) includes the chemically selective layer that would be applied for biosensing, thus altering the resonant frequency.

Stone, 1997). The reviews by Buttry (1991), Buttry and Ward (1992), Grate *et al.* (1993a,b) and Jossé (1994) present current theories of response in terms of the equivalent circuit models for conductive, liquid-loaded devices. Rather than relying solely on the resonant frequency changes accompanying changes in the surface layer(s), extended network analysis of the capacitive and resistive components of the circuit are now routinely made with commercially available analyzers.

1.6. FRONTIERS AND PROGRESS

In the introduction to her book, *Biosensors*, Elizabeth A.H. Hall began with the statement, "In the beginning was the word, and the word was *biosensors*" (Hall, 1991). The implication of a genesis was quite appropriate, but in actuality, research had moved to a fair point of maturity at the time Hall's book was prepared. Since then momentum has certainly increased due to new technologies and expanded applications. Progress is truly impressive, though many of the original goals for commercial viability are yet to be met. The future is still bright for anyone who wants to pursue an exciting area of research.

Creativity based in knowledgeable experience abounds in the quest for more effective bioanalytical sensors as well as alternate approaches in biosensing, in general. Over the years there have been multiple sensors developed for individual analytes—e.g., glucose, protons, oxygen, hydrogen peroxide, etc. There have been arrays designed for multiple analytes. Also, there have been some sensors designed only for specific analytes in specific sensing environments. If there is any degree of commonality among the various discrete devices it lies in the collective scientific cleverness in combining a wide variety of energy transduction modes with basic chemistry for achieving selective molecular recognition.

In sorting through what has been done and what is still to be done, it must be realized that the original multi-faceted goal of individual sensors that are accurate, reliable, selective, sensitive, rapid, miniaturized, reagentless, and stable devices has been realized only in some cases. When one combines the requirements for portability and possible implantation, the assessment of success to date merely points to more extensive research.

As time has passed, however, analysts have begun to acknowledge more realistically that not all of the above criteria must apply in all situations. While research in bioanalytical sensors was perking along over the last 10–15 years, complementary developments in some other areas—e.g. chemometrics and capillary electrophoresis for microseparations—have provided a new perspective on selective sensing by alternative strategies.

As current trends in biosensor research are assessed, there are some notable patterns emerging for shaping the near future of the field.

1. There are more sophisticated methods for chemical modifications which should improve immobilization methods, fine-tuning of molecular recognition for selectivity, and the use of new materials for both transducers and chemical transduction strategies. An example, illustrated in Figure 1-17, is Martin's polymer-coated nanoelectrodes which, for example, have been utilized for improved detection limits in ion-exchange voltammetry (Ugo, 1996).

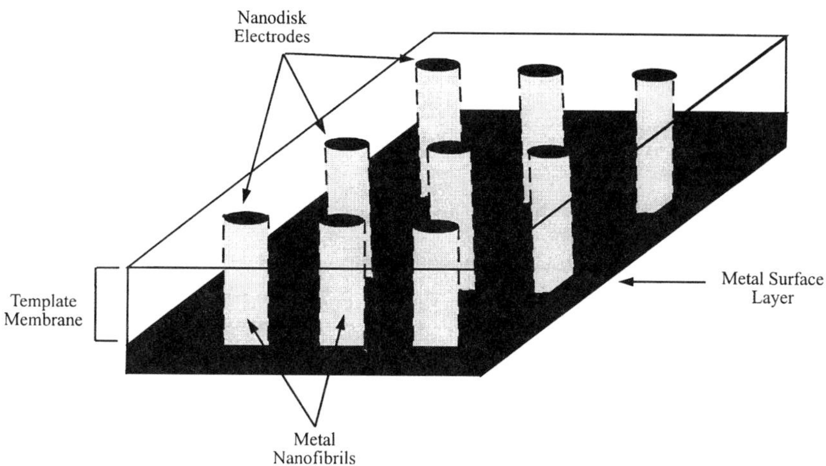

FIG. 1-17. Schematic of a nanoelectrode ensemble (NEE). Nanodisk electrodes are prepared using pores of a template membrane, and the method has produced gold nanodisks of 10 nm diameter. Adapted with permission from Martin, C. R. (1994) *Science* **266**: 1961–1966. Copyright 1994. American Association for the Advancement of Science, Washington, DC.

2. The perpetual challenges of real-world multicomponent samples have fostered some exciting innovations for multianalyte sensors, sensing arrays and chemometric approaches for non-selective or partially selective sensing. Figure 1-18 is adapted from the Cosofret (1995a) report of a multianalyte sensor used in the beating heart, which is a challenging analytical problem, to say the least.
3. In the realm of dimensional considerations, the original quest for miniaturized, portable—possibly implantable—devices is still a major challenge, though progress toward that goal is evident. Miniaturization and integration of components for combined separation and detection are major areas for future advancement. Advances in miniaturization are exemplified by the amperometric and optical arrays of Figures 1-19 and 1-20 as well as the intricate micromachining techniques illustrated in Figures 1-21 and 1-22.
4. As methods are adapted for miniaturization and extremely small sample volumes, such as single cells and samples beneath microscope tips, analysts must learn to use routinely microtechniques and also re-examine theories about physicochemical phenomena in dimensionally confined spaces (Bowyer *et al.*, 1992; Smith and White, 1992, 1993; Bard *et al.*, 1994; Beyer *et al.*, 1995; Gao, X. *et al.*, 1995; Takada *et*

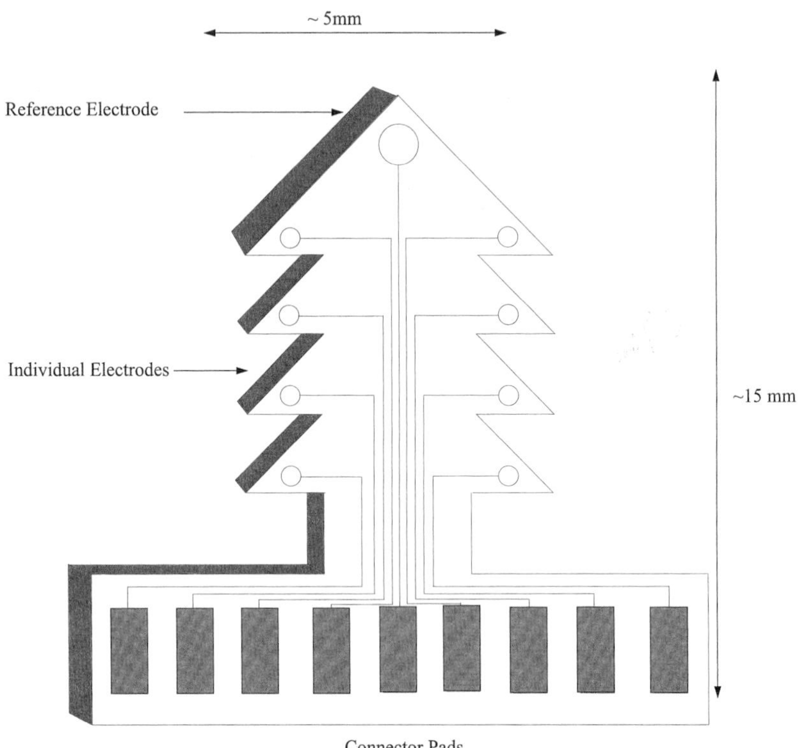

FIG. 1-18. Implantable sensor aray for sensing pH and K^+ in a beating heart. Adapted, with permission, from *Anal. Chem.* (cover) Vol. **67**(10). Copyright 1995, American Chemical Society, Washington, DC.

al., 1995; Lillard *et al.*, 1996, Tan and Kopelman, 1996; Bratten *et al.*, 1997; Clark *et al.*, 1997). Analytical determinations with spatial resolution on the nm to μm scale already contribute significantly to knowledge of surface modifications and chemical dynamics of living systems.

5. A whole new world has opened up regarding biochemical phenomena in non-aqueous media (Zaks and Klibanov, 1988; Guilbault and Mascini, 1993; Borzeix *et al.*, 1995; Guo and Dong, 1997). Just how far that will expand biosensing opportunities, particularly in industrial applications, remains to be seen.
6. Non-invasive sensing methods are now being developed with the advent of new transducers and chemometric treatment of data.

(a)

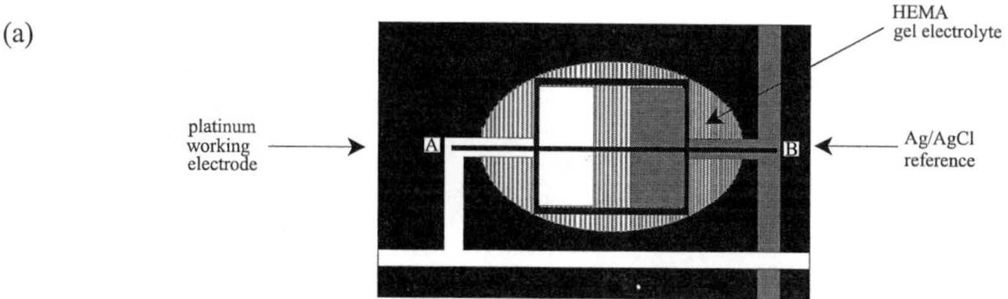

(b)

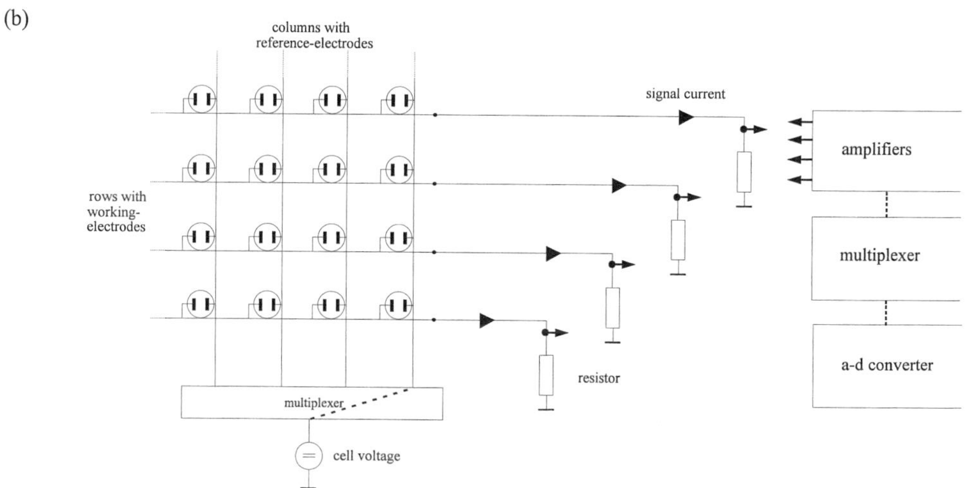

Fig. 1-19. Example of an amperometric sensing array for Clark-type oxygen sensing, with lower portion showing amperometric sensors on a silicon substrate. This was designed for 1024 cells on a 30 mm × 30 mm substrate. The upper portion shows an individual electrode of the array: A is the working electrode, and B the Ag/AgCl reference. The cell has immobilized poly-(2-hydroxymethylmethacrylate) (pHEMA) electrolyte gel on the surface. Adapted from *Sensors and Actuators* (B), Vol. **21**, Hermes, T. *et al.*, An amperometric microsensor array with 1024 individually addressable elements for two-dimensional concentration mapping, pp. 33–37. Copyright 1994, with permission from Elsevier Science Ltd., Oxford, UK.

Effective non-invasive approaches would go far in solving some of the most complex biosensing problems in natural environments.

1.6.1. Chemical Modifications

There are four major areas in which significant progress in chemical modifications has pointed to the things to come. There are now better methods for :

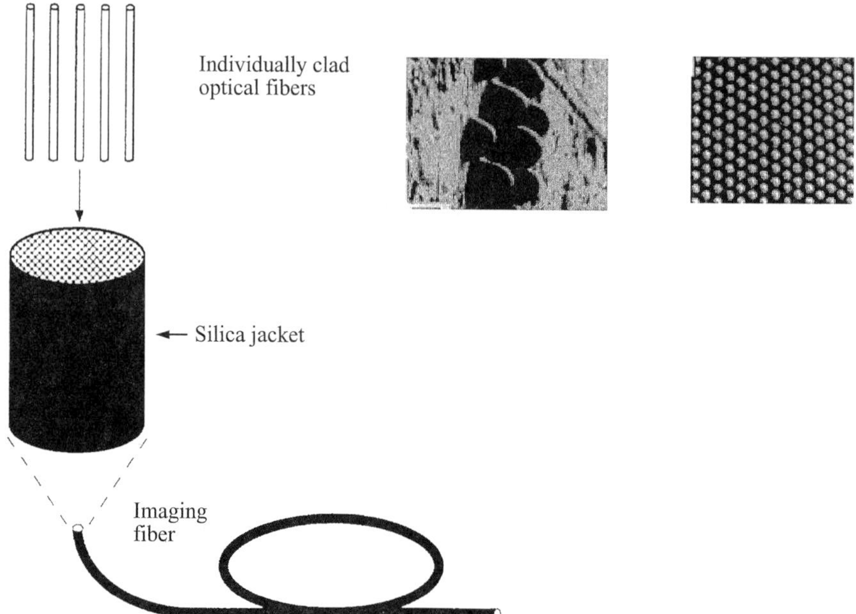

FIG. 1-20. Optical fiber bundle for imaging and sensing showing (left) a schematic of the 350 μm diameter bundle, and (right) an SEM of photodeposited polymethylsiloxane microstructure (sideview) and an SEM of a monodisperse polymer array. (left) Reprinted, with permission, from Pantano, P. and Walt, D. R., *Anal. Chem.* **67**: 481A–487A. Copyright 1995, American Chemical Society, Washington, DC. (right) Reprinted, with permission, from Healy, B. G. *et al.*, *Science* **269**: 1078–1080. Copyright 1995, American Association for the Advancement of Science, Washington, DC.

(a) engineering molecular recognition agents for more refined, biospecific selectivity tasks;
(b) designing the molecular architecture of immobilized layers for molecular recognition;
(c) studying and elucidating the actual interfaces that intervene between sample and energy transducer;
(d) designing the energy transducers, *per se*.

In the area of molecularly designing chemically selective agents, host–guest macrocyclics (Careri *et al.*, 1993; Beer, 1996; Dickert *et al.*, 1997), molecular imprints (Ansell *et al.*, 1996; Mosbach, 1996, Ramström *et al.*, 1996; Kriz *et al.*, 1997; Sellergren, 1997), nucleic acid ligands (McGown *et al.*, 1995; Kawazoe *et al.*, 1996) and genetically engineered components (Gilardi *et al.*, 1994; Sigal *et al.*, 1996; Schmid *et al.*, 1997; Scott *et al.*, 1997) may be expected to produce effective innovations for biosensors.

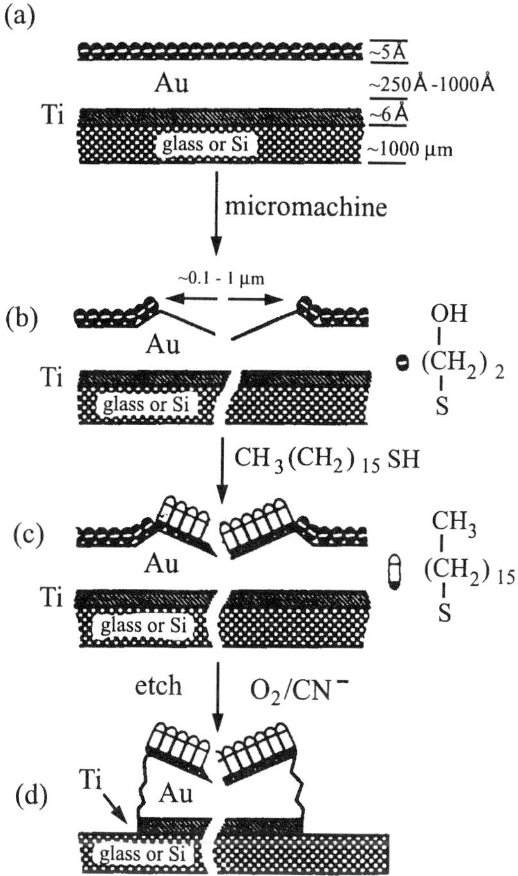

FIG. 1-21. Micromachining a self-assembled monolayer. Cleavage is created by a microscopic cut with a scalpel. Sulfhydryl groups are protected during etching. Reprinted, with permission, from Abbott *et al.*, *Chem. Mater.* **6**: 596–602. Copyright 1994, American Chemical Society, Washington, DC.

As discussed earlier, host–guest chemistry has been a long-standing approach in ion-selective electrode chemistry for cation selectivity. With newer options for tuning selectivity many of the more intractable problems of ion-selective sensing should be solved. Redox sensitive hosts have also been reported (Smith *et al.*, 1994; Beer, 1996). Zhou and Swager (1995) have provided an excellent example of using host–guest chemistry with fluorescence-quenching detection to good advantage. Using cyclophane acceptors,

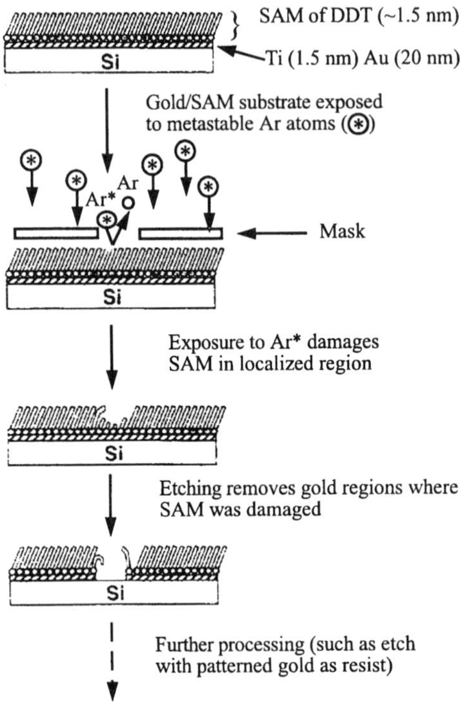

FIG. 1-22. Patterning SAMs with beams of neutral atoms in metastable excited states. Reprinted, with permission, from Berggren, K. K. *et al., Science* **269**: 1255–1257. Copyright 1995, American Association for the Advancement of Science, Washington, DC.

"wired" together by conjugated polymers, they were able to enhance significantly the sensitivity to fluorescence quenching observed with paraquat as analyte. The general use of fluorescent chemosensors should increase significantly in the future. The book edited by Czarnik (1993) deals with fluorescent chemosensors, and many of the sensing strategies discussed therein could be translated easily into a biosensor format. Host–guest chemistry, in general, should be a fertile area for further experimentation with regard to biosensors.

Molecular imprinting involves the synthesis of stable polymers which have a templated binding site for a molecule that is structurally complementary to the binding site. The polymer is referred to as a molecularly imprinted polymer, MIP. Molecular imprinting is newer to the biosensing world than it is to the worlds of chemical synthesis and affinity chromatography. Mosbach (1994, 1996), Sellergren (1997) and Kriz *et al.* (1997)

have provided reviews of the topic. Mosbach's group (Kriz *et al.*, 1995; Mayes and Mosbach, 1996) introduced biomimetic sensors based on molecularly imprinted polymers as recognition agents. Using a fiber-optic probe a dansyl derivative of phenylalanine was bound to the MIP. Although the MIP exhibited chiral selectivity, the prototype device suffered from poor signal/noise and long response time. The authors suggested that those problems can be overcome with design changes. Mosbach's group also used the MIP approach for an amperometric sensor for morphines (Kriz and Mosbach, 1995). More recently they have applied the MIP technique to produce "artificial" antibodies to corticosteriods (Kempe, 1996; Ramström *et al.*, 1996).

Matsui *et al.* (1995) developed an atrazine-sensitive MIP that should be applicable for sensing applications. The authors noted that while selectivity for atrazine relative to that observed for antibodies (Oroszlan *et al.*, 1993b; Tom-Moy *et al.*, 1995) was comparable, the atrazine affinity for the MIP seemed to be less than for antibodies. That may be improved with further developments. In a very different approach to using MIPs, Hutchins and Bachas (1995) reported a nitrate-selective electrode produced by an electrochemically mediated imprinting/doping method. Using a glassy-carbon electrode, polypyrrole was electrochemically polymerized in the presence of nitrate ion, thus forming nitrate sites in the polypyrrole network on the surface. Kugimiya *et al.* (1995) have also used an MIP for molecular recognition of sialic acid. Anderssen (1996) developed an MIP for s-propanolol which was used for a radioligand-binding study, but the method should be translatable to a sensing configuration.

Both the host–guest and MIP approaches involve synthetic species for molecular recognition. In the past only biocomponents were considered to be molecular recognition components for bioanalytical sensors. Yet, biocomponents are not always easily available, cost-effective or sufficiently stable. The opportunity for other alternatives which might lead to selective, stable devices presents definite avenues for further exploration.

One of the most exciting new developments in this area is that of nucleic acid ligands (NAL). They may be thought of as synthetically produced ligands based on naturally occurring molecular recognition agents. A chemical library of oligonucleotides (MW = ~3,000–40,000) is subjected to affinity binding to the target analyte. The oligonucleotides that retain the analyte are captured and amplified by PCR in an iterative process that enriches the affine species (aptamer). Then the enriched effective sequences are utilized as affinity substrates for the analyte for which they were selected. The method is referred to by several terms, including *in vitro* selection and systematic evolution of ligands by exponential enrichment (SELEX). McGown *et al.* (1995) have reviewed principles and applications of NALs and summarized some comparisons to antibodies. An application for patterned immobilization of folic acid with fluorescence detection of tagged NALs has also been reported (Kawazoe *et al.*, 1996).

While synthetic molecular recognition agents show great promise, there are still many options for biocomponents that need study. Genetic engineering of molecular recognition agents for biosensors is still in its infancy. Immobilization of cyt b_5 on a stationary support has been controlled by site-directed mutagenesis so that the proper cyt b_5–cyt *c* binding orientation is maintained (McLean *et al.*, 1993). The maltose binding protein was altered to optimize fluorescence effects near the maltose binding site (Gilardi *et al.*, 1994). Daunert's group has used genetically engineered bacteria for determining antimonite and arsenite (Scott *et al.*, 1997). Both Sigal *et al.* (1996) and Schmid *et al.* (1997) have used genetically engineered histidine-tagged proteins for studying oriented immobilization of proteins at surfaces.

The human genome project has created an explosive growth in sensors for nucleic acids and for studing hybridization. In the realm of genosensors high-density oligonucleotide arrays have been produced by different technologies and commercialized (Beattie *et al.*, 1995). For examples based on light-directed synthesis and ink-jet printing, respectively, see Pease *et al.* (1994) and Blanchard *et al.* (1996) plus the references therein. The Gette and Kreiner (1997) review updates applications of the device, described in the Pease report, introduced by Fodor *et al.* (1991). Several other types of sequence-specific sensors for DNA and hybrids have been developed (Millan and Mikkelsen, 1993; Hashimoto *et al.*, 1994; Millan *et al.*, 1994; Su *et al.*, 1994; Stimpson *et al.*, 1995; Ferguson and Walt, 1996; Wang, J. *et al.*, 1996a, 1997).

In the second category of advancements—improving immobilizations—it may be said that research has moved from the early days of "fence-it-in" with a membrane or "paint-it-on" the surface to sophisticated techniques for modifying transducer surfaces with:

- self-assembled Au–thiolate monolayers (Chidsey and Loiacono, 1990; Abbott *et al.*, 1994; Willner and Riklin, 1994, Willner *et al.*, 1993, 1996; Allara, 1995; Cheng and Brajter-Toth, 1995; Creager and Olsen, 1995; Frey *et al.*, 1995; Knichel *et al.*, 1995; Riklin *et al.*, 1995; Finklea, 1996; Sigal *et al.*, 1996);
- avidin-biotin binding (Pantano and Kuhr, 1993; Anzai *et al.*, 1994; Frey *et al.*, 1995; Caruso *et al.*, 1997; Dontha *et al.*, 1997);
- sol–gel preparations (Avnir *et al.*, 1994; Dave *et al.*, 1994; Narang *et al.*, 1994a,b; Lev *et al.*, 1995);
- phospholipids (Kallury *et al.*, 1992, 1993; Kallury and Thompson, 1994; Ong *et al.*, 1994; Pidgeon *et al.*, 1994; Shrive *et al.*, 1995; Wang, H. *et al.*, 1995);

- supported membranes (Nikolelis and Siontorou, 1995; Nikolelis *et al.*, 1996; Zhong and Porter, 1995; Sackmann, 1996).

With these methods, carefully designed, intricate molecular networks that bind or entrap analytes with better defined matrices on nanometric dimensional scales are possible. Thus, it is possible to control sensitivity and selectivity more closely by building layers with specific pore sizes, charge distributions and bonding characteristics.

Finally, chemical modifications for altering the actual energy transducers are being reported. Catalytic carbons (Newman *et al.*, 1995), new carbon composites (Alegret *et al.*, 1996; Santandreu *et al.*, 1997; Švorc *et al.*, 1997) and novel carbon paste formulations (Gorton, 1995; Kalcher *et al.*, 1995) have been reported. Miniaturized thermal sensors (Bataillard, 1993; Bataillard *et al.*, 1993; Xie *et al.*, 1994b; Shimohigoshi *et al.*, 1995) also fall into this category. But there are many other developments on the horizon—e.g., polymeric semiconductors and the so-called "smart materials." Martin's "nanomaterials", created by deposition of materials in nanoporous membranes (Figure 1-17), suggest many biosensing possibilities ranging from nanoscopic electrode arrays with a common biocomponent to multiple electrodes with multiple biocomponents (Menon and Martin, 1995).

It would not have been possible to reach the current level of structural design of interfaces had it not been for involvement of improved methods for spatial resolution in studying surfaces (Hubbard, 1995). Techniques using micropositioning, scanning probe microscopies, surface-enhanced spectral methods, mass-sensitive measurements, and electrochemical characterizations have enabled analysts to eavesdrop in the domains of these new molecular architectures with high resolution. Three approaches to scanning microscopy are schematized in Figure 1-23. The Hubbard (1995) book and Bard *et al.*'s (1994b) presentation of scanning electrochemical microscopy are rich with principles, applications and experimental details.

Perry and Somorjai (1994), Brecht and Gauglitz (1995), and Göpel and Heiduschka (1995) have recently reviewed analytical tools for characterizing thin films. The varied approaches to characterization are neccessary as informational guidance for the design of surfaces applicable for biosensors. The cited authors review factors related to depth profiling, lateral resolution, bonding forces and molecular dimensions of supramolecular complexes related to membrane transport, electron transfer, and photosynthesis. In order to move to the next level in designed thin films it will be absolutely necessary to achieve spatial resolution on the order of a few nanometers, typical of natural reaction sites in living systems. This goal is well in sight using the collective arsenal of spectroscopic and

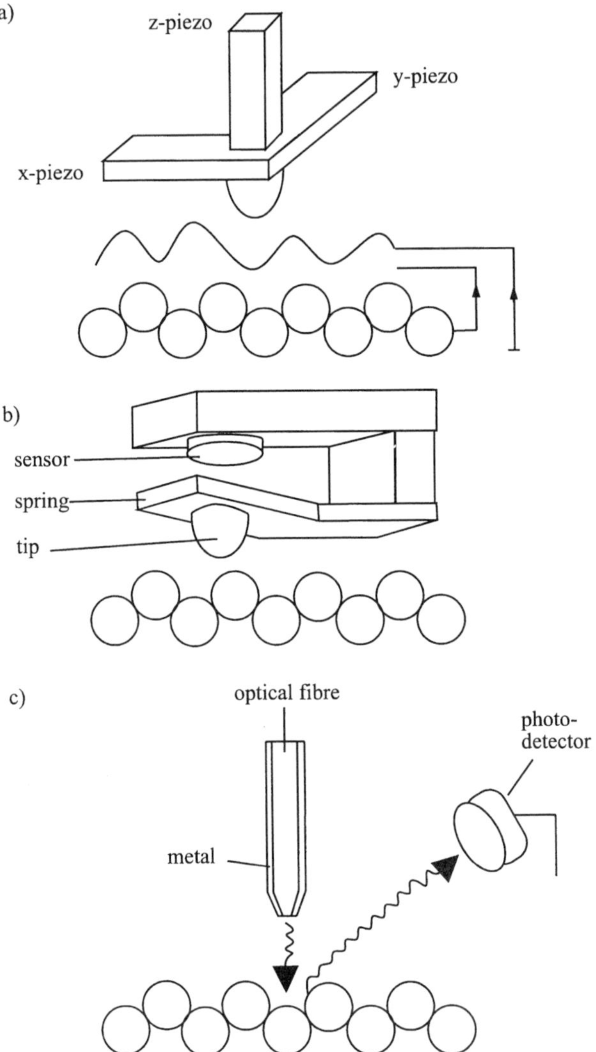

FIG. 1-23. Sketches of scanning probe microscopies: (a) scanning tunneling microscopy (STM), (b) scanning force microscopy (SFM), and (c) scanning near-field optical microscopy (SNOM). Reprinted from *Biosens. Biolec.* Vol. 10. Göpel, W. and Heiduschka, P., Interface analysis in biosensor design, pp. 853–883. Copyright 1995, with permission from Elsevier Science Ltd, Oxford, UK.

microscopic methods available. No doubt, much of the work in the next few years will be centered precisely in this area of interface characterization and design.

1.6.2. Multicomponent Challenges

Many devices that have resulted from the last thirty years of research are effective tools for selective sensing, though each and every one may be challenged to the utmost in complex sensing environments. Despite all efforts to design totally selective sensors for individual analytes in multi-component samples, a realistic assessment reveals that it cannot be done in every situation. With newer analytical technologies as options some practical decisions must be made about how much effort to expend in driving chemical selectivity of a specific device to its utmost perfection.

Some arrays for multicomponent biochemical analysis have been reported, but with miniaturization of devices, improvements in spatially resolved immobilization methods and more routine use of chemometrics, the development of arrays for all energy transduction modes should accelerate in the future (Hart and Hartley, 1994; Hart and Wring, 1994; Hendji *et al.*, 1994; Hermes *et al.*, 1994; Schmidt *et al.*, 1994; Meyer *et al.*, 1995; Pritchard *et al.*, 1995a,b; Sheppard *et al.*, 1995; Danzer and Schwedt, 1996; Ferguson and Walt, 1996; Gavin and Ewing, 1996; Healy and Walt, 1997). Some of the arrays, such as those reported by the Danzer, Sheppard and Hendji groups, begin the quest for turning characteristically non- or partially selective transducers into configurations amenable to chemometric analysis. Hermes' array of individually addressable amperometric sensors provides a different approach for spatial resolution of concentrations. Diamond (1993) has reviewed progress in electrochemical array research and Carey (1994) has reviewed multivariate arrays for industrial and environmental monitoring.

Arrays may be used to enhance a total signal output via an additive effect. That is basically the principle evident in metallized carbon-paste electrodes. The particles behave as multi-electrodes, thus enhancing the signal. As indicated earlier in this chapter, arrays may also be designed so that each element of the array responds to a different analyte. By impregnating carbon pastes with various metal oxides Chen *et al.* (1993) designed a four-element array for the simultaneous determination of carbohydrates and amino acids. That same group produced array strips for detecting glucose and lactate (Wang, J. and Chen, 1994a). Meyer *et al.* (1995) recently reported a 400-electrode array for oxygen, hydrogen peroxide and glucose. Applications for further multi-analyte sensing were projected. The miniaturized pH/K^+ sensor for implantation in the heart (Cosofret *et al.*, 1995a,b) and Walt's fiber-optic arrays (Pantano and Walt, 1995; Healy and Walt, 1997) are also examples of potentially powerful multianalyte sensors.

When discussing arrays, however, the next level of information content comes from the concomitant use of chemometric techniques. A natural synergy exists for complex bioanalytical samples and chemometric techniques. Except for the Chen work mentioned above, plus Danzer and Schwedt's (1996) and Slama *et al.*'s (1996) recent contributions, much of the research utilizing chemometrics and arrays has dealt with chemical sensors that do not ordinarily fall into the bioanalytical category. Consequently, application of chemometrics in the area of biosensing devices has not been extensive. Part of the reason for slowness in applying chemometrics to biosensing arrays has arisen from difficulties with designing arrays with spatially resolved immobilization of the biocomponents. Advances in micromachining and immobilization procedures for patterning biocomponents on surfaces should foster continued progress toward chemometric analysis of biosensing arrays.

The Chen array mentioned above (Chen *et al.*, 1993) was used in conjunction with partial least squares analysis for calibration/prediction. Their prediction error was 2.3%. Meyer *et al.* (1995) also suggest that their array can be adapted to a wide variety of applications for which pattern recognition methods in multianalyte situations will be useful. Recent work in using artificial neural networks (ANN) for pattern recognition of sensing array data should open many avenues of inquiry for multianalyte analysis (Blank and Brown, 1993; Zupan and Gasteiger, 1993; Wang, Z. *et al.*, 1995; White *et al.*, 1996; Sutter and Jurs, 1997). The discussion and references in Wang's report emphasize the fact that ANNs are particularly applicable to nonlinear data analysis.

Danzer and Schwedt (1996) developed three pH electrodes using acetylcholine esterase, alkaline phosphatase and acid phosphatase as the molecular recognition agents. Using chemometric methods for experimental design and simplex optimization they characterized the electrodes individually as a preliminary test before creating an array for pesticides and heavy metals. The enzymatic inhibition by the analytes was assessed on the basis of pH monitoring before and after exposure to the pollutants. This represents an application in which both the sensor (pH) and the sensing mechanism (enzymatic inhibition) are only partially selective for the analytes, but should be amenable to the combination of array format and chemometric analysis.

There is one more aspect of multicomponent analysis that must be examined when looking to the future—i.e. the microseparation systems introduced in Section 1.2.3. From the toolbox for chemical analysis, consider the conventional approach to analytical problems of multicomponent samples—separation and detection. Put into the equation the previously discussed needs for bioanalytical sensors. Rapid, reliable, selective measurements, some of which must be completed *in situ*, were to be accomplished without prior separations. That was because most of the conventional separation–detection methods of 10–15 years ago required sampling with off-line analysis. With the advent of capillary electrophoresis (CE), microdialysis (MD) and, more recently, miniaturized ultrafiltration (UF), the whole perspective on separation–detection has changed (Robinson and Justice, 1991; Baker, 1996; Schneiderheinze and Hogan, 1996).

The newer micromachined μTAS—i.e., capillary electrophoresis on a chip—poses truly exciting new possibilities (see Woolley *et al.*, 1997 and the references therein). In two 1994 editorials in *Analytical Chemistry* the Editor, Royce Murray, discussed precisely this dual-track development of chemical sensors and the separations-based sensors. With the capabilites of the new separation systems, there may be situations where development of a biosensor is either obviated by the optional approach or is impractical on a time and cost basis. Analysts of the future will enjoy the embarrassment of methodological riches—a wide variety of selective bioanalytical sensors and the separations chip. Choice of "best" method has always been one task of the analytical chemist. This new set of options just represents a new manifestation of that responsibility.

1.6.3. Small Volumes and Miniaturization

The term small volume can have multiple connotations, all of which imply miniaturization necessary for chemical analysis. For example, a natural source of a body fluid or intracellular analyte may be restricted in amount by practical considerations. Alternatives for acquiring and analyzing small volumes of sample must then be explored. Small volume can also mean that the *in situ* environment in which a measurement is being made is spatially confined to a small sampling region. Intracellular measurements would typify this second case. Medical applications and biochemical sources frequently present challenges arising from small volume constraints.

Situations for which a limited volume of bulk sample is available are exemplified by single-cell analyses as reported by the groups of Jorgenson, Yeung, and Ewing (Cooper *et al.*, 1992; Cheung and Yeung, 1994; Gilman and Ewing, 1995). In these analyses individual cells are usually introduced into a CE system, where lysis may occur with expulsion of the cell contents. The volumes of total sample for analysis usually range from nanoliters to femtoliters, depending on the cell from which the sample is acquired. Manipulations of the cells and sample handling must be done under microscopic observation.

Capillary electrophoresis has been the method of choice for these analyses. The technique is ideally suited to handling small volumes of samples. Electrochemical, fluorescence, and laser-induced fluorescence have been utilized as detection methods for the extremely low amounts measured. Microdialysis for sampling extracellular fluids also involves acquisition of an external sample for subsequent separation and detection, but the volumes are considerably larger, nanoliters to microliters (Justice and

Parsons, 1991; Newton and Justice, 1994; ; Zhou, S.Y. *et al.*, 1995; Niwa *et al.*, 1996).

A different connotation of small volumes is found in *in vivo* studies of the brain, intracellular events or the volume beneath a probe used for scanning electrochemical microscopy (SECM). In the last case a simple calculation indicates that attoliters (10^{-18} liters) are involved in the working space (Bard *et al.*, 1994). Electrodes implanted for studies in the brain sample from regions with micrometer dimensions, whereas intracellular measurements may reduce that value to well below a micrometer. Ewing's group (Chen, T. K. *et al.*, 1994) has presented a summary of calculations for amounts of material to be detected as a result of exocytosis from neuronal cells. In their study of exocytosis from rat PC12 cells a detection limit of 31 zmol (10^{-21} mole) was reported. That group and Bratten's have introduced micromachined picoliter "beakers" for some of the small-volume studies (Clark *et al.*, 1997; Bratten *et al.*, 1997).

The analytical challenges posed by these small volume situations provide some insight into the needs for miniaturization. There are other cases that may impose lesser constraints, yet require devices of greatly reduced dimensions relative to ordinary macroscopic tools. Several examples serve to illustrate dimensions required for different situations. The implantable ion–sensor array of Cosofret *et al.* (1995a) , shown in Figure 1-18, has outer dimensions of approximately 1 cm × 0.5 cm. Baker and Gough's (1995) implantable lactate sensors have a diameter of 2 mm with an electrode sensing region of 3 mm length. Wilson's latest design for a glucose sensor (Jung and Wilson, 1996) is for a device of 0.25 mm diameter with a sensing region of 3.5 mm. Meyer *et al.* (1995) designed their 400-electrode array with an overall 1 cm^2 area with each electrode of the array being 36 μm × 36 μm.

In contrast to these devices, Walt's fiber-optic bundle has an outer diameter of 350 μm and contains on the sensing surface ~6000 individual sensing units (Bronk and Walt, 1995; Healy and Walt, 1997). In the Healy report the authors propose possible spatial resolution of about 7 μm using the optical arrays. Kopelman's group (Tan and Kopelman, 1996) has developed several nanoscale fiber-optic sensors for bioanalytical applications. Ewing's group designed some of the first sub-micrometer carbon electrodes (Strein and Ewing, 1992) for studies in neuronal cells. These examples provide some basis for comparing dimensional demands of different sensing methodologies and environments. In a very innovative approach Zare's group took up the challenge of combining small sample volumes *and* small detectors. They used living cells as sensors, as depicted in Figure 1-24 (Shear *et al.*, 1995; Fishman *et al.*, 1996).

What the dimensions do not convey is the composite of problems in going from macroscopic prototypes for proof-of-concept to a miniaturized

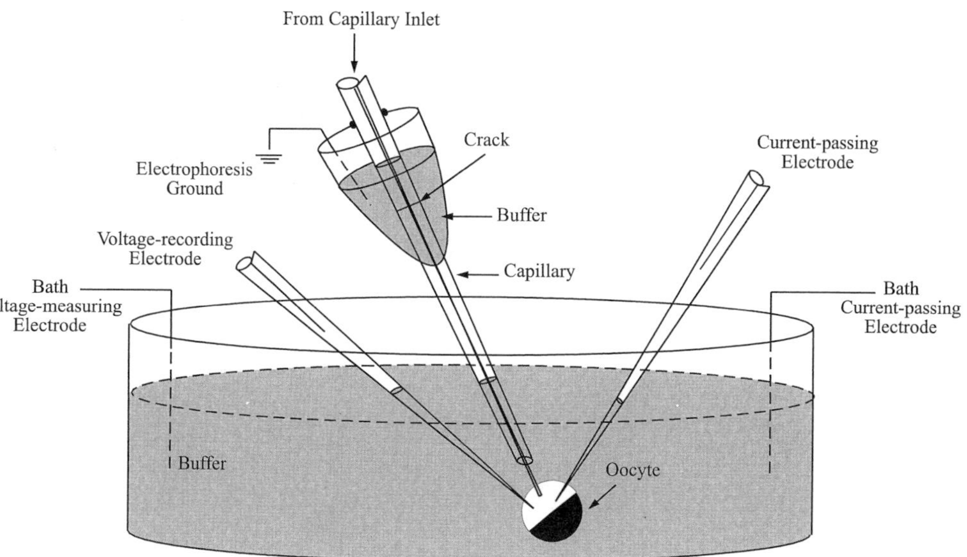

IG. 1-24. Single-cell biosensor. Single cell used as a biosensor for detection of capillary electrophoresis (CE) ffluents. Analytes binding to receptors produce changes in the plasma membrane ion permeability. Membrane urrent measurements were made with a two-electrode voltage clamp amplifier, an approach common to phy-ological measurements. Reprinted, with permission, from Shear, J. B. *et al.*, *Science* **267**: 74–77. Copyright 995, American Association for the Advancement of Science, Washington, DC.

version. A typical example involves developments in designing the implantable ion sensor of Cosofret *et al.* (1995b). In this report they point out that previous studies dealt with various developmental factors such as *in vivo* suitability, membrane optimization, reducing preconditioning time, improving adhesion of the membrane on the substrate, and improving biocompatibility. Buck *et al.* (1992) also reviewed the stages of development for that same device.

There are other challenges associated with miniaturization of devices. Furthermore, the specific hurdles in development are shaped by the eventual application for the sensor. The examples given have been from the biomedical and *in vivo* arena, but the challenges for environmental and bioprocessing sensors can be equally difficult, though different.

Some of the problems have to do with structural and fabrication considerations. There may be cross-talk between closely spaced electrodes in an electrochemical array. A miniaturized optical sensing unit might require a miniature source of radiation, such as an LED, and a miniaturized detector, usually a photodiode. Acoustic-wave devices suffer from extreme fragility of the components. These are all concerns when moving from macroscale prototypes to miniaturized devices.

Beyond fabrication challenges there are more subtle considerations for miniaturized devices. Theory and practice must agree if calibration and predicted responses are to be accurate. With more experience in the area of miniaturized devices the importance of understanding the uniqueness of confined spaces has become increasingly obvious. Many of the assumptions about macroscopic behavior are not valid for the microspace environment. Re-examination of theories about fluidics, electroneutrality, electroosmotic behavior, diffusion, and electrokinetic effects is being demanded by the prevalence of miniaturization in sensing devices. It is a theoretical field ripe for harvest.

1.6.4. Implantable Devices

Dimensional aspects of implantable devices were covered in the preceding section. There are other significant considerations in design of implantable devices. Biocompatibility and sensor stability are two of the major considerations. The review by Meyerhoff (1993) provides a good summary of factors to be dealt with in sensor design for *in vivo* applications, specifically intraarterial measurements of blood gases and electrolytes. Normal ranges of analytes, device sizes, flow-rate effects, stability, and sterilization requirements are discussed. The chapter by Claremont and Pickup (1989) and references therein also provide information on *in vivo* devices and the performance criteria.

When discussing biocompatibility of an implantable device, the encapsulation material is a major factor. This material normally consists of a biocompatible polymer such as polyurethanes or silicone rubber. Cosofret *et al.* (1995b) found that a poly-HEMA (2-hydroxyethyl methacrylate) encapsulation worked well for their potentiometric device. Neither duPont's Nafion nor calcium alginate worked as well. Espadas-Torre and Meyerhoff (1995) have reported a comparative study of thrombogenic characteristics of several polymeric materials utilized in ion-selective electrodes. They found that a polyurethane, treated with poly(ethylene oxide) (PEO), produced the best figures of merit for the potentiometric response. Brooks *et al.* (1996) studied the effect of immobilized heparin on a potassium-selective potentiometric electrode. They found that the response was unaffected and suggested that immobilized heparin might improve the biocompatibility of this type of sensor.

For implantable biosensors the requirements for tissue compatibility certainly must be met. Additionally, the encapsulating membrane must have permeability suitable for the sensing mode. For potentiometric devices this implies sufficient permeability of the ions for accurate and stable signals. Yet, the device cannot be subject to signal variations due to normal ionic or pH excursions. In the amperometric mode the diffusion patterns in the membrane are of prime importance. Thus, for implantable sensors the normal concerns for tissue compatibility are accompanied by a rather stringent requirement for transduction compatibility.

1.6.5. Non-aqueous Media

For a very long time it was thought that enzymatic activity was lost in non-aqueous media. Consequently, essentially all studies of enzymatic catalysis were performed in aqueous media. Logically, biosensors based on enzymatic catalysis were developed initially for the aqueous environment. More recently it has become apparent that not only is enzymatic activity retained in non-aqueous media, but some characteristics of the reactions have significant practical implications (Zaks and Klibanov, 1988; Dong and Guo, 1994; Yang and Murray, 1994; Borzeix *et al.*, 1995; Iwuoha *et al.*, 1995, 1997; Kamat *et al.*, 1995; Stancil *et al.*, 1995; Guo and Dong, 1997). In their study of horseradish peroxidase kinetics in organic solvents, Yang and Murray (1994) summarized the supposed advantages of working with enzymes in organic media, but pointed out that little is known about enzymatic reaction mechanisms in organic phases. The Kamat *et al.* (1995) review of enzymes in supercritical fluids provides a good perspective on the advantages and the problems yet to be solved. A great deal of research remains to be done in this area.

The first excursions into the world of enzymatic catalysis in non-aqueous media involved studies in organic solvents. More recently interest has developed in development of sensors for supercritical fluids (Kamat *et al.*, 1995). Microbial contamination may be lessened in non-aqueous media or separations may be facilitated. Favorable shifts in equilibria may favor synthesis in bioprocessing. Development of biosensing devices for non-aqueous media is particularly valuable for industrial and environmental applications. Frequently for those applications the analytes are not soluble in water.

The interest in organic-phase enzyme reactions spurred development of several organic-phase enzyme electrodes, OPEEs (Guilbault and Mascini, 1993; Dong and Guo, 1994; Deng and Dong, 1995; Mabrouk, 1996; Guo and Dong, 1997). Exactly what the role of water is for construction of those electrodes and use of them is still questionable. A microenvironment of water (microaqueous phase) is apparently necessary for activity of the enzymes (Zaks and Klibanov, 1988; Borzeix *et al.*, 1995), though the minimal amount is subject to more research (Mabrouk, 1996; Guo and Dong, 1997). Different preparations of the electrodes can produce behaviors dependent on the preparative methods. Predictability of response and standardization of fabrication cannot occur until there is agreement on the actual role of the water.

1.6.6. Non-invasive Methods

In previous sections there have been repeated references to situations in which invasive sensors either do not work properly or are harmful to the sensing environment. For these cases development of non-invasive methods is definitely of interest. Lifetime-based fluorimetry for sensing oxygen through skin has been utilized by Bambot *et al.* (1995a,b). There has been a series of reports on near-infrared measurement and chemometric analysis of data for glucose in complex matrices (Marquardt *et al.*, 1993; Small *et al.*, 1993; Shaffer *et al.*, 1996). Ito *et al.* (1995) have developed an ion-selective field-effect transistor (ISFET) for monitoring glucose transcutaneously. The infrared tomography method of Wienke *et al.* (1996a,b) may have real potential for non-invasive sensing if resolution can be improved.

Although much has been accomplished in each of the developmental areas discussed in this section, progress to date has merely provided stimuli for some incredibly interesting research projects in the future. A newcomer to almost any area of analytical sciences can be accommodated by future needs in understanding and developing bioanalytical sensors. The fun has just begun.

2. Designing for Performance

As indicated in Chapter 1, there are many combinations of chemical and energy transduction possible for designing discrete biosensors. These discrete biosensors—electrochemical, optical, thermal, or acoustic wave—may then be used as stand-alone devices, integrated into arrays of individual units or incorporated into sensing systems which involve additional components for sample handling and data acquisition. Because different types of sensors produce assorted distinctive response functions, there is no singular theory of response functions. Yet, there are aspects of design parameters and response characterization common to fabrication and performance of the various types of sensors.

Although biosensing devices are normally classified according to energy transduction, chemical transduction or analyte, another helpful categorization according to design considerations exists. Sensors or sensing systems may be designed for use as:

- disposable, one-shot devices, which require minimal or no added reagents and may be used by untrained personnel;
- portable, hand-held devices for repeated measurements, but may also have disposable or renewable sensing units and may be used by minimally trained personnel;

- sensing units for convective, batch mode or sample injection measurements;
- *in situ* devices for continuous monitoring, many of which must (a) be sterilizable, (b) require no introduction of additional reagents, (c) present no leachable components to sample, and (d) require either infrequent calibration or no re-calibration after introduction to the sampling site; or
- research devices for basic investigations of biochemical processes and sensing strategies, but not necessarily destined for mass production, portability or use by untrained personnel

The vast majority of bioanalytical sensors involve integration of chemical transduction and energy transduction to produce an overall response based on (a) ion-selective detection, (b) biocatalytic processes, or (c) reactions involving complementary ligand binding. Some of the minimally modified carbon-fiber electrodes used for *in vivo* measurements and acoustic-wave or optical devices which depend solely on vapor sorption provide examples of exceptions to the generality. For the most part, however, biosensors are fabricated with energy transducers modified with: ion-sensitive membranes; enzymes, cells or tissues immobilized in various types of films and membranes; or surface-bound complementary binding agents such as antigens, antibodies, oligonucleotides, molecularly imprinted polymers or nucleic acid ligands.

A general model of a discrete bioanalytical sensor is illustrated in Figure 2-1. Molecular recognition components are coupled in some manner to a transducer surface. In some cases there are auxiliary overlayers, shown in Figure 2-1 as the exclusionary or protective membrane. This may provide an additional element of selectivity, may be added for protection from fouling or may be required for biocompatibility. Analyte, usually accompanied by interferents, may be transported to the sensor by convection, diffusion or migration. Analyte partitions into and permeates the membrane or binds to a complementary molecular component of the modifying layer. On or within the membrane a molecular recognition event produces the detectable species which is responsible for signal generation. Diffusion within the membrane is shown in this sketch to emphasize two important points: (1) for some sensing membranes diffusion of chemical transduction products away from the sensor must be considered along with diffusion of analyte to the sensor; and (2) diffusion in the membrane differs from diffusion in the sample. The usual transducer surfaces on which these membranes or films are mounted are platinum, gold, various forms of carbon, glasses, semiconductors and quartz. It will be shown in Chapter 3 that several generally

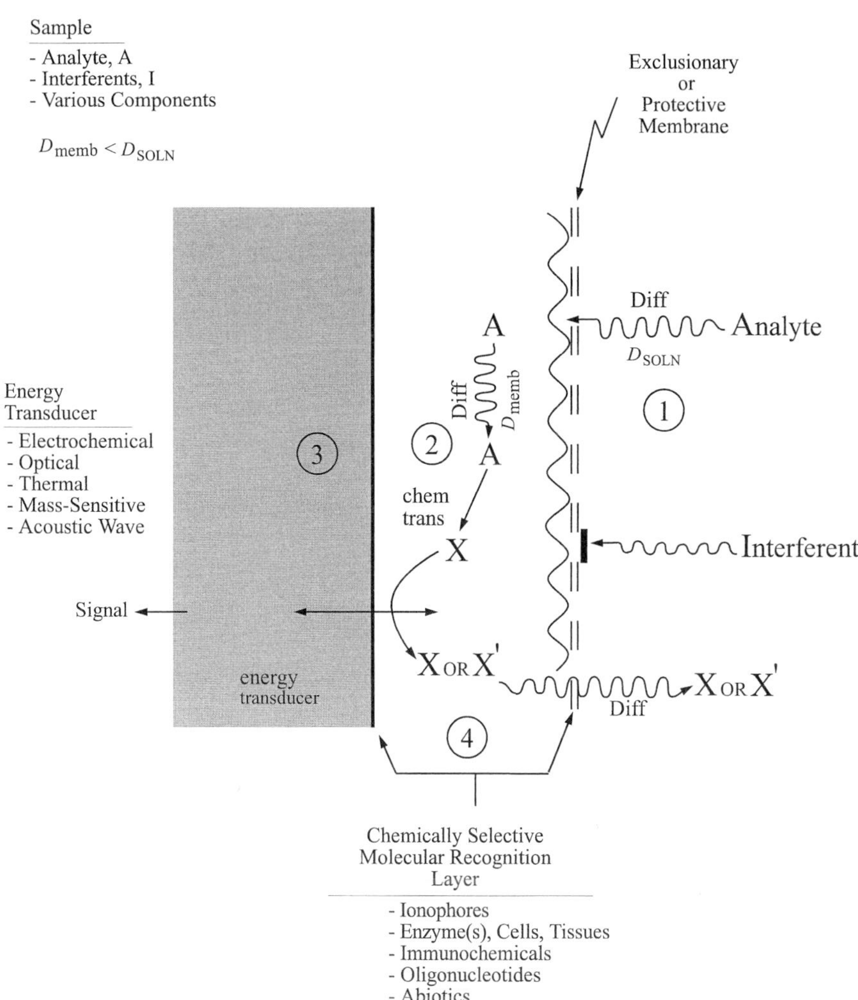

FIG. 2-1. Schematic of a generalized biosensor design: A, analyte; I, interferent; X, product of reaction in the membrane; D_{soln}, diffusion coefficient of A in the solution; D_{memb}, diffusion coefficient of A in the membrane. Not to scale; the thickness of the membrane is only nanometers to micrometers.

applicable immobilization schemes can often be customized with relatively little variation for most of the surfaces.

Although any of chemical transduction modes could be used, theoretically, with any of the ordinary energy transduction modes of detection, certain combinations for a given application may be more feasible than others. For example, mass-sensitive detection requires a chemical transformation resulting in a sufficient mass change to provide the necessary sensitivity. This requirement suggests that complementary binding processes between molecules of high molecular weights—e.g., antibodies with antigens or complementary nucleotides—may be more amenable to this type of detection than will reactions involving low molecular weight substances. Enzymatically catalyzed electron transfer reactions are best suited to amperometric measurements in which current flow due to the electron transfers is monitored. A basic sequence for this type of process is illustrated in Figure 2-2. The transducer→membrane distance represents the thickness *d* of the reaction layer. Potentiometric detection of potassium ions in a biological sample is illustrated in slightly different format in Figure 2-3 with both the sample/membrane and membrane/transducer interfaces emphasized.

The excellent detection limits afforded by various fluorescence modes often suggest this as a preferred detection strategy for species that have native fluorescence or those that can be tagged easily, even though other sensing approaches may also be applicable. A simple direct interaction of a fluorescently tagged antigen (Ag) interacting with immobilized antibody (Ab) is shown in Figure 2-4. In practice other strategies for using Ag/Ab interactions for quantitation are usually chosen. These will be discussed in later sections.

Of the four primary energy transduction modes, thermal sensing based on enthalpic changes is potentially the most universal, though enzymatically catalyzed processes, characterized by favorable heats of reaction, predominate even for that sensing mode. Table 2-1 summarizes some of the advantages and disadvantages of sensors, categorized by energy transduction.

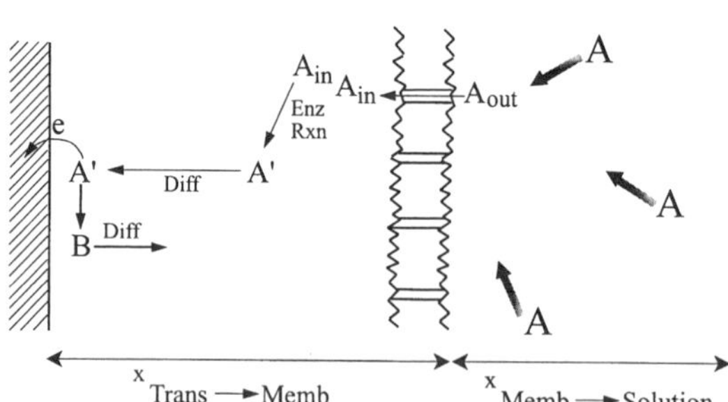

FIG. 2-2. General scheme for an enzymatically catalyzed amperometric sensor. A, a substrate for the immobilized enzyme, partitions into membrane where it reacts via enzymatic catalysis to produce A′, and this product of the enzymatic reaction diffuses the electrode where it is oxidized. The electrooxidized product B diffuses away from the electrode.

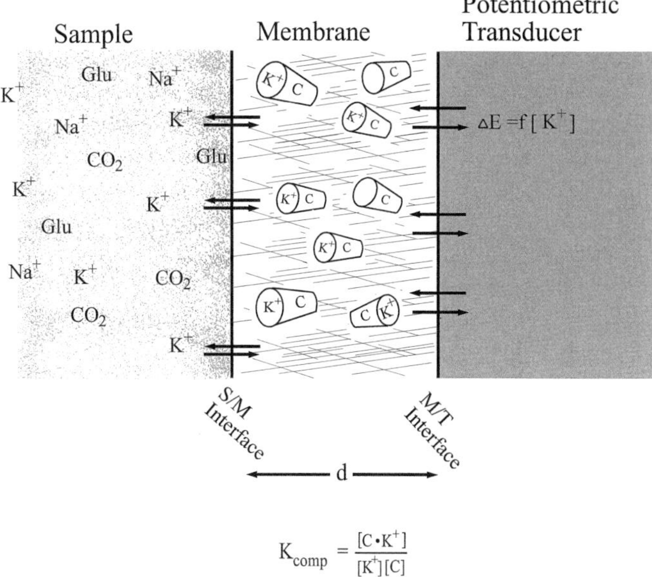

FIG. 2-3. Model of an immobilized ionophore for ion-selective sensing; ΔE is the potentiometric response.

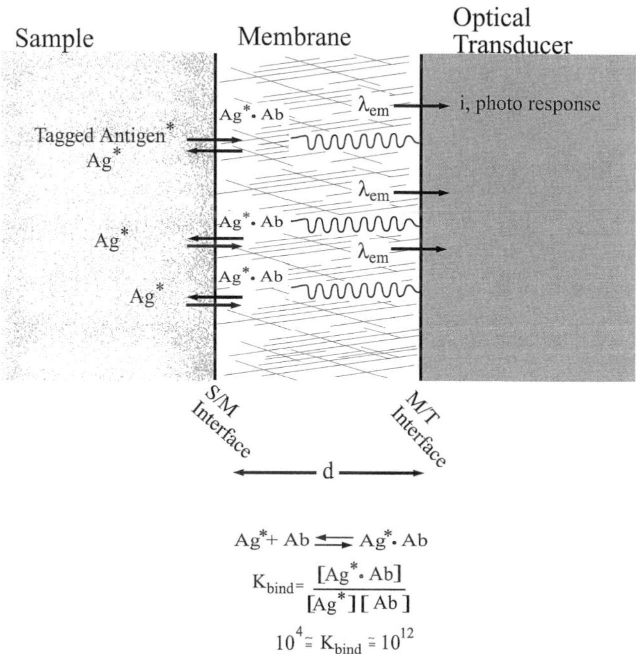

FIG. 2-4. Model of an immunosensor using light emission for detection. Ag^* is fluorophore-tagged antigen in the sample and Ab is immobilized antibody. The process indicated is a direct binding, chosen for simplicity in the illustration. Competitive and sandwich formats are more common immunoassay approaches.

TABLE 2-1 Comparative Qualities for Types of Biosensor

Type of Energy Transduction	Advantages	Disadvantages
Electrochemical		
Potentiometric	ISE translation relatively easy Easily miniaturized	Usually requires reference Limited linear range Often pH sensitive
Amperometric	Wide variety of biochemical redox mechanisms as basis Easily miniaturized Good dynamic range, controllable by membrane thickness Relatively good selectivity Can be designed for diffusion control of signal	Requires a reference Multiple membranes or enzymes may be necessary for required selectivity and sensitivity
Conductimetric	Easy to fabricate No reference required Low frequency source	Non-selective unless used in array format
Optical	No reference required, though multiple wavelength measurements help S/N Can be used in multiple modes: intensity, phase, frequency, polarization Evanescent wave interaction provides multiple designs, including real-time monitoring Multianalyte arrays easily fabricated Wide range of electromagnetic spectral range	Ambient light and scattering can be problem Limited dynamic range Miniaturization can affect magnitude of signal Limited selection of chromophores and fluorophores
Thermal	Applicable to all types of reactions Works for turbid or colored solutions Well-suited for off-line measurements Multiple enzyme formats easily accommodated	Inherently non-selective Thermal enzyme probes are inefficient in heat detection Flow-system required for most applications
Mass-sensitive (Acoustic Wave)	Good limit of detection for highly specific binding reactions Good for vapors Adaptable to arrays	Inherently non-selective unless used in arrays Wave propagation in liquid media can be a problem

The use of multiple sensing units to form arrays and the incoporation of discrete biosensors into sampling/detection systems represent a higher level of instrumental integration. Response functions in these cases not only involve the basic characteristics of the discrete units but also may reflect effects of miniaturization, more highly structured geometry, multiple contributors to a signal, convection and flow characteristics, or timing relationships due to tandem designs. Yet the same types of surfaces and chemical selectivity agents are utilized in the actual biosensing detection scheme.

Associating design and performance requires some understanding of the general operative factors affecting response. Identifying these factors for a particular application involves consideration of the entire analytical problem from requirements imposed by the sensing situation to reporting results in terms of the sensor's characteristics, including the figures of merit, FOMs. In this chapter several types of analytical problems for which discrete biosensors can be designed will be examined in general terms. Some of the connections between chemical and energy transduction will be discussed. Later chapters are devoted to more detailed discussions of specific types of energy transduction modes. Finally, the chapter concludes with a few examples of sensors designed for specific problems. Those sensors are examined in the context of their reported FOMs.

2.1. TYPES OF ANALYTICAL PROBLEMS

2.1.1. A Multitude of Analytes

Whether the application is in clinical/biochemical, environmental or process monitoring, there are several general categories of analytes for which bioanalytical sensors are useful:

nutrients and metabolites,
dissolved gases,
cations, anions and polyions,
enzymatic cofactors,
hormones,
drugs,
neurotransmitters,
heavy metals,
organic pollutants, and
toxins and noxious vapors.

While this may seem like quite an extensive menu, reminiscent of analytical sciences as a whole, sensor designs and detection schemes common to various types of sensors tend to lessen the expansiveness of the problem. Table 2-2 provides examples of a few important analytes and sensing alternatives that may be employed.

Clever chemical transformations of original analytes into fewer commonly occurring products reduce the problems considerably. Appendix A1 includes analytes referenced in the bibliography. Perusal of that list will show that a pattern emerges. As mentioned in Section 1.3, rather than developing a unique design for every situation, considerable effort has been directed toward particular types of sensors which are effective for multiple applications—e.g., those for alkali and alkaline earth metal ions, hydrogen peroxide, oxygen, NADH, and antibodies labeled with either fluorophores or enzymes. This consolidation among applications is quite evident in the sensors represented in the glucose list of Appendix A2. Examination of the actual research reports reveals the fact that most of those sensors are based on detection of hydrogen peroxide, oxygen or a few common oxidation–reduction cofactors and mediators.

Conversion of an analyte into a more easily detected species is not new to bioanalytical chemistry. For example, classical determinations of amino acids and sugars usually have involved derivatization or reaction with dyes for optical detection. Measurements without chemical conversions or derivatizations must be based on actual amounts of substances—e.g. microgravimetry—or on some inherent property of the analyte amenable to signal generation. For example, the molecule of interest might be fluorescent, or absorbent at a characteristic wavelength, or participate in an observable electron transfer process. Many biochemical analytes either do not possess these properties or provide insufficient native responses to be practical for most applications. For some applications proteins have been detected directly by the native fluorescence of the tryptophan and tyrosine residues, but discrimination among proteins in a mixture usually requires prior separation. Some, but not all, of the neurotransmitters exhibit electroactivity or native fluorescence.

There has been renewed interest in methods for direct determination of some of these analytes which have proven to be the most intractable in the past. For example, Yeung's group has been very active in developing methods based on laser-induced native fluorescence or particle counting to determine hemoglobin, enzymes and catecholamines at atto- and zeptomolar levels (Chang and Yeung, 1995; Lillard *et al.*, 1996 and references therein). Singhal *et al.* (1997) have devised a highly sensitive sinusoidal voltammetric technique for carbohydrates using 20 μm copper electrodes in a flow-injection analysis. Selective detection is based on discriminative responses at the higher harmonics of the frequency domain. With optimization of phase

TABLE 2-2 Selected Analytes and Sensing Alternatives

Analyte	Type of Sensor	Reference
Atrazine	Acoustic wave (STW)	Tom-Moy *et al.* (1995)
	Amperometric (PET cells)	Preuss and Hall (1995)
Biological oxygen demand	Fiber optic sensor	Preininger *et al.* (1994)
Botulinum toxin	Fiber optic–evanescent wave	Kumar *et al.* (1994)
Carbohydrates	Amperometric	Bilitewski *et al.* (1993)
		Colon *et al.*, 1993
	Amperometric (array)	Chen *et al.* (1993)
	AC voltammetry	Singhal *et al.* (1997)
PCBs	EC immunosensor	Sadik and Van Emon (1996)
Peroxides	Amperometric (enz. elec.)	Mulchandani *et al.* (1995)
	Amperometric (C-fiber)	Csöregi *et al.* (1994a)
Pesticides	FO immunosensor	Bier *et al.* (1992)
	Amperometric	Cagnini *et al.* (1995)
	Optical waveguide (immuno.)	Schipper *et al.* (1995)
Oxygen	Amperometric	Gooding and Hall (1996)
	Amperometric (array)	Hermes *et al.* (1994)
	Fluorescence lifetime	Bambot *et al.* (1995b)
	Near-field optical	Rosenzweig and Kopelman (1995)
Organic vapors	Acoustic wave (TSM)	Dickert *et al.*, (1996, 1997)
	Optical array	White *et al.*, (1996); Sutter and Jurs (1997)
	Acoustic wave (SAW)	Grate *et al.*, (1993a,b,c)
Metals	Cr(VI) electrochem–SAM	Turyan and Mandler (1997)
	Silver, mercury: IS optical	Lerchi *et al.* (1994, 1996)
Sucrose	ISFET	Tobias-Katona and Pecs (1995)
	Amperometric (flow system)	Filho *et al.* (1996)
TNT	Fiber optic–evanescent wave	Shriver-Lake *et al.* (1995)
Glucose	Amperometric	Jung and Wilson (1996)
	Amperometric, wired enzymes	Csöregi *et al.* (1994b)
	Amperometric, sol–gel	Narang *et al.* (1994b)
	Potentiometric	Lisdat *et al.* (1996)
	Thermal (*in vivo* with MD)	Amine *et al.*(1995)
Multianalytes		
Glucose/lactate/glutamate	Amperometric	Pandey and Weetall (1995b)
Glucose/urea/penicillin	Thermal	Bataillard *et al.* (1993)
Glucose/penicillin/ethanol	Thermal	Rank *et al.* (1993)
Glycerol/acetaldehyde/EtOH	Thermal	Rank *et al.* (1995)
Blood gases and electrolytes	Fluorometric	Bruno *et al.* (1997)
Oxygen/pH/carbon dioxide	Amperometric/potentiometric	Arquint *et al.* (1994)

angle and frequency selection they could obtain an LOD of 8 nM glucose. This technique should facilitate advances in the determination of glycoproteins, which have been relatively neglected in building the bioanalytical repertoire. Sierra, Galbán and Castillo (1997) have reported a glucose method based on the intrinsic fluorescence of the flavin group in glucose oxidase. The approach should be adaptable to both quantitation of substrates and kinetic investigations of any enzyme utilizing the flavin moiety as the enzymatic cofactor.

Most bioanalytical sensors are designed, however, for an indirect detection mode in which the analyte of interest is chemically converted to a more easily detected species. For example, reactions summarized in Table 2-3 (Sect. 2.1.2.4) illustrate common enzymatic conversions utilized in biocatalytic biosensors. A comparison of the reactions emphasizes the fact that certain detectable enzymatic products or coreactants predominate: protons, hydrogen peroxide, NADH, oxidation–reduction mediators, oxygen, carbon dioxide, and ammonia. Thus, the breadth of analytes in the previous section can be narrowed considerably to a few commonly occurring substances by incorporating these enzymatic reactions into the sensing mechanism. An analogous situation exists for immunoreactions for which a few enzymatic or fluorescent labels are common to many of the detection schemes. Horseradish peroxidase (HRP), alkaline phosphatase (AP), and fluorescein or fluorescein isothiocyanate (FITC) are examples.

2.1.2. Chemical Transduction Strategies

Based on the basic analytes shown in Section 2.1.1, it is instructive to consider some fundamental approaches to dealing with those categories of analytes. With regard to sensing strategies there is some practical overlap among the categories. For example, ion-selective detection may be applicable for a variety of situations beyond those which require determination of free ions present in samples such as natural waters or body fluids. Detection of ionic products of enzymatic reactions affords a good example. Enzymatic catalysis may be used when the enzyme's substrate is the primary analyte. Enzymatic catalysis may also be used merely as the "reporter" mechanism or tag for immunoreactions when the primary analyte is antigen to an antibody. Acidic and basic dissolved gases, some of which may result from enzymatic catalysis, may be determined by involvement of acid–base variations with either potentiometric or optical pH detection.

2.1.2.1. Ion-selective Sensing

Ion-selective detection is applicable to a great many bioanalytical problems: e.g.,

(a) free ions normally present in biochemical or environmental samples, e.g., H^+, Na^+, K^+, Ca^{2+}, Mg^{2+}, heavy metal ions, Cl^-, salicylate$^-$, PO_4^{3-}, NO_3^-, polyions such as heparin and protamine;

(b) ions that result from acid–base equilibria involving dissolved gases such as CO_2, NH_3, SO_2, NO_x; and

(c) ionic reactants or products of biocatalytic processes.

The importance of pH sensing as an ion-selective mode must not be overlooked in this overview. Development of durable, stable and accurate pH sensors has been an important chapter in the history of bioanalytical sensing.

The most widely used approaches to ion-selective sensing have been based on potentiometric or optical techniques. For potentiometric measurements solid-state conductors, polymer membrane-based electrodes and field-effect transistors are utilized. The experimental versatility of polymer membrane-based ion-selective devices has put them at the forefront of ion detection schemes. The basic potentiometric arrangement which was referred to in Figure 2-3, can also be represented as:

External ref	Sample a_x	Membrane	Internal filling solution a_{const}	AgCl/Ag

For optical sensing using polymeric membrane-based sensors, the sample/membrane segment may be incorporated onto an optical transducer.

Ion-selective membranes usually consist of polymeric matrices of varying polarities, ion carriers for ion exchange or complexation and additives which provide adjustments to the membrane resistance and ionic strength. Later sections will include discussion of various chemical modifications invoked for conveying ion selectivity, which results from a combination of ion partitioning effects and complexation equilibria within the selective membranes.

In Section 1.5.1, there was brief discussion of the fact that detection of anions has been an especial challenge, since effective discrimination among the species rests almost entirely on anion complexation within sensing membranes. As indicated in Chapter 1, the basic problem with anion selectivity arises from the tendency for anions to partition into organic membranes according to their relative lipophilicities (hydrophobicities), a phenomenon represented in terms of the Hofmeister series (Hofmeister, 1888):

$$ClO_4^- > SCN^- > I^- > Sal^- > NO_3^- > Br^- > Cl^- > HCO_3^- > OAc^- > SO_4^{2-} > HPO_4^{2-}$$

Hydrophilicity of anions in aqueous media is based on solvation energy which depends on the charge, basicity and ionic radii of the anions. If an anion is very hydrophilic then it will be less lipophilic. Hydrophobic anions are more lipophilic.

Lipophilicity of a species is usually stated as the logarithm of an octanol/water partitioning constant (log P) for that species, though Valkó *et al.* (1997) have recently suggested an alternative determination. They proposed a "chromatographic hydrophobicity index"(CHI) determined by a fast gradient reversed-phase HPLC method. The introduction to their report develops a good history of definitions and methods for lipophilicity, including references to databases and the recent IUPAC recommendations. The authors also provide the reasoning behind their CHI designation, rather than using the term, lipophilicity, based on binary solvent partitioning.

The relative tendencies to partition are the important factors when comparing two anions and achieving selectivity among ions. In order to shift the chemical order for preferred selectivity toward a given anion in the Hofmeister list, some chemical mechanism other than lipophilic partitioning must be invoked to facilitate a differently selective process. With greater understanding of the origin of selectivities, it is now known that if partitioning is complemented by anion complexation within the membrane the selectivity order can be manipulated for non-Hofmeister ordering of ion extraction into polymeric membranes.

Devising new complexing agents and membrane modifiers for anion selectivity is still underway. Principles of the various approaches have been reviewed by Yim *et al.* (1993). Barker *et al.* (1997) have applied optical sensing of anions using some of the metalloporphyrin complexors discussed in that review for general sensing of anions. Beer (1996) has summarized his group's efforts in synthesizing and characterizing several anion-selective agents which are largely based on organometallic macrocyclics such as modified calix[4]arenes. Anion interactions have been investigated in his studies using both optical and amperometric methods.

Xiao *et al.* (1997) adopted a different approach for anion selectivity. Rather than using organometallic interactions for selective complexation, they developed a very effective chloride-selective reagent based on hydrogen-bonding of Cl^- to a bis-thiourea ionophore. Bachas' group (Hutchins *et al.*, 1997) reported with exceptional thoroughness a guanidinium ionophore for salicylate, and potentially other oxoanions. The ionophore exhibits a high degree of selectivity, though the authors indicate that improvements in detection limits and selectivity must be made for applicability to clinical samples.

Although potentiometric detection has received an enormous amount of attention in past years, it is definitely not the only effective mode in ion-sensitive methodology. For example, by addition of ion-sensitive chromophores to the same types of membrane cocktails, chemical modifications developed for polymer membrane-based electrodes have been adapted for ion-selective optical devices that are also based on ion-selective membranes. Furthermore, pH-sensitive fluorescent dyes have become very useful for some applications related to bioanalytical problems. Bruno *et al.* (1997) have introduced a novel solid-state, multianalyte sensing array based on fluorometric detection. The proof-of-concept report dealt only with determination of pH, but the authors indicated that other analytes among blood gases and electrolytes had been tested successfully. The device, which is discussed further in Chapter 5, utilizes state-of-the-art components for every aspect of the instrumental design. The Bruno report also includes a good summary of comparisons between electrochemical and optical devices within the context of economic viability and mass production.

Since Tsien's pioneering work in cellular imaging by fluorescence (Grynkiewicz *et al.*, 1984; also see summary by Tsien, 1994) there has been increasing interest in the application of sensitive fluorescence measurements in real-time in confined spaces. Calcium, magnesium, sodium, potassium and phosphate ions are particularly important when elucidating biological processes. Shortreed *et al.* (1996b) have utilized fluorescent indicators in miniaturized optical sensors for *in situ* measurements. In the Shortreed study a Calcium Green amine fluorophore was bound to an optical-fiber surface. The ion-selective sensor exhibited a response sufficient for physiological measurements. Perhaps more significant, however, is the fact that the sensor had the required selectivity relative to magnesium, a problem that has plagued analysts in the past.

Although conductivity has been used for years for monitoring reaction kinetics, including enzymatic reactions, the method has been considered a relatively non-selective method for quantitation of ionic analytes in complex samples. This lack of selectivity arises from the fact that conductivity of a solution is based on the migration of all ions in the sample. The ability to couple conductimetric detection with selectivity conveyed by substrate-selective enzymatic catalysis, however, has renewed interest in the utility of that approach for biosensing. The Hendji *et al.* (1994) application for conductimetric determination of urea, glucose, acetylcholine and butrylcholine provides a good example.

Much of the work in ion-selective sensing has been driven by the needs for rapid, reliable clinical measurement of blood electrolytes and gases, determination of ionic balances in cells, and environmental monitoring of heavy metals and pollutants which produce excesses of ions such as nitrates and phosphates. Ion-selective potentiometric devices have an established record in clinical testing, but more recent developments, such as Bruno's fluorometric device mentioned above, have focused on different types of portable, miniaturized versions which can be micromachined and configured into multianalyte devices (Arquint *et al.*, 1994; Lemké *et al.*, 1992; Knoll *et al.*, 1994; Satoh and Iijima, 1995). Undoubtedly these and other types of sensing arrays will be the wave of the future in various applications of biosensing, including ion-selective detection.

One of the most challenging bioanalytical tasks has been that of preparing implantable ion-selective sensors. As is true of all *in vivo* applications, there must be no damage to the tissues and no harmful, leachable components. The sensor must continue to respond properly in the *in situ* environment and macromolecular adsorption or thrombus formation must be avoided or minimized. Meyerhoff (1993) reviewed many of the early developments in the area of implantable potentiometric sensors.

More recently, Cosofret *et al.* (1995a) have developed a solid-state potentiometric pH-K^+ sensor that can be implanted in the heart. Polymeric membranes for their device included ion-selective PVC matrices and a poly(hydroxyethylmethacrylate) (poly-HEMA) membrane for effecting good adhesion of the ion-selective membrane to the Ag/AgCl solid-state device. That same group reported separately on the successful encapsulation of their device with poly-HEMA (Cosofret *et al.*, 1995b).

In addition to the Cosofret studies of biocompatibility required for development of their potentiometric devices, Espadas-Torre and Meyerhoff (1995) addressed testing of biocompatible membrane materials for implantable ion-selective devices. They found that a polyurethane membrane covered with a poly(ethylene oxide) layer produced the best combination of response and biocompatibility.

During the last few years ion-selective detection, both potentiometric and optical, has been extended to polyions (Fu *et al.*, 1994; Wang, E. *et al.*, 1995; van-Kerkhof *et al.*, 1995). Meyerhoff *et al.* (1996) have reviewed developments in this direction. While emphasis has been on negatively charged heparin and positively charged protamine, the approaches are also applicable to other biochemically important polyelectrolytes such as polyphosphate and DNA. When dealing with implantable devices polyionic heparin is often administered to reduce the risks of thrombus formation. A logical question arises in this case: how will the presence of polyionic heparin in a sample matrix affect the response of an ion-sensitive device? Brooks *et al.* (1996) found that heparin on the surface of a potassium-sensitive electrode was not detrimental to performance. Indeed, they suggested that a surface coating of heparin on ion-selective devices might improve biocompatibility.

Unlike ion-selective responses to lower-charge ions based on sample/membrane equilibria, polyions produce a non-equilibrium response due to a different sample/polymer membrane partitioning process. For a polyionic-sensitive potentiometric device the phase-boundary potential for low concentrations of polyion is established by an equilibrium between ionic additives of the membrane and the concentration of the corresponding ions in solution. When the membrane is exposed to polyion in a sample, an exchange of polyion and membrane-bound counterions of lower charges occurs along with ion-pairing of the polyion with cationic additives in the membrane:

$$\mathrm{Pa}_{\mathrm{Aq}}^{-z} + z\mathrm{R}_{\mathrm{m}}^{+} + z\mathrm{A}_{\mathrm{m}}^{-} \rightleftharpoons \mathrm{R}_z\mathrm{Pa}_{\mathrm{m}} + z\mathrm{A}_{\mathrm{Aq}}^{-}$$

Before equilibrium is established, this reaction promotes a gradient of polyion concentration at the outermost boundary of the layer, thus creating a non-equilibrium situation between membrane and the sample. A steady-state signal based on flux of the polyion develops. Although this non-equilibrium response is a significant variation from the usual ion-selective potentiometric mechanism, Meyerhoff's group has shown that quantification can be quite sufficient with detection limits in the low $\mu g\ ml^{-1}$ ranges.

2.1.2.2. Dissolved Gases

Ion-selective sensing can be used as the detection mode for dissolved gases such as CO_2 and NH_3 by virtue of the aqueous phase acid–base properties of the analytes.

$$\mathrm{CO_2 + H_2O \rightleftharpoons H^+ + HCO_3^-}$$
$$\mathrm{NH_3 + H_2O \rightleftharpoons NH_4^+ + OH^-}$$

A typical sensor design includes an analyte (gas) permeable membrane which allows acid–base equilibration behind the membrane with detection by a cation sensing element. The original sensors for these types of measurements were based on the Severinghaus and Bradley (1958) design. In a Severinghaus sensor the gaseous species permeated a porous membrane and moved into an internal aqueous solution where the applicable acid–base equilibrium was established. The accompanying pH changes were monitored by a glass-membrane pH electrode in contact with the aqueous solution:

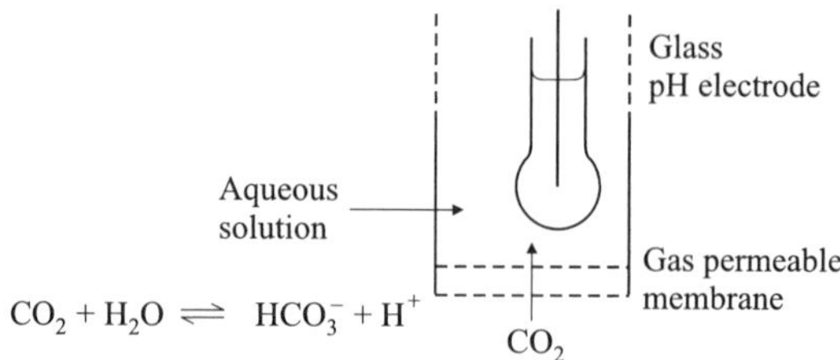

Newer designs utilize polymer membrane-based electrodes or ion-selective field-effect transistors (ISFETS) as the pH sensor, but the sensing principle is essentially the same. By changing to a buffered internal solution and ion-selective membranes for ionic species of the acid–base equilibria other than protons, selectivity can be improved:

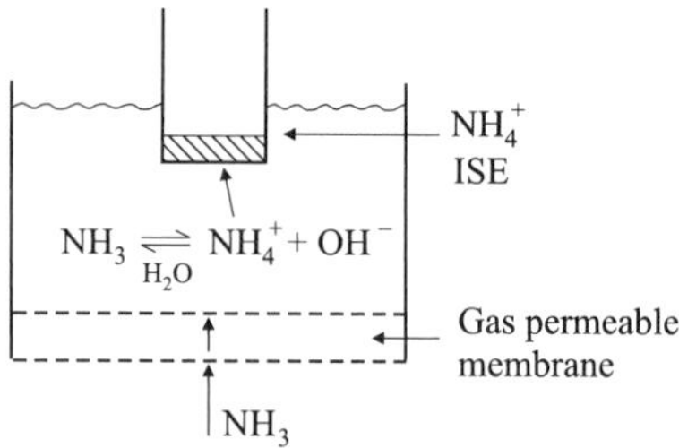

The extensive reviews by Yim *et al.* (1993) and Meyerhoff (1993) include basic principles of potentiometric blood gas sensors as well as several designs that have been tested for *in vivo* applications. Kar and Arnold (1992, 1994) developed fiber-optic devices based on essentially the same chemistries, but utilized an acid–base chromophore for optical detection of NH_3 .

Some of the most extensive work in bioanalytical sensing of dissolved gases has involved methods for dissolved oxygen. Detection of dissolved oxygen is of considerable importance in monitoring blood gases, in assessment of biological oxygen demand (BOD) and for determinations in which oxygen is a co-reactant of biological oxidation–reduction processes. Monitoring dissolved oxygen may be accomplished by any one of several schemes, but the most widely used methods are based on either the amperometric determination by a Clark-type oxygen electrode or on fluorescence quenching of metal complexes, commonly ruthenium or rhenium species (Xu *et al.*, 1994; Demas *et al.*, 1995; Li and Walt, 1995; Rosenzweig and Kopelman, 1995).

While many of the fluorometric methods rely on intensity variations, fluorescence lifetime measurements (Chapter 5) may also be used to good advantage (Lakowicz, 1994; Bambot *et al.*, 1995a,b; Hartmann *et al.*, 1995; Hartmann and Trettnak, 1996; Hartmann and Ziegler, 1996; Draxler and Lippitsch, 1996). Draxler's report, Demas *et al.*(1995) and Xu *et al.* (1994, 1995) all address the basic problem of understanding and modeling fluorescence quenching phenomena in polymeric matrices and on modified surfaces such as those used for biosensors.

While acidic–basic gases and oxygen have a long history in the development of dissolved gas sensing, a more recent challenge has arisen in the need for rapid, sensitive detection of nitric oxide (NO). There is increasing interest in determination of NO, primarily because of its role in intra- and intercellular signal transduction in tissues. Lehninger *et al.* (1993) and the introductory references in Maskus *et al.* (1996) and Zhou, X. and Arnold (1996) may be perused to survey developments in recent years. Nitric oxide is a radical with a very short half-life on the order of 1–6 s. It is a species that can be involved in a complicated series of redox reactions, one of which is oxidation by oxygen (Maskus *et al.*, 1996; Zhou, X. and Arnold, 1996):

$$
\begin{aligned}
2NO(aq) + NO_3^- + 2OH^- &\rightleftharpoons 3NO_2^- + H_2O \\
NO(g) + H_2O &\rightarrow HNO_2 + H^+ + e \\
2NO + 2e &\rightarrow N_2O_2^{2-} \\
2NO + e &\rightarrow N_2O + H_2O(\text{on Hg}) \\
4NO(aq) + O_2(aq) + 2H_2O &\rightarrow 4HNO_2(aq)
\end{aligned}
$$

A sensitive detection method, capable of both spatial and temporal resolution for *in vivo* measurements is required. Several sensing strategies, as indicated by Kikuchi *et al.* (1993), Ichimori *et al.* (1994), Lantoine *et al.*, 1995; Maskus *et al.* (1996), Friedemann *et al.*, (1996), and Zhou, X. and Arnold (1996), have been used for NO. Etches *et al.* (1995) have compared optical and electrochemical NO sensors for clinical applications.

Kikuchi *et al.* (1993) proposed an on-line chemiluminescent (CL) method which could be used for NO from perfused organs:

$$
\begin{aligned}
NO + H_2O_2 &\rightarrow ONOOH \ (\text{peroxynitrite}) \\
ONOO^- + \text{Luminol} &\rightarrow \text{luminescence}(h\nu)
\end{aligned}
$$

They reported pM concentrations in the perfusate which corresponded to a basal level of < 100 fmol per min per gram of kidney weight. The detection limit was reported to be $\sim$100 fM. In order to eliminate spectral interference of hemoglobin in the chemiluminescent reaction, Kikuchi added desferrioxamine to the final reaction solution.

Zhou, X. and Arnold (1996) converted the Kikuchi scheme into an optical-fiber sensor by confining the CL reagents to the fiber-optic tip behind a gas-permeable membrane. Based on the detection strategy employed, Zhou and Arnold were able to formulate a model of the response characteristics and propose a generally applicable mechanism for the effects of oxygen:

Optical fibre

$NO + H_2O_2 + Lum$

$NO \xrightarrow[\times 4]{O_2} 4HNO_2$

The authors pointed out that the oxygen involvement must be considered for all types of NO sensing modes inasmuch as the resultant signals will always be dependent on the relative rates of detection and the complicating oxidation.

Ichimori *et al.* (1994) introduced an amperometric NO selective electrode which is commercially available. The Pt/Ir (0.2 mm diam.) electrode is modified with an NO-selective nitrocellulose membrane and a silicone rubber outer layer. Their electrode was reported to be linearly responsive in nM concentration ranges with a time constant of ~1.5 s. Sensitivity was increased ~3-fold by raising the temperature from 26 °C to the physiological value of 37°C. Measuring NO in rat aortic rings under acetylcholine stimulation was reported as an example of the use of the electrode for *in vivo* applications.

Abruña's group developed a chemically modified glassy carbon electrode (GCE) for electrocatalytic reduction of NO (Maskus *et al.*, 1996). An electrochemically deposited film of vinylterpyridyl metal complex was used. The electrode, used in a solution of denitrifying bacterial cells, was coated with an outer layer of duPont's Nafion (3–5 μm thick) for improvement of selectivity. Submicromolar detection limits and "fast" response were reported, though the details concerning the latter FOM were not included. Characterization of the electrochemical behavior of NO at the modified electrode revealed oxygen effects, which are probably related to the side-reactions noted by Zhou and Arnold.

Friedemann *et al.* (1996) utilized a carbon-fiber electrode modified with an electrodeposited *o*-phenylenediamine (o-Pd) coating. They found that an underlayer of Nafion provided good sensitivity to NO and that selectivity against nitrite was optimized by a 3-layer overcoat of the Nafion. In the study their electrode was compared to a porphyrinic sensor of the type reported by Malinski and Taha (1992). The Friedemann o-Pd sensor exhibited better linearity, lower background charge and a better LOD than did their own version of the Malinski and Taha electrode.

2.1.2.3. Vapors

Detection of gas-phase analytes becomes an analytical problem in numerous applications. Vapors may be important in exhalation measurements and monitoring anesthetics, headspace accumulation in bioreactors, polluting organic vapors in the atmosphere or deadly vapors such as nerve or mustard gases. Much of the original work in vapor detection originated in industrial and environmental applications such as monitoring exhausts and developing the"artificial nose." A general approach is to provide a modified transducer which will respond to changes based on vapor interaction with chemically modifying layers. Sorption of the vapors is a major contributor to the sensing process. Chemical sorbents for the modifications have been adapted in many cases from proven materials for gas chromatographic substrates.

Essentially all energy transduction modes have been employed for vapor detection, oftentimes in applications totally unrelated to bioanalytical problems. Yet, the basic designs and sensing strategies are usually transferable into the bioanalytical realm. A considerable amount of advanced work in vapor detection has been completed with acoustic-wave (AW) devices, which are discussed as energy transducers in Chapter 6.

Some of the first and most extensive characterizations of multianalyte gas-phase samples have been accomplished using these acoustic-wave sensors (Kepley *et al.*, 1992; Grate *et al.*, 1993a,b,c; Martin *et al.*, 1994; Zellers *et al.*, 1995; Bodenhöfer *et al.*, 1996; Dickert *et al.*, 1996). The Kepley report introduces the use of a surface acoustic wave (SAW) device modified with an alkanethiol self-assembled monolayer tailored for selectivity to organophosphonates, which are nerve-stimulants sometimes used as warfare agents. The Grate and Zellers reports provide good introductions to the use of chemometric methods for analyzing and optimizing responses using AW sensing arrays.

The Bodenhöfer and Dickert reports cited above indicate that selectivity by sorption can be supplemented by host–guest inclusion using molecular recognition agents such as cavitands in the adsorptive layer. Grate *et al.* (1996) presented results that point to sorption as the operative mechanism, and they question the extent of contribution by molecular recognition. Dickert *et al.* (1996, 1997) provided spectroscopic evidence and molecular modeling correlations to support their proposal for the involvement of host–guest molecular recognition.

Walt's group (White *et al.*, 1996) has taken a different approach to sensing vapors. Using an optical array they immobilized on individual optical fibers the fluorescent indicator Nile Red in polymeric matrices of various polarities. The fibers were then clustered to form the array. Vaporous analytes were introduced to the array in 2 s pulses. The intensity–time analysis of vapor mixtures was completed by subjecting the results to pattern recognition using an artificial neural network (ANN) treatment. They were able to identify correctly individual analytes in binary mixtures and determine relative concentrations.

Sutter and Jurs (1997) extended the work of Walt's group by refining the ANN treatment of pattern recognition for the time-dependent fluorescence data. They optimized the number and types of descriptors necessary as ANN input for selectivity and quantitation. Success rates for classification of analytes were ~90–100% while quantification success was 97–100%. This type of optical array, the chemical modifications for such arrays and the ANN treatment could easily be translated into some very interesting bioanalytical applications.

Where toxic vapors are concerned another strategy actually employing a biosensor is that of enzyme inhibition. The physiological effects of numerous toxins, pesticides and nerve agents involve inhibition of enzymes, particularly acetylcholinesterase which hydrolyzes the neurotransmitter acetylcholine:

$$\text{Acetylcholine} + H_2O \overset{\text{AChe}}{\rightleftharpoons} \text{Choline} + \text{acetate} + H^+$$

As described by Smit and Rechnitz (1993) when discussing their tyrosinase sensor for toxins, sensing based on enzyme inhibition is a bit like the "canary in the mine" approach to sensing. It should be noted that when using enzyme inhibition as a sensing strategy, there may be an inherent lack of selectivity in the chemical transduction mode, since various inhibitors found in naturally occurring samples may affect the result. On the other hand, if the intent is to signal an alarm of potentially lethal concentations, as in the case of neuroactive compounds, then it matters little what inhibited the neurochemical processes. Hopefully, anything that is harmful will invoke a response. Mionetto *et al.*'s (1994) work illustrates an application of the acetylcholinesterase reaction to biosensing of neuroactive agents.

2.1.2.4. Biocatalytic Products

Biocatalytic sensors use isolated enzymes, often commercially available, or cells and tissues with intact enzymes as the selectivity and chemical transduction components. Table 2-3 includes several of the more common groups of enzymes used for hydrolysis, redox, and group transfer reactions. A great many of the ordinary analytes of Section 2.1.1 can be determined using these illustrative enzymatic reactions. Table 2-4 expands the information to

TABLE 2-3 Typical Enzymatic Conversions for Biosensing

Hydrolases

$$\text{Urea} + H_2O \xrightarrow{\text{Urease}} 2NH_3 + CO_2$$

$$\text{Penicillin} \xrightarrow[\beta\text{-Lactamase}]{\text{Penicillinase}} \text{Penicilloate}^- + H^+$$

$$\text{Acetylcholine} \xrightarrow[\text{cholinesterase}]{\text{Acetyl-}} \text{Choline} + \text{Acetate}^- + H^+$$

Oxidases

$$\text{Glucose} + O_2 \xrightarrow[\text{Oxidase}]{\text{Glucose}} \begin{matrix} \text{Gluconate}^- + H^+ \\ \rightleftharpoons \\ \text{Gluconolactone} + H_2O_2 \end{matrix}$$

$$\text{Glutamate} + O_2 \xrightarrow[\text{Oxidase}]{\text{Glutamate}} \alpha - \text{Ketoglutarate} + NH_3 + H_2O_2$$

$$\text{Amino acid} + H_2O + O_2 \xrightarrow[\text{Acid oxidase}]{\text{Amino}} \text{2-Oxo-Acid} + NH_3 + H_2O_2$$

Dehydrogenases

$$\text{Alcohol} + NAD^+ \xrightarrow[\text{dehydrogenase}]{\text{Alcohol}} \text{Acetaldehyde} + NADH$$

$$\text{Glucose} \xrightarrow[\text{dehydrogenase/PQQ}]{\text{Glucose}} \text{Gluconolactone}$$

Kinases

$$\text{Glucose} + ATP \xrightarrow[M_g^{2+}]{\text{Hexokinase}} \text{Glucose-6-Phosphate} + ADP$$

$$\text{Phosphocreatine} + ADP \xrightarrow[\text{kinase (CK)}]{\text{Creatine}} \text{Creatine} + ATP$$

TABLE 2.4 Examples of Sensing Options

Reaction	Sensing options
$\text{Urea} + H_2O \xrightarrow{\text{Urease}} \begin{matrix} CO_2 \\ \rightleftharpoons \\ HCO_3^- + H^+ \end{matrix} + \begin{matrix} 2NH_3 \\ \rightleftharpoons \\ NH_4^+ + 2OH^- \end{matrix}$	a) Ion-selective sensor b) Optical, using pH-sensitive dye c) Thermal
$\text{Glucose} + O_2 \xrightarrow{\text{GOx}} \begin{matrix} \text{Gluconolactone} + H_2O_2 \\ \rightleftharpoons \\ \text{Gluconate}^- + H^+ \end{matrix}$	a) O_2 Clark type electrode Fluorescence quenching b) H^+ Ion-selective sensor Optical, using pH-sensitive dye c) H_2O_2 Amperometric-direct Amperometric-HRP catalyzed Chemiluminescence Thermal (+catalase)
$\text{Glucose} + NAD^+ \xrightarrow{\text{GDH}} \text{Gluconolactone} + NADH$	a) NADH Amperometric Fluorometric

include some commonly occurring reactants and products for one-enzyme reactions, together with options for detection.

Enzymatic conversions may be sequenced for amplification of a signal or to simplify the detection strategy beyond that obtainable with a single enzymatic step. This strategy results in the design of bi- or tri-enzyme sensors. For example, thermal detection is commonly used for a variety of biosensing systems (Danielsson and Mosbach, 1989; Danielsson and Winquist, 1990). While many of the oxidoreductases catalyze reactions with a fairly good exothermic characteristic, amplification of the thermal signal can be achieved by coupling the catalase action on peroxide to the first reaction:

$$S + E \longrightarrow \overset{H_2O_2}{P} + E \qquad \Delta H_1$$

$$H_2O_2 \xrightarrow{\text{Catalase}} \tfrac{1}{2}O_2 + H_2O \qquad \Delta H_2 = -100\ \text{kJ mol}^{-1}$$

$$\Delta H_{Tot} = \Delta H_1 + (-100)$$

Sequencing of reactions in tandem mode may serve to increase the sensitivity of the method as above, or that approach can be used as a recycling scheme for enhancing the response. In a recycling mode an auxiliary "shuttle" molecule is used for constant re-stimulation of the sequenced reactions involving analyte, thus enhancing detectable signal above the response obtainable with one cycle. For all practical purposes, the sensor's response is amplified by the regenerative effect of recycled analyte or a product related to it. Scheller and Schubert (1992) provide numerous examples of recycling strategies and the analytical advantage gained from that approach. Scheller *et al.* (1995) reported pM detection limits using a recycling scheme with a bienzyme electrode. That report also includes a good comparison of the dynamic range and sensitivity improvements observed by invoking enzymatic recyling.

In a collaborative effort, Scheller has also been involved in devising a flow-injection analysis (FIA) using the recycling strategy to achieve a zeptomolar detection limit for alkaline phosphatase (Bauer, C. G. *et al.*, 1996). Glucose added to the FIA buffer acts as the "shuttle" in this case:

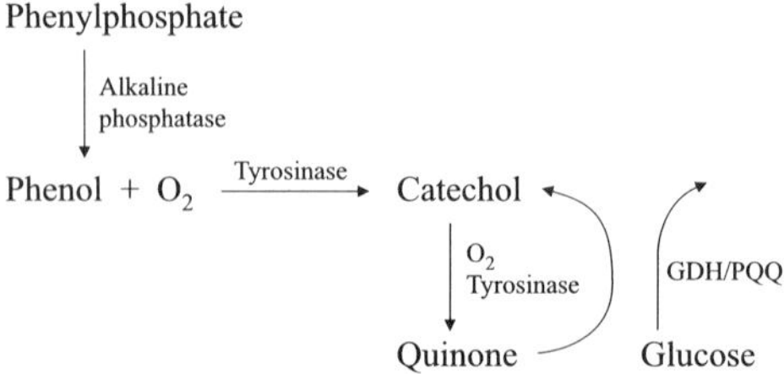

In this sensing scheme, oxygen is detected by a Clark-type amperometric sensor. Thus, recycling results in increased oxygen demand leading to a greater decrease in oxygen-dependent current. The alkaline phosphatase reaction was utilized for an environmental application, an electrochemical immunoassay for the pesticide 2,4-dichlorophenoxyacetic acid (2,4-D).

Using enzymatic sequencing for simplification of detection usually means further conversion beyond first-step products to more universal products such as protons, hydrogen peroxide, ammonium ion, etc. The need for this approach may arise because a more suitable sensor exists for the second-step product, but not the first, or there may be a difficult measurement interference for the first product and not for the second-step product. The bienzyme electrode for choline developed by Michael's group (Garguilo *et al.*, 1993) illustrates the former case. Whereas many of the oxidases can be used directly with some of the newer amperometric electrodes based on immobilized Os-based redox mediators, choline oxidase is not one of them. By coupling a horseradish peroxidase (HRP) reaction to the choline oxidation, however, Garguilo, *et al.* were able to effect a versatile sensor with submicromolar detection limits and good dynamic range for choline:

$$\text{electrode}\xrightarrow{e^-}\; Os^{+3}/Os^{+2} \;\rightleftarrows\; HRP_{red}/HRP_{ox} \;\rightleftarrows\; H_2O_2/H_2O \qquad H_2O_2 + \text{Betaine} \xleftarrow[\text{oxidase}]{\text{Choline}} \text{Choline}$$

Sequenced or parallel enzymatic reactions can also be used as "scrubbing" reactions to remove unwanted products and interferents—e.g., the ubiquitous interferent, ascorbic acid of tissues:

$$\text{Ascorbic acid} + O_2 \xrightarrow[\text{oxidase}]{\text{Ascorbate}} \text{Dehydroascorbate} + H_2O_2$$

In this example, ascorbic acid oxidase may be co-immobilized with other components to eliminate ascorbic acid as an interferent. One problem that arises is that the scrubbing reaction also produces H_2O_2 which also can interfere in some of the amperometric sensors operating at a potential sufficient for H_2O_2 electroactivity. For biocatalytic sensors which are adversely affected by hydrogen peroxide accumulation, catalase may be used as the cleanup or "scrubber" enzyme.

$$H_2O_2 \xrightarrow{\text{Catalase}} \tfrac{1}{2}O_2 + H_2O$$

Regeneration of oxygen in this reaction can actually be an advantage when using oxidases inasmuch as oxygen depletion in natural matrices often limits the utility of the sensor. The effective use of the sequenced and parallel enzymatic reactions relies on the creativity born of reasonable knowledge of enzymology and the compatibility of the various reactants and products of the reactions.

2.1.2.5. Bioligand Binding

Complementary binding based on non-covalent interactions between macromolecules forms the basis of several types of biosensors, most of which have been developed for immunoreaction determinations. The concerted effect of electrostatic, hydrogen bonding and van der Waal's forces in macromolcules can create very strong bonds, as evidenced by large association constants which may be as high as 10^{12}–10^{15} (see Alberts *et al.*, 1989). Antigen–antibody, lectin–carbohydrate, complementary oligonucleotides and nucleic acids comprise the major categories for applications. Biosensors based on these binding pairs often reflect the influence of affinity chromatography for the separation of macromolecular mixtures. The basic principles in both cases pertain to the strong, selective binding which plays such a fundamental role in molecular recognition.

While there is great advantage in the selectivity afforded by such high binding constants, there is also a downside to the strength of the complementary binding. Dissociation may be very difficult, thereby leading to regenerability problems for repeated usage of a biosensor. For example, consider an antigen–antibody interaction with an association constant of 10^{12}. In order to disrupt the associative binding to regenerate a refreshed sensing surface, washes using extreme pH's or chaotropic agents (hydrogen-bonding disrupters), such as urea or guanidinium chloride, may be necessary. Not only does this introduce additional processing steps, which are contrary to the basic philosophy of biosensors, but these necessary washes may also partially destroy the original molelcular recognition layers. Two particularly clever approaches to this problem have been introduced by Willner's group (Lion-Dagan *et al.* 1994a,b) and by Sadik and Van Emon (1996).

In the Willner approach, a photoisomerizable, antigenic dinitrophenyl spiropyran is attached to a self-assembled monolayer on gold. Antibody, as analyte, may be bound and detected when the spiropyran is in the correct form. In order to regenerate the antigenic surface, the surface is illuminated with 360–380 nm light, which inactivates the spiropyran, thus releasing the bound antibody. Then the spiropyran may be re-activated by radiation > 495 nm:

S—WWWO → (Ab, Y) S—WWWO>—

Active ON

λ_{act} λ_{inact}

S—WWWX + Y

OFF

The Sadik approach to the regenerability problem is analogous to Willner's, but electrochemical cycling between potentials performs the release function. Using a conducting polymer (polypyrrole) modified electrode, antibody may be bound to the surface. At an oxidizing potential an antigenic analyte is bound to the antibody and detected. Then a reducing potential is applied to release the antigen, thus regenerating the surface for repeated measurement.

2.1.2.6. Coping With Interferences

Throughout the discussions of previous sections there have been repeated references to different methods of coping with interferences. The same will be true of other sections of the book. Achieving adequate selectivity has been a major experimental odyssey in the development of biosensing strategies. Selective membranes have been mentioned. These membranes normally exclude unwanted substances by charge, size or a combination of charge and size, as indicated in Table 2-5. Edelman and Wang (1992) includes a brief overview of permselective membranes. Various forms of cellulose acetate, duPont's Nafion, electrochemically deposited polymers and phospholipids are among those cited. In that same volume, Fortier *et al.* reported on the use of the Eastman Kodak water-dispersed AQ polymers for enzymatic electrodes. They suggested these polymers as an alternative permselective matrix compared to Nafion.

In addition to the transport modifying membranes, "scrubbing" enzymes, ascorbic acid oxidase and catalase, were mentioned in the previous section as other strategies for coping with interferences. Sometimes the energy transducer itself may be be altered to improve selectivity. For example, elecrochemical pretreatments of carbon electrodes will sometimes separate redox potentials of analytes and interferents for sufficient selectivity. This approach has been widely used for selective detection of neurotransmitters in the brain (Justice, 1987; Wightman and Wipf, 1989; Boulton *et al.*, 1995).

If interferences cannot be eliminated by the usual chemical selectivity strategies, then one must choose either (a) a methodology that assesses interferences and faciltiates analysis in their presence or (b) a means by

TABLE 2-5 Typical Selective Membranes

Components	Comments
Cellulose acetate, CA	Size (and charge) exclusion
Nafion	Charge exclusion, repels anionic species
Cellulose acetate/Nafion	Mixed control
Polypyrrole, PPy	Formed by electropolymerization, porosity controlled by anion during polymerization
Polyurethane, PU	Biocompatible
Polyurethane/poly(vinyl alcohol)/polylysine, PU/PVA/PL	Biocompatible, permeable to ions
Poly(hydroxyethyl methacrylate), polyHEMA	Biocompatible
Poly(vinyl chloride), PVC, with ionophores	Permeable to ions
Self-assembled monolayers of thiols or thiotic acid on Au	Permeability can be controlled by head-groups
Polyvinylpyridine, PVP	Charge exclusion, repels cationic species
Poly(dimethylsiloxane) copolymer	Photopolymerizable, oxygen permeable
Sol–gel films	Allow diffusion of small molecules

which the interferents can be separated from the target analyte. In Chapter 1 the relationships between orders of instruments and extracting information about interferences were discussed. Thus, utilizing arrays comprised of partially selective sensors with chemometric extraction of the analytical information becomes an attractive option. Coupling separation methods, particularly CE and MD, to biosensing devices has also been introduced in the discussions of microseparations.

To state that the method of coping with interferences depends strictly on the nature of the analytical problem and the type of biosensors employed may seem like a sweeping, ambiguous statement. If there is, however, any aspect of bioanalytical sensors that cannot be generalized, it is the involvement of interferences in measurement. The interferences to sensitive detection vary widely among types of sensing situations and among different types of sensors. As the saga of bioanalytical sensors has unfolded, it has become obvious that desired selectivity for single analytes is achieved in only certain applications via sole use of a discrete biosensing element. Further, as the needs for multianalyte sensors increase, the issue of selectivity takes on a new dimension. Thus, research trends are definitely in the direction of coupling microseparation with detection by biosensors or increasing the options for sensing arrays, which can be operated as first- or second-order instruments and subjected to chemometric treatments. Examples throughout the remainder of this book will assume these trends as realities when discussing different sensing strategies that were introduced in Chapter 1.

2.1.3. Trace Concentrations

Across the spectrum of applications for biosensors concentrations and amounts of analytes may vary widely. Concentrations in the mM to μM ranges are frequently encountered, particularly in many of the bioprocessing or clinical problems. Some applications, however, involve much lower naturally occurring concentrations, thus requiring exceptional sensitivity and detection limits. Subnanomolar ($< 10^{-9}$M) concentrations are common for hormones or viral and tumor antigens, as well as some neurotransmitters. Environmental regulations for herbicides, pesticides or heavy metals often require determination of less than a part per billion (ppb). Several examples of previous sections have involved absolute detection limits of femto-, atto- or zeptomoles of analytes. These ultratrace amounts often occur as a result of using minute volumes of samples at very low analyte concentrations—e.g., CE of natural components of cells and tissues.

A variety of sensing strategies can be adopted for these trace-concentration problems. Fluorescence detection and amperometric detection using enzyme-recycling schemes have already been mentioned. Frequently laser-induced fluorescence is utilized with emitted radiation collected by an inverted microscope objective, i.e., chemistry under a microscope, which was mentioned in Chapter 1.

The extremely sensitive methods of enzyme-linked immunoassay (ELISA), usually conducted with optical detection, have been adapted to electrochemical immunoassay (ECIA) using redox-active enzyme-labeled antibodies. This approach was utilized in the Bauer, C.G. *et al.* (1996) study mentioned in Section 2.1.2 as an example of enzymatic recycling for amplification of signal. The extremely low LOD reported for alkaline phosphatase determination was attained by coupling ECIA with enzymatic recycling, a combination that becomes a powerful signal amplification strategy. Heineman's group in Cinncinnati has done much of the pioneering work in applications of ECIA (Heineman and Halsall, 1985).

Duan and Meyerhoff (1994) applied ECIA in an innovative procedure using a modified microporous gold electrode. Antibody was immobilized at the surface on one side of the electrode and analyte was "fed" from the other side, thus keeping the immunoreaction close to the detecting surface. Not only did the technique provide very good sensitivity and detection limits, but also it allowed elimination of washing steps that are usually necessary to remove excess antibody reagent in the bulk solution. The method was applied to whole blood samples for determination of human chorionic gonadotropin (hCG), which may be found in several types of malignant tissues. Detection limits were quite comparable to those of conventional ELISA.

Recently a different approach to dealing with an important analyte at trace concentrations was introduced. Based on foundational work in ECIA, Bard *et al.*'s (1994) scanning electrochemical microscopy (SECM), and Heineman's report of using SECM for imaging immobilized antibodies (Wittstock *et al.*, 1995), Shiku *et al.* (1996) developed an immunoassay which combines ECIA and SECM. Carcinoembryonic antigen (CEA) is a glycoprotein with a molecular weight of ~100,000. It is normally present in some embryonic tissues, but it is also a tumor antigen that occurs in a variety of carcinomas. Shiku's group microspotted (r = ~ 20μm, vol < 20 pL) horseradish peroxidase-labeled CEA–antibody and the antigen on glass slides, then imaged by SECM the reaction sites via a mediated hydrogen peroxide reduction. Their method detected $\sim 10^4$ molecules of CEA.

2.1.4. Temporal Resolution

Transient, steady-state and equilibrium-based responses can be used in biosensing. A transient response is the time-dependent signal observed initially upon analyte or energy transduction stimulus. The transient usually occurs before the sensor stabilizes to either a steady-state or equilibrium-based signal. A steady-state response develops when the rate of supply of detectable species to the sensor equals the rate of signal generation. An equilibrium-based response is one based on detection of an analyte that has attained chemical equilibrium in the analytical matrix, equilibrium with the energy transducer, or both. Immunosensors rely on the attainment of binding equilibria between antigens and antibodies. Some optical sensors are based on the acid–base equilibria of chromophores or fluorophores. Potentiometric methods are based on electrochemical equilibria described by the Nernst relationship. For steady-state and equilibrium-based responses the times for sample acquisition and attainment of the stable signal become the determining factors of both measurement timing and sample throughput. For other situations either continuous monitoring or response times faster than naturally occurring concentration changes are necessary. Continuous monitoring of glucose levels in diabetics and glutamate determinations related to neuronal activity may be used to illustrate considerations of temporal resolution.

Glucose is a common analyte for which sensors of almost every type have been designed. The clinical significance of monitoring glucose in diabetics has been a prime interest for many years. Glucose is also an important analyte in industrial applications ranging from foods to recombinant DNA biotechnology. Classical colorimetric determinations of glucose were based on the reactions of Fehling's solution with reducing sugars. In more recent years measurements using newer versions of color tests or biocatalytic sensors, usually operating in a steady-state amperometric mode, have predominated as preferred methods for glucose determination. The major advantages of the biocatalytic sensors are the selectivity, afforded by the choice of enzyme and modifying membranes, and the capability of *in situ* dynamic monitoring of concentrations.

Many glucose sensors are designed to have a linear range of response for normally occurring concentrations of ~0–25 mM. This range includes the normal (3.5–5.5 mM) value for blood glucose, usual pathological values in diabetic monitoring, and nutrient/metabolic values of biological fermentation processes. Concentrations may go above these values in severe cases of hyperglycemia and in certain other applications such as food processing. A sensitivity that allows concentration resolution of 1 mM or less is required for clinical monitoring, particularly at the hypoglycemic end of the linear range (Claremont and Pickup, 1989).

While time frames for biotechnological applications may allow for sampling and detection times on the order of minutes, clinical monitoring imposes more stringent requirements with regard to time. For example, Home and Alberti (1989) have indicated that for reasonable clinical efficiency, as well as therapeutic response, a glucose sensor should provide a result in 1 minute or less. Within this minute the sensor must be calibrated, a representative sample presented to the sensor and a reliable result made available for interpretation and action.

Working within this kind of time-frame immediately suggests a continuous monitoring strategy, which implies the use of either a rapidly responding transcutaneous sensor or an implanted device. Baker and Gough (1993,1996) have examined timing requirements in terms of response to dynamic changes in glucose concentrations. They formulated their testing and analysis on the basis of potential concentration excursions that might be as high as 0.25 mM min^{-1}. Baker and Gough presented theoretical treatments that define dynamic delay (time lag) and dynamic error in terms of sensor design and external mass transfer to the sensor. They also developed a computer-controlled device for simulating dynamic concentration challenges and testing sensor response times under varying conditions and patterns of concentration changes.

A second consideration of timing and response to normal fluctuations in glucose concentrations is reflected in the report of Thomé-Duret *et al.* (1996). This group presented results which indicate that an implanted subcutaneous glucose sensor detects rising glucose concentrations with a time *lag* between blood and subcutaneous levels, but that decreasing concentrations of glucose, such as those induced by insulin administration, show the opposite effect. They reported that subcutaneous glucose levels *lead*, rather than lag, decreasing blood levels. This represents a significant analytical result, particularly as it relates to hypoglycemic signaling. Moreover, the ability to distinguish different temporal relationships between the increasing and decreasing analyte concentrations attests to the effective, sophisticated designs that have emerged from years of glucose sensor research.

Another bioanalytical challenge involving both sensitivity and timing is that of determining in real time glutamate concentrations related to neuronal activity. Just as in the case of glucose, glutamate is an important analyte for applications other than living tissue measurements. Several groups have reported sensors designed for mM concentrations and relatively slow measurements, sometimes on the 4–10 min time scale (Arnold and Rechnitz, 1989; Karube, 1989b). Some of these sensors are very selective and are useful for a variety of applications in fermentation process control or in food industries.

Quite different challenges exist, however, for the nM to μM glutamate concentrations encountered in samples representative of neuronal activity in the brain or in brain slices. In these cases either *in vivo* modified electrodes (Pantano *et al.*, 1991; Pantano and Kuhr, 1993) or on-line flow systems—flow-injection analysis, microdialysis or capillary electrophoresis—are employed (Berners *et al.*, 1994; Niwa *et al.*, 1996; Cosford and Kuhr, 1996). Ultramicroelectrodes have the advantages of fast response times, aided by fast scan rates, and spatial resolution within tissues. On-line approaches, however, may provide more sensitivity and lower detection limits with adequate timing for some applications of real-time monitoring of extracellular levels of glutamate. For example, the Cosford and Kuhr (1996) capillary biosensor for glutamate has a 450 ms response time. As micromanipulations and microsampling devices are developed for more routine applications, some of the *in vivo* electrodes' advantages may be diminished on the basis of relative analytical efficiency.

2.1.5. Spatial Resolution

Some of the recent developments related to microvolumes and confined spaces were discussed in Section 1.6.3. There are essentially three perspectives when discussing spatial resolution in biosensing. Spatial resolution is a consideration for (1) *in situ* measurements in living systems, (2) designing sensing arrays which may be used for spatially resolved measurements, and (3) imaging biologically related reactions and sites.

(1) Significant efforts have been directed toward conducting measurements in confined spaces—e.g., intracellular measurements or within a single drop of sample. One might also be confronted with the task of monitoring species on or near cells or at particular sites in tissues. This challenge has been met ordinarily by fabricating micro- and ultramicrodevices that can be utilized without damage to the sensing site, yet are large enough to sample

adequately for reliable results. In most cases micropositioning and microscopic observation are required for proper manipulations.

Cahill and Wightman (1995), Cahill *et al.* (1996), Paras and Kennedy (1995), Pihel *et al.* (1994, 1995), Tan and Kopelman (1996), Shear *et al.* (1995), Fishman *et al.* (1995, 1996) and Niwa *et al.* (1996) provide several examples of electrochemical and optical detection schemes, as well as on-line microseparations (MD and CE) for these types of situations. In the Cahill and Pihel reports ultramicroelectrodes (~ 3–5 μm), both beveled and cylindrical, were used to monitor secretions from single cells by placing the electrodes within 1 μm of the cell surface. The Tan and Kopelman review (1996) includes numerous examples of the applications of nanoptodes for intracellular monitoring.

Shear *et al.* (1995) micropositioned the outlet of a 25 μm CE capillary on the surface of living single cells which were used as the sensing elements (Figure 1-24). Both fluorescence and electrochemical monitoring were used to detect the biological effects on the sensing cells of separated cellular components. The Fishman *et al.* (1996) study extended this approach to the use of cellular arrays for sensing. The entire detection scheme is designed for microscopic observation. For the cell-to-cell detection approach the group developed a motorized scanning microscope stage for moving the array beneath the CE outlet tip.

(2) Another consideration of spatial resolution arises when designing a sensor for a patterned or arrayed format. The spatial relationships of sensing sites and spaces between sites can be very important in interpreting signals and preventing unwanted interactions between sites. This is particularly true of electrochemical devices where electrical cross-talk or mass transport between sites may complicate measurements. Sreenivas *et al.* (1996) reported a sputtered carbon array that had minimal cross-talk. The sites were 150 μm × 36 μm with a 200 μm spacing between centers. The electrodes were fabricated so that the implantable shank tapered to a few micrometers at the tip. Spatially resolved monitoring in the brain is the ultimate goal for use of the array.

In a totally different application of spatial resolution Martin's group (Ugo *et al.*, 1996) took experimental advantage of the close spacing characteristic of their nanoelectrode ensemble (NEE), which was illustrated in Figure 1-17. The Au nanowire ensemble, fabricated by filling the pores of a nanoporous membrane, resulted in an electrode density based on 6×10^8 pores per cm^2. When the NEE was coated with an ion-exhange polymer and examined by voltammetric behavior of ferricinium and ruthenium species, greatly improved detection limits were observed because of mass-transport enhancement across the closely spaced redox sites. This is referred to as a "total overlap" condition because the diffusion fields of individual sites overlap to provide a uniform field across the whole ensemble, and thus an enhanced current signal. In contrast to the Sreenivas array, in which diffusional overlap must be avoided for spatially resolved measurements, that same phenomenon is used for analytical advantage in the NEE.

(3) The third aspect of spatial resolution important for biosensing applications involves imaging of surfaces and biologically active sites in tissues and cells. As was mentioned in Chapter 1, imaging methods are frequently utilized for elucidating reaction dynamics and physiological effects in biosensing rather than quantitative determination of specific analytes. A wide variety of imaging approaches have already proven useful to biosensing. Scanning electrochemical microscopy (SECM) (Bard *et al.*, 1991, 1994) and the light-addressable potentiometric sensor (LAPS) (Parce *et al.*, 1989; Owicki and Parce, 1992; McConnell *et al.*, 1992) were mentioned briefly in this context in Chapter 1. Walt's group has introduced several optically based sensors for imaging applications (Bronk *et al.*, 1995; Healy and Walt, 1997). One can only assume that applications will increase with improved technologies for miniaturization and microscopic observations. One of the most exciting developments in recent years has been that of near-field optical methods. Spatial resolutions in the tens of nanometers or better are possible using near-field optics. This approach is included in Chapter 5 on optical methods.

Before surveying some of the biologically related applications of imaging, it should be mentioned that some imaging approaches with potential for applications in biosensing have been reported, but have thus far been applied primarily to non-biochemical problems. Engstrom *et al.* (1992), Fiedler *et al.* (1995), and Bowyer, *et al.* (1996) have utilized fluorescence detection for spatially resolved imaging of electrode reactions. Engstrom's group generated hydroxide ion electrochemically at both platinum disks and a gold minigrid. Using a fluorescence microscope they imaged surface reactions via the fluorescein interaction in going from an acidic to basic form in the presence of hydroxide. For the minigrid electrode fluorescence was limited essentially to the 15 μm wide gold strips, thus indicating good localization with this technique. Engstrom's group has more recently extended their basic technique of fluorescence imaging to heterogeneous oxygen reduction on platinum, silver and glassy carbon electrodes (Bowyer *et al.*, 1996). They concluded that they could obtain micrometer spatial resolution in studies of electrode structure–activity relationships.

In a variation on the Engstrom approach, Fiedler *et al.* (1995) generated pH gradients at ultramicroelectrodes as a prelude to monitoring and creating pH gradients in picovolumes for biological studies. They also formulated mathematical modeling of the time-dependent pH gradients, and observed good agreement between theory and experiment.

One of the imaging techniques for biosensing, particularly important for oxidation–reduction phenomena, has been scanning electrochemical microscopy (SECM) developed by Bard's group (Horrocks *et al.*, 1993a,b; Pierce *et al.*, 1992, Pierce and Bard, 1993; Bard *et al.*, 1991, 1994). The SECM methodology is related to scanning tunneling microcroscopy with the variation that the scanning tip becomes an electrode. The tip is controlled either as a generator or collector of electrons for redox processes occurring in the imaged sample. Thus, various electroanalytical measurements made be made with high spatial resolution—micrometer or better for potentiometrically monitoring biologically related pH changes and enzymatic activity.

For enzymatic reactions the spatial resolution attainable is a mixed function of both the enzyme kinetics and the size of the SECM tip (Pierce and Bard, 1993). Alternative modes of SECM operation can be chosen for optimizing resolution and sensitivity. The Bard, Pierce, and Horrocks references cited above cover a variety of applications for cells, tissues and enzymatic reactions. The Bard, Fan, and Mirkin (1994) chapter on SECM is a complete treatment of theory and applications, including extraordinary details and instructions for the fabrication of SECM tips.

The SECM technique has been used by Heineman's group (Wittstock *et al.*, 1995). Though spatial resolution was less than that of Bard's studies, they were able to determine the density of *active* binding sites in immobilized antibody layers. The authors emphasized that this was an important outcome of the study since there is really no other way to reveal reasons for only partial activity of immobilized antibodies on sensing surfaces. Reduced signals may be due to inefficient energy transduction or to inefficient antibody–antigen binding. Being able to image the bound layers allows one to optimize the immobilization method. As mentioned in an earlier section, Shiku *et al.* (1996) combined results from the Bard and Heineman studies and utilized SECM for their microspotting determination of carcinoembryonic antigen (CEA).

The response of a light-addressable potentiometer (LAPS, Chapter 4) is based on photocurrents generated at a pH-sensitive silicon nitride surface. The device consists of a doped-silicon substrate overlaid with the insulator, silicon nitride, which interfaces with the sample solution.When the silicon is subjected to an alternating on–off cycle of LED illumination, a photocurrent proportional to sample pH at the sample/silicon nitride interface is generated. This type of sensor has been utilized by various groups for monitoring physiological changes for which spatially resolved pH variations can be used as the reporting mechanism (Parce *et al.*, 1989; Owicki *et al.*, 1990; Owicki and Parce, 1992; McConnell *et al.*, 1992; Rogers *et al.*, 1992). The Adami and Nicolini (Adami *et al.*, 1994, 1995a,b; Nicolini *et al.*, 1995) references in the Bibliography include designs and applications of a similar device referred to as the potentiometric alternating biosensor (PAB). Yoshinobu *et al.* (1996) have reported a variation on the LAPS using a scanning laser beam rather than LEDs for stimulation of the response, and spatial, as well as temporal, resolution of responses.

2.2. ION-SELECTIVE MEMBRANES

The measurement of ions in solution by ion-selective electrodes (ISEs) is one of the most widely used analytical methodologies with which all chemists are familiar. Commercially available potentiometric devices of varying selectivities for both cations and anions are commonplace in most laboratories. There is a vast literature covering theory and design of ion-selective devices. A computerized search for ISEs and authors will reveal an incredible number of seminal papers, even if the search is limited to the names E. Bakker, R. P. Buck, M. Lerchi, M.E. Meyerhoff, U. Schaller, W. Simon and Y. Umezawa. Although definitive theories of response and design have coalesced from combined efforts of recent years, important aspects of ion-selective sensing were largely empirically based for a very long time. Consequently, researchers in the field have been particularly attentive to experimental details in research reports.

Many ion-sensitive chemical transduction schemes are based on ion selectivity conveyed by ionophores—ion-exchange agents, charged carriers and neutral carriers—doped in polymeric membranes typical of ion-selective electrodes. In addition to the organic salts commonly used as dissociated exchangers, several macrocylics such as antibiotics, crown ethers and calixarenes have been used as neutral carriers which function by host–guest interactions (Cadogan *et al.*, 1992; Careri *et al.*, 1993; Suzuki *et al.*, 1995; Tsujimura *et al.*, 1995). Simon's group introduced originally a series of ionophores, now commercially available, which are designated by numbers as ETH ionophores (Bakker and Simon, 1992; Lerchi *et al.*, 1992, 1994; Seiler, 1991). The fundamental concept of ion-selective sensing using dissociated exchangers and neutral carriers with lipophilic additives was introduced in Chapter 1 (see Figure 1-15). The Chapter 1 summary can be extended to include charged carriers:

$$\underset{\text{Anion}}{A^-} + \underset{\underset{\text{Carrier}}{Y^-}}{L^+} + \underset{\underset{\text{Lipophilic additive}}{X^+}}{R^-} \rightarrow LA + Y^- + X^+ + R^-$$

As indicated in Section 2.1.2, ion-selective sensing may be implemented on the basis of either potentiometric or optical detection. Yim *et al.* (1993) have reviewed the various approaches to potentiometric sensing. The classic monograph by Morf (1981) includes a great deal of practical data along with theoretical discussions of ion-selective electrode behavior. Bakker *et al.* (1994) and Bakker (1997) present a comprehensive model for potentiometric responses with applications for selectivity in solutions of mixed ions. On a practical basis both potentiometric and optical devices are only partially selective among ions, but experimental comparisons of selectivity may be stated in terms of defined and predictable selectivity coefficients which characterize a particular device (Bakker *et al.*, 1994; Lerchi *et al.*, 1992, 1994; Schaller *et al.*, 1995).

The Swiss group of the late Wilhelm Simon presented theory and practice for a whole series of ion-selective optical probes (see Lerchi *et al.*, 1992 and references therein). Some of the selective membranes developed for potentiometric devices are also used for optodes, but in addition to the usual ionophores for ion selectivity, a chromoionophore is present. The optical response is a function of not only the target ion activity, but also the ratio of protonated to unprotonated chromoionophore concentrations and lipophilic counterions in the membrane. Tuning of response can be accomplished by choice of chromoionophore (via different K_a's) and counterion concentration. Kopelman's group has recently translated this approach into a miniaturized optical sensor for sodium (Shortreed *et al.*, 1996a). The Kopelman sensor uses a pH-sensitive chromoionophore for fluorescence rather than absorbance as the detection mode.

The responses of ISEs are based on measurement of a membrane–sample phase potential which is sensitive to the activity of the ionic analyte of interest. The ion-selective potentiometric response as an electrochemical tranduction mode is discussed further in Chapter 4. The basic experimental arrangement for a polymeric membrane-based device consists of an internal electrode and reference solution, the selective membrane across which an activity-dependent potential difference develops, and an external reference electrode to which the membrane potential is compared in the ΔE measurement:

$$\Delta E_{\text{cell}} = K_{\text{cell}} + \Delta\phi_{\text{memb}} = K_{\text{cell}} + \frac{RT}{zF}\ln\frac{a_{\text{A, sample}}}{a_{\text{A, int ref}}}$$

The response and selectivity of both potentiometric and optically based ion-selective devices are experimentally manipulated by the composition of the membrane. The most commonly used polymeric matrix is poly(vinyl chloride) (PVC). Various forms of PVC are available, including high molecular (HMW-PVC), carboxylated (PVC-COOH) and aminated (NH_2-PVC). A typical membrane "cocktail" for the usual cations and anions consists of polymer (~33 wt%), plasticizer (~65 wt%), ion carrier (~1.0–5.0 wt%), and ionic additives (~0–2 wt%) to provide low electrical resistance in the membrane and to satisfy electroneutrality conditions. Fu *et al.* (1994) noted that the optimum composition of the membrane for their polyion sensor had a significantly different ratio of components. In fact, the % composition of polymer and plasticizer were almost exactly reversed. Selectivity of the ion-selective membrane may be tuned by the absolute and relative values of the components. Table 2-6 lists some of the commonly used materials in each category.

An examination of examples in Table 2-6 indicates that there are various options:

(a) polymeric matrices may be one of several types, usually PVC, polysiloxanes, or methacrylates;

(b) ordinary plasticizers are bis(2-ethylhexyl) sebacate (DOS), bis(1-butylphenyl) adipate (BBPA) and 2-nitrophenyl octyl ether (o-NPOE), which cover a range of dielectric constants from ~2.5 to 25;

TABLE 2-6 Examples of Ion-selective Devices

Ion(s)	Membrane Components	Comment	References
H^+	PIP-PVC, KTpClPB, o-NPOE, ETH 5294, poly-HEMA	Used *in vivo*	Cosofret *et al.* (1995a,b)
K^+	Same, valinomycin		
K^+	Functionalized polysiloxanes, KTTFPB, poly-HEMA, 3 different carriers compared	Used for long-term monitoring under flow conditions	Reinhoudt *et al.* (1994)
Na^+	Room temperature vulcanizing silicone rubber, with oligosiloxane, modified calix[4]arene	Used in human body fluids, no additional plasticizer necessary	Tsujimura *et al.* (1995)
NO_3^-	Electropolymerized polypyrrole on glassy carbon, molecularly imprinted	non-Hofmeister selectivity	Hutchins and Bachas (1995)
Polyions	HMW-PVC, DOS, TDMAC, different ratio of components than usual	Non-equilibrium response reported	Fu *et al.* (1994); Meyerhoff *et al.* (1996)

(c) ion carriers may be dissociated exchangers, neutral carriers, charged carriers or reactive carriers such as the anion-sensitive organometallic complexes; and

(d) lipophilic additives are usually cationic long-chain alkylammonium salts or anionic tetraphenylborates—e.g., tridodecylmethylammonium chloride (TDMACl), sodium tetrakis[3,5-bis(trifluromethyl) phenyl] borate (NaTFPB) or potassium tetrakis(*p*-chlorophenyl) borate (KTpClPB).

If other components, such as a receptor or enzyme, are to be bound to the membrane, PVC is not necessarily the best choice due to the lack of an effective functional group for coupling. Polyurethanes (PU), poly(vinyl alcohol) (PVA), polylysine (PL) and combinations are useful in these cases. These polymeric materials also have the added advantage of being more biocompatible (Yim *et al.*, 1993).

Silicone rubber and a PU/PVC copolymer were reported by Goldberg *et al.* (1994) as good screen-printable ion-selective membranes for sensing arrays. Tsujimura *et al.* (1995) also used a silicone rubber-based membrane containing a modified calix[4]arene for an Na^+ electrode used in body fluids. Kim *et al.* (1997) have introduced a sol–gel ion-selective membrane based on what they called an organic–inorganic hybrid sol–gel. A triethoxysilane and 1,4-butanediol were used for forming the membrane. The electrode exhibited good response characteristics for determination of chloride ions. Selectivity among anions deviated from the usual Hofmeister sequence, even though a dissociated ion exchanger (TDMACl) was used as carrier. The authors did not speculate about the mechanism by which the selectivity differs from that observed with other dissociated carriers.

The preparation of conventional polymer membrane-based devices with internal filling solutions has changed little since the original work of Moody, Oke and Thomas (1970). Components are mixed in the appropriate proportions with a volatile solvent, usually tetrahydrofuran (THF). For macroelectrodes a film is formed on a flat glass or plastic surface for evaporation of THF. Polymeric disks are cut and mounted in commercially available electrode bodies. As mentioned above, membranes for smaller devices can also be screen-printed or drop-coated from a syringe. Reinhoudt (1992) described photopolymerization of some acrylates and polysiloxanes used as effective matrices on semiconductors.

Normally the membrane is conditioned several hours in a solution of the ion for which the membrane is to be selective. Bakker (1997) has suggested, however, a theoretically based alternative that should promote Nernstian behavior and better detection limits. For cation-selective, neutral-carrier membranes, he proposed pre-conditioning in solutions of the normally discriminated ion(s) rather than the primary ion.

Reinhoudt (1992) has summarized some of the problems related to PVC ion-selective membranes, which cannot always be adapted directly for solid-state devices and semiconductors. There may be adhesion problems to the sensor surface as well as spurious drifts. For semiconductors Reinhoudt solved the problems by interspersing a layer of poly-HEMA between the sensing layer and the semiconductor surface.

The adaptation to miniaturized devices poses challenges other than those of the chemistries involved. One problem involves application of the membranes to localized regions of the surfaces. Camman's group micromachined 1–500 μm "containment" wells on silicon chips (Knoll *et al.*, 1994). The PVC membranes could then be injected into the containments. They did, however, note the same types of adhesion problems reported by others. Goldberg *et al.* (1994) used a screen-printed epoxy "wall" to circumscribe the sensing area on their chips. Membrane cocktails were also applied by screen printing, though the formulations had to be adjusted for an acceptable rheology of the membrane pastes. Arquint *et al.* (1994) and Silber *et al.* (1996) have reported fabrication of miniaturized ion-selective multianalyte sensors which exemplify the technology for both solid state and FET transducers. The Pace and Hamerslag (1992) and Reinhoudt (1992) chapters in Edelman and Wang (1992) also illustrate many of the practical considerations of fabrication.

2.3. CHARACTERISTICS OF ENZYMES

Enzymology is a vast and fascinating scientific field. There are innumerable types of enzymes and mechanisms by which enzymes catalyze reactions. Multiple approaches to modeling and interpreting the kinetics of enzymatic reactions have developed over the years. There are extensive treatises, monographs, and textbooks which expound at length on the accumulated wisdom of scientists who specialize in enzymes. Fortunately, most undergraduate programs of study in chemistry or closely related fields now include, or require, introductory material on enzyme kinetics and mechanisms. The Abeles, Frey and Jencks (1992) and Lehninger, Nelson and Cox (1993) presentations of enzymatic catalysis provide good introductory-level treatment of enzyme kinetics. For analytical scientists with less experience in biochemistry, the Carr and Bowers (1980) monograph is still a first-rate primer of basic enzymology as applied to immobilized species. It is not the purpose of this section to repeat all of the fundamental relationships of soluble and immobilized enzymes. Rather, within the context of biocatalytic sensors, only the essential terminologies and characteristics will be distilled as a minimal foundation for further discussion of bioanalytical sensing.

2.3.1. Terminology

Enzymes are classified by a systematic numbering scheme according to the type of reaction catalyzed. They also have common names that are less specific, but universally accepted. Since all enzymes catalyze transfer reactions—electrons, atoms, or functional groups—the classification system is based on the type of transfer, the donor and the acceptor. There are four numbers of designation, the first of which indicates the major family:

1	Oxidoreductases	transfer electrons, H^-
2	Transferases	transfer functional groups
3	Hydrolases	transfer functional groups to water
4	Lyases	transfer groups to or from double bonds
5	Isomerases	transfer groups within molecules
6	Ligases	transfer by joining groups, accompanied by ATP cleavage

Within one category there may be optional routes to conversion of a substrate to product(s). For example, glucose may be oxidized by glucose oxidase, GOx (1.1.3.4), with oxygen as a cosubstrate to give gluconolactone and H_2O_2.

$$\text{Glu} + O_2 \xrightarrow{\text{GOx}} \text{Gluconolactone} + H_2O_2$$

Glucose may also be oxidized to gluconolactone by the action of glucose dehydrogenase, GDH (1.1.1.47). The mechanism involves NAD^+ as the required cofactor, rather than oxygen as a cosubstrate, and NADH is a product.

$$\text{Glu} + NAD^+ \xrightarrow{\text{GDH}} \text{Gluconolactone} + \text{NADH}$$

There is a third route to glucose oxidation. In addition to the NAD^+-dependent dehydrogenase, there is another glucose dehydrogenase, GDH (1.1.99.17), which is often used as the genetically engineered species. This enzyme requires neither oxygen nor NAD^+, but uses pyrroloquinoline quinone (PQQ) as the cofactor:

$$\text{Glu} \xrightarrow{\text{GDH/PQQ}} \text{Gluconolactone}$$

In recent years the importance of the quinoproteins has become increasingly obvious as they seem to be involved in many of the oxidation–reduction processes across all types of biological species (Davidson, 1993; Wang, S. *et al.*, 1996).

Other substrate–enzyme combinations may be used as examples of enzymatic options. Glutamate can be oxidized by glutamate oxidase (1.4.3.11) or glutamate dehydrogenase (1.4.1.3). Creatinine can be hydrolyzed to creatine by creatinine amidohydrolase (3.5.2.10) or it can be hydrolyzed to *N*-methylhydantoin and ammonia by creatinine iminohydrolase (3.5.4.21). The challenge becomes clear. It is absolutely necessary that the analyst is aware of the options among enzymes and the distinctions with regard to desired products as well as enzyme characteristics. Otherwise rational design of optimized sensors cannot be expected.

For the analyst an obvious question relates to when a particular enzyme should be used in preference to another. The glucose trio provides a good example of options and choices. The GOx catalyzed reaction for glucose oxidation is sensitive to ambient oxygen because oxygen is a cosubstrate. Consequently, as oxygen is depleted in a sample, performance degrades whether disappearance of oxygen or appearance of H_2O_2 is being monitored. One of the options for avoiding the oxygen problem is to change to another reaction—e.g., either of the GDH reactions. The GDH (1.1.1.47) reaction is an attractive option because the NAD^+/NADH couple is a well-established electrochemical or optical probe for monitoring biochemical reactions. These cofactors are not inexpensive, however, and the sensor design must include either addition of that reagent or a means of recycling the reagent internally in the sensor.

Finally, GDH (1.1.1.99.17) may be considered. This enzyme has not been widely available and it is expensive. On the other hand, the PQQ–GDH does have some advantageous characteristics for fabrication of enzymatic sensors. For example, GDH has a very high enzymatic efficiency. It will work with a variety of redox-active substrates and the electron transfer rate at electrodes is faster than that for a GOx electrode, thus providing higher current densities and sensitivity (D'Costa *et al.*, 1986; Ye *et al.*, 1993, 1994; Jin *et al.*, 1995).

There are other considerations in choosing an appropriate enzymatic sequence for a biosensor. Two examples will illustrate some of the considerations.

(a) Heller's group has used peroxidases in the design of several multi-enzyme amperometric sensors. One of the problems with their electrodes has been that they did not exhibit good long-term stability at the physiological temperature of 37 °C (Vreeke *et al.*, 1995b; Kenausis *et al.*, 1997). When they switched from the commonly used horseradish peroxidase (HRP,

1.11.1.7) to a soybean peroxidase (SBP) preparation, long-term stability was greatly improved even up to 65 °C.

(b) Creatinine is an important clinical analyte. There have been several sensor designs for determining creatinine. To understand the different design strategies it is informative to look at various enzymatic reactions involving creatinine and creatine which are present at comparable levels in blood:

$$(1) \quad \text{Creatinine} + H_2O \xrightarrow[\text{Iminohydrolase 3.5.4.21}]{\text{Creatinine}} N\text{-Methylhydantoin} + NH_3$$

$$(2) \quad \text{Creatinine} \xrightarrow[\text{Amidohydrolase 3.5.2.10}]{\text{Creatinine}} \text{Creatine}$$

$$(3) \quad \text{Creatine} \xrightarrow[\text{Amidinohydrolase 3.5.3.3}]{\text{Creatine}} \text{Sarcosine} + \text{Urea}$$

Most of the early sensors for creatinine were based on thermal detection of the hydrolysis reaction (Danielsson and Mosbach, 1989) or detection of the ammonia produced in (1). However, endogenous levels of ammonia exhibit a significant interference, so efforts were made to eliminate ammonia. Winquist and Danielsson (1990) describe an immobilized enzyme reactor that removes endogenous ammonia before potentiometric measurement of the enzymatically produced ammonia from creatinine hydrolysis:

$$NH_4^+ + \text{NADH} + \alpha\text{-Ketoglutarate} \xrightarrow[\text{dehydrogenase}]{\text{Glutamate}} \text{Glutamate} + \text{NAD}^+$$

Karube (1989a) used an entirely different approach. He described the use of the hydrolytic enzyme co-immobilized with nitrifying bacteria which consume the ammonia produced by hydrolysis. The bacteria consume oxygen in the process, and thus the detection method involves a conventional Clark-type oxygen electrode for monitoring decrease in oxygen:

$$\text{Creatinine} + H_2O \xrightarrow{3.5.4.21} NH_4^+ + N\text{-Methylhydantoin}$$

$$NH_4^+ \rightarrow NO_2^- \rightarrow NO_3^-$$

$$\text{Nitrifying bacteria}$$

$$\uparrow$$

$$O_2$$

A recent design which exhibits very good FOMs for creatinine determination in serum is one based on the multienzyme chemistry of Tsuchida and Yoda (1983).

$$\text{Creatinine} + H_2O \xrightarrow{3.5.2.10} \text{Creatine}$$

$$\text{Creatine} + H_2O \xrightarrow{3.5.3.3} \text{Sarcosine} + \text{Urea}$$

$$\text{Sarcosine} + H_2O + O_2 \xrightarrow[\substack{\text{Oxidase} \\ 1.5.3.1}]{\text{Sarcosine}} \text{Glycine} + \text{HCHO} + H_2O_2$$

The H_2O_2 of the third reaction is the detectable species. Mădăraş and Buck (1996) have modified the design for miniaturized, disposable amperometric sensors ("biochips") for creatine and creatinine. Their creatine sensor contains only the components of reactions 2 and 3. The creatinine sensor is the same except the third enzyme of reaction 1 is added. Then the creatinine assay can be conducted in a differential mode measuring the current difference between the two sensors. The Mădăraş and Buck device is an excellent example of accumulated experience with various facets of sensor design. The combination of optimized permselective membranes, multienzyme design, miniaturization and integration of components results in an effective device that has been well characterized (t_{95} < 60 s, DR of 0.9–1.2 mM in batch mode and 2.0 mM in FIA mode, LOD: 20 μM).

In addition to the classification of enzymes and options among reactions, products and detection schemes, there are some other terminologies which must be understood. Many of the enzymes require cofactors or coenzymes for the catalytic reactions of interest. Sometimes these are part of the native structure of the enzyme whereas in other cases the cofactors, such as metal ions, may be required as additional reactants for the catalysis. Examples of coenzymes are flavin moieties, NAD^+/NADH and PQQ . When a coenzyme is integrally bound to the enzyme in its naturally occurring 3-dimensional structure, the protein is referred to as the *holoenzyme* and the coenzyme is the *prosthetic group*. If the enzyme is isolated or produced without the prosthetic group, it is referred to as the *apoenzyme*. In some cases the holoenzyme can be reconstituted from the apoenzyme in the presence of the appropriate coenzyme. Such is the case for the genetically engineered GDH (1.1.99.17) mentioned earlier. Ye *et al.* (1993) reported that their reconstituted holoenzyme could be stored for 2 months at 4 °C without loss of activity.

Enzyme preparations are characterized according to the enzymatic activity. Preparations are designated as having *X* units of activity per milligram of the protein (U per mg) with a unit being defined as the amount that will convert 1 μmole of substrate to product per minute at a specified pH and temperature. Thus, if a preparation of alcohol dehydrogenase is characterized as having an activity of 200 U per mg of protein, using 3 mg of the preparation would provide 600 U of active enzyme. This should convert 600 μmoles of ethanol to acetaldehyde per minute at pH 8.8 and 25 °C, the standard assay conditions. The specific activity in U per mg rises with purification of the enzyme, and commercial preparations are available in a wide range of activities. Consequently, examination of original research reports will indicate that the amount of enzyme to be immobilized when preparing a sensor varies widely. A few milligrams are usually sufficient.

2.3.2. Kinetic Factors of Soluble Enzymes

Kinetic factors for an enzyme may be obtained from either equilibrium measurements or kinetic measurements based on the initial rate of reaction v. Carr and Bowers (1980) present a fairly thorough comparison of the two approaches, including predicted times for analysis. In practice the kinetic approach using initial rates is more common. Absorbance and fluorescence detection modes have been widely utilized for enzymatic assays, and many of the commercially available assay kits are geared to these instrumental approaches.

A typical plot of v, the initial reaction rate, versus [S], the substrate concentration, is shown in Figure 2-5. The data for such a plot are obtained from enzymatic assays in different experiments using the same amount of enzyme, same temperature and pH and varying substrate concentrations. This type of behavior, exhibited by most enzymes, is characterized by the plateau of maximum velocity V_{max}, under conditions of substrate saturation of the enzyme. Several points should be emphasized in anticipation of eventually relating these measurements to bioanalytical sensors. (a) Soluble enzymes are used. (b) Routine enzymatic assays are normally conducted at the pre-determined optimal pH for the enzyme. (c) The initial rate v must be determined before a significant portion of the substrate has been consumed—i.e., at the earliest possible time after mixing enzyme and substrate. (d) If a greater amount of enzyme is used than that represented in Figure 2-5, the V_{max} plateau will occur at a higher value—i.e., increasing the enzyme can be used to increase the initial reaction rate for better detection, but there is a practical limit imposed by the assumptions employed in analyzing these curves.

The interpretation of this type of enzyme behavior is presented in introductory textbooks of biochemistry or physical chemistry. Only the features most essential to enzymatically catalyzed sensors will be reviewed here. The hyperbolic relationship between v and [S] can be interpreted on either an equilibrium model proposed by Michaelis and Menten or by the steady-state model of Briggs and Haldane (Abeles *et al.*, 1992; Mathews and van Holde, 1990). The two approaches lead to the same saturation behavior shown in

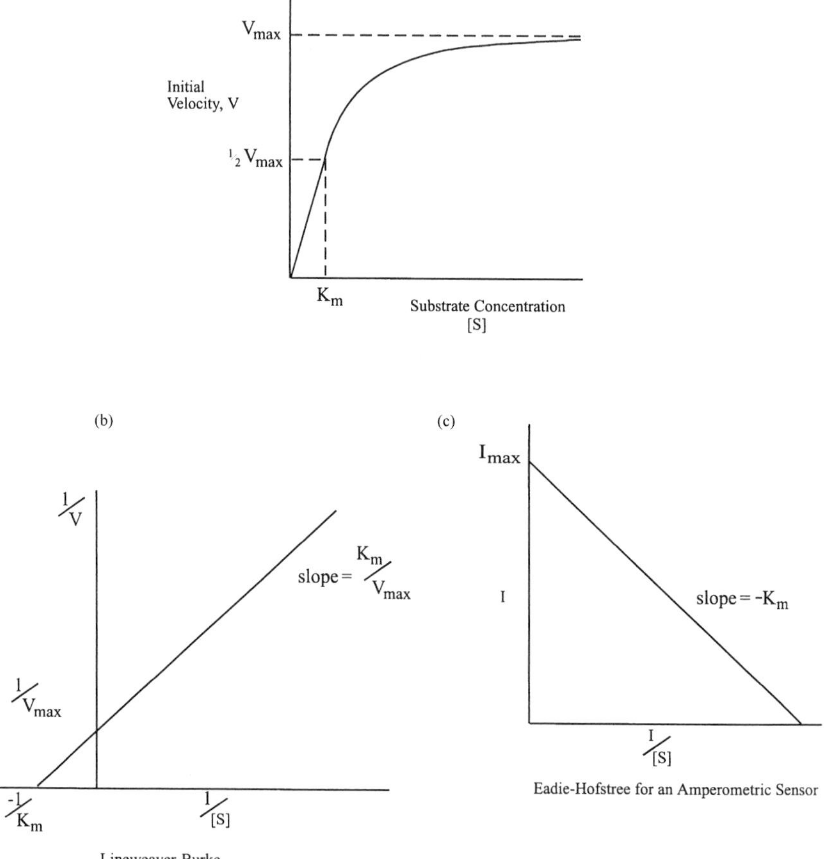

FIG. 2-5. Basic relationships for enzymatic catalysis. In (a) the usual hyperbolic curve for velocity of reaction as a function of substrate concentration is shown for one enzyme concentration. The velocity of reaction approaches a maximum rate for that enzyme concentration, but could be higher for a greater enzyme concentration. The Lineweaver–Burke plot of the data is shown in (b). An Eadie–Hofstree treatment of the data, shown in terms of amperometric sensor output, is illustrated in (c). See text.

Figure 2-5. Over the years the Michaelis–Menten descriptors have been conventional, even though the Briggs–Haldane model is widely used in interpreting the behavior.

The simplest enzymatic reaction mechanism of one substrate and one product can be used to define the characterizing factors of the Michaelis–Menten plots:

$$\mathrm{E} + \mathrm{S} \underset{k_{-1}}{\overset{k_1}{\rightleftharpoons}} \mathrm{ES} \overset{k_2}{\rightarrow} \mathrm{P} + \mathrm{E}$$

For this mechanism v is the rate at which the enzyme–substrate complex ES converts to product and free enzyme. The rate of product formation is then proportional to [ES]

$$v = \frac{\mathrm{d[P]}}{\mathrm{dt}} = k_2[\mathrm{ES}]$$

Total enzyme is always equal to the sum of free enzyme and that complexed with substrate:

$$[\mathrm{E}]_{\mathrm{tot}} = [\mathrm{E}] + [\mathrm{ES}]$$

At high substrate concentrations the enzyme becomes saturated with substrate, thus converting all of the enzyme into ES, and the V_{max} is proportional to only $[\mathrm{E}]_{\mathrm{tot}}$.

$$V_{\mathrm{max}} = k_2[\mathrm{E}]_{\mathrm{tot}}$$

In this region there is no dependence on [S] and the reaction is often described as zero order, referring to the lack of [S] dependence. In fact, the reaction is first order in this region because the rate of product formation is still dependent on $[\mathrm{E}]_{\mathrm{tot}}$.

The rate constant k_2 is determined under conditions of substrate saturation. It is the rate constant for maximal rate of the reaction. Usually the apparent rate constant for maximal rate is designated as k_{cat} because it may be comprised of multiple elementary reaction k's for complex mechanisms, and it represents the limiting rate of the reaction. For the simple Michaelis–Menten mechanism, $k_{\mathrm{cat}} = k_2$. The constant k_{cat} has the units of reciprocal time, and is designated as the turnover number. For a given unit of time with enzyme saturated with substrate, it represents the number of substrate molecules converted to product per enzyme molecule.

At the point indicated as $\frac{1}{2}V_{\mathrm{max}}$, the [ES] is also half its maximum value. The [S] producing this condition is defined as the Michaelis constant K_{m} of the enzyme for that substrate. For low [S] , i.e., $[\mathrm{S}] \ll K_{\mathrm{m}}$, the relative amount of enzyme complexed is low, and product formation increases linearly with [S]. Because the rate is linearly related to [S] in this region, it is often said that this is the first-order region, in keeping with the zero-order designation at saturation. The reaction in this region *is* first order with respect to substrate, but it is second order overall because the enzyme dependence is still there. It will become obvious below that the second-order rate constant for this region is k_{cat}/K_m. This is an important constant in that it allows one to compare specificities of an enzyme for different substrates (see, for example, Lehninger *et al.*, 1993). As described by Abeles *et al.* (1992) the "leveling off of the rate with increasing substrate concentration represents a transition from a second-order dependence on both reactants, [E] and [S], at very low substrate concentrations, to a first-order dependence on [E] alone at high substrate concentrations." While the distinction of orders in different regions of the curve has been emphasized here, there is common usage of the first- and zero-order terminology. It should be understood that when those terms are used the reference is with regard to [S] dependence only. The amount of enzyme still plays a dynamic role in the reaction.

The steady-state assumption may be used to obtain the rate equations, which lead to interpretation of the behavior shown in Figure 2-5.

For mass balance considerations:

$$[S]_{tot} = [S] + [ES]$$
$$[E]_{tot} = [E] + [ES]$$

However, based on natural biological systems, it is usually assumed that the relative amount of substrate bound to enzyme is small, therefore $[S]_{tot} = [S]$. In the design and modeling of bioanalytical sensors little attention has been given as to whether this is always a good assumption.

The rate of formation of ES is written as

$$\text{Rate of formation} = k_1([E]_{tot} - [ES])[S]$$

ES can either break down to product and free enzyme or dissociate back to substrate and free enzyme. Therefore the rate of disappearance of ES is

$$\text{Rate of breakdown} = k_{-1}[ES] + k_2[ES]$$

Applying the steady-state assumption,

$$k_1([E]_{tot} - [ES])[S] = k_{-1}[ES] + k_2[ES]$$

Rearranging:

$$k_1[\mathrm{E}]_{\mathrm{tot}}[\mathrm{S}] - k_1[\mathrm{ES}][\mathrm{S}] = (k_{-1} + k_2)[\mathrm{ES}]$$

$$k_1[\mathrm{E}]_{\mathrm{tot}}[\mathrm{S}] = (k_1[\mathrm{S}] + k_{-1} + k_2)[\mathrm{ES}]$$

$$[\mathrm{ES}] = \frac{[E]_{\mathrm{tot}}[\mathrm{S}]}{[\mathrm{S}] + \left(\dfrac{k_2 + k_{-1}}{k_1}\right)}$$

Defining $K_m = (k_2 + k_{-1})/k_1$,

$$[\mathrm{ES}] = \frac{[\mathrm{E}]_{\mathrm{tot}}[\mathrm{S}]}{K_m + [\mathrm{S}]}$$

Restating the formation of product:

$$V_0 = \frac{\mathrm{d}[\mathrm{P}]}{\mathrm{d}t} = k_2[\mathrm{ES}] = \frac{k_2[\mathrm{E}]_{\mathrm{tot}}[\mathrm{S}]}{K_m + [\mathrm{S}]}$$

Thus, the resulting equation "fits" the observed behavior. At low substrate concentrations ($[\mathrm{S}] \ll K_m$), the curve should be linear with a slope of V_{max}/K_m. For this region the rate constant is k_{cat}/K_m, which includes features relating the enzyme's activity and the stability of the ES complex, and thus its utility to compare effectiveness and activities with different substrates. At high substrate concentrations the rate is simply V_{max} or $k_2[\mathrm{E}]_{tot} = k_{cat}[\mathrm{E}]_{tot}$.

Determining values for V_{max} and K_m is usually approached from one of three types of linear plots which are for rearranged forms of the Michaelis–Menten equation. (1) The Lineweaver–Burke form is as shown in the lower left portion of Figure 2-5:

$$\frac{1}{V} = \frac{1}{V_{max}} + \frac{K_m}{V_{max}}\frac{1}{[S]}$$

(2) the Eadie–Hofstree plot is as illustrated in the lower right of Figure 2-5 in terms of amperometric data:

$$V = V_{max} - K_m\frac{V}{[S]}$$

(3) for the Hanes plot, which is not shown,

$$\frac{[\mathrm{S}]}{V} = \frac{1}{V_{max}}[\mathrm{S}] + \frac{K_m}{V_{max}}$$

The simple one-substrate mechanism used above is frequently operative in some biocatalytic sensors. There is, however, another mechanism that is important, particularly when using the oxidases as enzymes. The mechanism for oxidase activity is a two-substrate mechanism:

$$E + S_1 \rightleftharpoons ES_1 \rightleftharpoons E' + P_1 \overset{S_2}{\rightleftharpoons} E'S_2 \rightleftharpoons E + P_2$$

This is referred to as the ping-pong mechanism. It is only one of several ways for an enzyme to react with two substrates (Lehninger *et al.*, 1993; Mathews and van Holde, 1990).

For the example of glucose oxidase:

$$\text{Glucose} + \text{GOx} - \text{FAD} \rightarrow \text{Gluconolactone} + \text{GOx} - \text{FADH}_2$$
$$\text{GOx} - \text{FADH}_2 + \text{O}_2 \rightarrow \text{GOx} - \text{FAD} + \text{H}_2\text{O}_2$$

This mechanism may be treated with the general approach used for the simpler mechanism, but the basic Michaelis–Menten equation takes on a different form:

$$V = \frac{V_{\text{max}}[S_1][S_2]}{K_{\text{m,S}_2}[S_1] + K_{\text{m},S_2}[S_2] + [S_1][S_2]}$$

The K's in this generalized equation are composites of individual k's in the different elementary reactions of the mechanism. Due to the widespread interest in this type of enzymatic catalysis for sensors, a great deal of work has been published for modeling the responses, especially electrochemical sensors. If one sorts through the various models and assumptions, it will become obvious that sometimes the assumptions are either oversimplified or are inappropriate for the design of the sensor. Yet, when the enzymatic reaction variables are combined with the response variables of the detectors, the equations require simplification for solution (Section 2.5.5). Derivation of the response functions for sensors based on the ping-pong mechanism will not be included here, but the interested reader should consult among others Mell and Maloy (1975,1976), Carr and Bowers (1980), Turner *et al.* (1989), Cass (1990), Bartlett and Pratt (1993), Martens and Hall (1994), and Gooding and Hall (1996).

2.3.3. Effects of Immobilization

In the preceding section the characteristics of soluble enzymes were outlined. The V_{max}, K_m, k_2 or k_{cat}/K_m were shown to be important factors in characterizing an enzymatic reaction. When one immobilizes an enzyme at a surface the enzymatic behavior is altered. Putting the protein in a different type of environment and confining it to a microvolume can have a significant effect on the characterizing variables. Recognizing that there are always exceptions, it may be said that the heterogeneous behavior compared to that in homogeneous solution does show some general trends.

Depending on the method of immobilization the enzymatic efficiency may be considerably different from that in solution. This may be the result of random orientation of the molecules, thus shielding the active site for some of the molecules, or it may involve a degree of denaturation during immobilization. It may also reflect effects of the environment that shift reaction conditions away from the solution optimum.

Microenvironmental effects may be related to the inner-membrane's apparent pH, accessibility, ionic strength, polarity of the medium, and inhibitory action of accumulated products in the confined space.

It is instructive to examine the logic of microenvironmental effects on the pH optimum and Michaelis constant K_m for an enzyme-substrate interaction (see Carr and Bowers, 1980 and references therein). Consider first the pH effects that might be encountered in a matrix comprised of polyelectrolyte — e.g. polyamide, polyethylenimine or polysaccharide. If the immobilization matrix on a sensing device has an ionic atmosphere different from that of the surrounding solution, partitioning of positively charged protons or charged substrates into the matrix will be affected. This implies that the actual pH in the membrane may differ from that of the solution. Thus, an enzymatic reaction occurring in the membrane will experience a pH different from that of a comparable reaction involving soluble enzyme. The difference may be related to the electrostatic potential ψ in the membrane:

$$\Delta \mathrm{pH} = 0.43 z e \psi / kT$$

Here, z is the + charge on the proton; e is the charge on the electron; k is the Boltzmann constant and T is the absolute temperature. For a negatively charged membrane the inner pH will be lower than the pH of the solution. Thus, based on solution measurements of pH, a higher pH optimum for the enzymatic reaction would be ascertained.

An analogous argument can be made for effects on K_m of an enzymatic reaction, but in this case z represents the charge on the enzyme's substrate. The immobilized enzyme's net charge within the membrane would be an integral contributor to the ψ of the membrane.

$$K_m^{app} = K_m(ze\psi/kT)$$

Now the apparent K_m will be higher than that observed for the enzymatic reaction in solution if the substrate and matrix are of the same charge, but lower if they are of opposite charge.

Another factor to consider as an effect of immobilization is the tortuosity of the diffusional path within an immobilization matrix. This may affect directly the accessibility of enzyme to the substrate. For example, enzymes are sometimes immobilized in polymeric layers with cross-linking agents. This has the effect of creating a meshy network in the matrix. In modeling amperometric enzyme electrodes, Tatsuma *et al.* (1992b) compared the sensitivites of monolayers and cross-linked layers. The conclusion was that though the cross-linked layer provided an overall higher response because it contained more enzyme molecules, the enzymatic efficiency per enzyme molecule was greater in the monolayer case.

Pierce *et al.* (1992) observed in their scanning electrochemical microscopy study of glucose oxidase that excessively cross-linked hydrogels reduced the enzymatic activity, probably due to diffusional limitations. Hydrogels prepared with normal amounts of cross-linked protein, however, did not exhibit the same degree of diffusional limitation. Since the catalytic efficiencies of their preparations were shown to be lower than previous reports, there was also the possibility that the enzyme had been deactivated to some degree in the immobilization process.

Mădăraş and Buck (1996) also addressed the matter of cross-linking effects. In their report of a creatinine sensor they indicated that they had observed effects related to the degree of cross-linking. Their enzymatic layers with random orientation of the active enzyme were comprised of enzyme and inert bovine serum albumin (BSA) cross-linked with glutaraldehyde, which is discussed in Chapter 3. In order to examine systematically this effect, they defined a cross-linking factor f as the ratio of the amount of glutaraldehyde and total protein:

$$f = \frac{\%\text{GL}}{\%\text{TP}}$$

By varying the amount of active enzyme in the reaction layers, they varied the cross-linking factor. Increasing f did lower the sensitivity of the sensor. The authors reported that although there is a threshold degree of cross-linking for a viable reaction layer in their case, the apparent permeability of the layer did increase with decreasing values of f, as reflected in the higher sensitivity. For their further experiments they established a constant degree of cross-linking with $f = 0.008$–0.014 and total protein content of ~9–10% by weight.

2.3.4. Inhibition as a Tool

Discussion to this point has dealt only with the use of enzymes to determine concentrations of substrates for the given enzyme. Another sensing strategy is that of using enzyme inhibition as the mechanism for determining concentrations of the inhibitors *per se*. From elementary enzyme kinetics it will be recalled that inhibitor effects are reflected in changes of the soluble enzyme's characteristics such as V_{max} and K_m. Competitive inhibitors that bind at the same active site as the natural substrate will change the K_m for the substrate, as shown by a change of slope in the Lineweaver–Burke plots. Non-competitive inhibition, observed when inhibitor binds at another site on the protein, results in a change in the V_{max}.

For an inhibitor-based sensor, the response to the detectable species should be diminished in the presence of inhibitor. This has been demonstrated in several inhibitor-based sensors, as indicated in Section 2.1.2 for the Mionetto *et al.* (1994) acetylcholinesterase sensor.

Smit and Rechnitz (1993) used the inhibition of tyrosinase as a cyanide sensor. Wang, J. and Chen (1995a) also used tyrosinase as the enzyme, but monitored the inhibition by hydrazines. These examples point out precisely one of the problems with inhibition-based sensors. The selectivity advantage of using an enzyme with reasonable specificity for a substrate (analyte) is often lost once inhibition is used. That is because classes of compounds and even different types of compounds may inhibit the same enzyme. For example, acetylcholinesterase is inhibited by nerve gases, pesticides and some alkaloids. Therefore, selectivity of an inhibitor-based enzymatic sensor is, at best, restricted to a class of compounds. This is not to say that inhibition-based sensors cannot be successful. The very fact that inhibition occurs, regardless of the inhibitor, can be very useful analytically. Furthermore, with very well-designed experiments Preuss and Hall (1995) were able to determine selectively two herbicides by their differences in inhibitory characteristics during light and dark cycles of the photosynthetic electron transport chain in a cyanobacterium.

The Preuss and Hall study suggests another aspect of biocatalytic sensors. Cells and tissues are often used rather than the isolated enzymes (Chapter 3). This has significant advantages in that the natural matrix of membrane-bound catalysts with their accompanying cofactors are preassembled, so to speak, and organized according the biological function (Arnold and Rechnitz, 1989; Karube, 1989a; Cass, 1990; Karube *et al.* 1995b). There are two potential problems with this approach. First, the actual cells and tissues contain not only the bioselective enzymes of primary interest for the analysis, but also all others that might be present in the natural sample. These latter species may have deleterious effects on the primary sensing chemistry or may introduce different interferences than those encountered for isolated enzyme systems. Second, the mass transfer factors and reaction kinetics may differ significantly for cell/tissue sensors compared to those using the isolated and immobilized biocomponents (Ikeda *et al.*, 1996). This is attributable largely to the added feature of permeability of the cellular membranes. For example, response times may be impractically long or the response function may be difficult to model for guiding sensor design.

2.3.5. Sequenced Reactions

The use of sequenced reactions has been illustrated in several of the examples used throughout earlier sections. As indicated in Section 2.1.2, sequencing may be used for converting an enzymatic product to a more easily detected product, for eliminating interferences or for signal amplification. Sequencing two reactions for enhanced enthalpic output has been shown as one example. A recycling strategy, such as that described in Section 2.1.3, is a more complex example.

If sequenced reactions are utilized, the kinetic expressions take on the complexity of the chemical scheme. For example, for the simplest type of sequenced reaction:

(1) $$S \xrightarrow{E_1} P_1$$

(2) $$P_1 \xrightarrow{E_2} P_2$$

Carr and Bowers (1980) indicate that concentration of P_2, the detected species, is represented by

$$[P_2] = \frac{k_1'}{k_2'}[k_2't + \exp(-k_2't) - 1][S]$$

$$k_1' = \frac{k_1[E_1]}{K_{m,1}} \qquad k_2' = \frac{k_2[E_2]}{K_{m,2}}$$

Modeling of this behavior reveals a time lag in the production of the detectable species. This must be taken into consideration when attempting to make a steady-state measurement. Nonlinear behavior of $[P_2]$ can impart errors in measurement. If the entire system is allowed to equilibrate for an equilibrium-based measurement, then the lag phase of the reaction has little consequence.

2.4. THE BASICS OF BIOAFFINITY SENSORS

Although most of the work in sensors based on bioaffinity have involved immunoassay techniques, there are other sensing chemistries that fit the same category. In this section several of these will examined, though emphasis will be on antigen–antibody phenomena. The basic principles will be presented in descriptive form using only single-site binding examples. The Wyman and Gill (1990) book is quite thorough in dealing with various situations such as multi-site binding, allosteric binding and ligand linkages between macromolecular phases.

2.4.1. Characterizing Ligand Binding

Although biological macromolecules are involved, the reader will recognize much of the mathematical description of binding phenomena as being related to adsorption phenomena of introductory physical chemistry. Suppose there is a macromolecule, Mac, which can bind a ligand, L, at a single binding site. The reaction can be written simply as

$$\mathrm{Mac} + \mathrm{L} \underset{k_{-1}}{\overset{k_1}{\rightleftharpoons}} \mathrm{MacL}$$

The rates of formation and dissociation, frequently referred to as the on and off rates, of MacL are

$$V_{\mathrm{on}} = k_1[\mathrm{Mac}][\mathrm{L}]$$
$$V_{\mathrm{off}} = k_{-1}[\mathrm{MacL}]$$

At equilibrium the rates are equal, thus giving an expression for the K_{assoc} or K_{dissoc}:

$$\frac{[\text{Mac L}]}{[\text{Mac}][\text{L}]} = \frac{k_1}{k_{-1}} = K_{\text{assoc}} \qquad \frac{[\text{Mac}][\text{L}]}{[\text{Mac L}]} = \frac{k_{-1}}{k_1} = K_{\text{dissoc}}$$

The association and dissociation constants have units of M^{-1} and M, respectively. The fraction of macromolecular sites filled can be given as F and the fraction of empty sites by $1-F$:

$$F = \frac{[\text{Mac L}]}{[\text{Mac L}] + [\text{Mac}]}$$

Using the K_{assoc} expression :

$$F = \frac{K_{\text{assoc}}[\text{Mac}][\text{L}]}{K_{\text{assoc}}[\text{Mac}][\text{L}] + [\text{Mac}]} = \frac{K_{\text{assoc}}[\text{L}]}{K_{\text{assoc}}[\text{L}] + 1} = \frac{[L]}{[L] + \dfrac{1}{K_{\text{assoc}}}} = \frac{[L]}{[L] + K_{\text{dissoc}}}$$

The form of this equation shows that for low [L], $F \to 0$ and for large [L], $F \to 1$. Further, at half-saturation $F = 1/2$ and K_{assoc} [L] $= 1$. The [L] at half-saturation is the reciprocal of K_{assoc}. The hyperbolic curve is reminiscent of the Michaelis–Menten enzymatic behavior with analogous interpretations indicated. If the reciprocal of the equation for F is used a linear relationship results:

$$\frac{1}{F} = \frac{K_{\text{dissoc}}}{[\text{L}]} + 1$$

Thus, a plot of $1/F$ versus 1/[L] will have a slope of $1/K_{\text{assoc}}$. A more common plot is the Scatchard format, which involves plotting F/[L] versus F. The slope of this line is $-K_{\text{assoc}}$ and the y-intercept is K_{assoc}. Association constants on the order 10^4–10^{11} M^{-1} are common.

From the viewpoint of biosensing designs, there are two important facts to emphasize about the preceding relationships. The equations above apply to the binding phenomenon whether it occurs in homogeneous media or at a surface with one of the members immobilized. Second, the relationships assume an equilibrium situation in the final expressions. Consequently, most of the affinity-binding sensors are equilibrium-based sensors, and frequently require minutes to hours to exhibit an equilibrium signal.

In most bioffinity formats one of the members of the couple is immobilized at a sensing surface. There are several optical techniques which allow the analyst to determine both the amounts and effectiveness of immobilized species involved in binding reactions at sensing surfaces (Buckle *et al.*, 1993; Brecht and Gauglitz, 1995; Frey *et al.*, 1995; Lawrence *et al.*, 1996; Piehler *et al.*, 1996; Schipper *et al.*, 1996; Sigal *et al.*, 1996; Schmid *et al.*, 1997). Ellipsometry, optical waveguides, and internal reflection fluorescence have all been used in the works cited. Chapter 5 includes some discussion of these approaches. Film thicknesses on a nanometer scale can be determined. Surface coverages for monolayers of proteins are on the order of femtomoles per mm^2 of surface area. Piehler *et al.* (1996) should be consulted for a good example of the calculations for the avidin–biotin couple.

There is another very important point to be made about surface coverages of affine species. The method of immobilization has a drastic effect on the fractional sites available for binding. Ideally every immobilized molecular site would be available for ligand binding. If, however, the immobilization method results in randomly oriented molecules on the surface, crowds the molecules too much, or destroys some of the binding sites, then less than ideal binding will occur. It is not uncommon to have only 10–50% effectiveness in binding unless careful consideration is given to oriented binding, so the binding sites are projected toward the ligand. Some of the reported methods for achieving oriented binding will be included in the discussion of Chapter 3.

2.4.2. A Vocabulary of Immunoreactions

Because so many of the bioaffinity sensors are immunosensors, it is good to get the basic terminologies in order. A schematic of an antibody is shown in Figure 2-6. Antibodies are produced in the body in response to *anti*body *gen*erating species, antigens, such as viruses and bacteria. Antigens are macromolecular in nature with certain regions of the proteinic structure that will bind to the antibody. These binding regions on the antigen are called the antigenic determinants or epitopes of the antigen (Figure 2-7). Small organics, such as some hormones or pollutants, become antigenic by being conjugated to a macromolecular carrier. The unconjugated small organics, called haptens, can bind to antibodies but cannot themselves stimulate an immune response.

Antibodies are glycoproteins, immunoglobulins, with interesting symmetry in the structure. They are Y shaped molecules (MW $\sim$ 150,000) comprised of heavy and light chains of amino acids with a flexible hinged region. The disulfide bonds which connect the chains and create the hinged region are shown in Figure 2-6. The antigen-binding sites are located at the tips of the Y where variable 20–30 amino acid sequences make each antibody unique for a particular antigen. The base of the Y, referred to as the Fc region, is involved in activation of the complement system of the immune

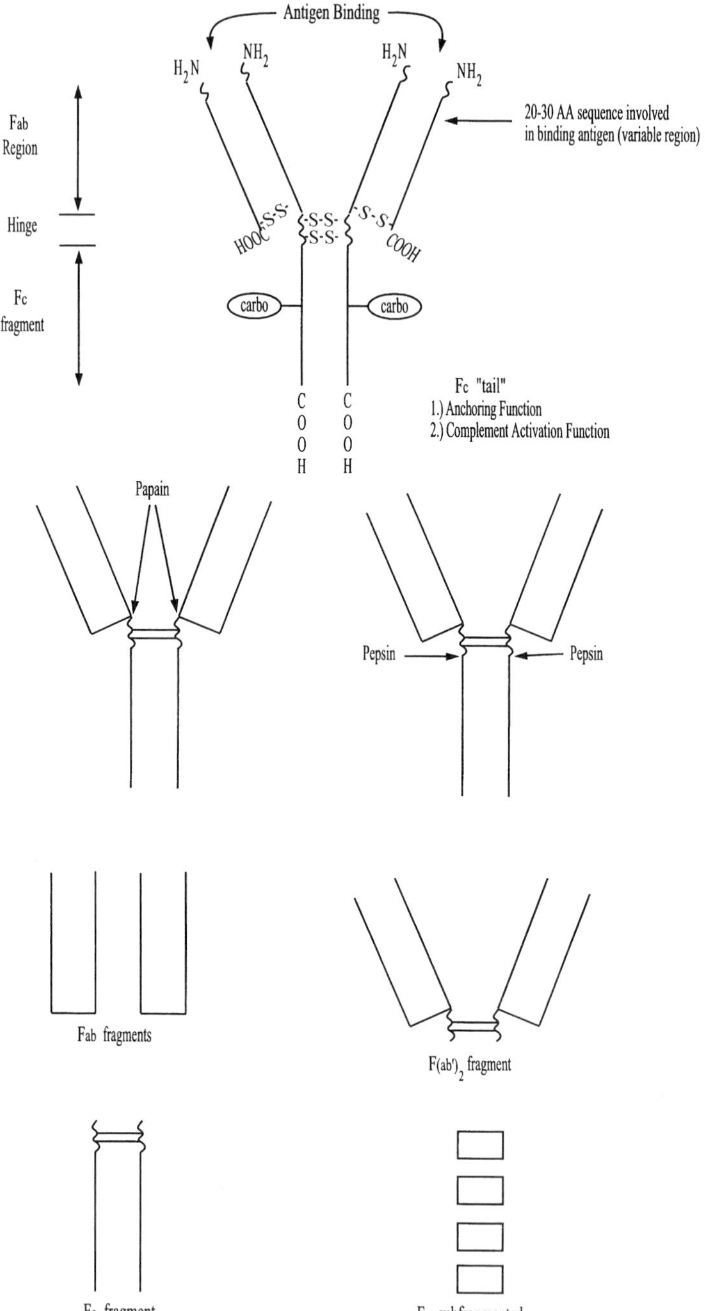

FIG. 2-6. Major features of the Y structure of antibodies or immunoglobulins (Ig). The Fab regions consist of heavy and light protein chains which distinguish the various Ig's. The lower portion of the illustration shows the usual fragmentation patterns with papain and pepsin (see text).

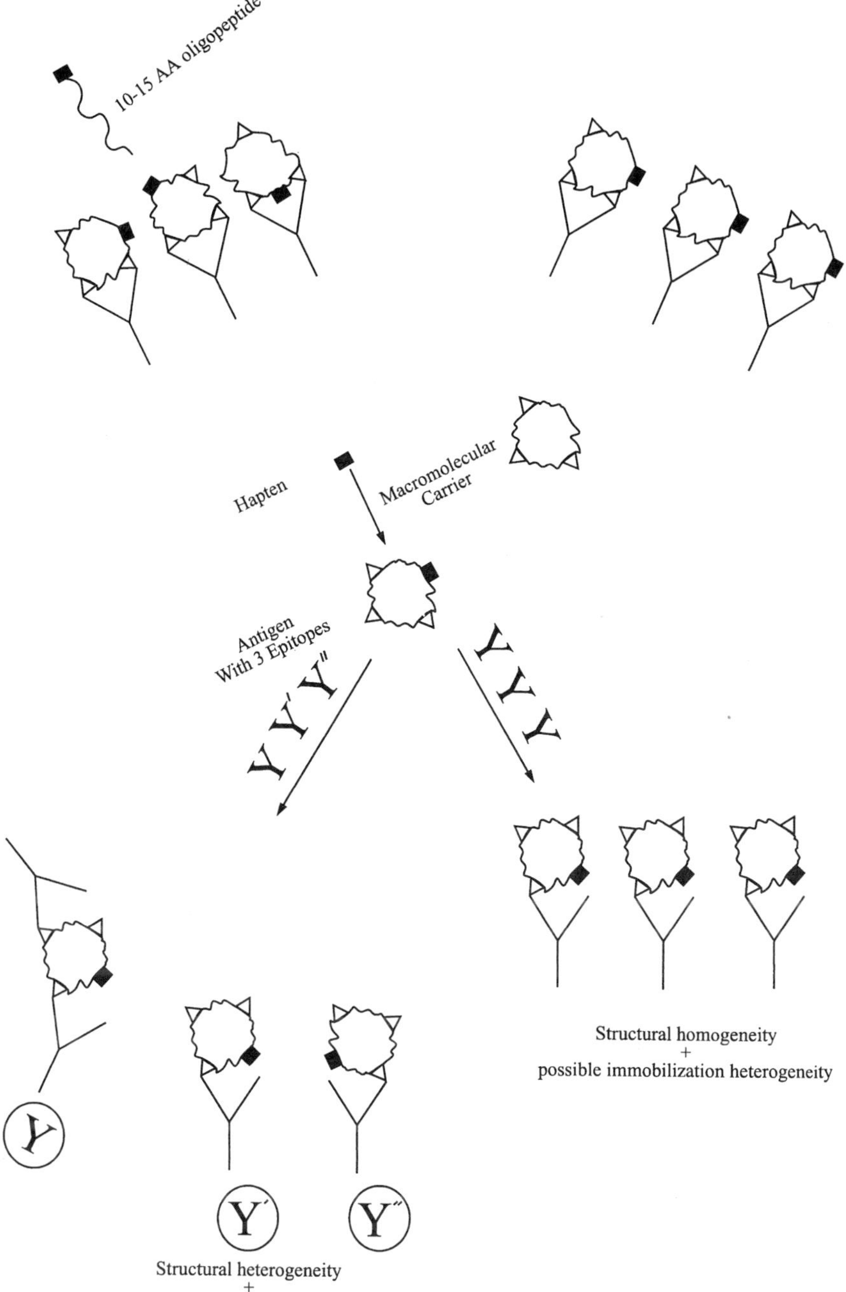

FIG. 2-7. Antigen–antibody binding. At the top a small organic molecule (hapten), shown as the dark block, binds to a macromolecule creating an antigen, which can then bind to an antibody. The triangular projections represent the epitopes, or binding determinant regions, of the antigen. Bispecific binding is illustrated in the scheme at the top, and the lower portion depicts the major difference between antigen binding to polyclonal (Y,Y′, Y″) and monoclonal (Y, Y, Y) antibodies.

response and binding to cell receptors. There are five classes of these immunoglobulins, IgM (pentamer), IgG, IgA (monomer or dimer), IgD and IgE. These forms are produced at different times during the immune response and differ in function, as well as location within the tissues of the body. An antibody may be fragmented by chemical treatment to create Fab or $F(ab')_2$ fragments which retain the binding function of the antibody. In Figure 2-6 the distinct fragmentation sites for papain and pepsin action on antibodies are shown. The retention of sulfhydryl groups on the Fab fragments plays a critical role in their oriented binding to sensor surfaces (see Lu, B. *et al.*, 1995, 1996).

When an antigen is injected into an animal different antibodies to the same antigen may be produced. Each type will bind to a different antigenic determinant or epitope of the antigen (see Figure 2-7). These antibodies are designated as polyclonal because they are produced by different types of cells. Monoclonal antibodies are produced when cells producing a specific antibody can be isolated and subjected to a cloning sequence so that the same antibody is produced by all the cells.

An immunoassay is an analytical procedure based on Ag–Ab reactions that can be used to determine either member as analyte. Immunoassays fall into two major classes: homogeneous and heterogeneous. In a homogeneous immunoassay the free Ag and bound Ag do not have to be separated. For example, an Ag may be labelled with a fluorophore or electroactive agent forming Ag*. Upon binding to Ab, the signal generated by Ab–Ag* can be distinguished from Ag*, and thus no separation is necessary. Heterogeneous immunoassays are more sensitive, but do involve separation of the bound and free antigen.

Heterogeneous assays may be designed in three major formats, which can be described most easily in terms of immobilized Ab, though that does not have to be the case.

1. Displacement—antibody is immobilized on a solid phase and labeled Ag* is added to Ab, filling all the available sites. When unlabeled Ag (the analyte) is introduced Ag* is displaced and the amount displaced will be related to the analyte concentration.
2. Competitive—antibody is immobilized. Both Ag* and Ag (analyte) are introduced to the Ab and compete for the binding sites. Bound Ag* is determined and its concentration is inversely related to the amount of Ag in the sample.
3. Sandwich—antibody, referred to as the capture Ab, is immobilized. Analyte, Ag, is added and it binds to the capture Ab. Then a second, labeled antibody, Ab*, is added, forming the "sandwich." Analytical determination of Ab* relates directly to the amount of analyte. It should be noted that this format can be used only for macromole-

cular antigens with at least two epitopes for the two binding processes involved in forming the sandwich.

Ordinary labels for these processes may be radioisotopes, fluorophores or enzymes for enzyme-linked assays (ELISA). The avoidance of radioisotopes for health and disposal reasons has prompted many of the newer approaches to immunoassay. The electrochemical immunoassay (ECIA) approach has attracted considerable interest over the past few years particularly among developers of sensors (Heineman and Halsall, 1985). In this procedure the labels are enzymes which produce an electroactive product that can be monitored. The Bauer, C. G. *et al.* (1996) sensor for alkaline phosphatase or 2,4-D provides a recent example.

Having presented the basic immunoassay formats in terms of immobilized antibodies, some exceptions should be mentioned to show the versatility of detection schemes. Tom-Moy *et al.* (1995) developed an acoustic-wave sensor which operated in a competitive format, but had an antigen analog rather than antibody immobilized. An atrazine analog was immobilized, then polyclonal antibodies and analyte, atrazine, were introduced to the sensor. A phase-sensitive detection method was used to determine the amount of antibody bound to the analog on the surface of the acoustic-wave device. A greater amount of atrazine in the sample provided less antibody bound to the surface.

Knichel *et al.* (1995) utilized as antigen an immobilized, synthetically produced epitope of the foot-and-mouth-disease virus. The authors pointed out that the use of the synthetic polypeptides is a safer, simpler method than trying to use the actual virus. In this direct binding assay with electrochemical detection, the immobilized epitope was exposed to monoclonal antibodies to the antigen. An electrical impedance signal that changed as antibody was bound was used for quantitation. The sensor was renewed for repeated measurements by urea washes, but the authors indicated that this aspect of surface renewal required more investigation.

There are several points to be emphasized about the foregoing introduction to terminologies and detection schemes. In the context of sensor designs, the immobilization, detection and regenerability aspects become paramount. Immobilization is the topic of Chapter 3, but some generalities about immunoassays should be emphasized at this point in the discussion. Antibodies are proteins. This means that the usual proteinic side-chains of the molecules can be used for immobilization. The carbohydrate regions on the Fc portion are also accessible for immobilization (B. Lu, 1996). Neither of these approaches assures an oriented immobilization with the Fab region projecting toward the incoming antigen. Thus, efficiency of binding can be affected. There are immobilization methods that "anchor" the Fc region to a surface using Protein A, Protein G or a genetically engineered ProtA/

ProtG (Chapter 3). This approach has been shown to produce much more efficient binding (Lu, B. *et al.*, 1995, 1996; Owaku *et al.*, 1995; Nakanishi *et al.*, 1996).

Whole antibodies do not have to be used for the immunoassay procedures. Both Fab and $F(ab')_2$ fragments may be utilized. These are produced by the proteolysis reactions that were shown in Figure 2-6 or they may be genetically engineered. Using these types of fragments, Shimura and Karger (1994) developed a highly sensitive biosensing technique they refer to as affinity probe capillary electrophoresis (APCE). Fluorophore-labeled antibody fragments (the affinity probes) were mixed with the antigenic analyte, then separated by CE with laser-induced fluorescence (LIF) as the detection mode. Detection limits in the pM range were obtained for determination of a recombinant growth hormone.

Finally, polyclonal and monoclonal antibodies each have their advantages. Polyclonal antibodies are cheaper and more easily acquired. Furthermore, as emphasized by Campbell (1989), the fact that polyclonal antibodies have binding sites for different eptiopes of the antigen can improve binding efficiencies in a sensor. Monoclonal antibodies all have the same binding characteristics and exhibit homogeneous behavior in solution. However, even monoclonal antibodies may exhibit some degree of heterogeneity in affinity constants when immobilized at a solid surface. The joint reports by Selinger and Rabbany (1997) and Rabbany *et al.* (1997) deal with these possible heterogeneities in affinity constants. Using a non-equilibrium flow immunoassay for TNT by a displacement protocol, the authors concluded that the heterogeneity was minimal, though it could be interpreted as significant if unmodified mathematical relationships for equilibrium binding were assumed. It should be noted that the Rabbany report involved a random rather than oriented immobilization method.

2.4.3. Oligonucleotides and Nucleic Acids

The nucleic acids are polyions comprised of chains of nucleotides which consist of pentoses, pyrimidine and purine bases and phosphate groups. The chain is linked by phosphodiester bonds. Using a common shorthand notation

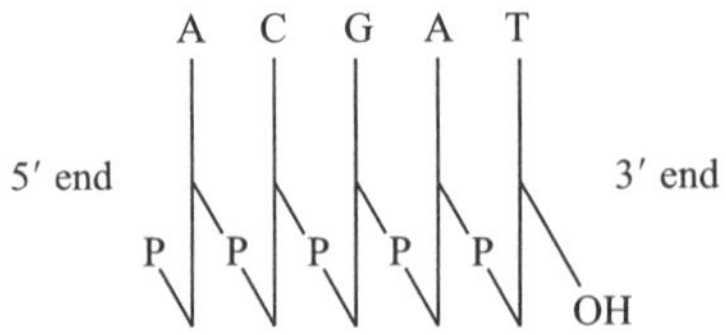

A C G A T

5′ end 3′ end

P P P P P OH

where P is phosphate and the 5′ and 3′ sites of the sugar rings are indicated. The purines adenine and guanine are represented by the A and G, respectively, while the pyrimidines, cytosine, thymine and uracil, are C, T and U. The complementary binding of double-stranded DNA (dsDNA) involves the C–G and A–T pairs. In RNA the thymine is replaced by uracil, giving C–G and U–A pairs. Ege (1989), Alberts *et al.* (1989), Mathews and van Holde (1990), Abeles *et al.* (1992), and Lehninger *et al.* (1993) should be reviewed for details of the binding characteristics. The phenomenon of single stranded DNA (ssDNA) binding to a complementary nucleotide sequence to form the double helix is probably one of the best known and remembered facts of introductory chemistry courses. It is, indeed, a fascinating chemical occurrence.

Many biosensors involving almost every mode of energy transduction have been developed for different types of applications using DNA sequences for complementary binding of oligonucleotides as the molecular recognition strategy. Dectection schemes have been based on essentially three phenomena related to the chemistry of the nucleic acids: (1) hybridization, which involves the immobilization of an oligonucleotide probe which can bind to the complementary sequence of a target oligonucleotide or DNA fragment; (2) detection of mutations or deletions, and (3) detection of genotoxins and drugs that either bind to DNA or are intercalators which insert between base pairs. A sampling of applications based on acoustic waves (Andle *et al.*, 1992; Su *et al.*, 1994), electroanalytical methods (Millan and Mikkelson, 1993, Hashimoto *et al.*, 1994; Millan *et al.*, 1994; Wang, J. *et al.*, 1995a, 1996a,b,d) and optical devices (Piunno *et al.*, 1995; Abel *et al.*, 1996) illustrates all three strategies. Mikkelson (1996) has provided a short review of electrochemical methods involving sequence detection, including summaries of commercial systems based on electrogenerated chemiluminescence and the light-addressable potentiometer. In that review she illustrates the extraordinary required sensitivity and detection limits by showing that 12 or 24 attomoles of DNA may be extracted from 7–8 million white cells per ml of blood. Mikkelson also includes an informative comparison of the FOMs for several types of DNA-based sensors.

The complementary binding for molecular recognition can be achieved by the use of nucleic acid fragments obtained by enzymatic digestion of the target DNA or by synthetic oligonucleotide sequences which are produced by automatic synthesizers. The sequences are available commercially, either individually or as arrays (Gette and Kreiner, 1997). Piunno *et al.* (1995) actually used the automatic synthesizer to build their oligonucletides on the surface of an optical sensor. These automated systems are based on sequential phosphotriester reactions to add one nucleotide at a time, analogous to solid-phase peptide synthesis (Abeles *et al.*, 1992; Mathews and van Holde, 1990; Lehninger *et al.*, 1993).

Certainly one of the most remarkable achievements in biochemical engineering in recent years has been the development of methods for creating oligonucleotide arrays. These arrays have fostered a revolution in sequencing by hydbrization (SBH, complementary binding of labeled DNA probes to targets). Borman (1996) and Gette and Kreiner (1997) have provided very readable introductory summaries of the techniques and some applications. Two of the major methods for creating these arrays involve light-directed synthesis (Fodor *et al.*, 1991; Pease *et al.*, 1994; Chee *et al.*, 1996) and ink-jet printing (Blanchard *et al.*, 1996). In the Chee *et al.* (1996) report the authors point out that they are able to identify a target sequence, quantitate by fluorescence measurement, and compare to a reference. Their arrays have 20 μm × 20 μm "features" which contain $\sim 4 \times 10^6$ copies of a specific probe with ~ 10 nm spacing between probes. They also reported sequencing of human mitochondrial DNA using these arrays.

In using oligonucleotide binding for molecular recognition labeled probes are utilized. The labels may be radioisotopes (Su *et al.*, 1994), fluorophores (Chee *et al.*, 1996; Abel *et al.*, 1996) or an enzyme (see Mikkelson, 1996). Stimpson *et al.* (1995) chose to use a light-scattering colloid with his optical waveguide. In the Su *et al.* (1994) study radioisotopes were used to validate results from acoustic-wave detection which does not require a label since it is a mass-sensitive technique.

As mentioned above, not all biosensing applications involve the complementary binding for sequencing purposes. Deletions and mutations may be investigated if the abnormal region of a sequence is known so that appropriate probes can be isolated or synthesized. Millan *et al.* (1994) used this approach for a model electrode to detect the sequence characteristic of cystic fibrosis. They found that the hybridization temperature, however, had to be raised to 42 °C in order to distinguish the normal and abnormal sequences of the target DNA. Viral DNA's have also been investigated with biosensors. Wang, J. *et al.* (1996a) proposed a possible electrochemical approach for detection of a sequence related to HIV, but the authors emphasized that the method would require considerable optimization for clinical use. Andle *et al.* (1993) determined polymerase-amplified genomic DNA using an acoustic-wave device. They could detect 200 ng per ml of the polymerase-amplified DNA.

A third category of biosensing strategies involves binding, intercalation and damage detection. Pandey and Weetall (1994) developed a general FIA protocol for detection of intercalators. Using a photochemically assisted electrochemical detection scheme they were able to detect the model intercalator, ethidium bromide, in sub-nanomolar concentration ranges. The method should be applicable without the photochemical step for electroactive intercalators.

Using acoustic-wave detection Su *et al.* (1995) reported an LOD of 10^{-7}M for platinum anticancer drugs. Kinetic analysis of their data indicated that for both *cis*- and *trans*-platin the DNA binding involves the hydrolysis products of the drugs. J. Wang's group has reported on electrochemical detection of radiation damage to DNA (Wang, J. *et al.*, 1997). The method involves changes in the guanine-DNA oxidation peaks after UV irradiation. Screen-printed solid-state electrodes were used.

2.5. GENERAL CONSIDERATIONS FOR DESIGNS

From the preceding discussions it becomes obvious that there is great variety in sensing designs and strategies. Regardless of the type of transduction mode and the combinations chosen, there are several general aspects of designing sensors that apply broadly. A brief examination of these considerations should provide a framework for attention to some subtle, and some not-so-subtle, details that affect design.

2.5.1. Aspects of Mass Transport

The supply of a detectable species at the surface of a sensor is usually separated into external transport, which relates to movement of a species to a sample/membrane interface, and internal mass transport, which pertains to transport into and through a membrane to the energy transducer. Not all sensors have modifying membranes. In those cases only aspects related to external mass transport would be considered. For sensors that do have membranes in which chemical transduction steps affect fundamentally the mass transport properties—e.g., biocatalytic membranes—the effects of those processes must be taken into consideration (Section 2.5.5). A basic physical chemistry textbook, such as Atkins (1990), should be reviewed for the basic definitions and relationships of mass transport. Bard and Faulkner (1980) present the fundamentals of mass transport with regard to a variety of electrochemical cases, including flow-through porous electrodes and rotating electrode designs. Valcárel and Luque deCastro (1994) cover theory and applications of flow-through biosensors.

Mass transport may occur by (1) diffusion related to concentration gradients, (2) convection which involves some form of hydrodynamic transport, and (3) migration which results from transport due to a gradient of electrical potential. The total flux (flow) J of a species i (mol $s^{-1}cm^{-2}$) to a surface (one dimension, x) may be stated in terms of these driving forces (Nernst–Planck equation):

$$J_i(x) = \underset{\text{Diffusion}}{D_i \frac{\partial C_i(x)}{\partial x}} - \underset{\text{Migration}}{\frac{z_i F}{RT} D_i C_i \frac{\partial \phi(x)}{\partial x}} + \underset{\text{Convection}}{C_i \nu(x)}$$

Here, D is the diffusion coefficient, z_i is the charge, $\frac{\partial C_i(x)}{\partial x}$ is the concentration gradient, $\frac{\partial \phi(x)}{\partial x}$ is the potential gradient and $\nu(x)$ is the velocity of solution (cm s^{-1}).

The relative importance of the individual contributors varies according to the type of sensor and the experimental design. For example, migration is the primary factor for conductimetric sensors in which charged species move toward electrodes according to their ionic mobilities, μ, with each ion carrying its own proportion of the current as indicated by its transference number, t. For most sensors other than those based on conductivity, migration is usually considered a negligible factor or the sensor is designed so that assumption will pertain.

Many types of sensors are utilized in a convective mode—stirred solutions, on-line reactor/detector units, flow-cells, FIA, rotating electrodes or even rotating reactors (Richter *et al.*, 1996). This means that external mass transport to the sensor will be governed by the hydrodynamic characteristics of the convective system. For example, stirring rates must be well controlled for reproducible results. Varying stirring rate can also be used as a means of determining whether external or internal mass transport controls the sensor's response (Turner *et al.*, 1989; Tatsuma and Watanabe, 1992; Tatsuma *et al.*, 1992). A response that shows no dependence on stirring or flow rate is being controlled by factors related only to internal mass transport, kinetics of chemical transduction in the membrane, or both.

When using flow-cells and reactor columns (Valcárel and Luque deCastro, 1994), the adjustment of flow rates must be determined in accordance with the timescale of chemical events occurring during chemical transduction. Residence time in the reactor or flow-cell, stated in terms of (cell volume/flow rate), becomes primarily important. If a stopped-flow design is employed, then the conditions revert to those of unstirred batch analysis.

Carr and Bowers (1980) have examined in detail factors of experimental design for ideal plug-flow and continuously stirred reactors in which immobilized enzymes are used. They designated these two cases as the extremes in the spectrum of convective systems. Several cases for interrelationships among cell volumes, flow rates, residence times and enzymatic reaction rates were treated. Basically the results indicate that for optimum sensitivity the fractional conversion of substrate to detectable product must be high.This implies adequate enzyme loading of the reactor/cell and a rate constant faster than the flow rate utilized.

The Xie, B. *et al.* (1992, 1993, 1994a,b, 1995) studies of thermal detection of enzymatic reactions with microflow systems requiring no more than 1 μl of sample provide good examples of the optimization of flow for maximum sensitivity. The Berners *et al.* (1994) design of tubular electrodes (i.d. = 400 μm) for an on-line glutamate sensor coupled to microdialysis involved test-electrode response times of < 1 s for the glutamate oxidase electrodes, and a calculated filling time of 20 s using a flow rate of 2 μl min^{-1}. Yet, the response time of the system was almost 1 min. The authors attributed the delay to probable diffusion and dispersion within the system.

If a flow-cell is used for an equilibrium-based chemical transduction such as antigen–antibody binding, then there must be sufficient time for the pertinent equilibrium to be attained before measurement. In a static cell, this may require a long time, even hours, to reach a stable signal (Dahint *et al.*, 1994; Renken *et al.*, 1996). Frequently the sample throughput of a flow-immunosensor is determined primarily by this necessary equilibration time, as well as the regeneration time, rather than the response time of the sensor itself. Stable signals may not be attainable for several minutes in these systems.

Rotating electrodes are particularly useful for characterizing the relative contributions of convection and diffusion to the response of amperometric sensors (Bard and Faulkner, 1980; Turner *et al.*, 1989; Gough and Leypoldt, 1979). The various mass transport and kinetic factors contributing to the amperometric response due to electron transfer can be separated and determined.

$$\frac{1}{i} = \frac{1}{i_{\text{diff}}} + \frac{1}{i_{\text{kinetic}}}$$

The angular velocity is ω, which is 2π times the number of revolutions per second. Plots of $1/i$ versus $1/\omega^{1/2}$ may be analyzed to determine the relative effects of convection and diffusion on the response.

The rotating reactor (RR) proposed by Matsumoto *et al.* (1993) and Richter *et al.* (1996) was mentioned above. This device has been proposed as an alternative to the IMER for continuous flow systems and rotating electrodes. The volume of the reactor is adjustable between 450 μl and 1.8 ml. Flow rates of about 1 ml min^{-1} were used with a rotation speed of about 900 rpm. Sample throughput was about 30 samples per hour. The authors did a thorough comparison of the efficiency of enzyme utilization under conditions of both continuous flow and stopped-flow. They concluded that the rotating reactor was more efficient than packed columns with regard to mass transfer/kinetics interrelationships, and that the RR permitted the use of smaller amounts of the enzymes.

Within sensor membranes or quiescent sampling sites, diffusion becomes a prime contributor to mass transport. As a reminder from introductory physical chemistry courses, diffusion involves the transport of matter in accordance with concentration gradients as indicated by Fick's laws of diffusion (see Atkins, 1990):

$$J(x,t) = -D\frac{\partial c(x,t)}{\partial x} \tag{1}$$

$$\frac{\partial c(x,t)}{\partial t} = D\frac{\partial^2 c(x,t)}{\partial x^2} \tag{2}$$

where D is the diffusion coefficient characteristic of a species in a given medium. As illustrated in Figure 2-8, D represents a diffusing species passing through D planes or slices of space per second, which may be visualized as a bullet moving through a deck of cards. The units for D are stated in

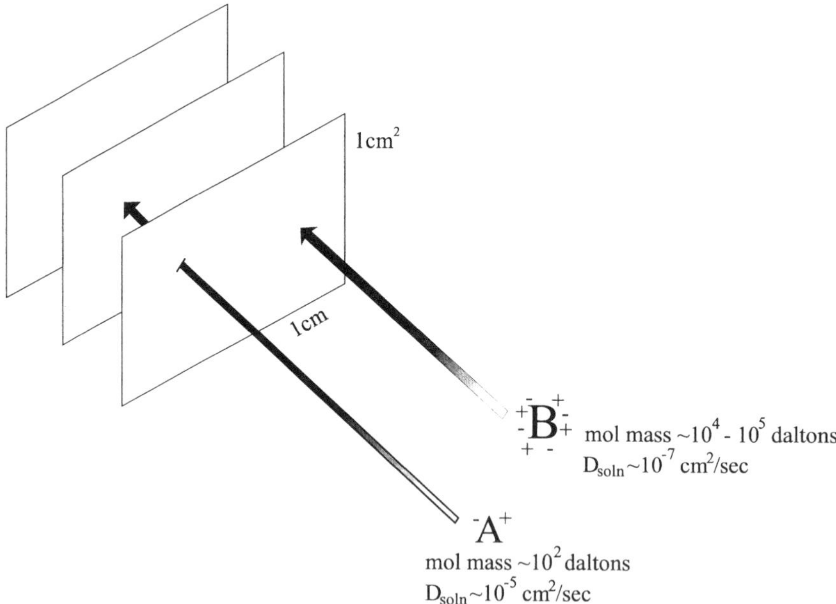

FIG. 2-8. Schematic illustrating definition of the diffusion coefficient—comparison of two molecules of different molar mass.

terms of $cm^{-2}\ s^{-1}$. In solutions of the same viscosity smaller molecules have larger D's than large molecules. Values of D are smaller in media of higher viscosities or under conditions of tortuous paths. The viscosity dependence, used especially in hydrodynamic studies of macromolecules, is reflected in the Stokes–Einstein equation:

$$6\pi a\eta = \frac{kT}{D}$$

Here, a is the radius of the molecule, η is the solvent viscosity, and D is the diffusion coefficient. k and T are the Boltzman constant and absolute temperature, respectively. Whereas D represents the number of planes a molecule can penetrate in a unit of time, the flux (flow) J is the number of molecules penetrating a particular plane in a unit of time. The units for flux are mol $cm^{-2}\ s^{-1}$. Flux is positive for a species moving toward a reference plane and negative if the species is moving away from the plane. Total flux depends on all the contributors to mass transport, as shown earlier. Flux due solely to diffusion is given by:

$$J(x) = -D\frac{\partial c}{\partial x}$$

This relationship indicates that it is the concentration gradient, not the actual concentration, that determines diffusional flux. In Figure 2-9 some of the considerations related to flux to a generalized sensor are schematized. Part (a) is for a situation in which the energy transducer consumes the incoming species—e.g. an amperometric sensor based on oxidation or reduction at the electrode. This type of sensor would be referred to as a nonzero-flux sensor and the concentration gradient of the detectable species will change with time, as shown in the family of curves. The concentration gradient region extending from the surface of the sensor is the diffusion layer of thickness δ. The thickness of this region—i.e., the extension of a graded region of concentration—changes with time for a nonzero-flux sensor, as shown. Even for a convective system, the physics of laminar flow indicate that there is a quiescent layer of solution adjacent to a surface which constitutes a diffusion layer. With efficient convective mass transfer the effect of this diffusion zone at the sample/membrane interface can be minimized. If the energy transduction mode—e.g., potentiometry or optical– does not involve consumption of the species (zero flux), then there will no concentration gradient at the transducer surface due to the energy-transduction process.

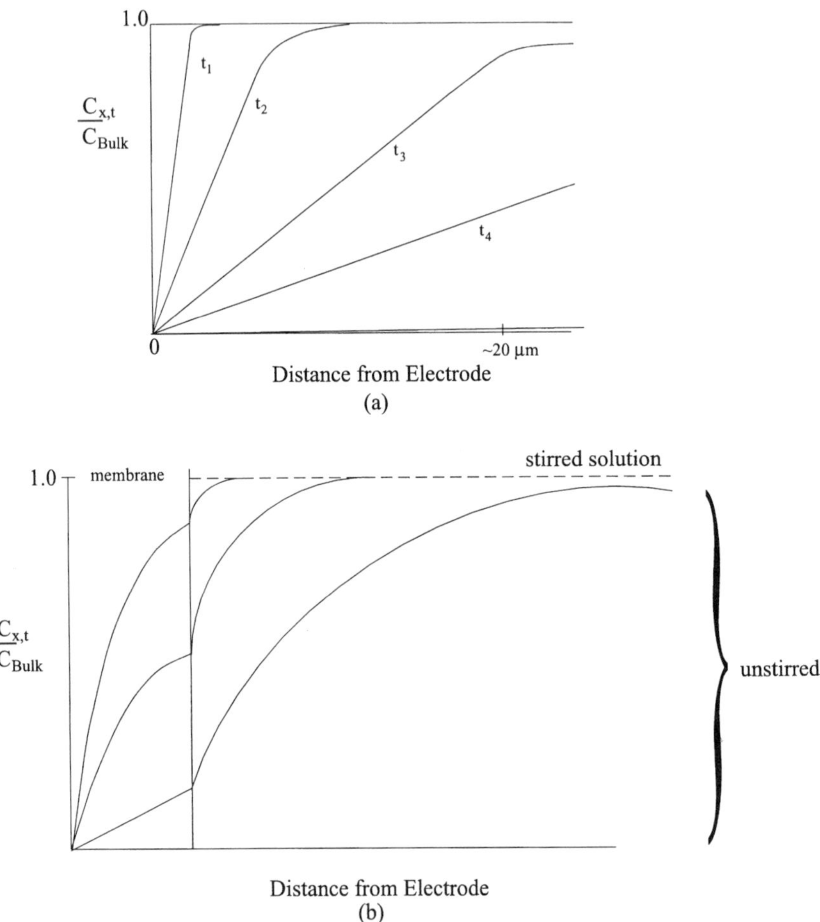

FIG. 2-9. Generalized pattern of concentration gradients at the electrode surface of an amperometric sensor. Linear diffusion to a planar electrode is assumed. (a) Unstirred solution. C is consumed at the electrode by an electron-transfer process (nonzero flux transducer). Increasing time is from $t_1 \rightarrow t_4$. (b) Generalized concentration gradient profiles with chemically selective membrane immobilized on the electrode.

Figure 2-9(b) relates to situations in which a membrane is placed over the energy transducer. If the bulk solution is stirred, the supply of analyte will be constant external to the sensor, but for a nonzero-flux sensor there will be diffusional gradient within the membrane. The discontinuity of the gradient at the solution/membrane interface is intended to illustrate that partitioning effects must also be considered (Justice, 1987; Meyerhoff *et al.*, 1996).

In Figure 2-10 the model of a biosensor is revisited as a pictorial summary of mass transport considerations related to external mass transport, partitioning/permeability and internal mass transport. Of the different energy transduction modes illustrated, only the top one for amperometric sensors represents a nonzero-flux example. The chemical transduction process is represented generally in the model by the A to B transformation. In Section 2.5.5 it will be shown that the concentration profiles across a membrane are dependent on the concomitant effects of diffusion and reaction.

A significant deviation from the convective mode dominance of external mass transport exists for certain *in situ* sensors. For example, consider the situation for *in vivo* measurements in the brain. In the unique physical environment in the extracellular space of the brain, corrections for localized environmental effects must be made. As has been discussed in full by Gardner-Medwin (1980) and Nicholson and Phillips (1979), the complex medium of the brain imposes new constraints on the basic mass-transport process. Measuring analytes in the restricted environment of interstitial spaces among cells requires some alterations to the basic mass-transport equations. Nicholson and Phillips cite typical values of 0.01–0.04 μm for interstitial distances between cells in the brain. For diffusional problems involving these distances and the usual sizes of *in vivo* sensing probes, it is

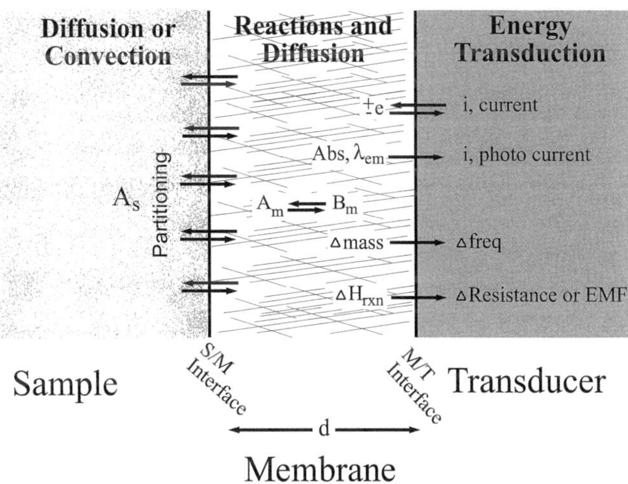

FIG. 2-10. Model of the basic factors for consideration of external mass transport, partitioning, internal mass transport and signal generation. Amperometric, optical, acoustic-wave and thermal energy transductions are represented.

not difficult to see that general derivations based on a number of fairly simplified assumptions may not apply.

Two of the most important corrective factors for *in vivo* measurements in the brain deal with (1) the tortuosity of path λ that alters the simple random walk approach in derivations of basic diffusion relationships and (2) the volume fraction α of extracellular fluid, which is the volume in which the analytes are actually measured relative to the volume of total tissue. Interstitial space—the actual sampling medium in this case—accounts for only about 20% of total tissue volume. The first of these factors, λ, results in a "corrected D", D^*, which is given by $D^* = D/\lambda^2$.

The overall effect of these two corrections, α and D^*, is that a lower rate of mass transport is observed *in vivo* relative to *in vitro* experiments with the same sensor. This is predicted by the $\alpha < 1$ and $D^* < D$. The corrections result in a revised diffusion statement for the time-dependent change in concentrations:

$$\frac{\partial C}{\partial t} = D^* \nabla^2 C + q/\alpha$$

where q is related to a point source of the substance being measured.

The original reference should be consulted for a full development and discussion. For a recent application of these considerations see the Friedemann (1996) research report of an *in vivo* NO sensor discussed in Section 2.1.2.2.

2.5.2. Partitioning and Permeability

Partitioning between phases is a concept familiar to all chemists. It has been discussed earlier in the context of lipophilicity of ions. The concept of permeability is probably newer to those less experienced in sensor designs. Permeability implies more than mere partitioning. It is the combined result of a species crossing an interface and moving through the new phase. Faster response times and higher sensitivities are generally the result of more favorable permeability of a modifying film on a sensor. The permeability of a species in a membrane is defined as $P_m = \alpha D_m/d_m$, where α is the partitioning constant, D_m is the diffusion coefficient of the species in the membrane and d_m is the thickness of the membrane. The units for P_m are cm s^{-1}. For gas-permeable membranes, the analogous expression is $P_m = (D_m)(S_m)$, where S_m is the solubility of the gas in the film. The unit in this case is (mol m^{-1} s^{-1} kPa^{-1}). Solubilities of gases in various polymers may be found in polymer handbooks.

Permeation of the sensing membrane is fundamental to the operational characteristics of most devices chemically modified with a thin film or membrane. For example, for the Zhou and Arnold (1996) optical NO sensor mentioned earlier in Chapter 2, several conventional sensing membranes were tested and compared, but they found that only a silicone material was effective. In the Tarnowski *et al.* (1995) study, discussed in Chapter 3 in the context of immobilizations on polymers, considerable effort was expended to characterize the permeability of their refunctionalized fluoropolymer with respect to oxygen. It was essential that the refunctionalization process did not diminish the permeability for the amperometric determination.

While many of the reported theoretical models of sensor response make simplifying assumptions about both partitioning and permeability, the reality of effects cannot be overlooked. Some practical considerations and examples of experimental approaches to determining permeability have been addressed for Nafion membranes (Fan and Harrison, 1992; Zook and Leddy, 1996), for overoxidized polypyrrole layers (Hseueh and Brajter-Toth, 1994; Palmisano *et al.*, 1995a) and self-assembled monolayers on gold (Cheng and Brajter-Toth, 1995, 1996). These reports provide good lessons in the experimental detail necessary to characterize the permeability.

Fan and Harrison (1992) examined the influence of casting solvent and temperature on the permeability of Nafion to neutrals, such as glucose. Rotating electrode studies (Gough and Leypoldt, 1979) were utilized to separate the current contributions of convection and transport through the Nafion films. They found that thermal curing of the films at 120–130 °C after casting reduced diffusion coefficients by as much as 4–8 fold compared to room-temperature cured films. Casting solvent also affected the permeabilties. Zook and Leddy (1996) compared three forms of Nafion with regard to wetted density. They used commercially available films, solvent-cast films and heat-treated cast films. They found that the unheated solvent-cast films had the lower wetted density. Heating brought the recast films to approximately the same wetted densities as the commercial films. They, too, noted the differences in films cast from different solvents.

Hseueh and Brajter-Toth (1994) undertook some of the definitive studies of permeabilities of ultrathin overoxidized ($>$ 0.7 V versus SCE) polypyrrole (oPPy) films. These films are of especial interest for amperometric sensors because electropolymerized films of pyrroles have been used as potential-controlled selective films. Carbon and platinum electrodes were (over)oxidized in pyrrole at +0.950 V versus Ag wire, and repeated coatings were applied for uniformity of the film. Using the rotated electrode appoach for measuring permeabilities, they determined that the films had good permeability to cations. The films would also discriminate between dopamine and ascorbic acid, an accomplishment always welcome to those involved in *in vivo* measurements.

Palmisano *et al.* (1995a) examined the same types of films, but in the context of using them for enzyme immobilization on amperometric electrodes. They performed extensive characterization of the films and found that the films seem to exhibit two domains, one hydrophilic and one hydrophobic. They tested the films against the usual interferents in brain fluids and found that permeabilities varied according to the hydrophilic/hydrophobic nature of the analytes. In the case of acetaminophen, the authors concluded that it partitioned into the membrane but could not diffuse through the membrane; thus they were unable to obtain a measurable permeability. Pihel *et al.* (1996) tested these types of electrodes in the brain using fast-scan voltammetry, and compared the behavior to Nafion-modified electrodes. They found stable, selective response to dopamine with oPPy exhibiting a higher sensitivity than Nafion, but some adsorption and loss of time-resolution were noted.

2.5.3. Activities, Concentrations and Amounts

Early on in a chemist's education it becomes common practice to invoke the thermodynamic assumption of either equal activities for two species or an activty coefficient of 1, thus equating activities and concentrations. For dilute solutions of ions and some other situations, this assumption can be quite valid. Unfortunately, the assumption is often introduced to simplify solutions to difficult equations, though sometimes it may be chemically inappropriate. There are several points to be made about activities, concentrations and amounts.

Just as is true of ordinary detectors, there are some energy transduction modes that are concentration-dependent (amperometry), some that are mass-dependent (microgravimetry) and some that are activity-dependent (potentiometry). For example, the acoustic-wave devices are utilized as mass-dependent devices, though other factors must be considered in interpreting responses. Most potentiometric sensors are, however, based on an activities-driven equilibrium that results in a measured electrode potential. As discussed in Section 2.1.2, Meyerhoff's group has demonstrated that polyion sensors provide an exception because a non-equilibrium situation can develop at the solution-membrane interface (Fu *et al.*, 1994). In this situation the potentiometric device becomes a nonzero-flux sensor.

The ion-selective optical sensors mentioned earlier in this chapter (Shortreed *et al.*, 1996a) have an optical response function that depends on the concentration of the chromophore, but the concentration of the chrompohore is determined by the activities-dependent acid–base equilibrium in the sensing membrane. Therefore, maintaining constant ionic strength in the membrane is essential and the sensor must be designed that way.

These examples serve as reminders that design must include consideration of the basis on which the response is to be based. A sensor design based on invalid assumptions will have a "built-in" deviation between theory and experiment. The corollary to this is that performance should not be assessed in terms of models formulated on assumptions that do not pertain to the real situation.

2.5.4. Effects of the Microenvironment

References have been made several times to performance considerations related to the microenvironment of a sensing membrane (section 2.3.3). For example, enzyme kinetics can be affected by internal gradients of charge or pH (Carr and Bowers, 1980). Michaelis constants change upon immobilization. Mădarăş and Buck (1996) and others have reported effects introduced by cross-linking within membranes. This last effect may involve similar tortuous pathways demonstrated for *in vivo* sensing in the brain, as well as orientational inaccessibility of active sites.

White's group has published several reports on the intricacies of microenvironmental variations at electrodes (Smith and White, 1992, 1993; Takada *et al.*, 1995; Lee, W.-W. *et al.*, 1993; Gao, X. *et al.*, 1995). They have shown that predicted electrochemical and acoustic-wave behavior can be altered drastically by incomplete or incorrect attention to the uniqueness of the microdomain. Everything from electroneutrality assumptions to microchanges in volume of the membranes and depletion effects should be considered for certain cases. The very different environment of membranes, compared to bulk solutions, has continued to be a challenge to the theoreticians and designers.

2.5.5. Terminology of Modeling Responses

Sensor response functions are highly dependent on the mass transport, equilibria and kinetics associated with the analyte. All factors that affect the total transduction process must be considered if theory and modeling are to inform design. This implies that all phenomena affecting transport to the sensor (external) and through any reaction layers (internal) to the surface of the energy transducer must be included in models for responses. In introductory physical chemistry textbooks—e.g., Atkins (1990)—the basic approach to problems of simultaneous consideration of diffusion, convection and reactions is presented in the context of material balance relationships.

During the last 20–25 years there have been numerous approaches to modeling various types of both biocatalytic and bioaffinity sensors. In some cases the mathematics is not simple, but available simulation and modeling software has evolved to make the whole task less tedious. For this introduction little more than the basic terminology and the fundamental logic of the methods will be presented. The bulk of work in modeling for analytical purposes has focused on biocatalytic sensors, particularly amperometric devices. Figure 2-11 illustrates a basic model for these sensors.

With emphasis on enzymatically based sensors, Carr and Bowers (1980) covered very thoroughly the work in this area through 1979. Their exploration and explanations of various situations are excellent. They also provided extensive compilations of applications—both potentiometric and amperometric—using immobilized enzymes. Eddowes (1990) develops with some mathematical detail the relationships and approximations for several examples of both zero-flux and nonzero-flux sensors for biocatalytic and bioaffinity cases. Although the mathematical formulations are not as extensive as Eddowes', Janata's (1989) treatment emphasizes the influence of some usual assumptions with caution about the importance of including the activity-dependent partitioning and pH dependent terms. Scheller and Shubert (1992) included mathematical details for several of the more common cases of enzymatic mechanisms, including considerations of both the num-

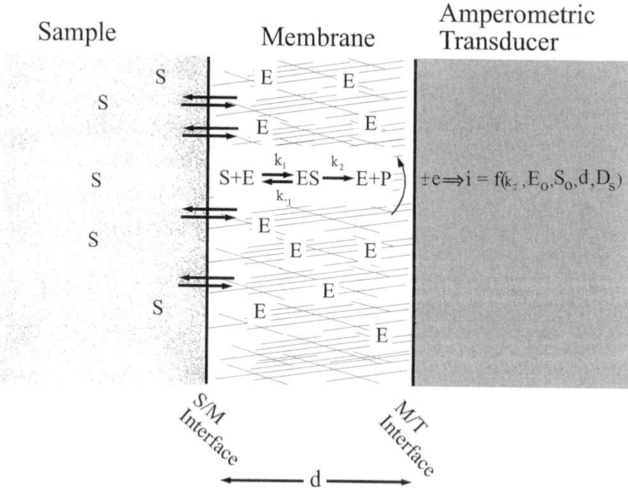

FIG. 2-11. Model of an amperometric sensor with chemical transduction by enzymatic reaction. Substrate (analyte) in the sample partitions into the immobilized enzyme-containing membrane. Product of the enzymatic reaction undergoes oxidation or reduction at the electrode, producing a detectable current which is a function of parameters related to both mass transport and the enzymatic reaction.

ber of layers, number of enzymes incorporated in the layers and sequential enzymatic catalyses. Albery and Craston (1989) combined their discussion of theory for several amperometric enzyme–electrode cases with multiple examples of systems studied. They show the results for different generations of amperometric electrodes, including the conducting salt electrodes. Schultz (1989) discusses fiber-optic sensors based on bioreceptors. Many of the relationships given for the different antibody–antigen situations are helpful for formulating the reaction terms of the diffusion–reaction equations.

Bartlett and Pratt (1993) have presented an extensive review of the various approaches to modeling processes for enzyme electrodes. Table 1 on pages 456–457 of that report provides a chronological accounting, with descriptive comments, of major developments in the field since 1971 and the various mathematical techniques that have been used for solving the differential equations using different assumptions and mechamisms. Tatsuma and Watanabe (1992b, 1993; Tatsuma *et al.*, 1992b) modeled the transient and steady-state responses and the oxidation–reduction mediation efficiency of enzymatic amperometric electrodes. In 1994 Martens and Hall presented the generally applicable case of an amperometric sensor in the presence of two oxidants, the natural substrate and a mediator. The 1996 Gooding and Hall report focused specifically on modeling responses for informed design. Ikeda *et al.* (1996) have extended the diffusion–reaction treatment to evaluation of the catalytic efficiency of a microbial cell sensor using a membrane-coated electrode.

In the following discussion, only biocatalytic sensors will be treated. There are analogous approaches for bioaffinity devices for which the primary concern is not usually the traverse of an analyte through a membrane, but is the ligand-binding reactions at the surface of the sensor. The reaction terms of the equations are introduced as the appropriate expressions for the bound versus free species or as the fractional coverage of available sites for binding. Examples are given in Eddowes' chapter (1990). Just as biocatalytic situations can become quite complex with multiple enzymatic steps, binding at multiple sites or separating specific from non-specific adsorption can complicate the bioaffinity calculations.

Considering biocatalytic sensors, optimization of sensor performance can be accomplished by careful analysis of variables in the solutions for diffusion–reaction equations and assessing their relative magnitudes (Martens and Hall, 1994; Martens *et al.*, 1995; Gooding and Hall, 1996). For example, it might be determined from the mathematical solutions that a different loading of enzyme in the layer should improve the sensitivity of the sensor or that a thinner layer might improve response time. The dynamic range of the sensor might be extended by altering the thickness of the reaction layer. Knowing the correct diffusion–reaction relationship can be a powerful tool in the quest for improved performance of a biosensing device. Yet there are many cases where a precise theoretical expression for concentration and time-dependencies of responses is, at best, elusive. Designers of sensors are the ones who must give attention to the details of the theory–modeling–performance triad. Oftentimes, probably most of the time, all of that planning and assessment is totally transparent to the end-user, as it should be in the majority of applications.

In the preceding sections it has been emphasized that external and internal mass transport can be considered separately and that while convection may play a big role in external mass transport, it will not be a significant factor in internal mass transport. For the purposes of this discussion it can also be assumed that migration is negligible. Inside a biocatalytic sensing membrane the factors of prime importance are the diffusional characteristics, the chemical transduction reactions, and the nature of the energy transduction process. Predictable patterns of responses can be derived from solutions to the appropriate diffusion–reaction equations with realistic, if sometimes approximate, initial and boundary conditions applied.

The goal is to describe accurately the response as a function of the analyte's activity or concentration, though the route to revealing this relationship often involves terms for a detectable species such as an enzymatic product. In order to treat both transient and steady-state signals, Fick's second law becomes the starting point. Fick's second law, the mathematical statement for rate of change in concentration, is in one dimension for planar diffusion:

$$\frac{\partial C(x,t)}{\partial t} = D\frac{\partial^2 C(x,t)}{\partial x^2}$$

As will be shown in Chapter 4, ultramicrosensors require the use of radial diffusion expressions due the minute radii of the sensors relative to the diffusional field around the tips:

$$\frac{\partial C(r,t)}{\partial t} = D\left(\frac{\partial^2 C(r,t)}{\partial r^2} + \frac{2}{r}\frac{\partial C(r,t)}{\partial r}\right)$$

where r is the radial distance from the center of the electrode. Those cases of radial diffusion will not be considered in this basic introduction. Following the usual logic in accounting for a total change in the concentration of detectable species, a term is introduced into the equation for the concomitant effect of kinetically controlled reactions, usually assuming Michaelis–Menten kinetics:

$$\frac{\partial[\mathrm{S}]}{\partial t} = D_{\mathrm{S}} \frac{\partial^2[\mathrm{S}]}{\partial x^2} - \frac{k_2[\mathrm{E}][\mathrm{S}]}{K_{\mathrm{m}} + [\mathrm{S}]}$$

$$\frac{\partial[\mathrm{P}]}{\partial t} = D_{\mathrm{P}} \frac{\partial^2[\mathrm{P}]}{\partial x^2} + \frac{k_2[\mathrm{E}][\mathrm{S}]}{K_{\mathrm{m}} + [\mathrm{S}]}$$

It is not difficult to find examples of some fairly elaborate enzyme mechanisms that present interesting challenges to solutions of the appropriate differential equations.

While each diffusion–reaction category is unique, there is an organized sequence of basically 3 steps in setting up the problem for mathematical solution or simulation. (1) First the diffusion–reaction equations are written, as above. (2) The initial and boundary conditions are stated. For example, if an enzyme electrode is being used in a convective mode—i.e., stirring, flowing stream, etc.—the concentration of analyte (substrate) at the sample–membrane interface may be assumed to be equal to the bulk concentration. For a zero-flux sensor there will be no gradient of concentration of detectable species at the transducer surface because the energy transduction process does not consume the species. On the other hand, for a nonzero-flux transducer—e.g., amperometric—the concentration of detectable species at the surface will be zero because it is being consumed in the transduction process. (3) Having stated the diffusion–reaction equations and established the initial conditions and boundary conditions, the decision must be made as to whether the transient or steady state response is of interest. When a steady state situation is being modeled, the assumption is that all the fluxes due to mass transport and reactions are balanced and can be equated.

Once the flux relationships, initial and boundary conditions are stated, then the solution may be sought. There are several options for solving the differential equations of diffusion–reaction and simulating the response functions—approximate analytical solutions, explicit finite difference, orthogonal collocation, etc. (Bartlett and Pratt, 1993). In many cases the differential expressions are complex and can be very difficult to solve. Numerical methods and digital simulations are widely used (Speiser, 1996).

This sequence is best illustrated by two specific cases outlined in Table 2-7. Assume that convective conditions are such that external mass transport is not a determining factor. Further, assume that a simple Michaelis–Menten mechanism pertains:

$$S + E \underset{k_{-1}}{\overset{k_1}{\rightleftharpoons}} ES \xrightarrow{k_2} P + E$$

The thickness of the modifying membrane is d and $0 < x < d$, with the surface of the transducer being $x = 0$:

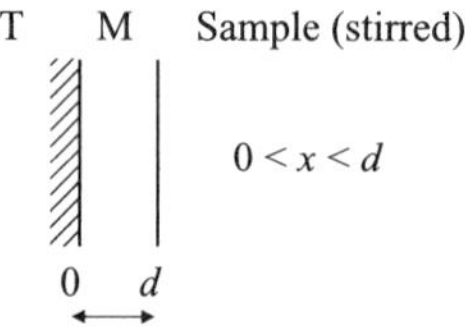

Case I is that of a zero-flux sensor such as a potentiometric device which responds to [P]. Case II is a nonzero-flux sensor at which product P is oxidized or reduced: P $\pm$ ne $\rightarrow$ P$'$. Table 2-7 contrasts the cases.

The interrelationship of diffusion and reaction in the membrane is not an easy thing to conceptualize because there are several interacting variables to consider. In Figure 2-12 the Thiele modulus and enzyme load terms, identified in cases I and II , are shown. Note that the expressions contain terms that reflect both the enzyme reaction rate and diffusion properties of the substrate. In the expression, d is the thickness of the membrane. The term, k_2/K_m, referred to as the catalytic efficiency of an enzyme, may vary over a wide range, even for one enzyme that is active toward several different substrates. The upper limit of efficiency would be approached at a diffusion-limited value of $\sim 10^9$ ((mol $l^{-1})^{-1}$ s^{-1}). This value would imply a diffusion-limited process for the totally effective collisions between substrate and enzyme. Thus, assuming there is enough enzyme in the reaction layer to turn over substrate molecules as they arrive, the Thiele modulus may be thought of as the available enzyme's efficiency in the confined space of the membrane relative to the diffusion characteristic of the substrate within that same space.

The Thiele modulus becomes the determining factor in the relative influence of diffusion and reaction terms on responses and it informs design (Carr and Bowers, 1980; Martens and Hall, 1994, Bacha *et al.*, 1995; Gooding and Hall, 1996). For a specific enzyme reaction the relative values of the original amount of enzyme in the membrane and the layer thickness determine the magnitude of the Thiele modulus. If the Thiele modulus term is substituted into the diffusion–reaction equation, as shown in Figure 2-13, we see that when ϕ^2 is large the reaction term will dominate the diffusion-reaction equation, meaning diffusion will be rate-determining. Thus, the sensor's response will be determined by the diffusion characteristics. This can occur with either a high enzyme load or a large thickness d, but increasing d will increase the response time of the sensor. If the term containing ϕ^2 is made small relative to the diffusion terms of the equation, then the enzyme reaction will determine the sensor's response pattern. This could occur if the enzyme load is low or if the reaction layer is very thin. Figure

TABLE 2-7 Logical Sequence for Modeling Responses Dependent on Diffusion-Reaction[a]

Basic diffusion–reaction equations for membrane:

$$\frac{\partial[S]}{\partial t} = D_S \frac{\partial^2[S]}{\partial x^2} - \frac{k_2[E][S]}{K_m + [S]}$$

$$\frac{\partial[P]}{\partial t} = D_P \frac{\partial^2[P]}{\partial x^2} + \frac{k_2[E][S]}{K_m + [S]}$$

Simplifying assumptions

1) Reaction in enzymatic membrane depletes S and generates P.
2) Steady-state response occurs when the rate of reaction within the membrane balances the rate of mass transport of S and P into and out of the membrane.
3) Mass transport of S to the sample/membrane interface is not a limiting factor.
4) S and P have equal permeabilities in the membrane and the partitioning constant for S is equal to one: $K_d = [S]_{memb}/[S]_{bulk} = 1$
(this is an unrealistic assumption widely used for simplification).
5) The enzymatic product P is the detectable species.

Definitions:

Thiele modulus $\Phi = d\left[\frac{k_2[E]}{K_m D_S}\right]^{\frac{1}{2}}$ Enzyme load $\Phi^2 = \frac{k_2[E]d^2}{K_m D_S}$

Initial conditions:

at $x = d$, S/M interface $[S]_d = [S]_{bulk}, [P]_d = 0$
at $t = 0$ for $0 \leq x \leq d$ $[S] = 0, [P] = 0$

For steady-state responses:

$$\frac{\partial[S]}{\partial t} = 0, \quad \frac{\partial[P]}{\partial t} = 0$$

$$D_S \frac{d^2[S]}{dx^2} = \frac{k_2[E][S]}{K_m + [S]} \qquad D_P \frac{d^2[P]}{dx^2} = -\frac{k_2[E][S]}{K_m + [S]}$$

For enzymatic reaction, if $[S] \ll K_m$, i.e. first order with respect to S

$$\frac{d^2[S]}{dx^2} = \frac{k_2[E]}{K_m D_S}[S] \qquad \frac{d^2[P]}{dx^2} = -\frac{k_2[E]}{K_m D_P}[S]$$

if $[S] \gg K_m$, i.e. zero order with respect to S

$$\frac{d^2[S]}{dx^2} = \frac{k_2[E]}{D_S} \qquad \frac{d^2[P]}{dx^2} = -\frac{k_2[E]}{D_P}$$

continued

TABLE 2-7 *continued*

CASE I Zero-flux transducer	CASE II Nonzero-flux transducer
Integrate differential equations with boundary conditions: $[S]_d = [S]_{bulk} \quad [P]_d = 0$ at $x = d$ $[S] = [P] = 0$ at $t = 0, \; 0 \leq x \leq d$	For $[S] \ll K_m$ $$\frac{d^2[P]}{dx^2} = \frac{\Phi^2}{d^2}\left(\frac{\cosh\left(\frac{x}{d}\Phi\right)}{\cosh\Phi}\right)[S]_d$$ at $x = 0, [P] = 0 \qquad$ at $x = d, \; [P] = [P]_d$
$$\left.\frac{d[S]}{dx}\right\|_{x=0} = 0, \; \left.\frac{d[P]}{dx}\right\|_{x=0} = 0$$ zero-flux condition	$$[P] = \frac{x}{d}[P]_d + \frac{[S]_d}{\cosh\Phi}\left(\frac{x}{d}[\cosh\Phi - 1] + [1 - \cosh(\frac{x}{d}\Phi)]\right)$$
Response $= f([P]_{x=0})$ For first order $[S] \ll K_m$ $$[S] = [S]_{bulk}\left(\frac{\cosh\left(\frac{x}{d}\Phi\right)}{\cosh\Phi}\right)$$ $$[P]_{x=0} = \frac{D_s}{D_P}[S]_d\left(1 - \frac{1}{\cosh\Phi}\right)$$ For zero order $[S] \gg K_m$ $$[P]_{x=0} = \frac{k_2[E]}{2D_P} \neq f([S])$$	For the nonzero-flux condition, need $$\left.\frac{d[P]}{dx}\right\|_{x=0} \text{ which is } \neq 0, [P]_{x=0} = 0$$ Differentiate expression for [P] and set $x = 0$ For current: $i = nFAD_P \left.\frac{d[P]}{dx}\right\|_{x=0}$ (A is the area of the electrode) For $[S] \ll K_m \quad i = \dfrac{nFAk_2[E]d}{K_m + \dfrac{Ak_2[E]}{P_{m,S}}}$ ($P_{m,s}$ is permeability of S) For $[S] \gg K_m \qquad i = nFAk_2[E]d$ Time to reach steady state: $$t_{ss} \leq 1.5\frac{d^2}{D_S}$$

[a] Mell and Maloy (1975, 1976), Carr and Bowers (1980), Eddowes (1990), Tatsuma and Watanabe (1992b), Tatsuma *et al.* (1992b), Bartlett and Pratt (1993), Martens and Hall (1994), and Gooding and Hall (1996) should be consulted for a variety of cases, assumptions and boundary conditions. For amperometric sensors the usual cases are for two-substrate or mediated reactions, extensions not covered in CASE II above.

2-13 depicts this balancing act between diffusion and reaction control of the response. Figure 2-14 summarizes some practicalities arising from the modeling process.

2.6. REPORTING PERFORMANCE—EXAMPLES

This chapter has dealt with many of the considerations of design that ultimately determine the performance of a sensor. As indicated earlier, the

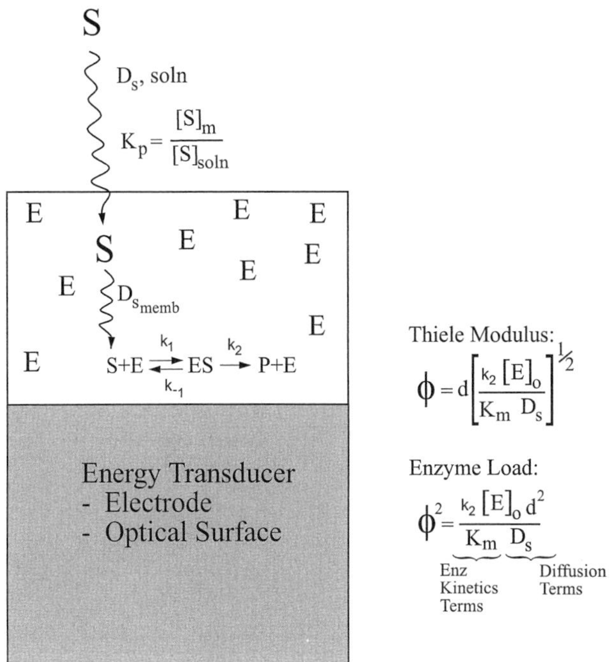

FIG. 2-12. Concept of enzyme load. E_0 is initial concentration of enzyme; K_m is the Michaelis constant for the enzyme reaction; D_S is the diffusion coefficient of substrate. Other terms are defined in the figure. In the formula for the Thiele modulus Janata (1989) includes appropriately a term for the enzyme's pH dependence. It is omitted here for clarity in emphasizing the counterbalance of diffusion and kinetic control.

performance is reported in terms of FOMs, sometimes with generally accepted conventions for reporting. In the following examples some variations of reporting performance will be examined.

2.6.1. An Amperometric Glucose Sensor

Because Clark's original glucose sensor has been the prototype for so many sensors, it seems appropriate to examine a newer version, the miniaturized, potentially implantable device of Jung and Wilson (1996). The sensor has a sensing area of 1.12 mm^2. For the eventual application of this sensor the dynamic range of the device should be $\sim$ 0–20 mM. The amperometric sensor is based on the same enzymatically catalyzed reaction used by Clark and others:

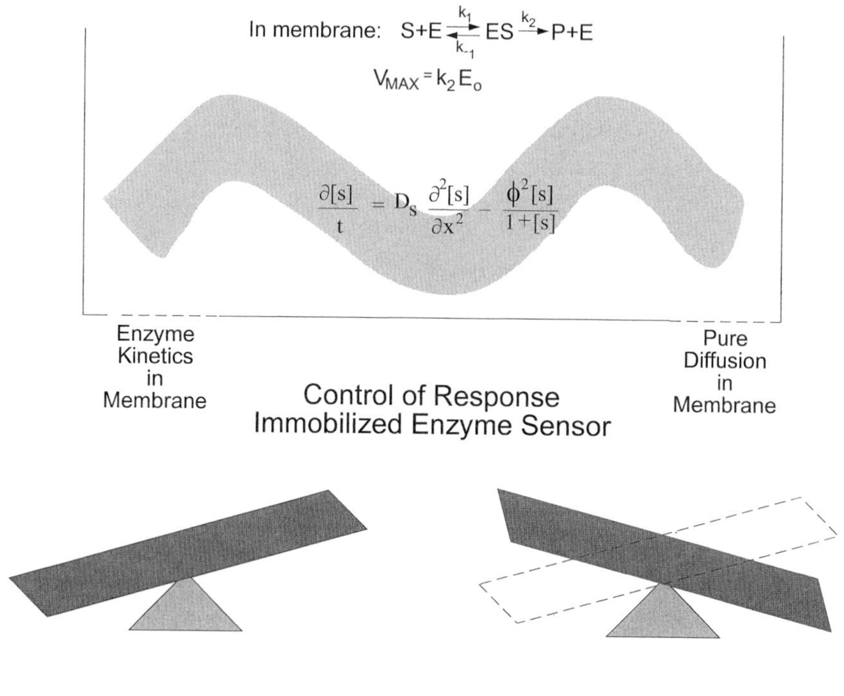

$$\text{Glucose} + O_2 \xrightarrow{\text{GOx}} \text{Gluconolactone} + H_2O_2$$

A molecular recognition layer of immobilized glucose oxidase (GOx) is covalently bound to a Pt wire. The authors point out that Pt continues to be the metal of choice in these types of measurement. Three different immobilization approaches were investigated: (a) entanglement of the GOx in a molecular network with bovine serum albumin (BSA) cross-linked with glutaraldehye, (b) as for (a) but with an outer layer of polyurethane (PU) applied over the enzymatic layer, and (c) covalent linkage of GOx to a self-assembled mercaptosilane film. In the presence of glucose as analyte, the diffusion-controlled amperometric signal due to oxidation of hydrogen peroxide at +600 mV versus Ag/AgCl is detected.

A perpetual problem for all glucose sensors based on GOx oxidation is that the signal is oxygen dependent. Oxygen is a cosubstrate for the enzyme. As oxygen is consumed in the enzymatic reaction it is depleted in the sample around the sensor and the linearity of response is lost. There have been numerous strategies for attacking this problem. In this sensor the outer

Zero-Flux Sensors

Unlike enzyme analysis in solution, linear response may be observed for $[S] > K_m$

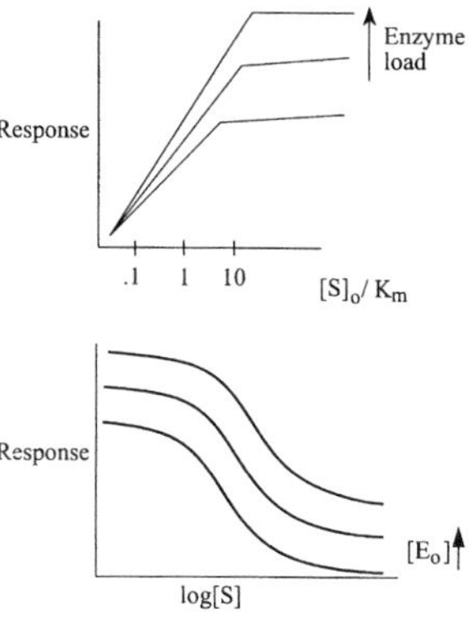

Response increases as d or E_o increases, if mass transport within the membrane predominates

Micro environment effects (pH, inhibitors, etc.) are more prominent at low enzyme load or $[S]_o >> K_m$ than at higher enzyme load or $[S]_o << K_m$

Thicker, less permeable membranes:
a) increase the linear dynamic range
b) make the response less dependent on external convection
c) increase response time, t_R

Nonzero-Flux Sensors

Highest sensitivity will be observed for membranes of high permeability and high enzyme load. External mass transport can limit sensitivity.

Higher enzyme loading increases the sensitivity

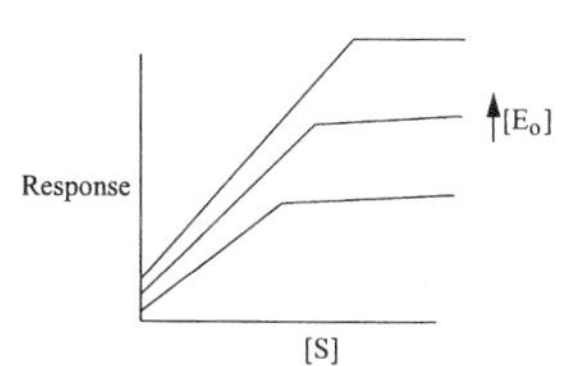

At high enzyme load linear response may be observed for $[S] >> K_m$

When mass transport predominates response increases with decreasing d, even if [S] is low

FIG. 2-14. General conclusions drawn from solutions of the diffusion–reaction equation for the one enzyme model.

PU membrane of design (b) above is more permeable to oxygen than glucose. Thus a diffusion-controlled signal over a wider glucose range can be accomplished by slowing, on a relative basis, mass transport of glucose into the layer. Polyurethane also prevents fouling of the sensor surface, blocks some interferents and has been proven as a biocompatible encasement for other devices.

In comparing the performance resulting from the three immobilization strategies, the GOx/BSA/GA electrode showed the highest sensitivity (as *d* response/*d* conc), but the dynamic range of the sensor was not sufficient. The sensor with covalently bound GOx covered the required dynamic range, but the sensitivity was lower than desired. The GOx/BSA/GA/PU membrane satisfied both criteria, as well as reducing oxygen dependence to a minimum. The sensitivity of the sensor was reported as 0.47 ($\pm$0.6) nA mM^{-1} while the dynamic range was $>$15 mM. Response time was determined in a flow-injection configuration. The time to reach 90% maximum signal (t_{90}) was $<$ 1 min for the GOx/BSA/GA/PU electrode.

Ascorbic acid, uric acid and acetaminophen are three of the most notorious interferents for glucose sensors in animal tissues and fluids. Jung and Wilson checked the sensor's selectivity with respect to these three substances. Selectivities based on the percentage error induced by physiological concentrations of the interferents and using a 5 mM glucose solution were reported. The largest error was found with acetaminophen. The conclusion was that it is possible to prepare glucose sensors for *in vivo* determinations with $<$ 3.1% interferent error at 5.5 mM glucose, the normal level.

2.6.2. An Optical Sensor for Glucose

Rosenzweig and Kopelman (1996) have reported an entirely different mode for sensing glucose. The same GOx catalyzed reaction as that of the previous sensor is used, but as the reaction proceeds the decrease in oxygen concentration is sensed optically rather than monitoring the hydrogen peroxide. They used 1–2 μm optical fiber tips with a fluorescent ruthenium complex immobilized with GOx in a photopolymerized acrylamide reaction layer. The metal complex is excited continuously (activated) by a laser source, fluoresces and is quenched by oxygen in the sample. As oxygen is consumed in the glucose reaction the emission detected increases due to a reduced level of quenching. Thus, the fluorescent metal complex becomes an indirect monitor of glucose concentration.

Several experimental innovations for manipulating microsized samples and devices are illustrated in this method. Kopelman's group has developed a series of sensors in the micrometer and sub-micrometer range for optical measurements in extremely small volumes. These miniaturized fibers for near-field spectroscopy are formed by drawing a fiber in a pipet puller while heating with a laser. Details of the spectroscopic considerations are discussed in Chapter 5.

The general technique for capturing emitted light from these tiny devices involves the use of an inverted fluorescence microscope objective. The sample is placed on a slide on the inverted objective. The optical-fiber tip is micropositioned at the sample interface. Laser excitation activates the fiber tip for fluorescence. The fluorescence intensity is inversely related to oxygen in the sample. Light—excitation and emission—captured by the inverted objective is passed to a dichroic mirror and filter which direct only the emitted light to a photon counting system.

The performance characteristics of this device are quite different from that of the Jung and Wilson (1996) amperometric sensor. The dynamic range was reported to be 0.7–10 mM glucose, less than the range for the amperometric device. The sensitivity, *per se*, of the optical device was not reported, though a calibration curve for the 0–10 mM range was presented. The limit of determination was 0.75 mM glucose, which does not compare favorably with other glucose optodes. Considering the microvolume of sample, however, the absolute detection limit is about 1 femtomole of glucose, which is much better than previous reports with larger optodes. The response time (t_{95}) was reported as 1.5 s.

2.6.3. An Acoustic-wave Sensor for Viruses

Acoustic-wave (AW) devices are increasingly important in the development of inexpensive sensors, and have been applied in various detection schemes for many years (Guilbault and Jordan, 1988). More details of the response mechanism are in Chapter 6. Briefly, the application of an alternating electrical field to the surface of a piezoelectric crystal results in the generation of acoustic waves of frequency f, characteristic of the crystal. Buttry and Ward (1992) describe the resonant frequency as that frequency which induces maximum particle displacement on the surface of the crystal. An oscillator for stimulation is coupled with a frequency counter or network analyzer for determining the oscillator circuit characteristics. Any modification to the surface of the piezoelectric material—e.g., adsorbed layers—induces a change Δf in resonant frequency. The change observed depends on several variables, including the mass of the modifying layer. A basic relationship between Δf and the mass is given by the Sauerbrey (1959) equation for quartz crystals:

$$\Delta f = -C_f \Delta m$$

$$C_f = \frac{2 f_q^2}{\rho_q \mu_q}$$

Here, ρ_q and μ_q are the density and the shear modulus of quartz, respectively.

König and Graetzl (1994) have utilized a piezoelectric device for quantifying several types of herpes virus. Their sensor provides an example of a third energy transduction option, perturbation of acoustic waves, and introduces a relatively straightforward application to immunosensing.

Antibodies (Ab) to five human herpes viruses were immobilized on the quartz crystal surface. Then the antigenic viruses (Ag) were introduced to the surface for Ab–Ag binding. The resulting mass variations upon ligand binding were determined as a function of frequency changes of the AW device in the resonant circuit.

The study demonstrated that this type of immunosensor could be used effectively for herpes viruses. The authors estimated that $5 \times 10^4 - 1 \times 10^9$ cells were immobilized on the crystal surface. Though not reported specifically, the sensitivity of the method appears to be about 50–150 Hz per 10-fold change in cell concentration. The detection limit was 5×10^3 cells.

Three different immobilization strategies were compared. The best results were obtained with a Protein A anchoring of the antibodies on the surface. Using the Fc-specific Protein A as the anchoring molecule allows for oriented Ab binding, which should promote optimal accessibility for the antigenic interaction. The most favorable result for this immobilization method is not surprising.

The inherently high selectivity due to Ab–Ag complementarity makes this immunosensor highly selective as an analytical device. Selectivity was not quantified in the report, but it was demonstrated that a specific Ab–Ag interaction is affected negligibly by the presence of other antigenic viruses. This is a significant result and exemplifies very well the ability to use molecular recognition as the primary source of selectivity. Acoustic-wave devices are among the most non-selective detectors in the arsenal of biosensing. Numerous factors—conductivity, viscoelastic effects of the surface layer, etc.—affect the resonant circuit characteristics of the method. Sorting mass-induced changes from other effects can be challenging. Anything that affects the surface affects the response. The König and Graetzl sensor shows exceptional selectivity considering the basic non-selective nature of the energy transducer.

There is another notable quality of their sensor. The modified crystal could be used 18 times without detectable loss of activity. Because Ab–Ag binding is so strong (K_{assoc} on the order of 10^{11}), regeneration of the sensor for repeated measurements is often a problem for these immunoassays. For the AW sensor described in this report, four regeneration methods were tested. The best results were obtained with elution of bound antigen by a competitive process. Antigen-specific synthetic peptides were used to "strip" the antigens from the Ab-modified surface. This avoided exposure of the Ab modified surface to harsh chemicals encountered in the usual washing steps.

Response time for the sensor depends on the time required for binding equilibration to occur between Ab and Ag. That time varied for each of the Ab–virus combinations, but all binding processes were complete in 60 min or less. Consequently, measurements were taken only after a standard 60 min incubation period.

2.6.4. A Thermal Sensor in Urinalysis

Karube's group in Japan has reported thermal microsensors that are useful for the determination of some analytes in urine (Shimohigoshi *et al.*, 1995; Shimohigoshi and Karube, 1996). The 1996 report dealt with the development of biothermochips for uric acid and oxalic acid in urine. Excesses of these substances in the urine indicate impending difficulties with renal calculi, kidney stones. These two sensors, as prototypes for an eventual integrated multianalyte sensor, illustrate very well the basic use of thermal sensing. Like AW devices, thermal transducers are basically non-selective. The selectivity is derived from the chemical transduction step, which is commonly enzymatic catalysis.

A potential problem to be solved for a combined uric acid/oxalic acid sensor is addressed in characterizing the individual devices of this report. Extremely low solubility of an analyte, uric acid, has to accommodated by choosing an operative pH which is also acceptable for sufficient enzyme activity in both reaction sequences.

The pertinent enzymatically catalyzed reactions are:
For uric acid

$$\text{Uric acid} + O_2 + 2H_2O \xrightarrow{\text{Uricase}} \text{Allantoin} + H_2O_2 + CO_2$$

$$H_2O_2 \xrightarrow{\text{Catalase}} \frac{1}{2}O_2 + H_2O$$

For oxalic acid

$$\text{Oxalic acid} + O_2 \xrightarrow{\text{OxO}} 2CO_2 + H_2O_2$$

$$H_2O_2 \xrightarrow{\text{Catalase}} \frac{1}{2}O_2 + H_2O$$

Enzymes were immobilized on the surface of the thermistors in a UV-polymerizing resin. The use of sequential enzymatic reactions for additive heats of reaction is a common approach for amplifying the thermal effect for detection. An insulated 1.5 ml sample cell, equipped with a sample injection port, was used. For this study the modified microthermistors were immersed in the sample/buffer solution.

A thermistor's resistance is a function of the temperature. The relationship between the two variables is a predetermined function. As the heat of the overall reaction raises the temperature of an adiabatic sample cell, the change in resistance of the thermistor is monitored electronically and converted to the corresponding ΔT. The temperature variation observed is a function of the concentration of analyte(s), thus providing calibration data.

Normally, the experimental conditions for an enzymatically catalyzed sensing strategy are chosen for optimum enzymatic activity. The pH optimum for urate oxidase (UO) activity is 8 and for oxalate oxidase (OxO) it is 5. Uric acid, however, is almost insoluble at pH's below 8. If an eventual dual analyte sensor were to be developed, it would be required to operate at pH = 8. The individual sensors, UO/CAT and OxO/CAT thermistors, were characterized at their respective pH optima, then the OxO/CAT sensor was checked for response at pH 8. It was determined that the response, though only 80% of the pH = 5 value, was sufficient for an eventual dual analyte device.

The biothermochips exhibited response times (t_{100}) of about 5 s. The UO/CAT device had a sensitivity of 8.1 mK mM^{-1} and a dynamic range of 1–6 mM at pH = 8. The sensitivity of the OxO/CAT sensor was 11.1 mK mM^{-1} and the dynamic range was 0.2–0.8 mM at pH = 5. Each device was tested at pH = 8 for an interference from the opposite acid—i.e., the OxO/CAT sensor against uric acid and the UO/CAT sensor against oxalic acid. The authors concluded no interference in either case. The lower sensitivity of the OxO/CAT sensitivity at pH 8 was attributed to the non-optimum pH rather than uric acid interference

3. Developing Bioselective Layers

Over the years a wide variety of immobilization strategies have been developed for creating selective reaction layers characteristic of bioanalytical sensors. Technologies for achieving selective reaction layers have evolved from those applicable only to hand-built prototypes to sophisticated methods for molecularly designed interfaces, mass-produced devices and spatially resolved patterning on a variety of surfaces. Progress is attributable largely to advances in assorted new materials for energy transducers and immobilization matrices, techniques for applying thin films to surfaces, and technologies for producing micromachined devices (Kovacs *et al.*, 1996).

It would be difficult, if not impossible, to catalog the myriad experimental protocols that have been reported for creating reaction layers of chemically selective agents on all types of sensing substrates. There are, however, three questions that suggest a scheme for fundamentals of chemical modifications. What are the species that must be immobilized to convey the necessary selectivity? What are the surfaces or matrices used for immobilizing these chemically selective agents? What are the major approaches for effecting the immobilizations? Some of these questions have been addressed partially in the preceding chapters. Table 3-1 includes examples of some preliminary answers to these questions of what?, where? and how?

TABLE 3-1 The What? Where? and How? of Immobilizations

Molecular Recognition Agents	Surfaces	Immobilization Strategies
Cells and tissues	Silicas	Adsorption
Proteins	Quartz	Entrapment
Enzymes	Metals	non-reactive gels
Antibodies	Platinum	Redox gels
Oligonucleotides	Gold	Conducting polymers
Nucleic acids	Silver, silver chloride	Photopolymerized matrices
Nucleic acid ligands	Metal oxides	Sol–gel networks
Lipids	Carbons	Entrapment with cross-linking
Langmuir–Blodgett bilayers	Glassy carbon	Glutaraldehyde linking
phospholipids	Carbon fibers	Polyethyleneglycol
		Glycidyl ether
Membrane receptors & transporters	Carbon pastes	Polyethyleneimine
Smaller organics	Carbon composites	Covalent coupling to
Haptens	Conducting salts	Surface (direct)
Redox mediators	Semiconductors	Organosilanes
Molecularly imprinted polymers (MIPs)	Polymers	Avidin–biotin anchors
	Nylon	SAMs, terminal groups
	Cellulose acetates	Supporting membranes
	Polyurethanes	Protein A or G anchors

Molecular recognition agents of biotic origin, as well as some abiotics, must be immobilized for selectivity toward many types of analytes. The surfaces and immobilization matrices shown in Table 3-1 represent a range of choices spanning the energy transduction modes commonly employed. It should be noted that some of the surfaces/matrices apply to more than one type of energy transduction. For example, gold may be used as an electrode material for electrochemical sensors, but it is also utilized as the optical reflection surface in surface plasmon resonance measurements and as the excitation electrode in many of the acoustic-wave devices. Another example is that of sol–gel inclusion of selective reagents, a technique that has been used for both electrochemical and optical sensors (Dave *et al.*, 1994; Lev *et al.*, 1995). As a result of certain commonalities across the spectrum of transducer modes, an immobilization method developed for one type of transduction mode may be applicable in other modes.

Immobilization protocols are quite diverse. While original research reports must be perused for experimental details, there are several general references that are widely used by most researchers. Weetall (1976, 1993) is considered one of the prime contributors to the science of immobilizing biocomponents on several types of inorganic and polymeric matrices. Carr and Bowers (1980) and Mosbach (1988) are excellent resources with good detail about both the immobilization considerations and effects of immobilization on enzymes. Taylor (1991) and Cabral and Kennedy (1991) are widely cited references for protein immobilization. Very practical examples of designing different types of sensors and sensing systems may be found in Cass (1990). Specific instructions for fabricating a variety of biocatalytic and bioaffinity sensors are included. Finklea (1996) has surveyed in good detail the theoretical and practical aspects of organized thin films, including copious material on self-assembled monolayers (SAMs) of organothiolates and disulfides on gold.

Many of the immobilization schemes and surface reaction chemistries presented by Weetall and others have been adapted from selective, durable chromatographic support modifications, then translated for transducer surfaces, sometimes with more intricate control of the molecular architecture of surface-bound species.While selectivity and durability imparted to chemically modified supports for chromatography are also essential for effective sensing strategies, a critical distinction between the two purposes of immobilization exists. For sensors there is the added factor of the supporting substrate's energy-transduction function. Therefore, immobilization strategies and characterization of the chemical modifications must be compatible not only with requirements for selectivity and durability, but also with generation of interpretable electrical signals.

For an excellent perspective on the state-of-the art with respect to control and elucidation of biofunctionalized sensing surfaces, the 1996 Sackmann report, and references therein, provides a concise, provocative summary of research in recent years. The scientific advantages accrued by coupling the power of modern surface characterization methods to biofunctionalization of surfaces is amply demonstrated by Sackmann's examples. Emphasizing lipid–protein bilayers, Sackmann classifies three types of supported membranes: (1) integrated membrane with a covalently fixed inner membrane, (2) supported bilayer separated from the substrate by an ultrathin water film or polymeric layer, and (3) ultrathin polymeric film of hydrophobic macromolecules which provides a cushion for monolayer receptors. Those examples also emphasize the necessity of a basic understanding of the various investigative approaches to elucidating surface phenomena because improved designs of biosensors rely on effective surface analysis.

In order to achieve the types of membranes to which Sackmann refers, there are five generalized approaches in routine usage:

1. *Deposition of mixed inner monolayers of long-chain, spontaneously adsorbed monolayers with terminal functional groups (amines, epoxy and hydroxy) for attaching biocomponents.* These monolayers relate primarily

to functionalized silane layers (Weetall, 1993) and SAMs of alkane thiols or disulfides on gold (Finklea, 1996). The strategy of including terminal hydroxyl groups in mixed monolayers of thiols reduces non-specific binding through hydrophobic regions of macromolecules by making the surfaces more hydrophilic.

2. *Use of biotin–avidin layers for anchoring biotin-labelled species.* This methodolgy is now so widespread that Sackmann refers to it as "already classical." Biotin, vitamin H, is a small organic of MW = 244. Avidin, isolated from egg whites, is a glycoprotein (MW $\sim$ 70,000). The avidin–biotin bond involves one of the strongest ligand–receptor interactions among biological species (Weisenhorn *et al.*, 1992; Wong *et al.*, 1997). Avidin has four binding sites for biotin. The streptavidin form, isolated from a strain of *Streptomyces,* exhibits a K_{assoc} of $\sim 10^{15}$ for the streptavidin–biotin bond with a bond energy of $\sim$ 35 kT (Wong *et al.*, 1997) where k is the Boltzmann constant and T is the absolute temperature.

There are several ways to effect the avidin–biotin bond for immobilization of species. Either member of the ligand–receptor pair may be affixed to a surface. Biotin is available commercially in a variety of derivatized forms, including a photoactive form with an azide functional group. Avidin may be adsorbed to electrodes and oxide surfaces, including glasses (Anzai *et al.*, 1994; Hoshi *et al.*, 1995; Polzius *et al.*, 1996). Also avidin may be coupled covalently to amino-derivatized surfaces through a conventional amide-bond formation which is discussed later. Utilizing surface-bound biotins, Abel *et al.* (1996) biotinylated their quartz surfaces either by adsorption of biotinylated bovine serum albumin at thiosilanized surfaces or by coupling an *N*-hydroxysuccinimide derivative of biotin to aminosilanized surfaces. Sundberg *et al.* (1995) biotinylated pre-treated glass surfaces with a photo-deprotectable biotin derivative which could be exposed for avidin binding upon light activation. By using a photolithographic mask with this scheme, they could photoactively pattern the surface with layers of macromolecules (Section 3.3.7). Dontha *et al.* (1997) reported a different method for depositing enzymes on carbon electrodes using a photoactive biotin and a laser interference patterning approach, rather than the photolithographic masks.

Utilizing a long-chain amino-terminal amine modification, Pantano and Kuhr (1993) tethered biotin to an electrode. Liu, Z. T. *et al.* (1995) adopted that approach for their organonitriles sensor using nitrilase. Utilizing surface force measurements and lipid supporting membranes, Wong *et al.* (1997) determined that a tethered biotin–streptavidin bond was sufficiently strong to "pluck" the tethered biotin from its supporting membrane—i.e., the biotin linked to a supporting lipid membrane was the weakest link in the chain of interactions. Their conclusion from the study was that flexible molecular tethers extend the effective range of ligand–receptor interactions and that the on-rate is a function of both the tether length and dynamics.

3. *Coupling proteins to functional groups of supporting polymeric layers via active succinimide esters.* An example of such a supporting layer is that of CM-dextran layers suggested by Löfas and Johnsson (1990) and utilized in both the Polzius *et al.* (1996) and Sigal *et al.* (1996) comparative studies of binding efficiencies. The same chemistry involving protein linking through succinimide ester intermediates also pertains to some coupling chemistries used in layer method 1, where terminal groups of the monolayers are utilized for linking proteins to the surfaces through either the amine or carboxyl groups on the protein. A water soluble carbodiimide, 1-ethyl-[3-(dimethylamino)propyl]carbodiimide (EDC), is usually used in conjuction with *N*-hydroxysuccinimide (NHS) to effect an amide bond between an amine group and a carboxyl group, one of which is on the derivatized surface and the other on the protein. The EDC activates carboxyl groups for ester formation (Ege, 1989).

4. *The use of monolayers with terminal chelating moieties complexed with metal ions that bind proteins reversibly.* This is exemplified by the Ni^{2+}/His-tagged protein approach utilized by Schmid *et al.* (1997) for sensing based on total internal reflection fluorescence (TIRF) and by Sigal *et al.* (1996) for surface plasmon resonance (SPR) studies of protein orientations. Although the application did not involve immobilizing a biocomponent, Kepley *et al* (1992) used essentially the same approach by chelating Cu^{2+} ions to a SAM for vapor-phase acoustic-wave detection of the warfare agent DIMP.

5. *The use of Langmuir–Blodgett films (Roberts, 1990) for anchoring and supporting biocomponents in an environment that mimics more closely that of natural membranes (Section 3.2.1).*

Several examples of surface pre-treatments and coupling chemistries referred to are given in Table 3-2.

Understanding the many aspects of modified surfaces for the immobilization of macromolecules requires some basic knowledge of surface phenomena, in general. Adamson (1990) and Masel (1996) present material basic to a generally applicable background in the physicochemical aspects of surface properties and adsorption. The Mottola and Steinmetz (1992) and Pecsek and Leigh (1994) books include varied coverage of chemically modified surfaces while the Vansant *et al.* (1995) reference emphasizes aspects of chemical modification of silica surfaces. Murray (1992) has presented a thorough overview of chemically modified electrodes. Much of that information is very helpful for similar surfaces used in other transduction modes. Because the chemical modifications of energy transducers span the dimensional domain of both thick and thin films, the Galan-Vidal *et al.* (1995) review of thick films for biosensors and the much more extensive Ulman (1991) book on ultrathin films provide foundational material.

TABLE 3-2 Examples of Pretreatments and Coupling Chemistries

Silanized surfaces

General $R_1(CH_2)_n\,Si(OR_2)_3$ + (surface)–OH, –OH → (surface)–O–Si(O)–$(CH_2)_n$–R_1 / O / (surface)–O–Si(O)–$(CH_2)_n$–R_1

(γ-Aminopropyl) trimethoxy silane, APTS

(surface)–OH, –OH + APTS → (surface)–O–Si–$(CH_2)_3$–NH_2 ≡ (surface)~~NH_2

3-Glycidoxypropyl trimethoxy silane, GOPS

(surface)–OH, –OH + GOPS → (surface)–O–Si–$(CH_2)_3$–O–CH_2–CH–CH_2 (epoxide, O) $\xrightarrow[NaIO_4]{H^+}$ (surface)~~C(H)=O

↓ H_2N–Prot: (surface)–O–Si–$(CH_2)_3$–O–CH_2–C(H)(OH)–CH_2NH–Prot

↓ H_2N–Prot: (surface)~~C(H)=N–Prot

Glutaraldehyde

(surface)~~NH_2 + Glutaraldehyde → (surface)~~N=C–$(CH_2)_3$–C(H)=O $\xrightarrow{H_2N\text{–Prot}}$ (surface)~~C(H)=N–Prot

Au surfaces

General

Au + H–S~~X $\xrightarrow{EtOH}$ (surface)–S~~X

Examples

Au + H–S~~NH_2 (Cysteamine) → (surface)–S~~NH_2

Au + (S–S ring)–$(CH_2)_4COOH$ (Thioctic acid) → (S S on surface ring)–$(CH_2)_4$–C(=O)OH $\xrightarrow[H_2N\text{–Prot}]{EDC}$ (S S on surface ring)–$(CH_2)_4$–C(=O)–N(H)–Prot

Some reports in the literature include helpful, comparative experimental results about different surfaces or immobilization methods. For example, Zhang, D. *et al.* (1994) compared the adsorption of cytochrome c_3 on three electrode surfaces—mercury, platinum and gold. They observed a complex set of circumstances in which the redox behavior of their form of cyt *c* varied among the three types of electrodes, and several irreversible conformational changes at the surfaces were proposed. Adsorption behavior at the mercury electrode was potential dependent, which suggests electrostatic interactions at the charged electrode surface. Yang, M. *et al.* (1993) also proposed that following electrostatic adsorption on an acoustic-wave device, proteins underwent conformational changes suggesting denaturation with exposure of the more hydrophobic regions.

Other comparative studies are useful for understanding surface phenomena related to immobilization strategies. Boitieux *et al.* (1990), Ahluwalia *et al.* (1991), and Lu, B. *et al.* (1995) have provided experimental comparisons of random versus oriented immobilization of antibodies and Fab fragments, a topic which has been summarized by Lu, B. *et al.* (1996). Whitesides' group has reported on a systematic study of the behavior of proteins at silica surfaces modified with various cationic polymers (Córdova *et al.*, 1997). The results of all of these comparative experiments are pertinent to a variety of sensor designs.

In a totally different application from that of most sensors, Pierce *et al.* (1992) examined by scanning electrochemical microscopy (SECM) enzyme kinetics on non-conducting surfaces using both membrane- and surface-bound immobilizations. Kinetic analysis was compared for Nylon-66, an Os-redox hydrogel and Langmuir–Blodgett films as sensing substrates. Cross-linking effects, mentioned in Chapter 2, were observed for the hydrogel preparation, with concomitant degradation of response.

Williams and Blanch (1994) compared five protocols for immobilizing antibodies on silicon nitride surfaces. Silanization of the surface followed by cross-linking proteins with glutaraldehyde proved to be the most effective of the methods applied. One of the major efforts of their study was directed toward finding ways to eliminate non-specific adsorption of sample–matrix proteins at the surfaces. They found that the problem was minimized if immobilizations were conducted in the presence of a detergent. The study did not reveal the effect of detergents on the conformation of the antibodies.

Polzius *et al.* (1996) conducted a thorough study of seven immobilization procedures for attaching antibodies to the surface of tantalum oxide waveguides. Protein G was used to anchor the antibodies through the Fc binding region, thus providing orientation for binding to antigen.

(a) Protein G adsorbed to silanized waveguides
(b) Protein G coupled via epoxy groups
(c) Protein G coupled via glutaraldehyde activation
(d) Protein G coupled directly via aldehyde groups
(e) Protein G coupled via *N*-hydroxysuccinimide activation
(f) Biotinylated Protein G bound to avidin adsorbed on the waveguide
(g) Biotinylated Protein G bound to avidin coupled covalently to CM-dextran coated waveguides

The authors were quite detailed in reporting the various immobilization chemistries. Their results indicated that electrostatic interactions played a decisive role in adsorptive binding and influenced non-specific adsorption. The authors concluded that adsorptive and covalent binding of Protein G resulted in equal sensitivities in the subsequent immunoassays. The avidin-mediated binding to Protein G produced higher responses whereas a CM-dextran support produced the most stable signal for $\sim$15 regenerative cyles.

Caruso *et al.* (1997) investigated the immobilization and hybridization of DNA using the quartz crystal microbalance (QCM or TSM) which produces a mass-dependent response. They used 30-mer oligonucleotides, one type biotinylated and one derivatized with a mercapto group. Single and multilayer films were prepared and compared. For the latter film they used successive deposition of avidin and poly(styrene sulfonate) which binds avidin through electrostatic interactions. While DNA hybridization was slower for the multilayer film, the sensitivity of the method was increased.

Progress in understanding the options for immobilizations and optimizing effective inclusion of biocomponents would not be possible without recent advances in surface characterization (Hubbard, 1995). While that is not a primary topic of this book, it should be borne in mind that molecular architecture of modifying layers cannot be understood or properly designed without knowledge of the means for characterizing the modified surfaces. Several relatively brief reviews and references therein are available for an introduction to this topic (Brecht and Gauglitz, 1995; Göpel and Heiduschka, 1995; Perry and Somorjai, 1994). The Göpel and Perry reviews cover a wide variety of methodologies. Brecht focuses on optical methods.

A variety of optical, electrochemical and acoustic-wave methods may be used to obtain information about the nature of reaction layers on sensors. Coulometric charge for electropolymerized layers and microgravimetry using acoustic waves provide mass-related and surface-coverage data. For example, both the Yamaguchi *et al.* (1993) and the previously mentioned Caruso *et al.* (1997) studies used the quartz crystal microbalance acoustic-wave device to investigate adsorption, immobilization and hybridization of DNA for sensing applications.

Ellipsometric and surface plasmon resonance data may be combined for data on thicknesses of films and dynamic changes within the films. For examples of elegant experiments using these methods for studying adsorption of proteins at surfaces, the reader is referred to Jordan *et al.* (1994), Jordan and Corn (1997), Frey *et al.* (1995) and Sigal *et al.* (1996). Bard and Faulkner (1980) have presented a very concise, lucid description of ellipsometry with examples from electrochemical applications. Briefly, a beam of linearly polarized light has electric vectors oriented the same direction, and such a beam may strike a surface at an angle for which all the electric vectors are oriented parallel to or perpendicular to the plane of incidence. At intermediate angles of incidence between zero and 90°, there will both parallel and perpendicular components all in phase. When the light is incident upon a surface at one of these intermediate angles, the parallel and perpendicular components will interact with the surface differently, thus altering the phase and amplitude relationships. The effect of this may be visualized as twisting the beam in such a way that an ellipse is swept out by rotation of the vectors. This is elliptical polarization.

Instrumentally the degree of ellipticity induced can be determined by a null-detection method which measures the amount of optical compensation necessary to correct for the elliptical polarization. The output provides data related to the index of refraction n of the surface layer and its thickness d. Dynamic changes in the layer may be monitored by time-dependent ellipsometric data. Thicknesses of films may also be monitored by surface plasmon resonance (SPR) measurements. When optical models are applied and the SPR and ellipsometric data are correlated, intricate characterization of submonolayer films and changes in films can be achieved (Jordan *et al.*, 1994; Frey *et al.*, 1995; Sigal *et al.*, 1996).

Adsorption. A major consideration for any type of biosensor is the role of adsorption as it affects responses. Adsorption of biomolecules plays a very big role in operational aspects of sensing devices. There is purposeful macromolecular adsorption as an immobilization technique, then there is nonspecific adsorption which must be minimized. Further, the development of many of the structural membranes of biosensors depend on spontaneously adsorbed layers for forming the silane, organothio- or avidin scaffolding to which molecular recognition species are bound (Zhong and Porter, 1995).

Previous sections have provided numerous examples of adsorptive phenomena in bioanalytical measurements. A composite picture emerges revealing that protein adsorption is a function of the type of surface, the pI of the protein, the size of the protein, and the solution conditions under which the adsorption occurs (Yang, M. *et al.* 1993; Zhang *et al.*, 1994, Córdova *et al.*, 1997; Jordan and Corn, 1997). In the last reference, regarding biopolymer adsorption at SAM-modified surfaces, the authors have presented a formulation for an overall adsorption coefficient K_{ads}, which is a combination of the K_a for the terminal group on a SAM and the K_{ip} for ion pairing during an electrostatically driven protein adsorption on avidin, $K_{ads} = K_a^n K_{ip}$ (n is the exponent for the general case of polyprotic SAM-terminal acidic groups, H_nX).

Extensive discussions of adsorption at surfaces may be found in Adamson (1990) and Masel (1996). Adsorption may be described by both the strength of the surface–adsorbate interaction and by the extent of the phenomenon at the surface. Adamson ascribes the term physisorption to those processes involving interaction energies of ~5–10 kcal mol^{-1} of adsorbate. These energies are on the same order as chain–chain interactions in polymers and cohesion forces in molecular solids. Physisorption involves primarily van der Waals forces. Chemisorption, according to Adamson, involves ~20–100 kcal mol^{-1} of adsorbate, an energy on the order of sublimation energies.

Masel assigns slightly different values to physisorption and chemisorption, i.e., 2–10 kcal mol^{-1} and 15–100 kcal mol^{-1}, respectively. Masel also provides a further distinction. A species is physisorbed if it adsorbs without undergoing significant changes in electronic structure. If electronic structure is perturbed, then the process is chemisorption. This latter distinction applies particularly to Au–S–R interactions of self-assembled monolayers (Finklea, 1996). Finklea indicates that the Au—S bond is on the order of 40–50 kcal mol^{-1}, a value which definitely puts it into Adamson's and Masel's chemisorbed categories.

Beyond the energy of the sorption process, the extent of adsorption may be distinguished as monolayer or multilayer. For a monolayer all adsorbate is "on" or in close proximity to the solid surface. Corn's group has conducted extensive studies of the previously mentioned SAM-supported monolayers of poly(L-lysine) to which proteins may be adsorbed or covalently attached (Jordan *et al.*, 1994; Frey *et al.*, 1995; Frey and Corn, 1996; Jordan and Corn, 1997). Multilayer adsorption results from attractive forces between adsorbate molecules which cause them to "condense" into a film, a process referred to as condensation (Masel, 1996). An example of the latter process has been mentioned earlier—i.e., the adsorption study of Caruso *et al.* (1997) in which the multilayers were based on electrostatic interactions between layers of avidin–biotinylated DNA and poly(styrene sulfonate).

3.1. DEPOSITING FILMS AND MEMBRANES

Before proceeding to the chemical aspects of designing bioselective layers some of the technologies, other than adsorption, for depositing these films and membranes should be considered. A minimal vocabulary for some of the immobilization techniques, together with scaling of layered thicknesses, was introduced in Chapter 1. In some cases the method for application of the molecular recognition layers may be critical for achieving the desired performance. In other cases there may be more latitude in options for deposition, though one or more may be preferable.

Lateral resolution for miniaturized devices has been discussed in the spatial resolution section of Chapter 2. Photolithography, screen-printing, ink-jet printing and light-directed synthesis on substrates were introduced. Whereas lateral resolution is critical when working with well-defined miniaturized components, Gooding and Hall (1996) have emphasized the equally important necessity for good depth resolution, especially when designing membranes for which diffusional characteristics are fundamentally important. They compared generally some methods with regard to reported depth resolution in depositing enzymatic membranes: screen printing, ~100–500 μm; photolithographic patterning, ~150 μm; and ink-jet printing, ~1–5 μm. Because electrodeposition can be so closely controlled by coulometric monitoring, Gooding and Hall proposed that it may be a method of choice for many applications.

While electropolymerization for amperometric sensors was the focus of Gooding and Hall's suggestion, electrochemically polymerized layers have been used for a variety of applications. For example, electropolymerized layers for anion-selective sensing have been investigated by several groups (Edelman and Wang, 1992; Yim *et al.*, 1993). In Chapters 1 and 2 different approaches to ion-selective sensing and the necessary selective membranes were discussed. One aspect of anion sensing not included in that discussion involves the use of organometallic complexes for achieving selectivity. Electropolymerized membranes may be used for this type of application.

The use of organometallic-based anion sensing may be thought of as having its origins in the biochemical behavior of some naturally occurring metalloporphyrin complexes with anions. Bachas' group applied this approach early on using Vitamin B_{12} derivatives (Palet *et al.*, 1993). Based on the fact that metalloporphyrins, interact biologically with a variety of anions, one could predict that these species should make good anion-selective ionophores. Porphyrins doped in the usual manner into PVC matrices and those immobilized in electropolymerized films were compared. Non-Hofmeister selectivity was observed through the use of the metalloporphyrins, and a very effective nitrite sensor was reported. The result indicated that there might be good advantage in the stability of ion-selective electropolymerized films as compared to PVC matrices.

Meyerhoff's group also investigated the use of electropolymerized matrices for anion-selective electrodes (Yim *et al.*, 1993). They found that different mechanisms probably pertain to the observed potentiometric responses. A pictorial summary of their proposal for multiple contributors to the response is shown in Figure 3-1. Ion-exchange and redox interactions with both the matrix and ionophore were proposed.

Two of the oldest methods for application of immobilizing matrices are dip- or drop-coating. An electrode or optical fiber may be simply dipped into a liquid form of the pre-mixed coating, then allowed to dry by evaporation of solvent. Alternatively, there are some matrices—e.g., some acrylates— that work best when subjected to photopolymerization after being dip-coated (Healy *et al.*, 1995).

Drop-coating is self-explanatory. A surface may be mounted surface-side-up for simple application of one or more drops of the coating solution. Repeated application results in multilayer modification. Immobilization

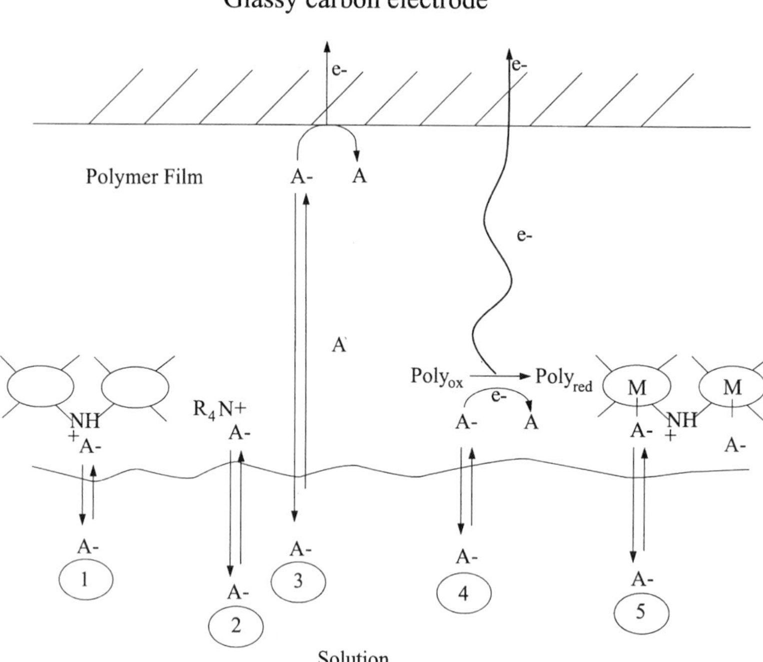

FIG. 3-1. Possible mechanisms for potentiometric response (anion selectivity) of an electropolymerized metalloporphyrin film modified electrode: (1) ion-exchange process with positively charged polymeric backbone; (2) ion exchange due to entrapped quaternary ammonium ion of supporting electrolyte; (3) redox response at the underlying electrode surface; (4) redox response due to redox reaction with conducting polymer; and (5) metal–anion interaction through axial ligand exchange. Reprinted from *Biosens. Bioelec.* Vol. **8,** Yim, H-S. *et al.*, Polymer membrane-based ion-, gas- and bio-selective potentiometric sensors, pp. 1–38. Copyright 1993, with permission from Elsevier Science Ltd., Oxford, UK.

layers formed by either dip- or drop-coating may be several μm in thickness. For microdevices that are manually fabricated, drop-coating via use of a syringe may be necessary to form sufficiently small drops for the localized active areas on the surface. While dip- and drop-coating are perhaps the simplest methods for modifying surfaces there are inherent problems of controlling layer thicknesses and reproducibility.

Commercially available spinners may be used to spin-coat a surface. This technique provides a relatively reproducible modifying layer that can easily range from thin to ultrathin ranges of membrane classifications. Spinning speeds of 2000–2500 rpm are common. Frank *et al.* (1996) have presented an informative report dealing with the structure of spin-cast thin and ultrathin films. Petit-Dominguez *et al.* (1997) have reported spin-coated, sol–gel modified graphite electrodes designed for ion selectivity via ion exchange. They used spinning speeds of 3500 rpm and amperometric characterization of ion permeability and pre-concentration by using electroactive ionic species.

In order to create defined features and active sites on miniaturized devices, more controlled methods of application must be employed. Screen printing is one of the more widely used methods for fairly good resolution and reproducibility (Kissinger and Heineman, 1996; Goldberg *et al.*, 1994; Hart and Wring, 1994). A patterned mask for the surface layout is created either by a photographic emulsion technique or by etching. The latter approach provides better spatial resolution. The layering mixture, paste or ink, is then applied through the mask with the use of a flexible squeegee blade. Goldberg *et al.* (1994) provides considerable detail about screen printing within the context of several types of solid-state electrochemical sensing devices. (note: on p. 177 of the report thermic seal should read hermetic seal.). They gave particular attention to practical factors related to pasting composition for adequate spreading and feature definition. Rohm *et al.* (1995) used screen printing with UV polymerization for their enzyme electrodes. Screen printing has been broadly applicable in the production of disposable devices. Several of the Gilmartin, Hart, Hartley and J. Wang entries in the Bibliography involve screen-printing applications.

Ink-jet deposition, using either solenoid or piezoelectric dispensing mechanisms, has become increasingly applicable to the creation of miniaturized reaction sites. Lemmo *et al.* (1997) have described their system for use in combinatorial synthesis. They evaluated dispensing factors affecting the deposition of 180 nl–1200μl of solution into 2304 microwells arranged on a $8\frac{1}{2}$ inch × 11 inch polypropylene sheet. Blanchard *et al.*, (1996) reported a piezoelectric dispensing method that produces high-density oligonucletide arrays on an oxidized silicon wafer. They were able to define droplets 100 μm in diameter, separated from each other by 30 μm hydrophobic spaces. One of the major advantages emphasized in both reports is the speed with which these defined surfaces can be created.

In evaluating some new catalytic carbon materials for amperometric biosensors, Newman *et al.* (1995) used ink-jet printing for applying an outer Nafion membrane to screen-printed enzyme electrodes. They compared dip-coating, spin-coating and ink-jet printing for the application of the layers and decided that the ink-jet application provided the best signal /noise ratio combined with the desired selectivity offered by the layer.

Polymeric modifications may be deposited on surfaces in a controlled manner by any one of several polymerization techniques. Walt's group, and later Kopelman's group, have produced several types of optical biosensing devices using Walt's controlled photopolymerization on optical fibers, including optical fiber arrays (Tan *et al.*, 1992a,b; Bronk and Walt 1994;

Bronk *et al.*, 1995; Healy *et al.*, 1995; Healy and Walt, 1997). A silanized fiber is placed in a solution of precursors to an acrylamide copolymer. Light of the appropriate wavelength is directed through the fiber to initiate the photopolymerization process on the distal end of the fiber.

Electrochemical devices are often modified with different types of polymeric matrices, some of which may be electropolymerized *in situ*, as mentioned earlier. Several types of polymeric species utilized are commonly classified as redox polymers, ion-exchange or coordination polymers and conducting polymers (Martin and Foss, 1996). Redox polymers include an electroactive group that is usually a side-group on the polymeric backbone. For example, Abruña's group (Pariente *et al.*, 1994–1996) has investigated a variety of electrodeposited dihydroxybenzaldehydes which result in redox-active films containing quinone groups. These films have been useful for the electrocatalytic oxidation of NADH on glassy carbon electrodes. They found the ortho-isomers to be quite effective. Ion-exchange and coordinative polymers lack electroactivity, *per se*, but may become electroactive by permeation of redox species into the layer.

Conducting polymers are usually deposited with coulometric monitoring of the quantity, for determining membrane coverage. The films are comprised of polymeric chains of conjugated aromatic rings (e.g., polypyrrole) or aromatic rings connected by polarizable linkages (e.g., polyaniline or poly-(*o*-phenylene diamine)). These polymers are distinguished by the fact that they are usually conductive in an oxidized state and insulating in a reduced state. This means that charge transfers and permeation by ionic species can be controlled by the applied potential. Polyaniline presents an additional control parameter in that its characteristics are also pH sensitive. Bartlett has reviewed electropolymerized films as the inclusion matrix for enzymes (1993). Bätz *et al.* (1991) summarized the structure–behavior characteristics of polypyrrole, *per se*.

Other types of polymerization schemes have been employed, such as the Rohm *et al.* (1995) use of UV-polymerizable pastes for screen-printing enzyme electrodes. Bu, English and Mikkelson (1996) used a one-step, UV-initiated polymerization of a charged redox gel containing glucose oxidase to produce an improved glucose sensor. Nakanishi *et al.* (1996) used plasma polymerization to immobilize antibodies and Fab fragments on a QCM surface.

3.2. SURFACES FOR IMMOBILIZATIONS

Typical surfaces of energy transducers include silicas and quartz, platinum, silver, gold, metal oxides, several types of carbon and carbon composites, conducting salt electrodes and organic polymers such as nylon or cellulose acetates. Each type of surface has distinctive characteristics related to immobilization chemistries for biosensors. For example, platinum is characterized by PtO moieties that are always present. Thus, in devising a chemical modification for Pt surfaces, it is not the Pt which will be involved in the binding, but the oxide film (Gilaman, 1967; Murray, 1992; Aschauer *et al.*, 1995). A similar situation of oxygen functionalities exists for carbon surfaces, but there is a greater variety in the types of functional groups on carbon. Hydroxyl, carbonyl and carboxyl groups may be present, depending on the nature of the pre-treatments for cleaning and activating the surfaces (McCreery, 1991; Pantano *et al.*, 1991; Chen, P. and McCreery, 1996).

Gold is used not only for electrochemical studies, but also as an immobilization surface for some of the optical methods, such as surface plasmon resonance (SPR), and acoustic-wave devices which may include gold electrodes as part of the energy transduction process. As has been mentioned above, gold surfaces have the distinguishing property of exceptional affinity for sulfur groups. Thus, the self-assembly of RSH or RSSR groups by chemisorption at gold surfaces has become an increasingly important approach to creating well-controlled, intricately designed sensing layers (Kepley *et al.*, 1992; Abbott *et al.*, 1994; Berggren *et al.*, 1995; Cheng and Brajter-Toth, 1995, 1996; Frey *et al.*, 1995; Finklea, 1996; Sigal *et al.*, 1996; Willner *et al.*, 1996; Caruso *et al.*, 1997). While thiols are used in a great many of the SAM studies, thioctic acid and cystamine have also been widely used (Cheng and Brajter-Toth, 1995, 1996; Willner *et al.*, 1996; Schmidt and Schuhmann, 1996). These monolayers are extremely durable, exhibit very useful electrochemical properties, and provide a supporting membrane for covalent attachment of molecules away from the sensing surface. One of the essential features of SAMs for sensing applications is the presence of a terminal group, X (Au–S–X), which can be used either for tuning the hydrophilicity or permselectivity of the layer (Cheng and Brajter-Toth, 1995, 1996), or for covalently binding biomolecules.

Silica and quartz surfaces are characterized by the presence of Si–OH groups, but with a pK_a of $\sim$ 2–4 (Córdova *et al.*, 1997) the surfaces are negatively charged at pH's above 4 due to $Si—O^-$. The negatively charged surface of silica has important implications for the adsorption of macromolecules. As mentioned earlier, several reports, including those of Sigal *et al.* (1996) and Polzius *et al.* (1996), indicate the strong dependence of adsorption on the pH and ionic strength of the solvents, as well as the pI of the proteins (Table 3-3). The immobilization of biocomponents at silica surfaces has one of the longest histories in biosensing, paralleling developments in affinity chromatography. Weetall (1976, 1993) and references therein are particularly valuable as resources for these types of immobilizations. For many of the semiconductor-based devices, silicon nitride (Si_3N_4) surfaces are involved. As discussed in Vansant *et al.* (1995), the types of

TABLE 3-3 pI's of Selected Proteins[a]

Protein	pI[b]	Protein	pI[b]
α-acid glycoprotein	3.0	Glutamate oxidase	6.0
Protein G	3.5	Streptavidin	6.4
Ovalbumin	4.7	Hemoglobin	6.8
Polyphenol oxidase	4.7	Human IgG	7.0
Serum albumin	4.9	Myoglobin	7.0
Urease	5.0	Avidin	10.0
Superoxide dismutase	5.0	Cytochrome *c*	10.7
		Lysozyme	11.0
Insulin	5.5		

[a]Compiled from various sources.
[b]Isoelectric pH (no net charge).

usual pretreatments of Si_3N_4 often alter this surface by oxidation to SiO_2, so that it may be considered as an Si—OH surface for chemical modification (see also Williams and Blanch, 1994).

A report on the interesting modification of silica surfaces for oriented binding of proteins is that of Schmid *et al.* (1997). The basic immobilization chemistry through chelation, mentioned as one of the major approaches to biofunctionalization, was developed originally for affinity chromatography, and had been adapted earlier by Sigal *et al.* (1996) for use at a gold surface in SPR studies of oriented protein binding. In the Schmid study of optical transduction by fluorescence a quartz surface was silanized with 3-(mercaptopropyl)trimethoxysilane, then a metal chelator, a derivatized nitrilotriacetic acid (NTA-maleimide), was attached. After chelation with Ni^{2+}, the surface promotes the oriented binding of histidine-tagged proteins.

Polymeric supports and matrices for immobilizing biocomponents are varied and the functional groups available for binding depend on the side-chain groups of the polymeric backbone. For example, cellulose triacetate, polyamides and polyesters offer functional groups susceptible to immobilization chemistries. Nylons have been widely used. Carr and Bowers (1980) summarized some of the reactions used for nylon modifications. The *O*-alkylation method, used by Pierce *et al.* (1992) for immobilization of glucose oxidase on nylon disks, provides an example.

Polyvinyl chloride (PVC), which is so important in ion-selective sensing, is not amenable to protein immobilizations by covalent binding, due to the lack of functional groups for linking macromolecules. By using the aminated or carboxylated forms of PVC, however, immobilization may be accomplished, though usually with low loading (Yim *et al.*, 1993). Meyerhoff's group has investigated a variety of polymers for use with potentiometric enzyme electrodes. Cellulose triacetate (CTA) and polyurethanes were found to be effective for enzyme immobilizations. The CTA reaction chemistry, creating asymmetric binding of enzymes on one side of the membrane is summarized in Yim *et al.* (1993).

Another example of linking to polymeric substrates is that of the 1996 Kawazoe study using nucleic acid ligands, NALs. For the immobilization of folic acid to be determined by binding to labeled nucleic acid ligands, they used a poly(ethylene terephthalate) base-film with an N_3-derivatized poly (allylamine) layer which could be photochemically attached on the surface of the base-film. Folic acid was then immobilized by a conventional amide-formation chemistry using a water soluble carbodiimide (Table 3-2).

The importance of understanding the behavior of biomolecules at polymer surfaces should not be underestimated. Many of the disposable sensing devices could be made more simply and cost-effectively using polymeric substrates. The work of Xia *et al.* (1996) in making elastomeric replicas of optical components illustrates some of the current technological capabilities using polymeric materials. Lawrence *et al.* (1996) recently described disposable Perspex diffraction gratings for an optical device. The binding surface for optical interaction was coated with a gold film to which biocomponents could be bound, though the application in their study did not involve that aspect of the device.

Roberts *et al.* (1997) recently described photoablated polymeric μ-CE devices for microdiagnostic systems. The photoablation technique was used to produce polystyrene, polycarbonate, cellulose acetate, and poly (ethylene terephthalate) microdevices. The authors noted that bovine serum albumin dampened electroosmotic flow, an observation that must be taken into consideration when designing liquid handling systems for the devices. The interaction of biomolecules with various polymeric materials has been widely investigated with respect to biocompatibilities. Adsorption and covalent binding phenomena at these surfaces, which would also be pertinent to sensor applications, have received far less attention. A broad menu of immobilization methods analogous to those for other surfaces has not yet emerged. For example, Kallury *et al.* (1993) were able to develop a scheme for immobilizing enzymes to a phospholipid layer on fluoropolymers. Osborne and Girault (1995), however, indicated difficulty in immobilizing enzymes on the same type of polymer, which was used as a gas-permeable membrane in their creatinine sensor.

This type of problem involving immobilization of biocomponents on polymeric substrates has been examined by Tarnowski *et al.* (1995) with regard to fluoropolymer membranes used for Clark-type electrodes. In order to facilitate the immobilization of enzymes and cells on these membranes, that group refunctionalized the membranes by glow-discharge plasma. That treatment produced surface hydroxyl groups which could be used for silanization with subsequent protein or cell coupling. This tactic raises an important question about effects of the treatment on the gas permeability of the membranes, since that function is fundamental to their utility. Using a variety of electrochemical characterizations of the permeability and diffusion coefficent for oxygen, the authors were able to demonstrate that the modifications did not have an adverse effect on permeability. Both whole-cell and glucose sensors fabricated with the modified membranes performed essentially the same as those with unmodified membranes.

3.2.1. Cleaning and Pre-treatments

Before immobilization procedures can be used successfully the tranducer surfaces must be cleaned carefully and pre-treated for the coupling chemistries of immobilization. Cleaning may be a simple process such as boiling glass surfaces in hot nitric acid, followed by copious rinsing with high-quality water (Weetall, 1993) or ultrasonic cleaning in commercially available detergent solutions, followed by aqueous rinses (Schmid *et al.*, 1997). There have been varying opinions on the use of detergents with surfaces that will be used with proteins, due to the possible effects of protein denaturation, but with newer detergent products and careful attention to removal of the surfactant material it has now become a widely used method. Indeed, Williams and Blanch (1994) found that detergent added to the immobilization solutions minimized non-specific adsorption on their immunosensors.

Cleaning solid metal surfaces is frequently accomplished by repeated polishing with diamond pastes or alumina slurries using high-quality water. McCreery has emphasized the problems created by additives in some of the commercial alumina preparations (Chen, P. and McCreery, 1996; Kissinger and Heineman, 1996). Taylor *et al.* (1996) also mentioned problems with carbonaceous impurities in aluminas as they compared several methods for cleaning silver surfaces to be used for surface-enhanced Raman spectroscopy. They reported that the Ar^+ sputtered surfaces were the cleanest they obtained, but that the highly active surfaces created by this technique adsorbed carbonaceous impurities from the atmosphere upon handling. Their best results came from a chemical polishing procedure specific for silver, followed by an electrochemical treatment.

Anderson and Winograd (1996) have discussed metal film electrodes and emphasized the point that the usual cleaning procedures are not practical for many of those devices. Therefore, film electrodes are frequently designed as one-shot or disposable devices. The exception they mentioned is a film layered between substrate plates and used in an end-on geometry. Even for these the authors suggest a cleavage process to renew a surface rather than polishing.

Earlier sections have emphasized that gold surfaces are among the most widely used in sensing designs. Finklea (1996) describes with extraordinary thoroughness the use of gold as a substrate for organized thin films of thiols and disulfides. His chapter includes not only behavior and distinguishing features of the gold films, but also considerations in cleaning and characterization. Finklea, and Anderson and Winograd, note that because some gold films (and other metals) are freshly prepared by vapor deposition, cleaning is sometimes omitted. When cleaning is utilized it may involve heating in hot acid. Many laboratories report—e.g., see Caruso *et al.* (1997) and references therein—the use of H_2O_2/H_2SO_4 (piranha) solution which is *highly oxidizing, thus requiring the utmost caution in its use, storage and disposal.*

Carbon electrodes are used in various forms—glassy carbon, carbon fibers, carbon pastes, carbon films and carbon composites. Some of the newer carbon composites for sensors have been described by Newman *et al.* (1995), Alegret *et al.* (1996), Švorc *et al.* (1997) and Santandreu *et al.* (1997). The chapter by McCreery and Cline (1996) is one of the latest general and very useful references on carbon electrodes. Considering primarily glassy carbon and carbon fibers, they explain in detail the types of carbons, cleaning and activation procedures, and electrochemical characteristics of each type. McCreery's (1991) contribution is more extensive with emphasis on electrochemical behavior. Wightman and Wipf's (1989) chapter on ultramicroelectrodes also includes details of fabrication, surface activation approaches and characterization of carbon-fiber electrodes.

It should be noted that with carbon-fiber electrodes there are limited options for cleaning in the usual sense. Polishing and beveling can be accomplished, very carefully, but even then further treatments may be necessary for good performance (McCreery and Cline, 1996; Wightman and Wipf, 1989).

Activating the surface-oxygen functionalities by electrochemical pretreament of carbon serves more than one purpose. This may involve no more than a few seconds of cycled potential variations, usually about 0–2.0 V versus SCE (McCreery, 1991). This electrochemical pretreatment not only offers functional groups for eventual immobilizations, but also has the effect of changing the heterogeneous electron transfer kinetics for electroactive species (McCreery, 1991; Chen, P. and McCreery, 1996). Thus, it may be used as a means of separating overlapping peaks for more selective responses. This approach has been very useful in studying neurotransmitters (Justice, 1987; Wightman and Wipf, 1989; Kissinger and Heineman, 1996). As an alternative to electrochemical pre-treatment, Strein and Ewing (1994) also reported a laser activation process for carbon-fiber electrodes.

Immobilizations of biomolecules may be effected with direct binding to a surface, for example, with cyanogen bromide or cyanuric chloride chemistries (Murray, 1984, 1992; Weetall, 1993). In practice, however, for several reasons this is almost never done. This sort of immobilization leads to molecular crowding of biomolecules on the surface. Also reproducibility of the amount of immobilized material is difficult to control. For some acoustic-wave devices and electrodes there is also the problem of putting the biomolecular species within the charged environment of the electrode double layer. Modern methodologies almost always involve preparation of the surfaces of sensors with some type of intermediary spontaneously adsorbed films or covalently bound tether to which the bioactive components will be bound (Pantano *et al.*, 1991; Zhong and Porter, 1995; Sackmann, 1996; Wong *et al.*, 1997). Figures 3-2, 3-3 and 3-4 illustrate the general approaches. These supportive layers not only enable one to effect controlled molelcular designs at the surfaces, but also they extend the immobilized molecules away from the surface. This provides a "molecular reach" and spatial flexibility toward analytes in a sample, as discussed in the avidin–biotin studies of Wong *et al.* (1997), thus enhancing the sensing function (Pantano *et al.*, 1991; Wong *et al.*, 1997).

In the Zhong and Porter (1995) review of organized thin films the authors discuss Langmuir–Blodgett films (Roberts, 1990) and three major preparative strategies for spontaneously adsorbed films: chemisorption of long-chain carboxylic acids at metals with native oxides, organosilane-based monolayers, and organosulfur-derived monolayers chemisorbed at gold. Perusal of the literature will reveal that the last two are universal in their application. Weetall (1993) has presented in good detail the use of organosilanes to pre-treat inorganic matrices. Aminosilanes, epoxysilanes, glycidoxysilanes and sulfhydrylsilanes are widely used (see Table 3-2). Weetall points out that his experience has been better with the trifunctional silanes which are more stable on inorganic supports. The silanizations may be conducted in aqueous media, in organic media, such as toluene, or they may be evaporated onto the surfaces from a volatile solution. Polzius *et al.* (1996) include several different silanization procedures in their comparative study of seven immobilization methods.

The application of the RSH and RSSR groups to gold is a simple procedure, sometimes involving no more than immersion of the substrate in an ethanolic solution of the adsorbate. For example, Kepley *et al.* (1992) used a 1 mM ethanolic solution of 11-mercaptoundecanoic acid. Caruso *et al.* (1997) used 5 mM 3,3′-dithiodipropionic acid in ethanol. Cheng and Brajter-Toth (1995) used a mixture of thioctic acid and 1-hexanethiol for their study of permselectivity of the SAM films. Willner *et al.* (1993–1996) used 0.02 M aqueous cystamine dihydrochloride solution.

The deposition of Langmuir–Blodgett films in the vertical mode involves transferring a pre-compressed amphiphilic layer to a substrate at the air–water

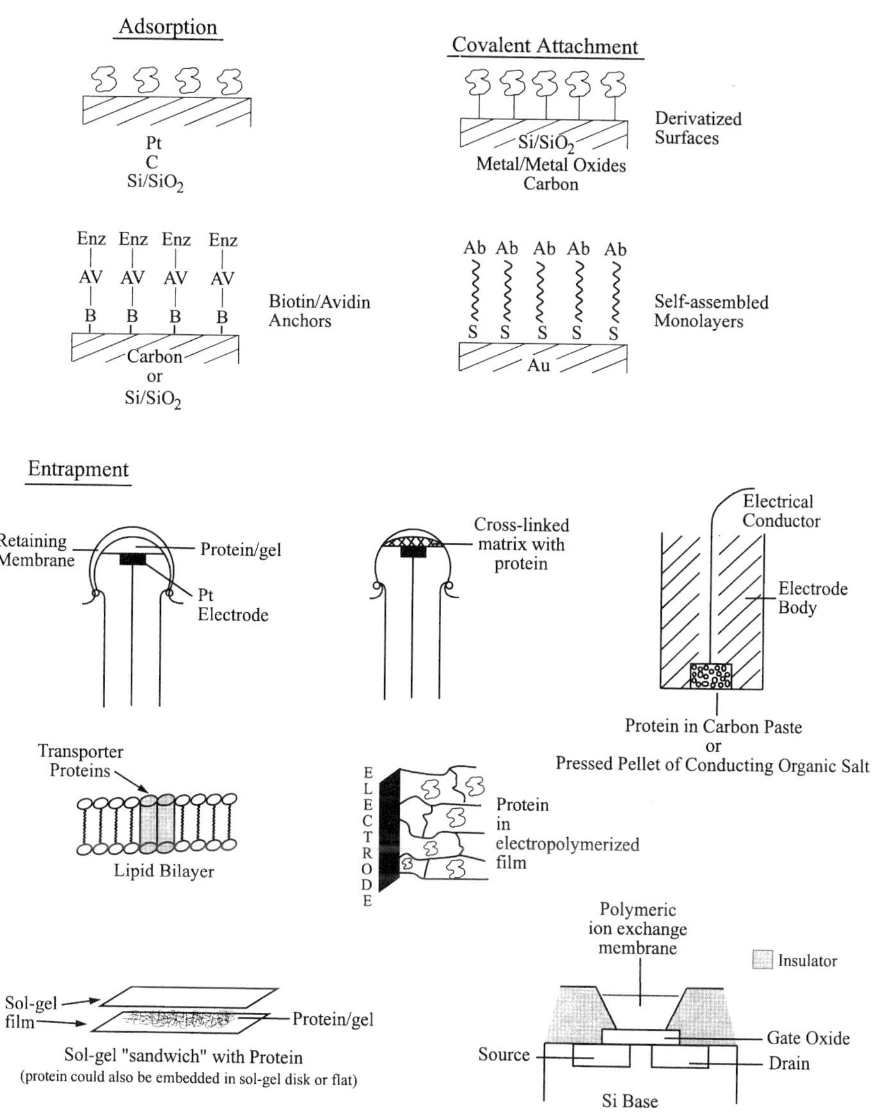

FIG. 3-2. Various immobilization strategies for chemical transduction

interface (Roberts, 1990; Zhong and Porter, 1995). A good analogy is holding a microscope slide by the edges and dipping into a solution with oil on top, then slowly withdrawing the slide. There are commercially available units for producing the L–B films. Pierce *et al.* (1992) describe their procedure for a comparative SECM study. Owaku *et al.* (1995) deposited an L–B film of Protein A for anchoring antibodies for an fluorescence-based immunoassay.

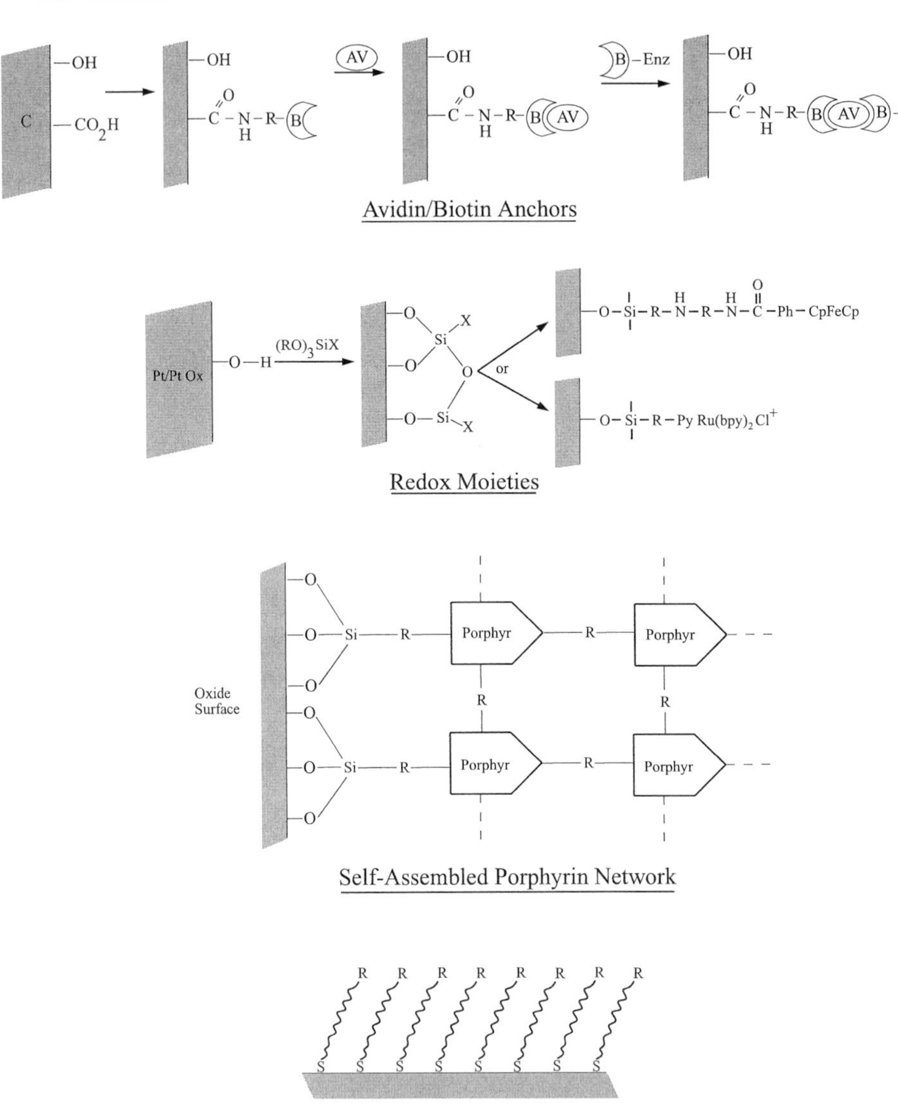

FIG. 3-3. Examples of more highly controlled immobilizations: (top to bottom) biotin–avidin linkages (tethering chain can be adjusted in length by choice of R); chemically modified electrodes using silanization; self-assembling porphyrin network and self-assembled thiols on gold. R is an end-group that can be varied.

3.3. BIOSELECTIVE AGENTS

In Chapter 2 major approaches to selectivity and chemical transduction strategies were explored generally. Proteins, primarily enzymes and antibodies; smaller molecules such as metal complexes, enzymatic cofactors, med-

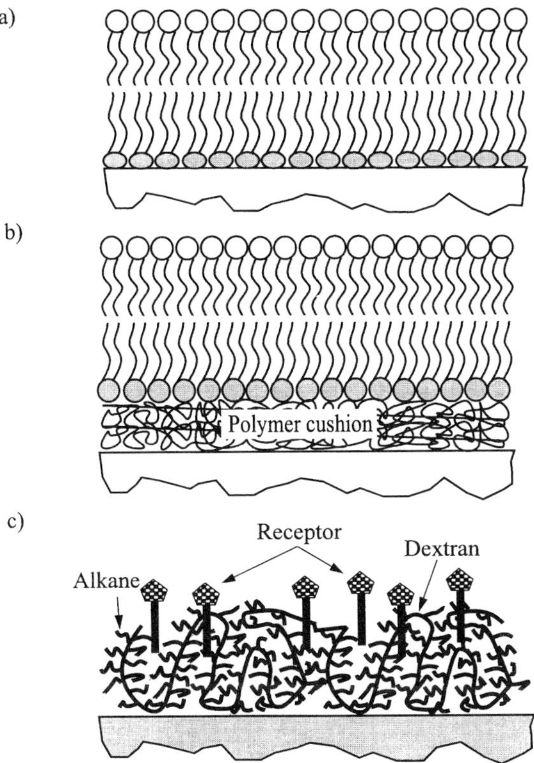

FIG. 3-4. Supported membranes: (a) integrated membrane with covalently bound inner monolayer; (b) supported bilayers; and (c) ultrathin polymer film of hydrophobized macromolecules forming a cushion for monopolar receptors. Reprinted, with permission, from Sackmann, E., *Science* **271**: 43–48. Copyright 1996, American Association for the Advancement of Science, Washington, DC.

iators and some antigens, oligonucleotides and nucleic acids, cells, tissues and membrane transporters are the usual molecular recognition species or auxiliaries for effecting the required selectivity. While each component may require unique conditions of immobilization for optimal response, there are some general approaches for each category or type of chemically selective reagent. Without giving details of procedures, which may be found in the original literature, some of the major biocomponents and immobilization methods will be presented.

3.3.1. Reactive Sites on Proteins

Enzymes and antibodies constitute the largest groups of proteins used and are fundamental for to the operation of biocatalytic and bioaffinity sensors. Other proteins, such as bovine serum albumin, casein and avidin, may be used as auxiliary membrane components, as surface conditioners to prevent non-specific adsorption, or as intermolecular linkers. In these cases the auxiliaries might affect sensor response, but not be primary reactants in the response-determining chemical-transduction steps.

Because all proteins have the same types of sidechains and functional groups, many of the protein immobilization techniques are generally applicable to both enzymes and antibodies. Some of these common methods involve adsorption, intermolecular cross-linking or covalent binding via amide bonds using the sidechains (Table 3-2). Sulfhydryl-group linking reactions are applicable to both the RSH side-chains of enzymes and the RSHs which result from pepsin digestion and disulfide reduction to produce Fab fragments of antibodies (Chapter 2). There are, however, distinguishing characteristics of each type of protein which may require customization for particular situations. Each type has its own set of requirements for immobilization with retention of optimal bioactivity. Enzymes and antibodies are quite different in their macromolecular structures. Thus, an immobilization protocol for achieving accessibility to an enzyme's active site might require a variation in techniques for the purpose of orienting binding sites of an antibody. The usual categories for approaches to protein immobilization have been: 1, adsorption on a surface; 2, entrapment in a membrane; 3, intermolecular cross-linking; and 4, covalent attachment to a support.

Various aspects of adsorption have already been introduced. Entrapment in a membrane may be as simple as mixing components and applying to the sensor surface (Shul'ga *et al.*, 1994) or it may involve a more elaborate approach such as photo- or electropolymerization (Pantano and Walt, 1995; Bartlett and Cooper, 1993).

Entrapment in sol–gel matices has also become a very effective immobilization procedure which can be used for both optical and electrochemical sensing (Lev *et al.*, 1995; Dave *et al.*, 1994; Narang *et al.*, 1994a,b; Tsionsky *et al*, 1994; Tsionsky and Lev, 1995; Sampath and Lev, 1996; Petit-Dominguez *et al.*, 1997). For example, enzymes may be immobilized different ways using the sol–gel films. In preparing electrochemical sensors for glucose, lactate and L-amino acids, Sampath and Lev (1996) prepared the gel with the enzyme present during gelation. The sensors exhibited good FOMs and could be used with a single drop of analyte on the surface. Narang *et al.* (1994a,b) took a different approach. They deposited the enzymatic layer between two sol–gel films in a sandwich structure (see Figure 3-2). Their sensor could be used in either an optical or electrochemical transduction mode.

The sol–gel process is a relatively simple preparation. Low molecular weight alkoxysilanes, such as tetraethoxysilane (TEOS) or tetramethoxysilane (TMOS), are mixed with water as the precursors to gel formation. A solvent, either ethanol or methanol, is usually added for the purpose of improving miscibility, though Sampath and Lev found that the alcohol produced in the hydrolysis was sufficient. The mixes are stirred for several hours to days for the hydrolysis to occur. Lev *et al.* (1995) describe the stages of polycondensation as hydrolysis, silanol–silanol condensation and silanol–ester condensation.

Cross-linking refers to the use of bifunctional agents to create intermolecular links between biomolecules for purposes of stabilizing the layers and preventing leaching from the reaction layers. A universal homobifunctional cross-linking reagent is glutaraldehyde, but Heller's group and others use PEG, poly(ethyleneglycol diglycidyl ether), for their redox gels (Ohara *et al.*, 1993; Garguilo *et al.*, 1993).

$$R_1\text{—}NH_2 + OHC\text{—}(CH_2)_3\text{—}CHO \longrightarrow R_1\text{—}N{=}CH\text{—}(CH_2)_3\text{—}CHO$$

Glutaraldehyde

$$\downarrow R_2NH_2$$

$$R_1\text{—}N{=}CH\text{—}(CH_2)_3\text{—}CH{=}N\text{—}R_2$$

$$\underset{\text{O}}{\triangle}\text{—}CH_2\text{—}O\text{—}(CH_2\text{—}CH_2\text{—}O)_n\text{—}CH_2\text{—}\underset{\text{O}}{\triangle}$$

PEG

The glutaraldehyde chemistry may also be used as a covalent binding strategy to link proteins via their amino groups to an amino-derivatized support. Unfortunately, any time glutaraldehyde is used for intermolecular cross-linking there is also the possibility of intramolecular cross-linking in the structure of the protein. Thus, proteinic activity may be affected.

The cross-linking factor of Mădrăş and Buck (1996) was introduced in Chapter 2:

$$f = \frac{\%\ \text{GL}}{\%\ \text{Total protein}}$$

They performed a systematic study of their creatinine sensor's sensitivity as a function of f, varying both the glutaraldehyde (GL) percentage composition and total protein composition. In the latter case the % GL was held constant at 0.27%. For the experiments designed to hold f constant at variable total protein concentrations, low total protein concentrations in the membrane required such low amounts of GL that the enzyme washed from the membrane. In going from $f = \sim 0.01$ to < 0.05 the sensitivity of the sensor dropped 30-fold, thus indicating the degree of influence of cross-linking within the membrane. Their optimum values seemed to be f = 0.008–0.014.

The established categories of adsorption, entrapment, cross-linking and covalent binding (compared in Table 3-4) are useful for conceptualization of the processes involved, but not necessarily indicative of all types of biosensing strategies. For example, some bioanalytical problems involving the use of single cells or cellular suspensions sometimes do not require immobilization at all (Ewing *et al.*, 1992; Burnette *et al.*, 1996; Niwa *et al.*, 1996). Figure 3-5 illustrates the diversity of biosensing challenges. Immobilization of enzymes and antibodies is depicted with additional possibilities of membranes and transporters, and of whole cells.

A fact not emphasized in Table 3-4, but mentioned in earlier discussions, is that of random versus oriented immobilization. For all practical purposes randomly oriented immobilization will occur unless chemistries for binding to specifically chosen sites on the protein are implemented. This can have a

TABLE 3-4 Traditional Immobilization Methods

Method	Advantages	Disadvantages
Adsorption	Simple, quick	Desorption, leakage Difficulty with reproducibility Difficulty with controlling thickness
Gel entrapment	Simple, quick Many variations Relatively mild	MRA[a] leakage from matrix Polymerization process may "harm" MRA Permeability to interferents can be a problem
Cross-linking	Simple, quick Mild conditions Many options of bifunctional organics MRA strongly held, reducing leakage	Difficult to control inter- and intramolecular linking MRA may be deactivated Introduce sizing restrictions on substrates due to cross-linking
Covalent attachment	Many options for functional groups for reaction with MRA Strongest bonding to prevent leakage Relatively mild Good history of success	Difficult to protect active sites in binding reactions Enzyme activities difficult to predict Monolayers and highly organized molecular architecture difficult to achieve

[a]Molecular recognition agent.

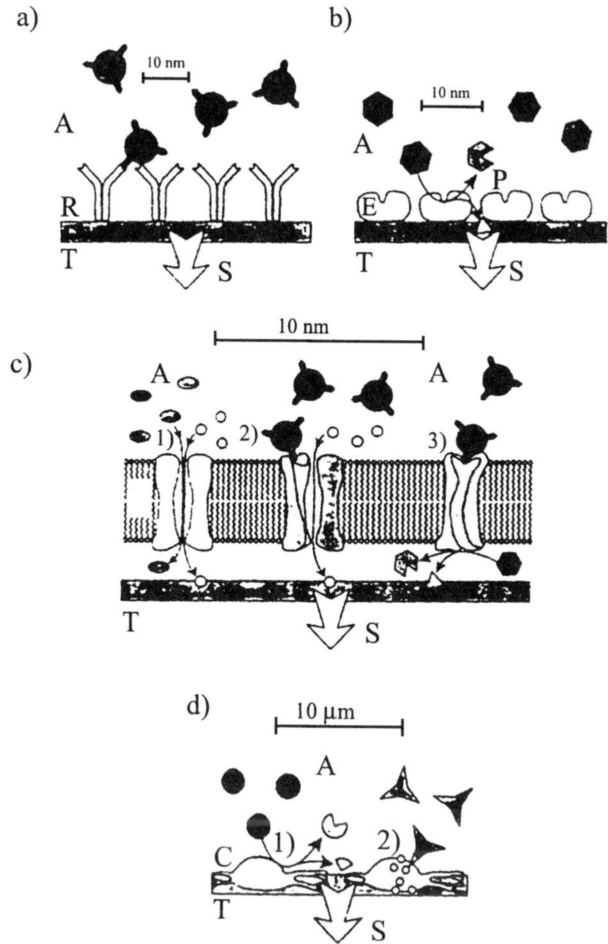

FIG. 3-5. Examples of biosensing principles and interfaces: (a) antibodies; (b) enzymes; (c) transmembrane components (1, transporters; 2, ion channeling; 3, signal receptor showing subsequent enzyme activation); and (d) whole cells. A is analyte, T is the transducer and S is the signal output. Reprinted from *Biosens. Bioelec.* Vol. **10**, Göpel, W. and Heidusehka, P., Interface analysis in biosensor design, pp. 853–883. Copyright 1995, with permission from Elsevier Science, Ltd., Oxford, UK.

drastic effect on the sensing efficiency of both biocatalytic and bioaffinity sensors, a point that bears repeating.

Reactive linking sites on proteins include the $-NH_2$, $-COOH$, $-SH$ and aromatic amino acids in the protein sequence. These are the groups that largely determine adsorption characteristics, primarily in accordance with different ionization states of the proteins, and they are the groups that participate in appropriate coupling chemistries. Exactly which groups are exposed to the surface of the protein, as well as the ionization state, depend on the pH, ionic strength and polarity of the solvent in which the immobilization occurs. Consequently, solution conditions must vary for the controlled adsorption or immobilization of different proteins. The pI of a protein (see Table 3-3) is defined as the pH at which there is charge balance between the positive and negative charges. Below the pI the protein will carry net positive charge. Above the pI the protein will have a net negative charge. Control of solution conditions or microenvironmental charge atmospheres must be considered in sensor designs.

Many of the coupling chemistries using either the $-NH_2$ or $-COOH$ groups involve the formation of an amide bond between an amino group on a prepared surface and the carboxyl group of the enzyme, or vice versa (Weetall, 1993; Ong *et al.*, 1994; Pidgeon *et al.*, 1994). One of the most widely used methods for accomplishing this type of bond is that of activating carboxyl groups on a support with water-soluble 1-ethyl-[3-(dimethylamino)propyl]carbodiimide (EDC) which promotes an acyl intermediate, then treating with *N*-hydroxysuccinimide (NHS) (Ege, 1989) to form the succinimide esters mentioned earlier. The protein can then be coupled to form the amide bond.

$$R_1-N{=}C{=}N-R_2$$

Carbodiimide (EDC)

$$\text{Prot}-\text{COOH} + \text{NHS} \xrightarrow{\text{EDC}} \text{Prot}-\text{C(=O)}-\text{O}-\text{N(succinimide)}$$

$$\xrightarrow{\text{surface}\sim\sim NH_2} \text{surface}\sim\sim\text{NH}-\text{C(=O)}-\text{Prot}$$

The same chemistry could be used for activating a carboxyl group on the support, then coupling to an amino group of the protein.

The EDC–NHS method may be used also for derivatizing proteins. Iwuoha *et al.* (1997) derivatized directly the ε-lysines of horseradish peroxidase (HRP) as succinimide esters for an organic phase enzymatic electrode (OPEE). They found that the NHS–HRP was more effective for sensing in organic media than was the native enzyme. Using an imidazole complex of microperoxidase-11 in dimethyl sulfoxide, Mabrouk (1996) found that the rate of electron transfer to the enzyme was independent of water content in the solvent between 0.1% and 1.8% water content. These examples illustrate not only the use of derivatized proteins, but also some of the variances between aqueous and organic sensing strategies.

For attaching to sulfhydryl groups of a protein the formation of a disulfide bond with an RSH prepared surface or the attachment through a heterobifunctional linker to an amine-derivatized surface may be used (Weetall, 1993):

N — $OCOCH_2CH_2$ — S — S — (pyridyl)

Succinimidyl 3 – (2 – pyridyldithio) proprionate
SPDP

$$\text{—}\!\sim\!NH_2 \xrightarrow{\text{SPDP}} \text{—}\!\sim\!NHCOCH_2CH_2\text{—S—S—(pyridyl)} \xrightarrow{\text{HS–Prot}} \text{—}\!\sim\!NHCOCH_2CH_2S\text{—S—Prot}$$

Lu, B. *et al.* (1995) utilized the SPDP method for attaching the –SH portion of Fab fragments to an aminated silica surface. That binding approach was compared with attachment through a glutaraldehyde linkage, which would involve more random binding. The GL linkages provided more coverage of fragments on the surface, but the SH-linked fragments exhibited much greater inmmunobinding activity due to oriented binding.

Some special precautions must be taken in using the –SH derivatized supports and RSH compounds, due to the extreme susceptibility of RSH groups to oxidation. Weetall suggests using dithiothreitol reduction of an RSH surface prior to effecting the immobilization. The SPDP approach using aminated surfaces avoids this problem with regard to the support oxidation, but the handling of the HS–Fab fragments still requires attention to the oxygen susceptibility. In the Lu study, the Fab's were isolated and prepared with additions of the dithiothreitol and in deoxygenated solutions. The Fab immobilization was conducted under an argon atmosphere.

There are also techniques for linking proteins to supports through the aromatic amino acids of the proteins. This approach is used less frequently primarily because the hydrophobic regions of the protein are not on the macromolecular surface in aqueous solution. Linkages to the hydrophobic regions will take on new importance with increasing use of organic media for enzymatic reactions (Zaks and Klibanov, 1988; Guilbault and Mascini, 1993; Borzeix *et al.*, 1995; Iwuoha *et al.*, 1995, 1997; Mabrouk, 1996).

3.3.2. Examples of Engineered Molecules

While the basic immobilization strategies above apply to both enzymes and antibodies, there are other important considerations related to the uniquely active portions of the molecules. Suppose an immobilization method is chosen whereby a protein's amino groups are involved in the bonding. That means that all amino groups on the molecule may be susceptible to the immobilization chemistry. If these sites are proximate to the active site for substrate or antigen binding, then the efficiency of molecular recognition, and thus the sensing process, may be impaired.

How does one choose a particular site for an enzyme immobilization? If the structure of the enzyme has been elucidated, then databases of structures may be consulted and logical sites may be chosen based on the structure. Alternatively, based on structural and cloning data, an immobilization site may be genetically or chemically engineered away from the active site of the molecule (Hong *et al.*, 1993, 1994; McLean *et al.*, 1993; Gilardi *et al.*, 1994; Lu, B. *et al.*, 1995; Witkowski *et al.*, 1995; Sigal *et al.*, 1996).

Gilardi *et al.* (1994) used site-directed mutagenesis to engineer a cysteine moiety into maltose binding protein. The purpose was to create a protein-bound fluorophore as the basis for a reagentless sensor for maltose. By putting a microenvironmentally sensitive fluorophore near the active site, the fluorescence behavior should change in accordance with substrate binding, and thus be dependent on substrate concentration. Lacking cysteine in the native structure of the protein, the mutation allowed the introduction of a thiol-active fluorophore near the active site of the protein. They did experience RSH oxidation problems as mentioned above, manifested as dimers of the protein, but found that additions of mercaptoethanol maintained the monomeric structure. Acrylodan was found to be the best of several fluorophores tested.

Witkowski *et al.* (1993) also introduced a genetic engineering procedure for creating enzyme–ligand conjugates that would be more effective in enzyme-linked immunoassays. When the engineered conjugate was compared to the usual EDC–NHS-produced conjugates, the assay was significantly more sensitive. The authors also noted that the total enzymatic activity of the engineered product was much higher on the solid substrate for the assay, thus implying a more oriented immobilization with the engineered product. That same group has extended the approach to developing biotinylated enzyme–ligand conjugates which can be used on avidin-treated surfaces (Witkowski *et al.*, 1995).

For oriented binding of cytochrome *b*, McLean *et al.* (1993) also introduced a biotin binding site by genetic engineering so the active site of cyt *b* would be properly aligned with its electron transport partner, cytochrome *c*. The biotin moiety on the "opposite" side of the protein from the active site allowed biotin–avidin immobilization while optimizing the projection of the active site away from the surface. The effectiveness of the approach was demonstrated by chromatographic evidence, but the application to sensing strategies by controlled molecular recognition at surfaces is an obvious extension.

As will be discussed in Chapter 4, for some sensors based on oxidation–reduction processes it is necessary to include in the sensor design an enzymatic cofactor or an electron transfer mediator. The major problem with heterogeneous electron transfer to enzymes lies in the fact that electron transfer sites for some enzymes are buried deep into the tertiary proteinic structure, often blocking direct electron transfer from an electrode. Mediation of the process may be invoked by the addition of electron shuttles (mediators) that will interact simultaneously with the enzyme and the electrode. The saga of achieving electrical contact between an electrode and buried redox-active sites of oxidoreductases has been ongoing for years. There have been numerous reports of ways to achieve the necessary electrical connection (Turner *et al.*, 1989; Heller, 1990; Lötzbeyer *et al.*, 1994; Blonder *et al.*, 1996; Schmidt and Schuhmann, 1996; Willner *et al.*, 1996; Iwuoha *et al.*, 1997).

Blonder *et al.* (1996) utilized the electron-transfer blocking effect purposely by allowing an interceding layer of antibody, as analyte, to interrupt electron transfer. The design was developed for an electrochemical immunoassay method. They used a mixed monolayer of antigen and electron-transfer mediator on the electrode to effect an electrocatalytic current for glucose in the presence of glucose oxidase. With addition of antibody, specific for the bound antigen, the electrocatalytic current decreased because the electrical connection between glucose oxidase and the electrode was disrupted. They were able to determine labeled antibodies in the concentration range of 1–50 $\mu g\ ml^{-1}$.

Du *et al.* (1996) have recently presented evidence that a cyanometalate-modified nickel electrode exhibits an electrocatalytic current for direct electron transfer to glucose oxidase. For those cases where direct electron transfer from an electrode surface to a buried cofactor cannot be accomplished, there are several options for providing synthetic pathways for electron transfer: (1) include an electron mediator as a free species in the reaction layer from which it might leach; (2) include the mediator in either a polymeric redox network on the surface of an electrode or in the electrode material, *per se,* such as the carbon paste electrodes; (3) attach it directly with flexible spacers to the enzyme thus effecting better contact with the electrode; or (4) "wire" the enzyme's electron transfer cofactor through a mediator to the electrode, then connect the apoenzyme to complete a "wired" circuit from the interior of the reconstituted holoenzyme to the electrode (Heller, 1990; Bu *et al.*, 1995,1996; Willner *et al.*, 1996; Schmidt and Schuhmann, 1996).

Au –S—WWW NH$_2$ —(PQQ, EDC)→ Au –S—WWW NH— PQQ

Cystamine

↓ H_2N–FAD

–S—WWW NH— PQQ— NH— FAD

↓ Apo–GOx

–S—WWW NH— PQQ— NH— FAD (2e, 2e) Glucose → Gluconic acid

The Schmidt and Schuhmann review of reagentless oxidoreductase sensors provides a good introduction to some strategies for actually attaching mediators and cofactors to enzymes for sensor applications (1996). The extensive references in Willner *et al.* (1996) also provide a chronology of that group's efforts toward promoting electron transfer from an electrode to an oxidoreductase. Their 1996 report of reconsituting at the electrode surface the holoenzyme, glucose oxidase, through a PQQ mediator includes data for one of the highest current densities ever reported, thus indicating very efficient heterogeneous electron transfer.

Antibodies, whether polyclonal or monoclonal, present a totally different set of opportunites for binding strategies (Lu, B. *et al.*, 1996). While the same side-chain groups as those of enzymes are found in antibodies, the latter have some additional options due to their structure. For example, in addition to the usual side-chain reactions of proteins, it is possible to utilize the integral carbohydrate moieties of the antibodies (see Figure 2-6). This may lead to different, but not necessarily better, oriented binding than do most of the general methods mentioned above.

An antibody has antigen-specific sites (Chapter 2) which can be oriented toward the incoming antigen by anchoring the Fc tail of the antibody to the surface of a sensor. Protein A, Protein G or the genetically engineered form of ProteinA/ProteinG may be used (Lu, B. *et al.*, 1996). Protein A, which may be isolated from baterial cell walls, binds the anchoring Fc portion of antibodies. Protein G exhibits the same function, but with different specificity according to the sources of antibodies. Protein A/G is a genetically engineered form which spans the specificities of both A and G (Polzius *et al.*, 1996; Lu, B. *et al.*, 1996, Sigma Chemical Company catalog). Considering the commercial availability of these anchoring agents, the ease of the immobilization process and the obvious need for oriented binding of antibodies, there is really little reason to continue random immobilization of antibodies unless a sensor's sensitivity and limit of determination are sufficient with a random immobilization procedure. As discussed in Chapter 2 and above, the use of –SH binding using the Fab fragments may also be adopted for oriented binding (Nakanishi *et al.*, 1996; Lu, B. *et al.*, 1995).

3.3.3. Cells and Tissues

As indicated earlier, not all sensing strategies utilizing cells and tissues involve immobilization. For measurements using single cells or cellular suspensions maintaining the viability of the preparations is the major factor. For experiments conducted in Petri dishes, for example, the cells are submersed in biological buffers or artificial body fluids to mimic the natural environment (Ewing *et al.*, 1992; Paras and Kennedy, 1995; Schultz *et al.*, 1995; Niwa *et al.*, 1996). Burnette *et al.* (1996) were able to obtain good data on norepinephrine transporter kinetics of genetically engineered cells simply by using a rotating electrode in a cellular suspension.

For actual sensor designs cells and tissues may be bound to an energy transducer by something as simple as a piece of dialysis tubing and an O-ring, or they may be distributed in biologically compatible matrices such as calcium alginate, agar, biological buffers, etc. (Arnold and Rechnitz, 1989; Karube, 1989a; Karube *et al.* 1995b). Brewster *et al.* (1995) suggested a novel method for the storage and immobilization of photosystem II photosynthetic electron transport, (PET) cells. The goals were to improve long-term storage and provide an active cell preparation that could be used for quick semiquantitative screening for herbicides. They suggested a technique for immobilizing the PET cells in calcium alginate microbeads that remained viable for 1 week when stored at 4 °C. Preuss and Hall (1995) used a calcium alginate matrix for immobilization at a glassy carbon electrode in elegant experiments designed to study the effects of herbicides on PET cells. The cellular matrix was covered with a dialysis membrane for containment.

Rather than immobilizing genetically engineered plant tissue on a sensor surface, Wang, A. J. and Rechnitz (1993) constructed a packed column, a tissue reactor, then used it in a flow system with external fluorescence detection. The authors commented that the slow response time could be overcome by adapting the immobilization to the tip of a optical fiber.

Preininger *et al.* (1994) did use an optical substrate for a microbial sensor designed for improved biological oxygen demand (BOD) measurements. They mixed yeast cells with poly(vinyl alcohol) (PVA) and spread it on a sensor surface modified with an oxygen-sensitive fluorescent ruthenium complex. They observed a response time of 3–10 min, significantly faster than conventional BOD protocols, and the upper LOD was 100 mg per liter BOD.

For a totally different type of application, Watts *et al.* (1994) designed an optical biosensor for actually monitoring *Staphylococcus aureus* (Cowan-1 strain) cells. These cells express Protein-A on the surface. By immobilizing human immunoglobulin IgG with glutaraldehyde linking to an aminosilane surface of a resonant mirror device (Chapter 5) or through coupling to CM-dextran, they were able to monitor cells in the 4×10^3– 1.6×10^6 range in milk samples. This is an example for which random binding of the IgG with some Fc regions projecting away from the sensor surface is actually advantageous, since accessibility to the Protein A on the cell walls of the sample is basic to the detection method.

For their "dynamic microbial sensor" that was described in Chapter 2, Slama *et al.* (1996) immobilized their cells in a polyurethane hydrogel. Prepolymer was mixed with buffer, then cross-linked with poly(ethylene imine). That solution was mixed with the previously harvested suspension of cells. The preparation was spread on a gas-permeable polyethylene membrane for attachment to a Clark-type oxygen electrode. A second gas-permeable protective membrane was also part of the assembly. The sensor was used in an FIA system for chemometric analysis of two-component mixtures.

3.3.4. Oligonucleotides

The art and science of producing highly ordered oligonucleotide arrays have been raised to a high level with the commercial availability of automated devices for hybridization studies (Fodor *et al.*, 1991; Pease *et al.*, 1994; Blanchard *et al.*, 1996; Chee *et al.*, 1996). In addition to these high-throughput systems, there is also the need for simpler DNA sensors that could be used for purposes other than massive sequencing problems. Several groups have developed such devices using different kinds of energy transduction and immobilization chemistries.

Millan and Mikkelson (1993) and Millan *et al.* (1994) utilized an electrochemical method for monitoring hybridization, including detection of the cystic fibrosis sequence deletion. In the 1993 study they bound DNA to a glassy carbon electrode through the deoxyguanosine groups using conventional EDC–NHS chemistry. For the cystic fibrosis study they used the same immobilization chemistry, but applied it to immobilization of ssDNA at a carbon paste electrode.

Abel *et al.* (1996) developed an optical-fiber sensor for DNA. The immobilization scheme was based on avidin deposition on the fiber surface, with subsequent attachment of the 5′-biotinylated 16-mer oligonucleotides. They used a polymeric sodium salt and Tween detergent to prevent non-specific adsorption. They reported an LOD of 132 pmol of the complementary olionucleotides.

Piunno *et al.* (1995) took an entirely different approach to immobilizing oligonucleotides on quartz fibers. They adopted the same strategy as that used in commercial synthesizers. The first structural unit was applied to the surface as a nucleoside tethered to a long-chain linking molecule (see the original report for the relatively simple chemistry of that process). From that point the fiber was subjected to the usual automated synthesizing process using phosphoramidite chemistry for sequential additions of bases to build the oligonucleotides.

Caruso *et al.* (1997) devised a method, described earlier in other contexts, for covalently binding DNA to gold surfaces for their acoustic-wave studies of hybridization. The Au was first prepared with a –COOH terminal SAM. Then avidin was covalently bound to the surface through EDC–NHS chemistry. This supporting membrane of avidin could then be used for attaching 5′-biotinylated DNA. This group's use of multilayering through successive deposition of electrostatically attracted layers has already been discussed in the context of adsorption.

3.3.5. Lipids

Many of the natural biological processes involving species for which sensors have been developed occur within the hydrophobic regions of membranes rather than in an aqueous environment. Logically it would be advantageous to be able to study these bioreactants or design sensors for them based on a more natural microenvironment. The use of Langmuir–Blodgett bilayers as supporting membranes was introduced earlier in the chapter. They provide one avenue to creating a more biochemically realistic sensing situation.

Kallury *et al.* (1992, 1993) and Kallury and Thompson (1994) adopted a different approach. In an attempt to develop enzymatic sensors with enhanced stability for storage and sterilization, they designed covalently bound phospholipid films on silica, tungsten and polytetrafluoroethylene.

While the chemical steps are numerous, they were thoroughly reported in the original literature, and involve the creation of amino-derivatized supports coupled to phospholipids. The enzyme urease was then bound to the long-chain phospholipid tether by EDC–NHS chemistry either through the protein's amino groups in the first report, or through the –COOH groups in later reports. Extensive surface characterization using spectroscopic and ellipsometric methods was performed. Film thicknesses were on the order of 90–100 Å. The remarkable feature of these phospholipid-bound enzymes was the enhanced thermal stability. The enzymes were stable to temperatures up to 100 °C and the modified polymeric substrates could be boiled in aqueous soultion for 1 hour with minimal loss of activity.

Pidgeon's group has also been quite active in the immobilization of phosholipids on solid surfaces, though their main goal is that of immobilized artificial membranes (IAM) for chromatographic supports, particularly in drug binding studies (Ong *et al.*, 1994; Pidgeon *et al.*, 1994 and references therein). In the latter report they developed mixed ligand membranes consisting of a phosphatidyl choline (PC) layer and layers containing a series of other PC derivatives. The effectiveness of these membranes for drug binding studies was demonstrated through the strong correlation to results from analogous experiments using liposomes. The authors noted extended applications that might be possible, including sensing membranes which might be especially stable in organic media.

3.4. MOLECULARLY IMPRINTED POLYMERS

Although MIPs, introduced in Chapter 1, do not fall into the category of immobilized biocomponents, the rising importance of these very useful biomimetic materials requires attention to the potential for biosensing (Kriz and Mosbach, 1995; Kriz *et al.*, 1995; Mosbach, 1996; Ramström *et al.*, 1996; Sellergren, 1997). For example, Kriz *et al.* (1995) developed a distal-end fiber-optic sensor comprised of an MIP confined to the fiber by a conventional nylon net/O-ring arrangement. The polymer had been imprinted with dansylated phenylalanines. The optical sensor exhibited the chiral selectivity characteristic of MIPs when tested in acetonitrile solutions. The time required for a stable signal was 4 hours, due to diffusional mass transport into the polymer and binding kinetics.

Levi *et al.* (1997) devised a sensing method for the determination of chloramphenicol (CAP) in the 3–1000 μg ml^{-1} range. An MIP was prepared with a Methyl red derivative of CAP as the imprint molecule. In an HPLC system the solvent containing CAP-MR was equilibrated in the HPLC system. Injected samples of CAP in acetonitrile, some having been extracted from serum, then displaced the CAP-MR which could be detected optically in the flow stream. By changing to a column length of 5 cm and using a 2 ml min^{-1} flow rate, they were able to complete a measurement in 5 mins. In the report the authors defined a molecular imprint factor which related the chromatographic capacity factor for the imprinted polymer to the capacity on a non-specific polymer prepared in an identical manner: $I = k'_{sp}/k'_{n}$.

Most of the developmental work in MIPs has been related to chromatographic and extraction materials that exhibit high loading capacities and exceptional selectivity, even for chiral separations (Kriz and Mosbach, 1995; Matsui *et al.*, 1995; Kempe, 1996; Mayes and Mosbach, 1996; Muldoon and Stanker, 1997). The translation of the technology to sensors is an obvious extension, which has been reported in several of the references above and references therein, but some obstacles such as inefficient energy transduction, slow mass transfer in rebinding of the template and highly solvent-dependent behavior must be overcome for general usage.

One can conceive of almost any relatively low molecular weight analyte as the template molecule. Primary emphasis has been on the use of the MIPs as antibody mimics since the smaller haptens may be imprinted, thus precluding the necessity of raising antibodies against them, and binding constants rival those of some antigen–antibody pairs (Sellergren, 1997). It should be noted that McGown *et al.* (1995) and Kawazoe *et al.* (1996) have suggested the same sort of versatility and broad applicability to nucleic acid ligands (NALs). If, in fact, synthetic molecular recognition agents of these types can be shown to have comparable selectivities and transduction compatibilities of naturally occurring biocomponents, then they may prove to be ideal alternatives for the sometimes more fragile, unstable and expensive natural materials.

3.5. PATTERNED SURFACES

Arrays of discrete sensing units, microdevices of various types and chemical arrays have been discussed in previous sections in the context of frontiers of research, as well as spatial and depth resolution. Full treatment of all the ways to achieve patterned sensing surfaces is beyond the scope of this introduction. It is important, however, to emphasize the distinction between patterned energy transducers on which the same or different molecular recognition agents might be deposited, and chemical arrays patterned on a common transducer. Kovacs *et al.* (1996) and Karube *et al.* (1995a) and references therein may be perused for an introduction to the basics of micromachining devices, as schematized in Figure 3-6.

There have been many approaches to patterning surfaces with arrays of chemicals. Bhatia *et al.* (1993) used UV-patterning of silanized surfaces to create hydrophobic regions. Corn's group has used the photosensitivity of

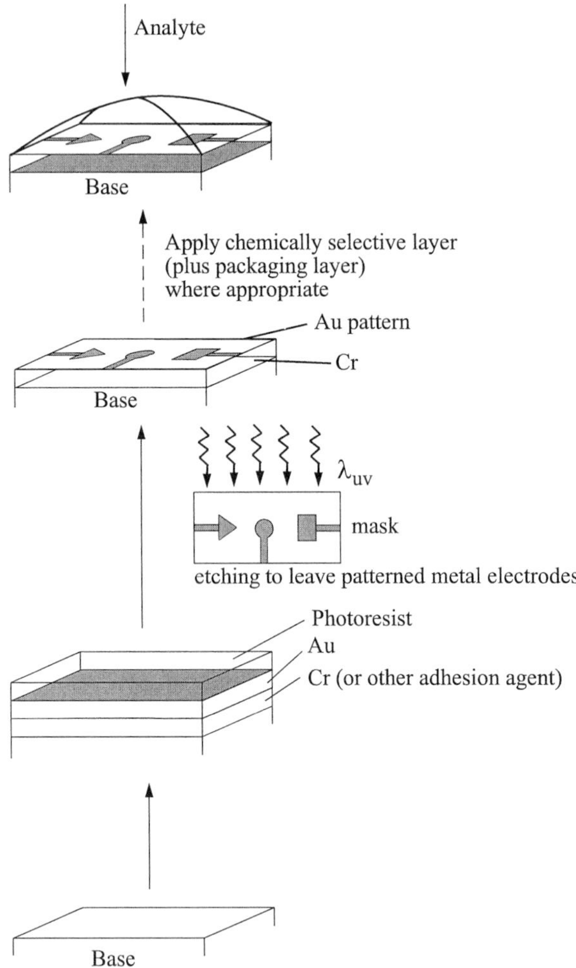

FIG. 3-6. Schematic of the process for patterning a microsensor using a photolithographic technique. After cleaning, 200 Å Cr followed by 2000 Å Au layers are sputtered on the base. A ~1.5 μm layer of photoresist is then spin-coated over these. Next a mask is applied and the unit is selectively exposed to UV radiation. Etching reveals the metal electrode pattern and the chemically selective layer, etc. may be applied.

the Au–S bond to photopattern gold surfaces for regions of controlled protein adsorption (Jordan and Corn, 1997). Shiku *et al.* (1995) used scanning electrochemical microscopy to pattern surfaces with diaphorase. Prichard *et al.* (1995a,b) and Morgan *et al.* (1995) developed a photolithographic method for patterning antibodies by using an avidin-treated surface with an azide-derivatized photobiotin.

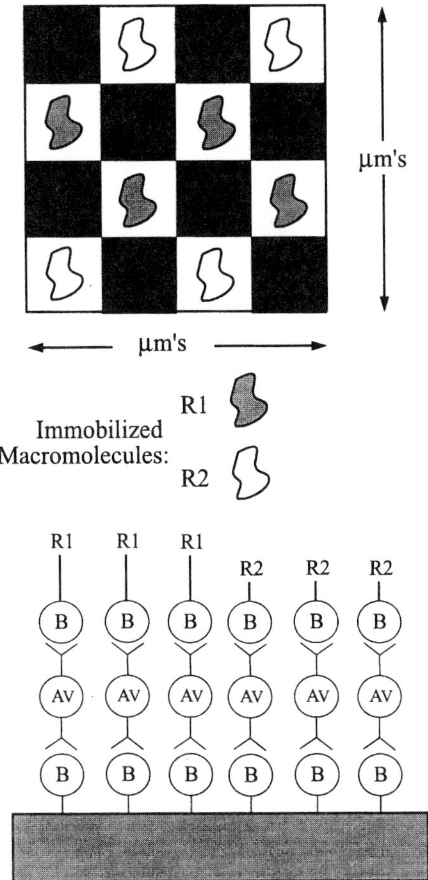

FIG. 3-7. Representation of micro-patterning immobilized biocomponents with avidin/biotin technology. For examples with antibodies, see Sundberg *et al.* (1995); for enzyme immobilizations see Dontha *et al.* (1997).

The patterned chemical array of Sundberg *et al.* (1995) was produced by using a photolithographic mask to pattern a surface covered with a photoactive biotin species (see Figure 3-7). Once the biotins are exposed by light activation, a layering process based on avidin–biotin chemistry can be initiated. Any biocomponents that can be biotinylated could be used for attachment to this type of prepared surface. They demonstrated the process with antibodies on glass slides, but it could be used for other biocomponents as well. Kuhr and coworkers (Dontha *et al.*, 1997) have extended this general approach to a "maskless" method, using a laser interference pattern for activating photobiotin for immobilization of enzymes at microsites on the surfaces of carbon fibers.

These various types of patterning technology, along with other methods that have been discussed in previous sections, have become routine in many laboratories. As recently as the 1960's chemists were desperately trying to develop techniques that would allow them to "see" what was happening at a surface during a dynamic process. Now they are micropatterning surfaces with nanoscale molecular architecture that is truly impressive.

4. Fundamentals of Electroanalytical Biosensors

Although any of the basic phenomena of charge and electron transfer may be invoked for biosensing applications, potentiometric and amperometric sensors have been the most widely studied and developed for practical devices. Conductimetric and impedimetric modes (Section 4.4), which are inherently less selective, have received more attention is recent years as high-quality instrumentation, better methods for achieving molecularly selective modifications and chemometric techniques have improved. Before surveying various types of electrochemical devices used in bioanalytical problems, a few notes of review and nomenclature may be helpful. Because there are so many experimental combinations of the variables—potential, current, concentration, and time—some analysts are often bewildered by the assortment of methodologies as outlined in Figure 4-1.

General references to electrochemical theory and applications include Bard and Faulkner (1980), Rieger (1994), Wang, Joseph (1994), Oldham (1994) and Kissinger and Heineman (1996). The Turner *et al.* (1989) contributed volume includes chapters on most of the electroanalytical applications in the context of biosensors. Cass (1990) provides practical

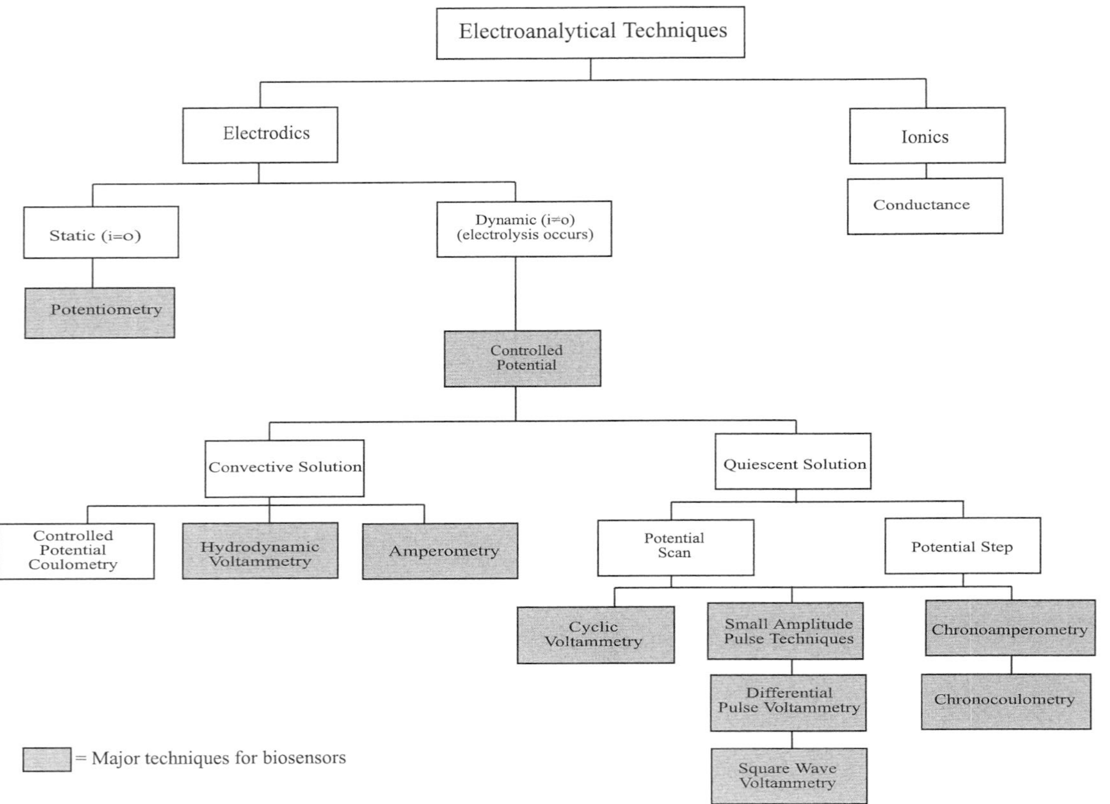

FIG. 4-1. Outline of electroanalytical methods. Shaded boxes are major techniques for biosensors.

information on the fabrication and characterization of several types of electrochemically based biosensors. There is also in the last reference development of the basic theory of responses for biocatalytic and bioaffinity sensors.

4.1. REVIEW OF TERMINOLOGY

The simplest form of classification of electroanalytical methods simply indicates the variable measured—e.g., potential, current, conductance or impedance. Modifiers are added according to the variable controlled in the measurement. For example, amperometry is the measurement of current, whereas voltammetry is the measurement of current as a function of variable applied potential. Chronoamperometry is the measurement of current as a function of time at a constant value of the applied potential. The shaded boxes in Figure 4-1 indicate the modes that have been used most commonly for elucidating biosensing mechanisms and developing biosensors. The resulting devices almost always operate is one of the four basic modes indicated above.

It should be recalled from basic chemistry that electrochemical cells are classified as Galvanic or electrolytic. A Galvanic cell is one in which a chemical reaction occurs spontaneously with generation of energy which is measured either as the current flow or as the electromotive force in units of volts between two electrodes, one of which is a reference. Two Galvanic examples from the world of biosensing are ordinary potentiometric sensors (zero current) and the less common microbial fuel-cell sensors. In the latter, immobilized microbes consume nutrient, which is the analyte, and as a result of the internal microbial oxidation–reduction processes a detectable current (and potential) proportional to the amount of analyte is generated at an electrode.

In contrast to Galvanic cells, which generate their own energy, an external source of energy in the form of a power supply or potentiostat is used to drive an electrolytic cell. Commonly the cell may be forced to adopt an applied potential, and as the ratio of oxidation–reduction species changes to comply, current flows in proportion to the amount of material being oxidized or reduced. The current may be integrated over time to obtain the number of coulombs, and thus the number of chemical equivalents transformed. Alternatively, though rarely used for biosensors, an external source of current may impress a current flow, in which case the potential of the cell will change accordingly. An applied potential or current may be constant or may be varied in any one of several time-dependent patterns. For example, cyclic voltammetry refers to the method of scanning applied potential forward and back between two limits and measuring the resulting current (refer to Figure 1-11).

The fundamental relationship that governs the types of behaviors just described is the Nernst equation, which, for the reaction

$$\mathrm{O} + ne \rightarrow R$$

is

$$E = E^{\circ} + \frac{RT}{nF} \ln \frac{[\mathrm{O}]}{[\mathrm{R}]}$$

where E is the potential of an electrode versus a reference and E° is the standard potential, R is the gas law constant, 8.31441 J mol^{-1} K^{-1}, T is the absolute temperature, and F is the Faraday constant, 9.64846×10^4 C per equivalent.

Implicit in the above description is a reference electrode against which potentials are either measured or applied. The reference electrode consists of a stable, reversible oxidation–reduction couple at equilibrium. The primary reference for stating standard electrode potentials is the normal hydrogen electrode (NHE), but this is rarely used routinely due to its impracticality. The more common secondary references are the saturated calomel (SCE)and silver/silver chloride (Ag/AgCl) electrodes:

$$\mathrm{Hg_2Cl_2} + 2e \rightleftharpoons 2\mathrm{Hg} + 2\mathrm{Cl^-} \quad \mathrm{E}^{\circ} = +0.2847\mathrm{V}\ \text{versus NHE}$$
$$\mathrm{AgCl} + e \rightarrow \mathrm{Ag} + \mathrm{Cl^-} \quad E^{\circ} = +0.22\mathrm{V}\ \text{versus NHE}$$

The choice of reference electrode has proven to be a challenge for designing miniaturized and solid-state devices and for some applications where diffusion of the reference's internal components into a sample may be detrimental. The maintenance of the reference's equilibrium potential is essential. In order to reduce liquid junction potentials, a porous fiber junction is often used for SCEs. Over time diffusion of species from the reference, e.g., chloride ion, into a sample can contaminate the sample, and should disturb the equilibrium of the electrode if enough diffuses out. For the Ag/AgCl electrode one problem is that the potential is dependent on constant chloride ion activity, which cannot be assured in natural matrices such as body fluids or processing broths. Moussey and Harrison (1994) found that Ag/AgCl references implanted subcutaneously failed unless protected by a polymeric coating of either thermally cured Nafion or polyurethane. Electrodes treated in this manner were stable *in vivo* for 2 weeks.

Several approaches have been proposed for acceptable reference electrodes to be used with miniaturized and solid-state devices. If the sensor is to be used in a flow system, such as FIA, the reference may be put in contact in-line in the stream rather than being incorporated onto the sensing unit. Goldberg *et al.* (1994) suggested that a reference electrode might be built into the electronic control unit which would accept plug-in devices containing the working electrodes. Pötter *et al.* (1995) designed a perchlorate reference field-effect transistor (REFET). They micromachined a hole into which they could deposit $KClO_4$ and $CaSO_4$ hydrate. That was then covered with a polymeric film punctured with a 30 μm hole for making contact with the sample solution. The authors reported that the device would last only 4 days. Nolan *et al.* (1997) reported a solid state Ag/AgCl reference that had the NaCl electrolyte immobilized in a PVC matrix which was covered with either Nafion or polyurethane. The polyurethane coat was found to provide better stability; the electrode was stable for up to 90 days. Lvovich and Scheeline (1997) incorporated a tungsten rod as reference into the electrode body of their superoxide/hydrogen peroxide sensors (Section 4.3).

Cells may be operated in a two-electrode configuration, commonly designated as an indicator and reference for potentiometric measurements and as a working electrode and reference for amperometric sensors. In the latter case, the two-electrode design will suffice if (a) currents are low enough to avoid disturbing the reference's equilibrium potential and (b) the *iR* drop through the solution does not alter significantly the applied potential from the potentiostat. To guard against both problems a three-electrode configuration is often adopted using a working electrode, a reference and an auxiliary electrode through which the current passes. The reference in this potentiostatic arrangement is operated in a non-current mode for electronic comparison of set potentials to actual potentials. Any detected difference is fed back electronically to the potentiostat for adjustment of the potential applied between the auxiliary and working electrodes (Kissinger and Heineman, 1996).

It should be noted that micro- and ultramicroelectrodes are often used in the two-electrode configuration because the currents are so low that *iR* drop is not a problem. For example, a current of 1 microampere passing through a solution of 100 ohms resistance will drop only 0.1 millivolt across the solution. Ultramicroelectrodes pass even smaller currents than 1 microampere. This is one reason they have been especially useful for non-aqueous work where solution resistances are considerably higher.

A reminder about electron flows is in order. Electrons roll down the energy hill. That is for spontaneous processes with ΔG's < 0, ΔE must be positive. There is a convention used throughout the literature for depicting the electron flow using biocomponents as electron transfer agents. Not all authors use the same directional notation, but the outcome is the same. An oxidative process may be illustrated with the following, where M refers to a mediating electron transfer "shuttle" molecule which is discussed in Section 4.3:

A_{red} B_{ox} C_{red}
e e e
A_{ox} B_{red} C_{ox}

M_{red} GOx_{ox} Glucose
e e e
M_{ox} GOx_{red} Gluconolactone

4.2. POTENTIOMETRIC SENSORS

As indicated in earlier chapters, potentiometric methods have proven to be particularly good for ions and dissolved gases which produce ions in many types of bioanalytical problems. The apparent simplicity of potentiometric methods, as usually presented in undergraduate courses, is deceptive. Obtaining stable, selective signals in complex media is far more challenging than determining one or two ions in synthetic solutions. Furthermore, miniaturized and solid-state devices introduce design problems unlike those of macroelectrodes. The discussion of reference electrodes in the previous section illustrates that point.

The fundamental concepts of potentiometric measurements and some typical macroelectrodes are summarized as a review in the illustrations of Figures 4-2 through 4-4. Essentially the same elements of the pCa macroelectrode (Figure 4-4) and others analogous to it can be translated into miniaturized versions. For the basic potentiometric measurement there is the usual arrangement of indicator and reference electrodes with a very high impedance voltmeter for measurement of the response under conditions of zero current flow (Figure 4-3). It should be noted that the internal reference element of the ion-selective electrode (ISE) is distinct from the external reference to which the potential is compared in the measurement. The potentiometric measurement is a zero-flux, or non-consumable, mode of measurement.

The guiding principles of potentiometric measurements are found in the Nernst equation, which states the fundamental relationship involving the

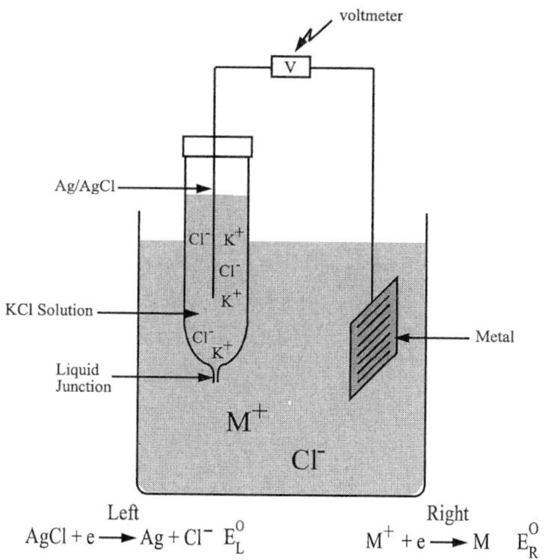

$$M^+ + e \longrightarrow M \qquad E_R^O$$

$$AgCl + e \longrightarrow Ag + Cl^- \qquad E_L^O$$

$$M^+ + Ag + Cl^- \longrightarrow M + AgCl \qquad E_{cell}^O = E_R^O - E_L^O$$

$$E_{cell} = E_{cell}^O - \frac{RT}{n\mathcal{F}} \ln \frac{\overset{1}{a_M}\ \overset{1}{a_{AgCl}}}{a_{M^+}\ \underset{1}{a_{Ag}}\ a_{Cl^-}}$$

$$E_{cell} = E_{cell}^O - \frac{RT}{n\mathcal{F}} \ln \frac{1}{a_{M^+} a_{Cl^-}}$$

or

$$E_{cell} = E_{cell}^O - \frac{RT}{n\mathcal{F}} \ln \frac{1}{[M^+][Cl^-]}$$

or

$$E_{cell} \cong E_{cell}^O - \frac{2.303RT}{n\mathcal{F}} \log \frac{1}{[M^+][Cl^-]} = E_{cell}^O - \frac{.059}{n} \log \frac{1}{[M^+][Cl^-]} \text{ at 298K}$$

FIG. 4-2. Fundamental potentiometric measurement utilizing a metal electrode in a solution of its metal cation.

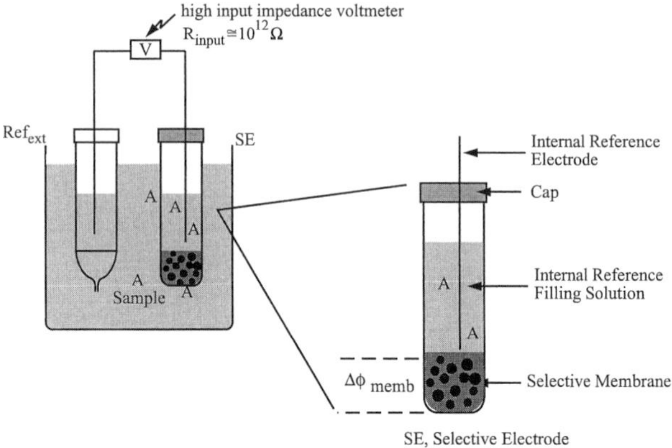

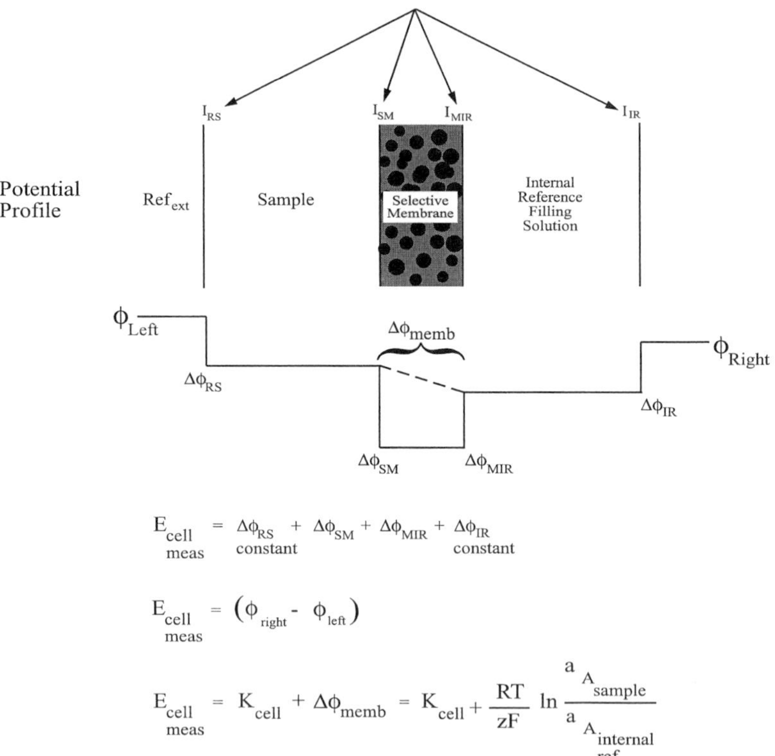

FIG. 4-3. Potentiometric measurement with membrane separating the cell compartments. The resulting potential is a function of the potential across the membrane, and is comprised of a phase boundary potential and a diffusion potential.

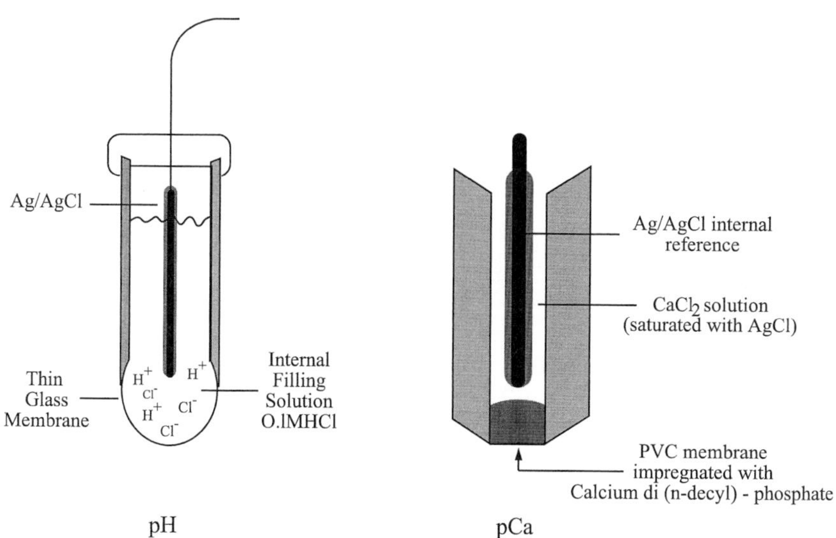

FIG. 4-4. (Left) Typical glass membrane electrode for pH measurements, and (right) polymer membrane-based ion-selective electrode for determining calcium ions.

potential difference between electrodes and the activity of the species responsible for the response. The applicable relationships are shown in a simplified format in Figure 4-3. Comparable expressions may be written for each ion in a mixture. For thorough treatments of ISE theory and practice Morf (1981), Amman (1986) and Umezawa (1990) should be consulted. The IUPAC (1995) report, discussed below, also includes major references in the historical background of theories regarding selectivity and its assessment.

4.2.1. Polymer Membrane–based Sensors

Many common potentiometric biosensors are designed around the basic concepts of polymer membrane-based ISEs, which were introduced in Chapters 1 and 2. The membranes are doped with ion-exchangers, neutral or charged carriers. Lipophilic additives, RX, are included to reduce the membrane resistance and maintain charge blance in the membrane. The additives are not just negligible species—they directly affect response characteristics which may be partially tuned by adjustments in the composition. The zero-current potentiometric measurement has its origin in the phase potentials developed across membranes designed for selectivity for a given species. As shown in Figure 4-4, the measured potential for the cell depends on the potential difference ϕ_{memb} across the membrane. Fundamental thermodynamics of electrochemical cells indicate that the potential is determined by the relative activities of the ion on each side of the membrane and the charge on the ion, as shown in Figure 4-3.

The Nicolsky–Eisenman equation has been the usual relationship for comparing in a quantitative way the selectivities of ion-selective devices Nicolsky *et al.* (1967).

$$E = \text{Const} + \frac{RT}{z_A F} \ln \left[a_{\text{A}} + \sum_{\text{B} \neq \text{A}}^{\text{B}} k_{A,B}^{\text{pot}} (a_{\text{B}})^{z_A/z_B} \right]$$

A lower value of $k_{\text{A,B}}{}^{\text{pot}}$ for the interfering ion indicates higher selectivity for the primary ion. When comparing two electrodes, both designed to measure A, the one with the lower $k_{\text{A,B}}^{\text{pot}}$ for an aniticipated interferent B would be preferred. This indicates that interfering ions have a lesser effect on response. Normally analysts use one or more of three ways to determine $k_{\text{A,B}}^{\text{pot}}$.

1. Fixed interference method. A constant activity of B is maintained as the activity of A is varied. *E*'s are plotted versus the activity of the primary ion, A. Then $k_{\text{A,B}}^{\text{pot}}$ is calculated from the data at an activity for A determined from the intersection of the two linear portions of the curve
2. Separate solution method. Measurements of the potentials for two separate solutions are made. The first solution has only the primary ion at an activity *a*. Then the interferent is measured at that same activity in a separate solution.
3. Matched potential method. In a reference solution the activity of the primary ion is changed from *a* to a$'$ and the corresponding potential change ΔE is recorded. Then in an identical reference solution with the same initial activity of primary ion, the interfering ion B is added until the same potential change ΔE is attained. Then $k_{A,B}^{\text{pot}}$ is defined to be $\Delta a_{\text{A}}/a_B$ (see Figure 4-5).

There have been many discussions of potentiometric results which suggest either deviation from the basic N–E equation or inconsistencies when comparing various electrodes, fabrication techniques or methods of determining *k*. An IUPAC recommendation (Umezawa *et al.*, 1995) has provided a thorough discussion of the usual anomalies and causes. Basically, there are three sources of the problems with application of the N–E equation. (1) Primary and interfering ions of different charges frequently lead to deviation from the basic relationship. (2) The equation assumes that both the primary

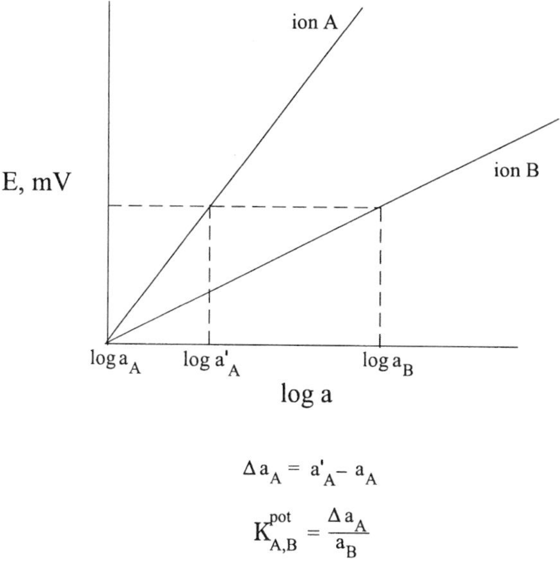

FIG. 4-5. Illustration of the matched potential method for determining selectivity coefficients. A is the primary ion, and B is the competing ion. See text for more details of the method. Also refer to Umezawa *et al.* (1995) for a full discussion of different approaches for determining $K^{pot}_{A,B}$ and for a full report of IUPAC recommendations.

ion and interfering ion(s) demonstrate Nernstian behavior—i.e., a $(0.059/z)$ slope for the E versus log (a) plot. There are many cases where this is not the real situation. (3) Values of k often have obvious dependence on the absolute activities at which k is determined. The practical result, however, is that cross-comparisons for differently prepared electrodes or measurement methods lead to widely varying k's in some cases. The IUPAC report suggests that the matched potential method of determination probably avoids most of the apparent inconsistencies, inasmuch as it is a method independent of the N–E equation. Only in the limiting cases of equal charges for primary ion and interferents and Nernstian behavior of all the ions do the results from the matched potential method correspond to the other measurement methods.

Bakker *et al.* (1994) have proposed a selectivity formalism for polymer membrane ISEs. Their model provides an approach to a more accurate prediction of k and reinforces the validity of the matched potential method for determining k. Perhaps more significant is the fact that the theory establishes a firm basis for designing the composition of polymeric membranes for required selectivities. The original report should be consulted for a complete treatment of the response functions for dissociated ion-exchangers, neutral and charged carriers in the membranes. The membrane formulations with carriers and lipophilic additives, R, as introduced in earlier chapters are utilized. Some of the results may be summarized.

1. For polymer membranes based on dissociated ion exchangers, selectivity is determined by the difference in lipophilicity of two ions.

2. For neutral and charged carriers the selectivity depends on the differences in solvation energies of the two ions of interest as well as the stability constants for complexation with the carrier. This ionophore complexation is the additional factor discussed in Chapters 1 and 2 which makes it possible to obtain non-Hofmeister selectivities for anion sensing.

3. Their selectivity coefficient k_{AB}^{sel} is different from the N–E definition, but may be related to it:

$$k_{AB}^{sel} = a_A^{1-(z_B/z_A)} (k_{AB}^{pot})^{z_B/z_A}$$

4. The required $k_{A,B}^{pot}$ for a relative error of $p_{A,B}$ in assessing the primary ion's activity is given by

$$\underset{\text{Required}}{k_{AB}^{pot}} = \frac{a_A(AB)}{A_B(AB)^{z_A/z_B}} \left(\frac{p_{AB}}{100}\right)^{z_A/z_B}$$

where the AB terms in parentheses refer to the ions in mixed solution. The significance of this result is that even for ions of different charges—e.g., usual concentrations of intracellular potassium and magnesium ions—a required selectivity coefficient can be calculated and the membrane composition adjusted to achieve the desired selectivity.

5. The 1994 work of Bakker *et al.* confirmed the matched potential method of determining selectivity coefficients as the natural approach based on the theory of responses.

Two extremely thorough reports applying the principles of the above treatment are those of Hutchins *et al.* (1997) for characterizing a salicylate sensor based on a guanidinium ionophore and Xiao *et al.* (1997) for their highly selective chloride sensor. Interestingly, in the Hutchins work the authors noted that at low concentrations of salicylate chloride as an interfering ion produced a response too low to measure with the matched potential method. Thus, they had to use the fixed interference method instead.

4.2.2. Solid State Devices

In the preceding discussion it was assumed that the medium on each side of the membrane was a liquid phase, though the reference ion may actually be in a gel for convenience of handling. In this case ionic charge transfer occurs at the membrane interfaces. A totally different situation exists if the sample solution contacts a membrane cast on the surface of a solid contact such as a noble metal wire or carbon surface. These solid contacts once covered with the membrane become the analogy of a coated-wire electrode (CWE). The electrical contact for these situations, membrane–solid rather than membrane–solution, presents some designing challenges in order to obtain performance comparable to other types of ISE's. Because the membrane–solid interface of the CWEs involves transitional control from ionic to electron charge-transfer, there have been performance difficulties with stability and reproducibility.

Janata (1989) has pointed out some of the difficulties with several approaches to this problem—e.g., use of aqueous gels between the solid contact and ion-selective membranes. These layers may be permeated by extraneous species, including water, which change the nature of the layer. Metal–metal salt contacts are a choice, perhaps a very good choice, but there is some difficulty if they are unprotected when used *in vivo*. The use of electropolymerized films for ion sensing is also a choice—e.g., the approach to anion sensing discussed in Chapter 3 (see Edelman and Wang, 1992). Cadogan *et al.* (1992) developed a sodium-selective electrode with $NaBF_4$-doped polypyrrole (PPy) solid contact. For comparison they also used platinum as a CWE with a PVC ion-selective membrane. Their impedance measurements indicated that the PPy layer definitely lowered the charge transfer resistance to facilitate the necessary ionic to electronic conductivity transition. The device with the PPy solid contact exhibited essentially no oxygen sensitivity, in contrast to the Pt/PVC CWE.

Platinum, copper and silver wires, and also carbon, have been used as solid contacts. Carbon-based CWE's sometimes incorporate the ion-exchange material into a membrane mix with carbon paste. For the implantable ion-selective device of Cosofret *et al.* (1995a,b) Ag/AgCl contacts were used with polyHEMA layers between the contacts and the ion-selective membranes. Meyerhoff's group used an Ag–epoxy formulation for their contact with a polyurethane/functionalized polyvinyl copolymer (Yim *et al.*, 1993; Goldberg *et al.*, 1994). The device could be used as a stand-alone ion-selective device, or the polymeric overcoat could be used for immobilizing biocomponents, thus producing a well-behaved biosensor.

In a thorough, but succinct, report on some thick-film ion sensors, Pace and Hamerslag (1992) described the design and characterization of a multi-analyte device for determination of blood electrolytes. A ceramic–metal composite (Ag) was used with an interlayer of carbon in a polymeric matrix. Then the ion-selective PVC membrane was deposited over the interlayer. This arrangement provided a kind of graded transition from ionic to electron charge-transfer with very good results. The sensors were used in a split-flow system with a reference sensor in one line, and a differential signal was measured. The authors gave particular attention in the report to identifying and characterizing the various interfacial potentials which might introduce measurement error. Though they demonstrated the excellent response function of the device, they also noted a 100–300 milliseconds transient which was not accounted for in their model and a slight, probably acceptable, drift of a few millivolts over a period of several hours. The sensing system could be used with as little as 50 μl of undiluted serum.

4.2.3. Field-effect Transistors

When thinking of miniaturization of instrumentation a logical connection is that of semiconductors. For potentiometric measurements the field-effect transistor (FET) has proven to be an effective device in the area of biosensing (Blackburn, 1989; Karube, 1989a; Karube *et al.*, 1995 a,b; Reinhoudt, 1992; Reinhoudt *et al.*, 1994). Where miniaturization is required, these FET-based devices should continue to hold a prominent position. It should be noted, however, that just as there have been problems with solid-contact devices, there have been similar problems with FETs. Adhesion of membranes to the surfaces, bothersome pH sensitivity due to the SiO_2 nature of the surfaces, and drifts have been noted by many groups. Reinhoudt's group (Reinhoudt, 1992; Reinhoudt *et al.*, 1994) solved several problems at once with the use of chemically modified surfaces and polyHEMA interlayers. The SiO_2 surfaces were silanized, then overcoated with a photopolymerized layer of acrylate or siloxanes. The polysiloxanes provided superior performance, even in the absence of the polyHEMA interlayers.

Although it is not the intent of this monograph to provide a substantial background in the physics of semiconductors, the FET's characteristics that make it so adaptable for biosensing should be examined, at least, qualitatively. Any introductory physics or electronics textbook can be consulted for a general background in FET principles and operation. For the essentials of FETs specifically within the context of biosensing, Janata (1989) and Blackburn (1989) may be useful for fundamentals of the devices. The reports by deRooij and van den Vlekkert (1991), Reinhoudt (1992) and Reinhoudt *et al.* (1994) contain details of fabrication procedures and chemical modifications.

Referring to Figures 4-6 and 4-7, the most basic chacteristics of the insulated gate field-effect transistor, IGFET, which make it useful for ion sensing can be summarized. Conduction in intrinsic silicon is negligible at

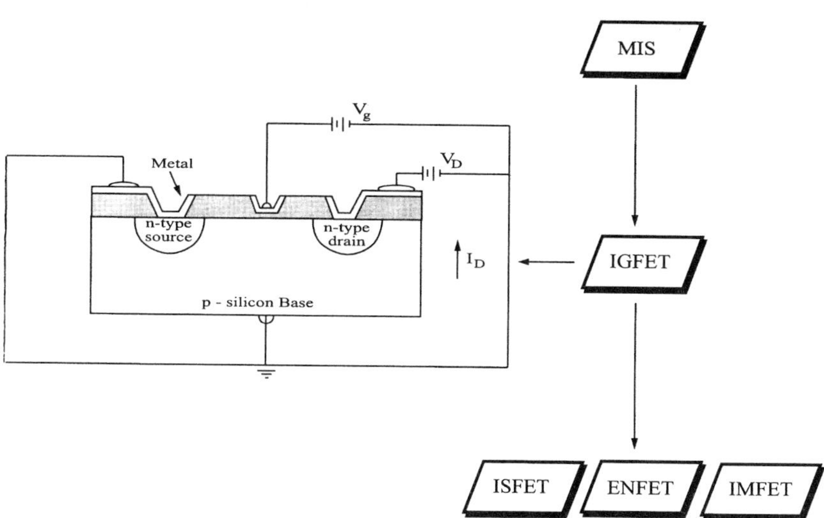

FIG. 4-6. Generic field-effect transistor (FET), showing the conceptual transition to related devices

room temperature. In order to provide for the possibility of electron mobility in the lattice the silicon can be doped with group V elements—e.g., P or As—which have one more electron than silicon, thus producing an n-type Si. Alternatively the silicon can be doped with group III elements—e.g., B or Al—which have one less electron than silicon, thus producing a p-type Si. For illustrative purposes in Figure 4-6 the p-type has been chosen.

For electron conduction to occur in a semiconductor, electrons must move in response to an electric field. For the insulated gate field-effect transistor, IGFET, this can be accomplished by applying a voltage between the metal gate and semiconductor substrate, separated by an insulator. If the positive applied voltage V_g is sufficient to promote a charge separation within the semiconductor lattice, then it is possible to create an electron dense region near the Si surface in contact with the insulator, while an electron deficient region, comprising the "holes" of the p-Si, develops toward the opposite region of the substrate.

Prior to the application of V_g, the region between the n-type Si drain and the source is a non-conducting region which cannot accommodate a current flow. Effectively there is an open circuit between source and drain. As a result of the electron-rich field created at the semiconductor–insulator surface by application of V_g, a conducting circuit path is created for the flow of a current I_D between drain and source. The amount of current I_D which flows is a function of V_g, as long as V_g does not go so high as to saturate the

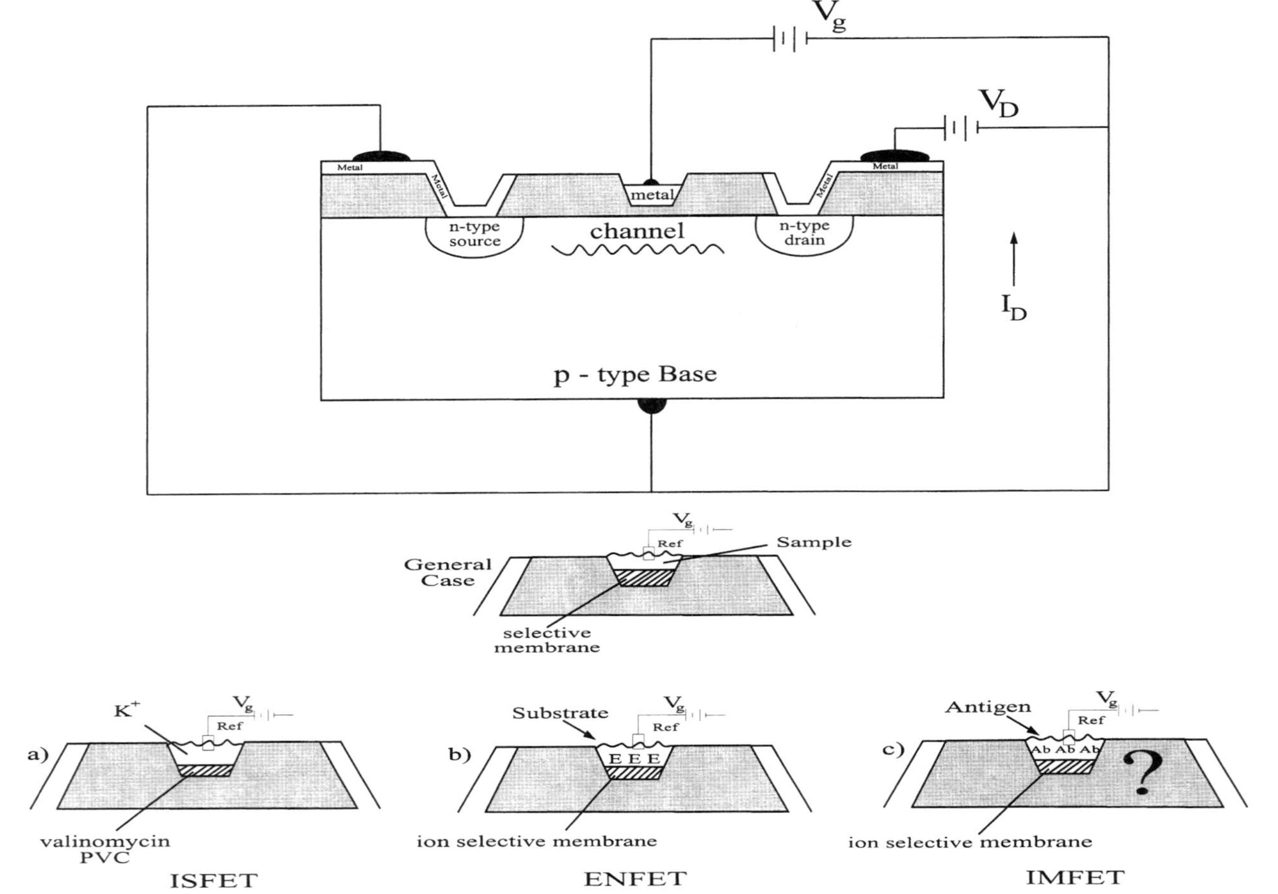

FIG. 4-7. FET families for ion-selective responses arising from inclusion of ionophores, enzymes and antibodies in the selective membrane.

surface field and produce a constant current. Consequently, between a certain threshold voltage for inducing charge separation and a saturation voltage, the IGFET produces a current signal that is dependent on the voltage V_g, applied between gate and substrate.

A very logical transition can be made to a chemically modified IGFET, the CHEMFET. If the metal gate is changed to a chemically selective membrane and the device is put into a solution, as shown in Figure 4-7, essentially the same possibilities exist for response as that observed in the generic IGFET. Now, however, the partitioning of species and chemical reactions in the membrane will modulate the applied voltage V_g, which again determines the magnitude of the drain current I_D. Blackburn (1989) discussed in some detail the two basic approaches for signal processing—i.e., either in a constant voltage, V_g mode or a constant current I_D mode. The constant current mode allows for direct readout of potentials, more akin to the conventional ISE potentiometric response.

With some minor modifications, essentially the same sensing membranes that are useful in other potentiometric devices can be used for the CHEMFET. This assumes the use of modifications and stabilization strategies such as those discussed earlier. This means that most of the ion-selective, enzymatically catalyzed and gas-selective sensing schemes should be adaptable to the modified FETs. Furthermore, there is also the possibility for some very useful miniaturized devices that can be encapsulated for biocompatibility.

Depending on the type of chemistry employed in the membrane the nomenclature extends CHEMFET to ISFET for ion-selective and ENFET for enzymatic molecular recognition (Figure 4-7). Most of the ENFETs actually function as ISFETs with ionic changes arising from enzymatically catalyzed reactions involving ionic species. IMFETs have been less successful because they do not exhibit the same ionic sensitivity. Blackburn (1989) presented interesting calculations showing that although the IMFET is, in theory, another extension of the CHEMFET, the double layer capacitance of the adsorbed molecules in an immunoreaction would require a device with a charge transfer resistance so high as to be impractical, if not unrealizable.

Janata (1989) has summarized much of the early work of his group in designing ENFETs for a variety of analytes, including glucose and penicillin. He provides a good analysis of design in terms of the diffusion–reaction equations (Chapter 2) that pertain. Data are provided demonstrating nonlinear concentration gradients in the membrane due to diffusion–reaction, thus negating some of the assumptions made in earlier treatments of diffusion–reaction for zero-flux sensors. Required thicknesses, based on the calculations for the nonlinear concentration profiles, were 150 μm for the glucose sensor and 20 μm for the penicillin device.

Karube (1989b) and Karube *et al.* (1995a) have described numerous ENFETs, including those using urease and ATPase as the catalysts. Shul'ga *et al.* (1994) introduced a very interesting ENFET for glucose. They added to their GOx membrane the redox mediator potassium ferricyanide, $[Fe(CN)_6]^{3-}$. They were able to demonstrate that at the onset of the sensor response to glucose, the glucose prefers the oxygen as the cosubstrate. When the usual oxygen depletion effect begins to take effect, the mediator shifts in as the cosubstrate of the reaction. The purpose of using the mediator originally was to achieve an amplification of the proton-sensitive signal at the FET surface. The mediator reaction with the reduced enzyme produces 3 protons per glucose molecule instead of 1 generated by the oxygen cosubstrate. The authors' schematic of the diffusion–reaction considerations for this ENFET provides a good lesson in the various factors that must be considered.

4.2.4. Gas Sensors

The classical Severinghaus electrode, introduced in Chapter 2, was one of the first to be developed for the measurement of blood gases, and there have been numerous variations of the original design (Yim *et al.*, 1993; Meyerhoff, 1993; Telting-Diaz *et al.*, 1994). The Telting-Diaz device, illustrated in Figure 4-8, was a dual lumen, implantable catheter for simultaneous determination of pH and CO_2. One compartment contained a CO_2-sensitive bicarbonate solution and the other compartment was pH-sensitive. The sensor exhibited good accuracy and stability for 30–65 h in blood loop experiments. As an *in vivo* test for using the device in non-heparinized animals, the catheter was coated with a biocompatible tridodecylammonium–heparinate complex. The coating, however, had an adverse effect on the pH function, but not the CO_2 function, of the sensor.

The Severinghaus-type sensor is basically an ion-sensing electrode equipped with a gas-permeable membrane on the exterior. Gas passes through the gas-permeable membrane into an aqueous solution where the acid/base equilibria involving the dissolved gases are established. The accompanying pH or anion variations are then detected by an internal ion-selective electrode or ISFET. The cell potential is given by the Nernst equation, with the logarithmic term dependent on the partial pressure of the gas in the sample.

$$E_{\text{cell}} = \text{Constant} + 0.059 \log P_{\text{gas}}$$

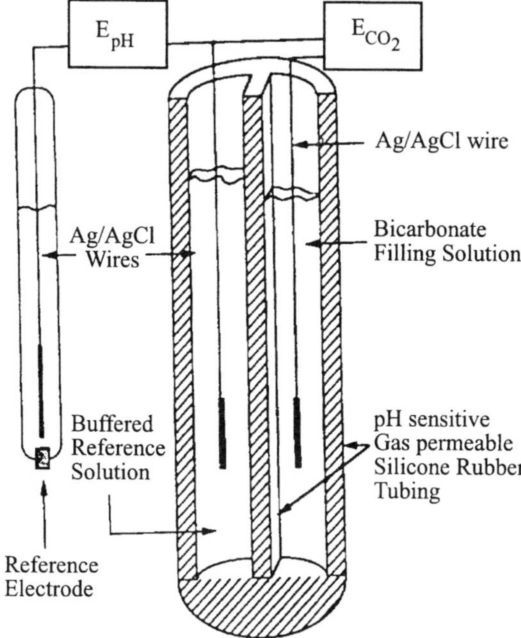

FIG. 4-8. Dual analyte device for measuring intravascular K^+ and CO_2 *in vivo*. Outer membrane is impregnated with ionophores for the potassium-ion selectivity. Outer membrane is also permeable to carbon dioxide which effects a pH variation in the bicarbonate solution. Reprinted from *Biosens. Bioelec.* Vol. **8**, Yim, H-S., Polymer membrane-based ion-, gas- and bio-selective potentiometric sensors, pp 1–38. Copyright 1993, with permission from Elsevier Science Ltd., Oxford, UK.

There are several characteristics of the classical Severinghaus design that have prompted continued research into more effective ways for measuring dissolved gases, particularly for *in vivo* situations. Some problems centered on the use of a glass electrode, which is not practical for miniaturized devices. The use of various ISE designs have eliminated that as a practical problem. The Severinghaus design also has a fundamentally slow response due the multistep sequence necessary for response and recovery. The Telting-Diaz catheter mentioned above exhibited a response time of ~80 s for a silicone tubing wall, but the time increased to ~8 min for a polyurethane wall. One of the most troublesome aspects of many Severinghaus designs is the lack of selectivity. Because the response is based on the ionic changes accompanying the acid/base equilibria of dissolved gases, any species that can diffuse through the gas-permeable membrane and affect H^+ activity can be a significant interferent. Meyerhoff (1993) has discussed progress made in several areas of sensing dissolved gases, including improvements in selectivity by use of some innovative polymer membrane chemistries. His group has been able to improve both selectivity and detection limits used in clinical applications involving blood gases.

4.3. AMPEROMETRIC SENSORS

Although all amperometric sensors involve the measurement of current flow as the primary transduction phenomenon, there are many different sensing applications based on that principle. For example, in direct fuel-cell sensors the current generated spontaneously by microbial action can be detected by simple measurement tools involving the iR drop across a standard resistor. For indirect fuel cells and some enzymatically catalyzed electrodes, however, Clark-type oxygen electrodes or hydrogen peroxide amperometric detectors are operated in the more common mode of measuring current at a preselected applied potential for the working electrode. For electrochemical immunoassay methods (ECIA), enzyme labels on the immunocomponents are probed by amperometric detection of products of enzymatically catalyzed reactions. In contrast to the examples with immobilized biocomponents, some of the *in vivo* sensors for studying neurotransmitters by amperometric methods are minimally modified with nothing more than deposition of permselective membranes on the surfaces. Even though these examples vary widely, they have in common the basic operational principle of amperometric detection.

Three types of electrodes are shown in Figure 4-9. The sketch in (a) is a typical Clark-type electrode, introduced in previous chapters. The biocomponent may be enzyme, cells or tissues, or a mix including cofactors and mediators. The matrix may be some type of biological gel medium, a UV- or photopolymerized layer or a cross-linked macromolecular network. In miniaturized format, this same type of device might have micropatterned electrodes with deposited membranes, as discussed in Chapter 3. The needle-type electrode in (b) represents a possible configuration for an implantable amperometric device. The chemical and energy transduction may be almost identical to that of (a), but in miniaturized form the application is quite different, as is the required biocompatibility. A variation on this type of device was used in Chapter 2 as one of the performance examples (Jung and Wilson, 1996). One type of multiple-electrode device, the interdigitated array (IDA) is illustrated in (c). Other examples of arrays might be the circular array of eight electrodes for wall-jet operation (Niwa *et al.*, 1996) or the probe-array of sputtered-carbon microelectrodes of Sreenivas *et al.* (1996). Although all of these have different geometries from those in (a) and

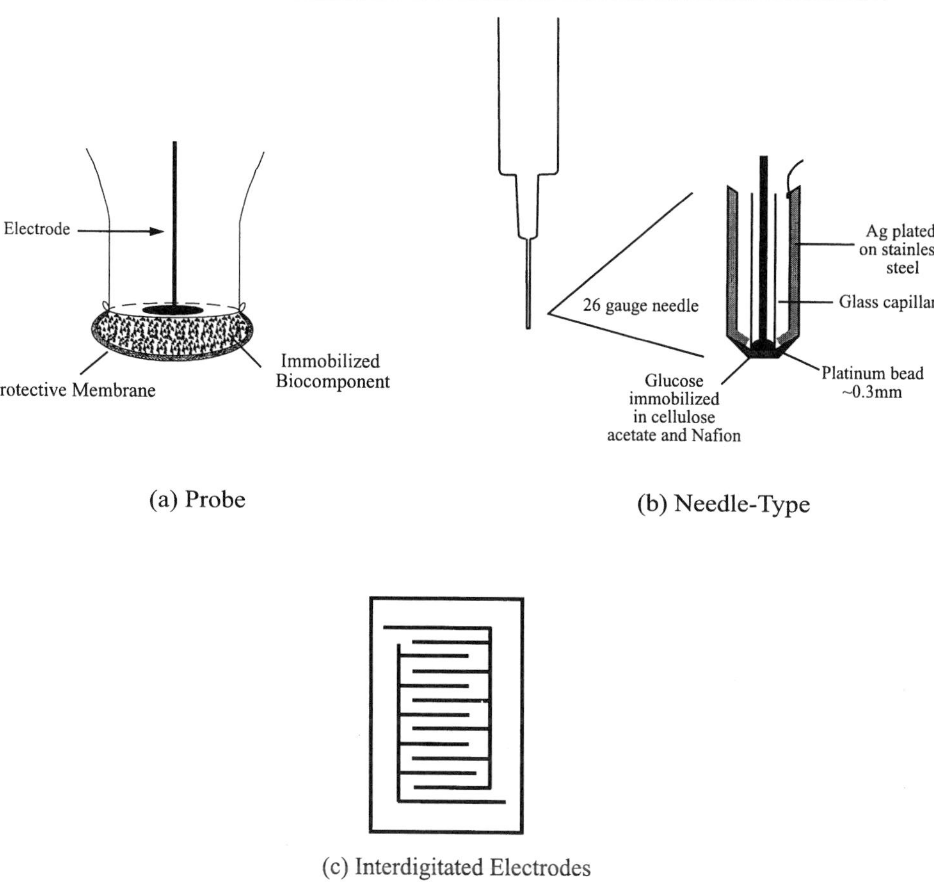

FIG. 4-9. Three amperometric sensor designs: (a) immobilized biocomponent in gel with protective membrane, (b) needle-type electrode, and (c) interdigitated electrodes.

(b), fundamentally the same chemical modifications and transduction mechanisms can be employed.

The carbon-fiber electrode of Figure 4-10 is an ultramicroelectrode (UME) of the kind commonly used for *in vivo* brain studies (Justice, 1987; Wightman and Wipf, 1989; Boulton *et al.*, 1995). The surface may be modified with only a permselective membrane for selectivity (Chapter 3) or it may be treated so that the device becomes an enzyme electrode (Pantano *et al.*, 1991; Dontha *et al.*, 1997). The carbon fiber of Figure 4-10 is shown with the fiber tip flush with the casing to form a disk electrode. It may also be fabricated with a part of the fiber protruding from the casing, thus producing a cylindrical electrode (Wightman and Wipf, 1989). Not only does the diffusion characteristic of these UME's, in general, differ significantly from

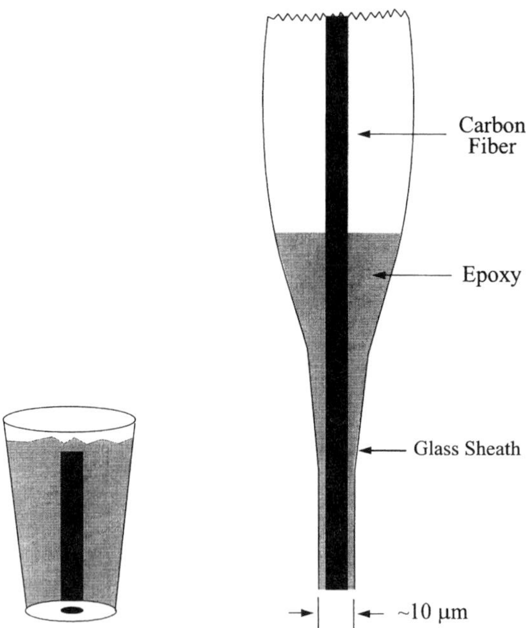

FIG. 4-10. Typical carbon fiber electrode.

that for a macro- or microelectrode, but the theoretical expressions for current for these two UME geometries — disk and cylinder — also differ (Heinze, 1993).

In Chapter 3, several electrode surfaces—Pt, Au and C—were discussed in the context of basic surface behavior, cleaning and pre-treatments. Two important types of electrodes used for amperometric sensors were not included in that discussion—conducting salt electrodes and some of the newer carbon matrices and composites. The chapters by Albery and Craston (1989) and by Bartlett (1990) provide good background in the properties, fabrication methods and applications of conducting salt electrodes. These materials are comprised of crystals of planar organic molecules that have high delocalized π-electron density above and below the plane of the molecules. They consist of a donor–acceptor pair, such as the donor, tetrathiafulvalene (TTF), and acceptor, tetracyanoquinodimethane (TCNQ). Bartlett has summarized the necessary characteristics to make the salt structures, such as TTF–TCNQ, conducting: (1) there must be segregated stacking of molecules—i.e., donors and acceptors segregated—(2) the molecules must form new aromatic sextets upon gain or loss of electrons, thus creating the electron mobility, and (3) there must be charge transfer between the stacks. These conducting salt electrodes have a limited potential range (~ +0.3 to ~ −0.2 V versus SCE), but they have some very favorable properties. For example, the electrodes are quite stable as long as the potential limits at which the materials themselves are oxidized or reduced are not exceeded. Second, flavoproteins are readily adsorbed to the electrodes, thus providing a simple method of preparing many of the oxidoreductase electrodes. Third, although the mechanism has been subject to controversy, conducting salt electrodes represent one of the few types of electrodes for which a direct catalyzed current is observed for the flavoproteins. In most other cases, the process must be mediated by another component.

Several types of what may be termed mixed-matrix carbon electrodes have been utilized. Some of the original electrodes in this category were the carbon-paste electrodes in which graphite was simply mixed with a paraffin oil as a pasting liquid (Kissinger and Heineman, 1996). Various components may be mixed into the paste, including enzymes (Gorton, 1995), enzymes and conducting salts (Pandey and Weetall, 1995b), and metal complexors (Wang, J. and Chen, 1993). There are also several types of metallized catalytic carbons (Wang, J. *et al.*, 1995d; Newman *et al.*, 1995) and graphite–epoxy composites that have been used for determining neurotransmitters (Justice, 1987) and for other biosensing applications, including immunosensors (Alegrets *et al.*, 1996; Santandreu *et al.*, 1997).

From the preceding examples, it may be ascertained that there is no shortage of electrode materials on which different designs of biosensors may be based. Each type of electrode has its own potential window of operation, usually governed by potentials at which either the solvent itself or some component in it is electroactive or the electrode material is oxidized (Table 4-1). The potential limits are often sensitive to the solvent's pH, supporting electrolyte or polarity. The electrode's potential limits define the potential range over which a compound may be investigated, for example by potential scanning techniques. The limits are analogous to wavelength cutoffs of optical materials, though they arise from a totally different phenomenon.

For a typical amperometric experiment, a pre-determined potential relative to a reference is applied as a potential step to a working electrode. The potential is chosen for facile electron transfer to the species of interest and is applied for a time t_{step}, which usually ranges from a few milliseconds to seconds. Upon potential impulse the charge on the electrode changes and there is a transient current comprised of both a non-Faradaic component for charging the solution double layer at the surface and a Faradaic component involved in the electron transfer. The charging current decays rapidly and

TABLE 4-1 Approximate Potential Limits of Common Electrode Materials (volts versus SCE or Ag/AgCl)

Electrode Material	Anodic Limit	Cathodic Limit
Pt	1.3 to 0.8	0– −0.8
Au	1.5	0
C (depends on type)	1.2–1.5	−1
Conducting Salt	0.3	−0.2
SnO_2	1.8	−0.5

for a planar electrode with semi-infinite diffusion of analyte the Faradaic component is given by the familiar Cottrell equation:

$$i_{(t)} = \frac{nFAD_0{}^{1/2}C_0^*}{\pi^{1/2}t^{1/2}}$$

If current is recorded as a function of time for the duration of the potential step, the method is referred to as chronoamperometry. If the current is integrated over that period of time ($\int_o^t i(t)\mathrm{d}t = Q(t)$), it is chronocoulometry. If, instead of monitoring the current for the entire period of applied potential, the current is sampled at some particular time t, the method is usually referred to simply as amperometry or potential-step amperometry. Such sampled currents taken at the same time t for different concentrations of analyte may be seen to be linearly related to concentration. For monitoring dynamic processes, such as neurotransmitter levels, amperometric signals may be acquired in the sampled mode over a period of minutes, or even hours, to ascertain the time-dependent behavior of the electroactive species *in vivo* (Justice, 1987).

For a modified electrode such as an enzyme electrode, the mass transport characteristics and concentration profile for detectable species at the electrode surface are governed by the diffusion–reaction conditions described in Chapter 2. In this case the sensor's current is described by a current function different from that above,in accordance with behavior described in Chapter 2. The current is usually allowed to decay initially to a stable, background level, then analyte is introduced. There is an immediate transient as the enzymatic reaction begins, then the current attains a steady-state value when all the fluxes are balanced in the membrane.

The presumption of a predetermined value of the applied potential was mentioned above. This becomes a primary question for the experiment to be successful. While one might assume that tables of thermodynamic standard potentials could be used for such determination, that is not the case in practice. Oftentimes heterogeneous electron transfer becomes kinetically limited for many of the typical biocomponents or at modified surfaces. This means that a potential greater than the thermodynamic standard potential—i.e., an overpotential—may be needed to effect electron transfer. Furthermore, the overpotential necessary on one type of electrode surface may be different from that for another surface. Consequently, the values for applied potentials are determined experimentally for the specific conditions under which the analysis is to be conducted.

In order to determine the necessary applied potential for the operation of an amperometric sensor, current behavior as a function of scanned potentials is investigated. An example of a result using a cyclic scan of potential at a stationary electrode was shown in Chapter 1 (Figure 1-11) in the general discussion of inherent selectivities. Other types of scanned methods, some in convective modes, may be utilized. In Figure 4-11 the current–potential behavior of a hypothetical stirred solution of glucose and glucose oxidase is shown. The illustration describes both the reduction of the cosubstrate, oxygen and the oxidation of the product hydrogen peroxide. In practice an applied potential corresponding to one or the other would be used. The Clark-type amperometric electrode with an applied potential of about −0.6 V versus SCE monitors oxygen depletion in samples and is widely used for a variety of sensing problems. Most of the glucose electrodes are designed to monitor the oxidation of hydrogen peroxide at about +0.6 V versus SCE because the linear concentration dependence is affected adversely by the depletion of oxygen in samples. The two options are schematized in Figures 4-12 and 4-13, with projected calibration curves and notes concerning experimental considerations.

Cyclic voltammetry is often used for characterizing the oxidation–reduction behavior of samples. One of the advantages of this technique is that effects of homogeneous side-reactions can be investigated and rates of heterogeneous electron transfers may be determined if that information is helpful in the development of a sensor. The general approach is shown in Figure 4-14. The example chosen is more complicated than the simple two-component mixture that was illustrated simply as a qualitative method in Figure 1-11.

A linearly variant pattern of potential versus time is applied to a stationary electrode. *It should be noted that both the potential scale and current scale have been inverted in this example.* This has been done purposely to emphasize the fact that various authors utilize different conventions for displaying their results. The same is true even for commercially available simulation software (see Speiser, 1996). The reason for this has historical roots in the early development of electroanalytical methods using mercury electrodes at

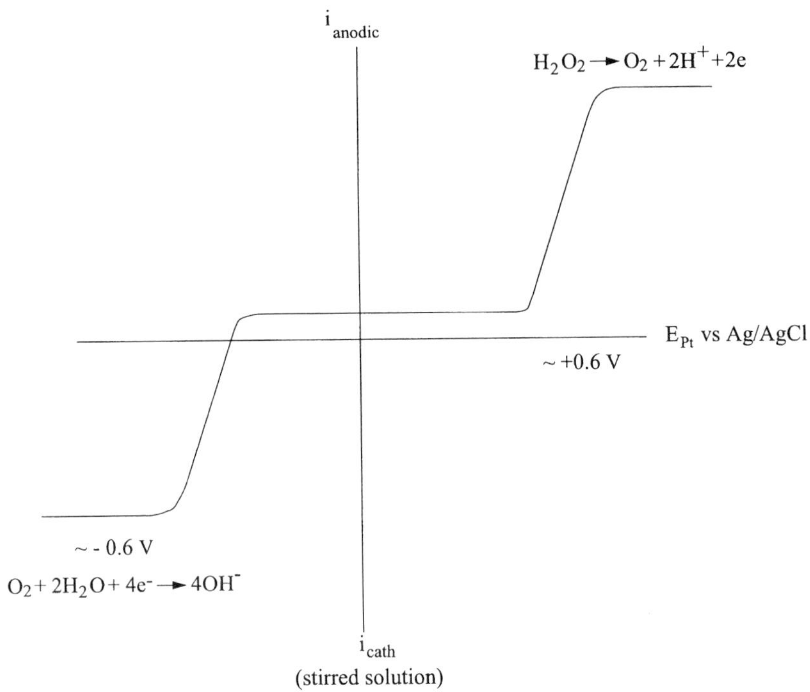

$$Glu + O_2 \xrightarrow[\text{Oxidase}]{\text{Glucose}} H_2O_2 + \text{Gluconolactone}$$

Gluconolactone ⇅ (hydrolysis) Gluconic Acid

<u>Sensing Options:</u>

- Follow disappearance of O_2

- Follow appearance of H_2O_2
  - Electrochem → $H_2O_2 \rightarrow O_2 + 2H^+ + 2e$
  - Optical → $H_2O_2 + \text{Lum} \xrightarrow{\text{HRP}} \text{Lum}^* + h\upsilon$

- Follow pH changes accompanying the Gluconic Acid equilibrium

FIG. 4-11. Schematic of limiting currents: stirred solution for oxygen reduction and hydrogen peroxide oxidation related to the enzymatically catalyzed oxidation of glucose. The potential is scanned in time. Alternate detection schemes are also summarized.

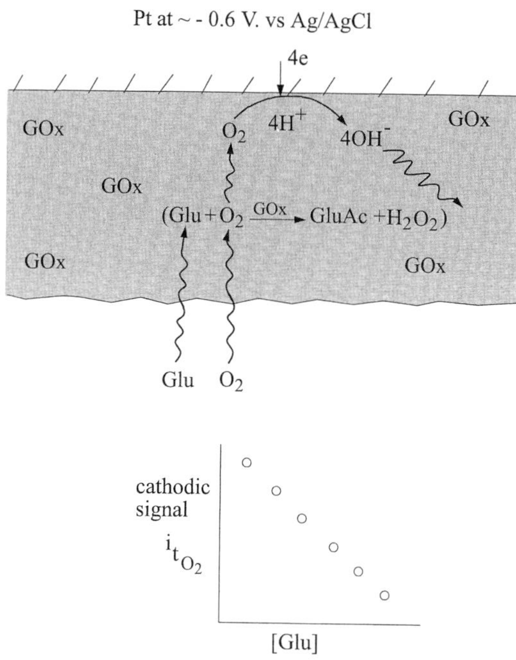

FIG. 4-12. Linear portion of amperometric calibration curve for glucose with oxygen the electroactive species utilized for detection. Three problems can arise. (1) When monitoring O_2 the determination becomes very dependent on the partial pressure of O_2 in the solution. O_2 is depleted from small volumes or low reserve. (2) Reactions show pH effects that can be even more significant in localized environments. (3) H_2O_2 tends to inhibit GOx, so catalase is sometimes added.

negative potentials versus SCE and cathodic currents printed as positive displays.

The process illustrated in Figure 4-14 is one in which the species X is not in the original solution, but is produced in a homogeneous reaction of the first anodic peak product. As the potential is scanned back toward 0 versus SCE not only does the cathodic peak for A^{2+} resident near the electrode appear, but also the new peak appears for the reduction of newly produced X. While this particular mechanism may not be a general one, it is not uncommon to encounter follow-up chemical reactions coupled to electron transfer processes. This is especially true of many of the compounds encountered in bioanalytical problems, as will become apparent is some later sections.

There are several points to be made about current–potential curves like those of Figure 4-14. It should be obvious that, unlike the stirred solution curves of Figure 4-11, peaks rather than plateaus are observed. This arises

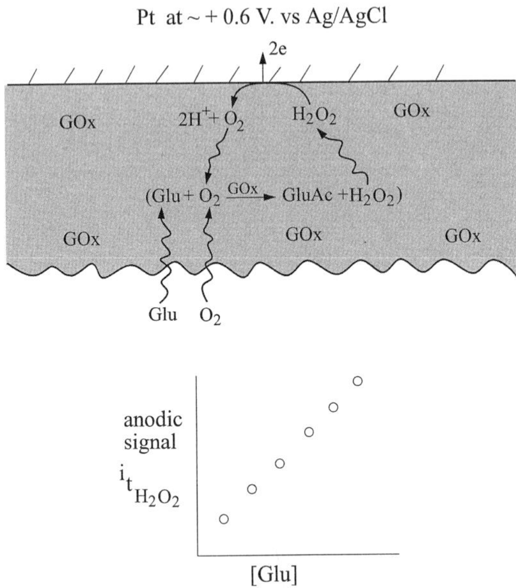

FIG. 4-13. Linear portion of amperometric calibration curve for glucose with hydrogen peroxide the electroactive species utilized for detection.

from the fact that in quiescent solution there is an exponential fall-off of the analyte's concentration at the electrode surface once the peak voltage for the microelectrolysis has been reached. For a general reaction $0 + ne \rightleftarrows R$

$$\frac{C_{ox}}{C_{Red}} = f(t) = \exp\left[\frac{nF}{RT}(E_i - \nu t - E^o\right]$$

Here, ν is the scan rate. Also, without the complication of the follow-up reaction in Figure 4-14, one would expect essentially the same amount of current for reduction of A^{2+} as that observed for oxidation of A, assuming that the scan rate is not too slow to allow A^{2+} to diffuse away from the electrode. Thus, one expects $i_{pa}/i_{pc} = 1$ for an uncomplicated case. For a stationary, planar electrode of area A in unstirred solution the basic peak current response is (Bard and Faulkner, 1980)

$$i_p = (2.69 \times 10^5)n^{3/2}AD_0{}^{1/2}\nu^{1/2}C_0^*$$

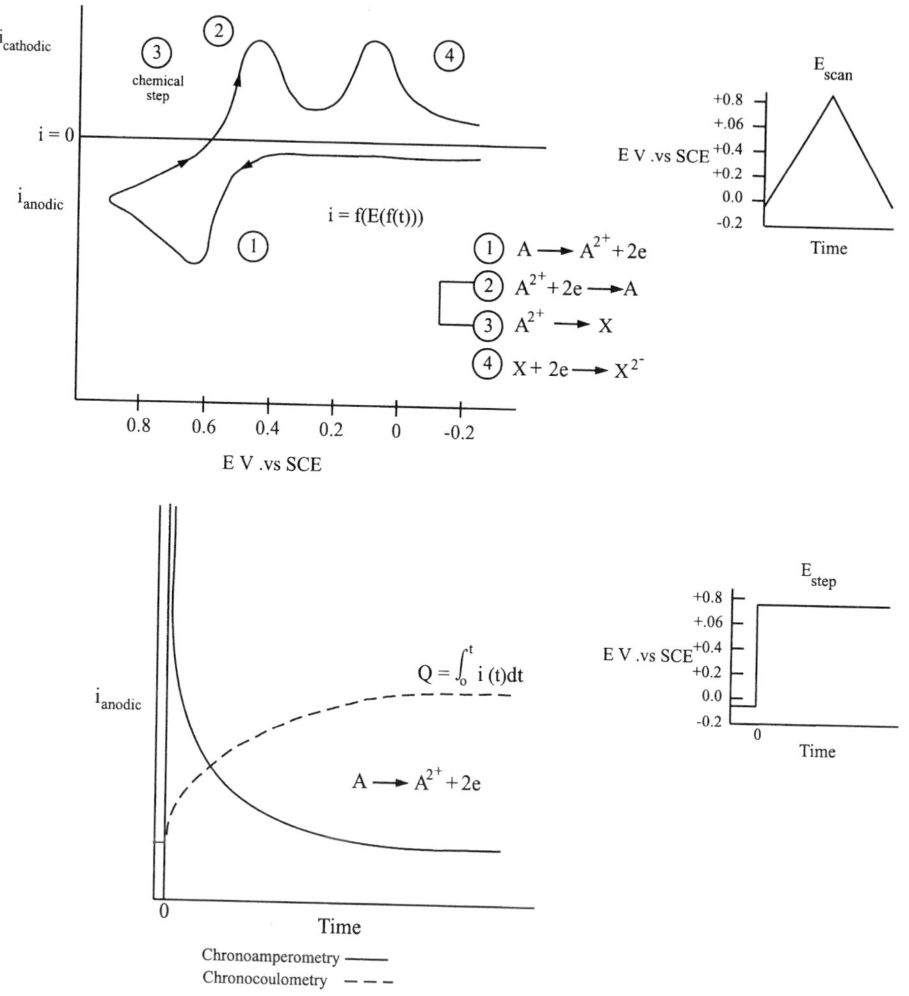

FIG. 4-14. (Top) Cyclic voltammetric curve. This example is for electron transfer reactions coupled with a homogeneous chemical reaction. The applied potential input pattern is shown top right. (Bottom) Basic chronoamperometric and coulometric outputs expected for a potential step input (shown right).

for $0 + ne \rightleftarrows R$. A is the area of the electrode, D is the diffusion coefficient of the electroactive species, ν is scan rate and C_0^* is the bulk concentration of O.

There are established diagnostic criteria using these relationships for determining the mechanism of coupled heterogeneous and homogeneous processes (Bard and Faulkner, 1980; Kissinger and Heineman, 1996). Most of the criteria involve comparing cathodic and anodic peak currents, the shifts of peaks, and peak separations as a function of scan rate. Much of the tedium of that process has been removed by the (belated) commercial availability of simulation software for that purpose (Speiser, 1996). An analyst developing a biosensor may very well have to invoke various diagnostics for elucidating the mechanism of a sensor's response function and optimizing the design. The end-user will be unaware of the extent of that exercise if a properly operating device is the result.

The current–potential behavior of ultramicroelectrodes (UMEs) is very different from that of macro- or even microelectrodes (Heinze, 1993; Wightman and Wipf, 1989). Several types of miniaturized electrode designs are illustrated in Figure 4-15, and the Pt UME of Heinze (1993) and Tschuncky and Heinze (1995) is shown in Figure 4-16. This latter device is included for two reasons. It shows some of the intricacies involved in the construction of an electrode other than the carbon fibers used for illustration thus far. Second, the particular design was proposed by the authors as

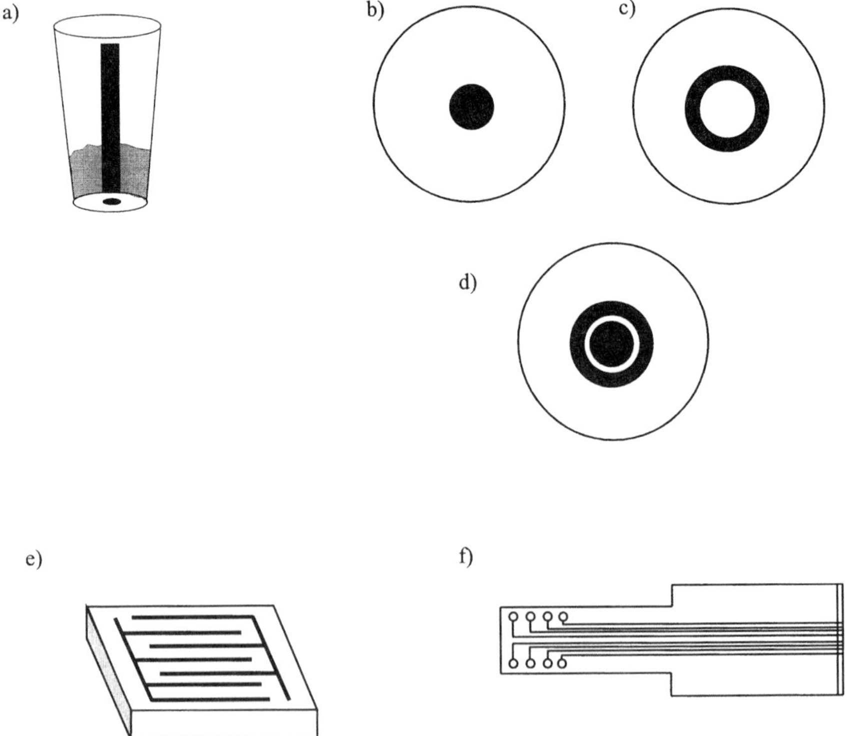

FIG. 4-15. Various configurations of ultramicroelectrodes: (a) carbon fiber; (b) disk, (c) ring; (d) ring–disk; (e) interdigitated electrodes; and (f) stiletto design for array on tip.

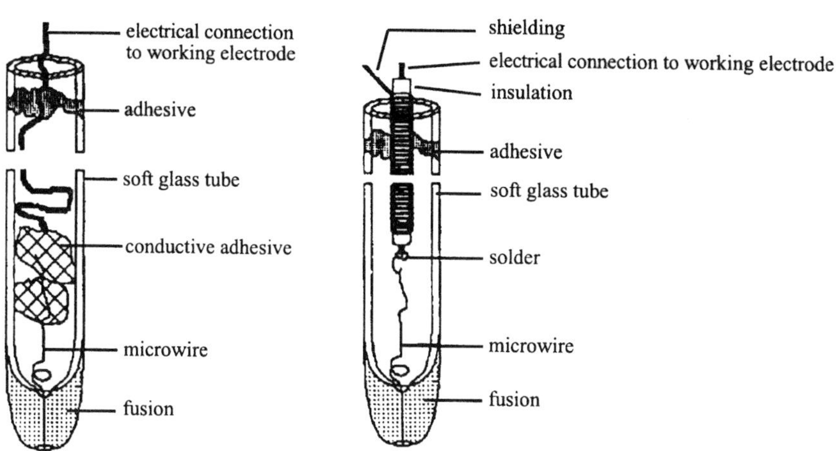

FIG. 4-16. Improved design reported for UME construction. The connections and shielding have been improved to lower capacitive effects by a reported factor of 5. This allows the study of fast heterogeneous rate constants using fast-scan voltammetry, e.g. 20,000–120,000 V s^{-1}. Reprinted with permission from Heinze, J. (1993) *Angew. Chem. Int. Ed. Engl.* **32**: 1268–1288. Copyright 1993, Wiley-VCH, Weinheim, Germany. For further improvements using a slightly different approach, see Tschunky and Heinze (1995).

an improved UME shielded for the purpose of reducing capacitive currents at high scan rates. The Tschuncky and Heinze (1995) report includes several constructional approaches for lowering stray capacitance associated with a working UME. Some of the designs require alteration of instrumentation for electronic implementation of subtractive methods.

The basis for the unique behavior of UMEs is summarized in Figures 4-17 through 4-20. Because the electrode is so small relative to the surrounding solution, radial rather than planar diffusion occurs. This has the effect of an increased supply of analyte to the sensor. Steady-state rather than peaked currents are observed (Figure 4-19). The expressions for current must be adjusted accordingly, as shown. It should be noted for the diffusional pattern of Figure 4-17(b) that for very short times the diffusional field is similar to the planar case. In Figure 4-18 another difference between macroelectrodes and UMEs is shown—i.e., the different decay times of capacitive currents.

Figures 4-19 and 4-20 describe the different current patterns when potential scans are used. In Figure 4-20 it may be seen that very fast scan rates are the equivalent of "short" times, as mentioned above, and the effects of planar diffusion begin to be manifested in the high scan rate curves. Peaked curves are observed for very fast scan rates. The scan rates necessary

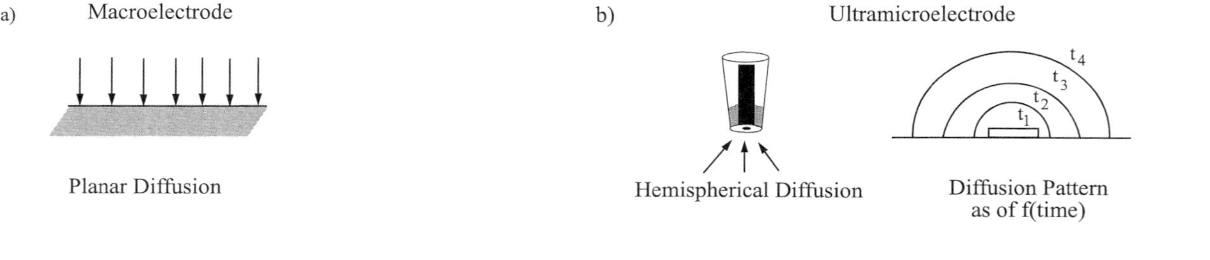

c)

Current-Time Curve:

$$\frac{it^{1/2}}{nFAD^{1/2}C^*} = \frac{1}{\pi^{1/2}}$$

$$\left(1 + \pi^{1/2}\left(\frac{Dt}{r_o^2}\right)^{1/2}\right)$$

Correction Term

Current-Time Curve:

$$\frac{it^{1/2}}{nFAD^{1/2}C^*} = \frac{1}{\pi^{1/2}}\left(1 + \pi^{1/2}\left(\frac{Dt}{r_o^2}\right)^{1/2}\right)_{spherical}$$

(d) Limiting Current: $i_{lim} = 2nFC^*Dd$

hemisphere	sphere	disk
$d = \pi r_o$	$d = 2\pi r_o$	$d = 2r_o$
↓	↓	↓
$i_{hem} = 2\pi r_o\, nFC^*D$	$i_{sph} = 4\pi r_o\, nFC^*D$	$i_{disk} = 4r_o nFC^*D$

FIG. 4-17. Current–time behavior for (a) planar diffusion and (b) radial diffusion: *i*, current; *D*, diffusion coefficient; *A*, area of electrode; *t*, time; *F*, Faraday constant; r_0, radius of the UME; and *d*, electrode surface diameter.

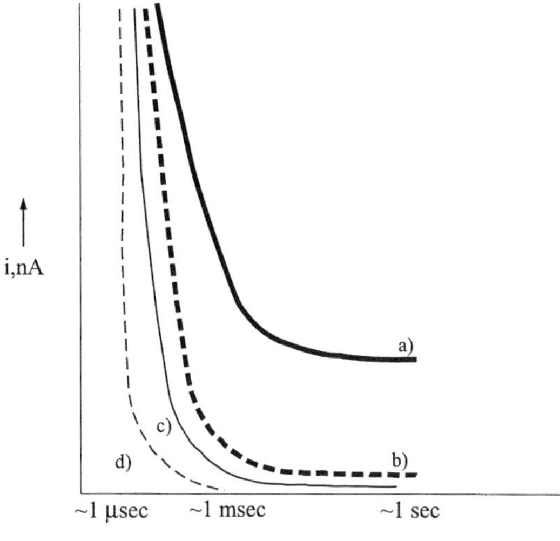

FIG. 4-18. Comparison of current–time behavior for macroelectrode and UME. Note that the capacitive current falls more rapidly for the UME. The relative currents are not to scale. The UME would produce nanoamps to picoamps of current whereas the current at the macroelectrode might be on the order of microamps. Curve (a) is a Faradaic current with planar diffusion, $i_{f,p}$; curve (b) is a capacitance current with planar diffusion, $i_{c,p}$; curve (c) is a Faradaic current with spherical diffusion, i_{fs}; and curve (d) is a capacitance current with spherical diffusion, $i_{c,s}$. Note that (1) at $t = 0$ the current is theoretically ∞, but limited by the electronics, and (2) i_{cap} falls much faster for UMEs.

for observing this effect are much higher than ordinary ones of conventional CV, but they are used for monitoring fast processes *in vivo* (Wightman and Wipf, 1989; Pihel *et al.*, 1996).

One of the advantages of ultramicroelectrodes lies in the fact that very small currents passed by ultramicroelectrodes—nano- or picoamperes range—make the *iR* drop through a solution a lesser problem than with macroelectrodes, thus allowing two-electrode configurations routinely. This makes the entire process of miniaturization much simpler and less expensive than it would be otherwise when designing bioanalytical sensors. A second advantage is the lower capacitive effects at UMEs, as shown in Figure 4-17. This provides an electrode system with a reduced RC time constant, and thus much faster scan rates can be used in potentiodynamic experiments. For measurements in dynamic biological systems this opens up the possibility of studying many of the more interesting natural processes that occur with high rate constants.

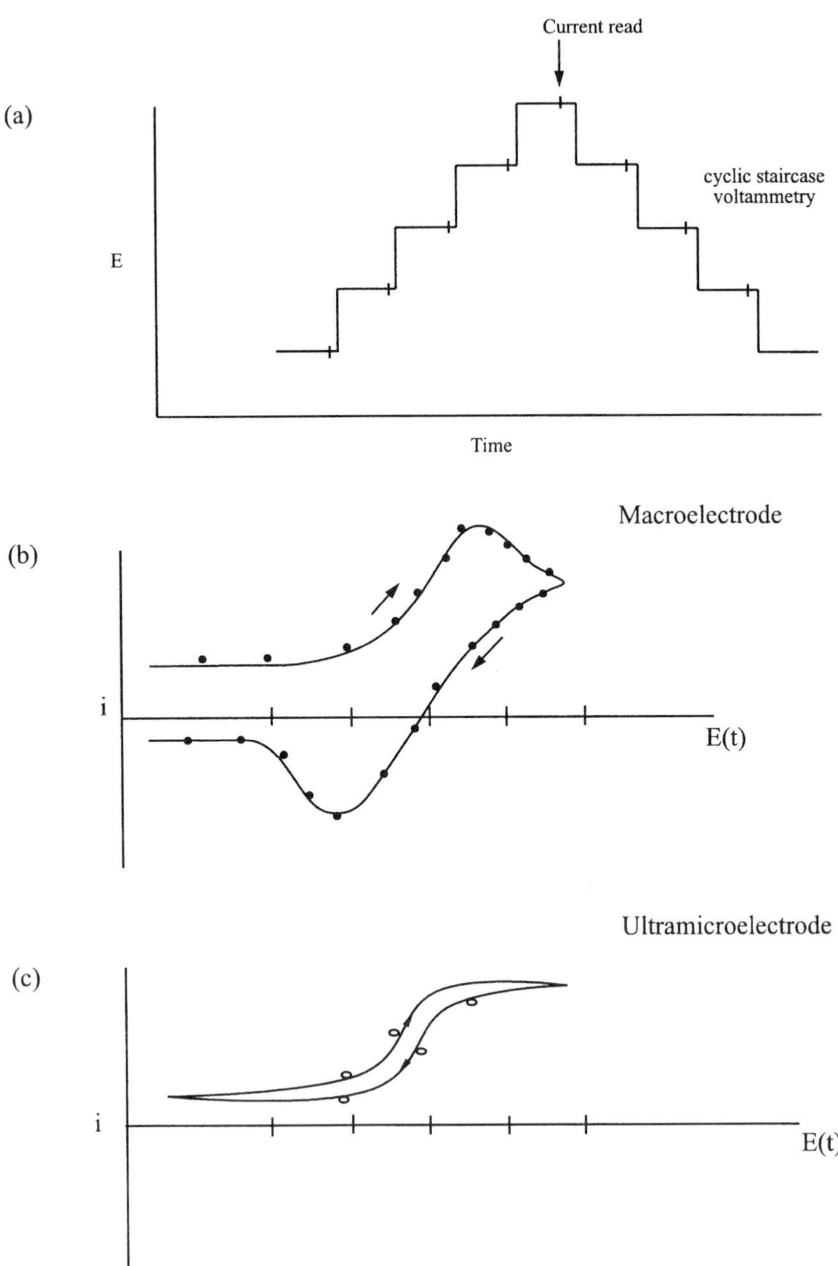

FIG. 4-19. (Top) Cyclic staircase potential input as a function of time, (center) general pattern for the current–potential behavior of a macroelectrode compared to (bottom) that for a UME under the same conditions of potential input.

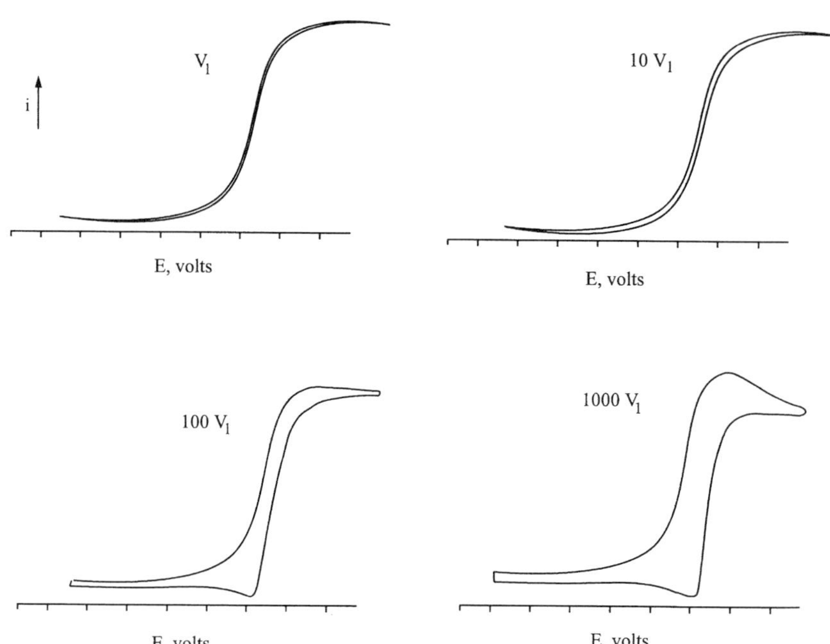

FIG. 4-20. Influence of the time–radius ratio on current–potential patterns for varying scan rates. For fast scan rates (short times) the radial diffusion term is minimized. See Heinze, 1993.

Some of the designs used for illustrative purposes involve more than one electrode. This additional factor presents new possibilities not only for arrays but also for investigating the coupled heterogeneous and homogeneous reactions mentioned earlier. Some of the first uses of dual working electrodes were introduced in rotating ring–disk electrode combinations (Bard and Faulkner, 1980). In those cases product generated at one electrode, for example the disk, was examined at the second working electrode, the ring, to determine the influence of intervening homogeneous reactions in the sample medium. Alternatively, two products, one at each working electrode, can be generated and the resulting reaction of those products studied. This was the experimental design for some of the first studies of electrogenerated chemiluminescence (ECL) (Faulkner and Bard, 1977). Radical cations generated at a electrode reacted with the radical anions generated at second electrode. The homogeneous reaction of the products produced an excited state species which would then decay to the ground state, emitting light characteristic of the energy difference between states. The basic principles of those early experiments are still utilized in some bioanalytical sensors based on the ECL phenomenon. In Chapter 5 examples of ECL-based biosensing systems will be discussed.

Another case for using dual electrodes is for the purpose of improving signal/noise by compensating for effects of matrix interferents which affect the current signal of amperometric sensors. Basically, one fabricates an enzyme electrode using in the reaction layer of the primary working electrode an enzyme that is active toward the analyte of interest. Then a secondary working electrode, fabricated with an identical surface modification minus active enzyme, is added into the circuit. The instrumental output is then taken as a differential reading of current between the two electrodes.

Other applications for multiple working electrodes include amperometric monitoring with small electrodes for spatial resolution, arrays of electrodes and interdigitated electrodes (IDEs). Theoretically the instrumental problem of acquiring simultaneously several current outputs without signal interference is the same in all cases. With microarrays and IDEs, however, there are some very stringent design criteria to avoid difficulties due to capacitive effects and "crosstalk" between closely spaced microelectrodes of an array.

As discussed in Chapter 2, models and simulations of commonly encountered diffusion–reaction mechanisms have been presented (Bartlett and Pratt, 1993; Speiser, 1996). Scheller and Schubert (1992), Eddowes (1990) and Albery and Craston (1989) present some of the applicable cases for amperometric sensors using many of the electrodes discussed herein. The Scheller and Schubert reference includes step-by-step details for modeling some of the more complex chemical and mathematical sequences. Ultramicroelectrode responses and applications have been presented in detail in the Wightman and Wipf (1989) chapter. Hall's group has studied the basic parameters for designing sensors using oxidases as enzymatic components. Their reports include extensive derivations of diffusion–reaction equations for amperometric devices (Martens and Hall, 1994; Martens *et al.*, 1995; Gooding and Hall, 1996). Throughout this and the preceding chapters, numerous amperometric sensors have been used as examples in the context of various topics. Rather than attempting to describe in this section the multitude of amperometric sensors that have been developed over the years, a summary of some more recent ones using state-of-the-art designs is provided in Table 4-2. The original reports should be consulted for details. It should, however, be instructive to consider in a more integrated manner some of the general bioanalytical problems that have been confronted using different types of sensors. The remainder of this section addresses some of these topics.

TABLE 4-2 Selected Recent Amperometric Applications

Reference	Distinguishing Features	Analytes
Cosnier *et al.*, 1997	Glutamate oxidase, polyphenol oxidase immobilized electrochemically using amphiphilic polymer; promotes immobilization of negatively charged enzymes	Glutamate, dopamine
Guo and Dong, 1997	Organic phase enzyme electrode, organohydrogel, works with variety of sensing media and analyte solubilities; tested with HRP, PPO, BrbOx	Peroxides, phenols, bilirubin
Bu *et al.*, 1995	Redox gel, ferrocene-containing, polyacrylamide layer with enzyme; one step polymerization using ternary catalyst	Glucose
Willner *et al.*, 1996	Electrically-wired GOx on Au–SAM; reconstituted holoenzyme on surface; highest current density reported to date	Glucose
Kenausis *et al.*, 1997	Electrically wired thermostable HRP using Os3/2 mediation in redox gel	Glucose, lactate
Du *et al.*, 1996	Cyanometalate modified Ni electrode reported to effect direct electron transfer to GOx without mediators	Glucose
Bauer *et al.*, 1996	Sophisticated enzyme-recycling scheme with zeptomolar LOD for alkaline phosphatase. Used in ECIA for 2,4-D	Alkaline phosphatase, 2,4-D
Lvovich and Scheeline, 1997	Dual sensor for simultaneous monitoring of superoxide ion and hydrogen peroxide; tungsten reference; tested with xanthine system.	Superoxide and hydrogen peroxide
Invitski and Rishpon, 1996	Disposable, one-step immunosensor using enzyme channeling design	hLH
Duan and Meyerhoff, 1994	Au–SAM mesh electrode for ECIA, separation-free sandwich assay; Back-fed analyte	hCG
Filho *et al.*, 1996	TCNQ mediator with GOx and invertase in CPE for FIA; inexpensive, works with high concentrtion of sucrose without dilution	Sucrose

4.3.1. Mediators

There has been repeated reference to the fact that direct electron transfer from an electrode to oxidoreductases is frequently a major problem. The cyanometalate modified electrode reported by Du, *et al.* (1996) is a definite exception in that regard. The authors provide extensive evidence for the direct transfer. For other types of electrodes, however, the difficulty may be due to the location of the electroactive center deep within an enzyme's structure or it may be due to electrode surface effects related to adsorption and orientation of the macromolecule. In Chapter 3 several chemical approaches to effecting electron transfer in these cases were mentioned—e.g., engineering molecules for more facile electron transfer, immobilizing mediating molecules with enzymes, coating surfaces with mediators so they will become catalytic for the electron transfers, or covalently binding mediators to enzymes or "wiring enzymes" to electrodes through mediating molecules (Turner *et al.*, 1989; Cass, 1990; Heller, 1990, 1992; McLean *et al.*, 1993; Blonder *et al*, 1996; Schmidt and Schuhmann, 1996; Willner *et al*, 1996).

A mediator also becomes important in attempts to eliminate oxygen-depletion effects on signals of oxidoreductase sensors by becoming a surrogate cosubstrate for oxidases. Mediators are often included in designs of microbial cell sensors. Though the reality of using mediators is often complex, the fundamental basis of mediated electron transfer is a simple thermodynamic principle:

$M_{ox} + Enz_{red} \rightleftharpoons M_{red} + Enz_{ox}$	ΔE^0, solution or membrane reaction
$M_{ox} + ne \rightleftharpoons M_{red}$	E^0_M facile heterogeneous process
net result $Enz_{red} \rightleftharpoons Enz_{ox} + ne$	$-E^0_{enz}$, kinetically hindered electrode process
Enzyme has been oxidized indirectly	$\Delta E^0 = E^0_M - E^0_{Enz}$

Even though the biological reaction of interest may be thermodynamically favorable, the kinetically controlled heterogeneous electron transfer requires an inaccessible overpotential. A properly chosen mediator will simultaneously come to equilibrium with both the biocomponent and the electrode. The electron transfer will occur at the mediator's potential without the requirement of a high overvoltage for the biocomponent. If a species is to be oxidized by a mediator, then the mediator redox couple must have a more positive E° than that of the species. Once this concept is grasped, various applications seem to have more in common.

In a fuel cell designed for energy production rather than sensing, the difference between the E° of the mediator and that of the biocomponent may be chosen to be large so that the energy derived from the cell is as large as possible. In the case of mediated enzyme sensors, E° will be chosen to be close to the biocomponent's E° to avoid introducing side-reactions with other redox components that may be present. Additionally the mediator must, of course, exhibit fast heterogeneous electron transfer kinetics.

Several of the common mediators employed are shown in Table 4-3. Various chapters in Turner *et al.* (1989) and Cass (1990) deal with mediated enzyme sensors, including many based on microbial preparations. Gorton *et al.* (1992) provide not only an excellent history of NADH chemistry, but also extensive information on various mediators that have been used for designing electrocatlytic surfaces for NADH oxidation.

Some of the definitive work on the use and efficiencies of mediators for oxidase electrodes has been from Hall's group (Martens and Hall, 1994; Martens *et al.*, 1995).

TABLE 4-3 Approximate Formal Potentials for Some Common Mediators[a]

Mediator	Approximate $E^{0'}$ (pH 7) V. versus NHE
$Ru(bpy)_3^{3+/2+}$	1.27
$Os(bpy)_3^{3+/2+}$	0.84
1,1-dicarboxylic acid ferrocene	0.64
$Ferrocene^{+/0}$	0.44
$Fe(CN)_6^{3-/4-}$	0.36
1,4-benzoquinone/hydroquinone	0.28
N,N,N′,N′-TMPD	0.27
2,6-dichloroindophenol	0.22
1,2-naphthoquinone/hydroquinone	0.14
Phenazine methosulfate	0.08
Thionine	0.06
Methylene Blue	0.01
Phenosafranine	−0.25
Benzylviologen	−0.36
Methyl Viologen	−0.44

[a] Compiled from various sources: Bard *et al.* (1994) and references therein; Bartlett *et al.* (1991) and Eggins (1996).

Consider the following problem, using the glucose reaction as an example.

$$\text{Gluc} + O_2 \xrightarrow{\text{GOx}} H_2O_2 + \text{Gluconolactone} \qquad \text{overall}$$

(1) $\text{Gluc} + \text{GOx} \longrightarrow \text{GOx}_{red} + \text{Glucono}-$

(2) $\text{GOx}_{red} + O_2 \longrightarrow \text{GOx} + H_2O_2$

The linearity of the calibration curves for glucose concentrations is affected adversely by the depletion of oxygen in the samples. One very widespread solution has been to invoke the mediated approach where the mediator becomes, supposedly, the primary oxidant, thus removing oxygen dependence for the sensor's function.

$$\text{Gluc} + \text{GOx} \rightarrow \text{GOx}_{red} + \text{Glucono}-$$
$$\text{GOx}_{red} + M_{ox} \rightarrow \text{GOx} + M_{red}$$

In the real-world a sensor designed with a mediator will still be used in samples that contain oxygen. Thus, the enzymatic reaction now operates in the presence of two oxidants. The question becomes one of the efficiency of the mediation coupled with the effectiveness in removing oxygen dependence.

Using diffusion–reaction models with extensive analysis of their models, Hall's group was able to demonstrate that oxygen interference for a sensor designed for mediated electron transfer was particularly significant at low substrate levels below $K_{m.}$ Further, they showed that the rate of reaction for the mediator–enzyme reaction must be much higher than that of the oxygen–enzyme reaction for the mediation to be effective for the performance of the sensor. Their conclusion was that the rapid diffusion characteristics of oxygen make it very difficult for a mediator to compete. It should be recalled that Shul'ga *et al.* (1994) detected this same effect in their FET-based glucose sensor designed for ferricyanide mediation.

4.3.2. Oxygen Species

Dissolved oxygen and hydrogen peroxide have been used repeatedly as examples of analytes for which biosensing devices have been developed. In fact, the whole of oxygen electrochemistry is important in developing bioanalytical methods (Mabrouk, 1996; Postlethwaite *et al.*, 1996; Iwuoha *et al.*,

1997). The oxidation-reduction sequences involving oxygen species are numerous:

	E° v. versus NHE
$O_2 + 2H_2O + 4e \rightarrow 4OH^-$	+0.401
$O_2 + 2H^+ + 2e \rightarrow H_2O_2$	+0.695
$H_2O_2 + 2H^+ + 2e \rightarrow 2H_2O$	+1.763
$4H^+ + 2O_2^{\cdot-} \rightarrow O_2 + 2H_2O$	Disproportionation

O_2 and H_2O_2 have received major attention as analytes, but H_2O_2 and the reactive oxygen intermediates resulting from the incomplete reduction of oxygen are among the most toxic substances that can accumulate in tissues (Mathews and van Holde, 1990; Lehninger *et al.*, 1993). Mutagenic alterations of DNA have been attributed to oxidative damage by reactive oxygen species. Lvovich and Scheeline (1997) have introduced a dual-analyte sensor for superoxide and hydrogen peroxide. The two-channel device has a built-in WO_3 reference electrode. The H_2O_2 sensing unit is comprised of a glassy carbon electrode with an electrodeposited horseradish peroxidase/PPy layer. For the second electrode the same material was used with an overlayer of superoxide dismutase (SOD) for detecting superoxide.

$$H_2O_2 + 2H^+ + 2e \xrightarrow{\text{HRP}} 2H_2O$$

$$2O_2^{\cdot-} + 2H^+ \xrightarrow{\text{SOD}} O_2 + H_2O_2$$

Both sensors demonstrated linear behavior in the nM to low μM range and the response was sufficient for one second simultaneous determinations of $O_2^{\cdot-}$ and H_2O_2. The sensor was tested under the stringent conditions of the sequential xanthine/xanthine oxidase reactions, which involve numerous steps. The model developed for the sensor's behavior corresponded to experiment within 10%, which is really quite remarkable considering the complexity of the reaction sequence.

4.3.3. NADH

The saga of NADH oxidation has been ongoing for almost half a century. It should be obvious to the reader at this point that far more has been done with amperometric enzyme sensors based on oxidase catalysis than any other type. Part of the reason for this has been problems with other enzymes for accomplishing the same task. In Chapter 2 the other types of enzymes for oxidizing glucose (e.g., NAD^+-dependent dehydrogenases, DH) were introduced as alternatives to oxygen-involved sensing with oxidase electrodes.

$$\text{Glucose} + \text{NAD}^+ \xrightarrow{\text{GDH}} \text{Gluconolactone} + \text{NADH}$$

The use of dehydrogenases is of special interest because there are hundreds of NAD^+-dependent dehydrogenases which catalyze a huge number of biologically important reactions.

Several problems arise in the effort to utilize NAD^+-dependent dehydrogenase electrodes. The soluble coenzyme must be supplied to the sensor as an external reagent, a requirement which is not desirable for mass-produced devices and ease of use. Furthermore, the coenzymes are not inexpensive. If they are used for any application of enzymatic catalysis it would be good to be able to recycle them.

In principle the NADH produced in the DH reactions could be recycled in a sensor if reoxidation of the molecule could be accomplished at accessible potentials. The problem with this is not the principle, but the reality. The electrochemical reduction of NAD^+ and oxidation of NADH, which require rather high overvoltages, produce some intermediates that halt the process through rapid formation of dimers and accumulation of products at the electrode surface.

$$\text{NADH} \xrightarrow[\sim 0.7\text{V}]{\pm e} \text{NADH}^{+\cdot} \xrightarrow{-\text{H}^+} \text{NAD}^\cdot \underset{\sim 1.16\text{ V}}{\overset{\pm e}{\rightleftharpoons}} \text{NAD}^+$$

$$\text{NADH}^{+\cdot} \xrightarrow{\text{NAD}^\cdot} \text{NADH} + \text{NAD}^+ \qquad \text{NAD}^\cdot \xrightarrow{\times 2} \text{NAD}_2 \xrightarrow{\sim -0.1\text{V}} \text{NAD}^+$$

Underwood and Burnett (1973) summarized much of the early work to that date. The 1992 summary by Gorton *et al.* brings the history forward with a thorough review of the strategy of mediated NADH oxidation using modified solid electrodes and carbon paste electrodes. Gorton (1995) has also reviewed the use of these carbon paste electrodes modified with enzymes, including the dehydrogenases utilized with mediators. In this case the coenzyme is mixed into the paste formulation with the mediator and the enzyme. These electrodes show good response over a wide substrate range and could be operated with a significantly lowered overvoltage. The authors reported, however, that stability of the dehydrogenases was a problem.

Nowall and Kuhr (1995) reported a chemically modified carbon surface that stabilized electrocatalytic peaks for NADH, as well as several neurotransmitters. Without speculating on the exact nature of the surface or the mechanism, the authors described a process whereby NADH is first adsorbed to the electrode using a sine wave excitation (600 mV bias) at 38 °C. Then the electrode is exposed to ascorbic acid, thus completing the formation of the catalytic surface. The elevated temperature is essential; modification without it was ineffective. The remarkable result of this process is that a very stable electrode with excellent response to NADH is obtained. The LOD for NADH was improved by an order of magnitude at the catalytic surface and the linearity of calibration extended from 1 μM to 1 mM NADH. The other very useful property of this new surface was found in its ability to shift peak potentials for a variety of compounds, including neurotransmitters. Furthermore, the observed shifts were unique for each compound, thus providing a route to better selectivity among the compounds. While these shifts probably reflect compound-specific effects on the heterogeneous electron transfer rates for each species, no effort was made to elucidate that phenomenon.

Pariente *et al.* (1994, 1996) also reported effective modified electrodes for the electrocatalytic oxidation of NADH. Their electrodes were introduced in Chapter 3 as examples of using electropolymerized films for modifying electrodes. They used a series of dihydroxybenzaldehydes electrodeposited on glassy carbon at an oxidizing potential. The 2,3-DHB and 3,4-DHB were the most effective due to the *o*-quinone functionality which had been shown previously as an effective catalytic agent for NADH. Catalytic currents in the presence of NADH were linearly related to NADH concentrations from about 10μM to 1 mM. The authors suggested that the films would be useful for biosensors based on dehydrogenases. This approach still would not solve problems of a requirement for an external source of the coenzyme.

4.3.4. Ascorbic Acid

Ascorbic acid is the nemesis of analysts trying to do determinations in natural body fluids and tissues. The species is ubiquitous and often present in concentrations (μM to mM) considerably higher than some other analytes. This is particularly true in the brain, where basal levels of neurotransmitters might be in the nM range. Ascorbic acid is electroactive at approximately the same potential as many of the neurotransmitters (~0.4 V versus SCE at carbon), thus selectivity in brain studies has continued to be a problem. By electrochemically pretreating the carbon surfaces the peak may be shifted to eliminate the problem, at least partially. Selective membranes and coatings, discussed in Chapter 3, have also been used routinely. The use of ascorbic acid oxidase as a co-immobilized enzyme for removal of ascorbic acid introduces a problem with the oxygen dependence of a peroxide detector because the two oxidoreductases compete for the oxygen.

For analyses involving the determination of hydrogen peroxide a different type of problem arises. Although there has been some controversy about the origin of the effect, it has been observed that the presence of ascorbic acid introduces a negative bias in determining H_2O_2 (Palmisano and Zambonin, 1993; Lowry *et al.*, 1994). The reason proposed by the Lowry group is that a homogeneous reaction occurs between the peroxide and ascorbic acid (AA)

$$H_2O_2 + AA \xrightarrow{M^{+z}} 2H_2O + \text{Dehydro AA}$$

They presented evidence of the kinetic complications, as well as data indicating that trace metal ions catalyzed the reaction. The effect could be eliminated with addition of EDTA. The Palmisano group presented some evidence that the problem was due not to the homogeneous reaction, which has a slow rate, but to electrode fouling by oxidized ascorbic acid products.

There have been several strategies for coping with the ascorbic acid problem. As mentioned above, sometimes something as simple as electrochemically pre-treating a carbon electrode will solve the problem by moving the ascorbic acid peak. Cahill and Wightman (1995) actually performed simultaneous measurements of ascorbic acid and catecholamines by using the electrochemical pre-treatment strategy for one electrode. In other cases, both for brain studies and in H_2O_2 detection schemes, a variety of selective membranes have been used, particularly Nafion, though both overoxidized polypyrrole and electrochemically deposited *o*-phenylenediamine have also been used (Pihel *et al.*, 1996; Palmisano and Zambonin, 1993).

Berners *et al.* (1994) took an entirely different apprroach in their glutamate determination involving peroxide detection. Using an on-line system with microdialysis sampling, they installed a pre-electrolysis electrode to oxidize the ascorbic acid before the sample reached the glutamate oxidase modified electrode. Thus, the ascorbic acid was "scrubbed" before contacting the H_2O_2 producing part of the system.

4.3.5. Rotating Electrodes

The use of rotating electrodes has been introduced earlier, in the discussion of permeability studies in Chapter 2. For the novice in the field the idea of a rotating electrode being useful in biosensing may seem a bit foreign. The commercially available units are simple to use and in most cases have a more precision-controlled rotation rate than one can achieve with stirrers. In addition to the availability, there is also great utility in the basic theory of the responses. From discussions of the diffusion–reaction equations, it should be remembered that certain solutions were obtained with the assumption of steady-state currents produced in convective mass transport. Rotating electrodes provide one type of convective mode. Further, the very nature of membrane electrodes suggests that it would be instructive for designs if one could sort the various effects controlling the responses. Rotating electrodes provide one way of doing that sorting.

The basic theory of rotating electrodes has been discussed in detail by Bard and Faulkner(1980). Based on consideration of the diffusional and convective conditions of mass transport, an expression for the steady-state current can be developed (Levich equation):

$$i_{\text{lim}} = 0.620\ nFAD_0{}^{2/3}\omega^{1/2}\nu^{-1/6}C_0^*$$

Here $\omega = 2\pi f$ where f is the number of rotations per second and ν is the kinematic viscosity (specific viscosity/density). If potential is scanned during the rotation the current at any potential can be related to the limiting current. The useful aspect of the relationships is that by using an inverse form it is possible to separate the terms into the individual kinetic and diffusional terms:

$$\frac{1}{i} = \frac{1}{i_k} + \frac{1}{i_{\text{lim}}}$$

Thus, plots of $1/i$ versus $1/\omega^{1/2}$ should be linear with an intercept of the kinetic term. This approach may be used to determine heterogeneous rate constants in basic electrochemistry.

For membrane electrodes, the same sort of separational approach is used by applying the steady-state flux balance conditions to all the transport and reaction rate relationships. In this case each parameter is related to internal transport, external mass transport or kinetic terms related to the membrane and enzymatic characteristics:

$$\frac{1}{i_{\text{total}}} = \frac{1}{i_{\text{internal}}} + \frac{1}{i_{\text{external}}} + \frac{1}{i_{\text{kinetic}}} + \frac{1}{i_{\text{adsorption}}}$$

There are established plotting formats to extract from the data the individual effects that might control the response (Turner *et al.*, 1989; Cass, 1990 and references therein). This provides one route to determining design parameters. Hall's group uses a different approach based on the diffusion–reaction equation and the Thiele modulus (Martens and Hall, 1994; Martens *et al.*, 1995; Gooding and Hall, 1996).

4.3.6. Biocatalytic Efficiency

Biocatalytic efficiency is of prime importance in the use of microbial preparations as immobilized biocomponents (Turner *et al.*, 1989; Hall, 1991; Cass, 1990) . The efficiency aspect is analogous to the concept of enzymatic efficiency stated as k_2/K_m values. Analytes may include compounds such as alcohols, sugars, amino acids, various nitrogen species, or vitamins. The basis is that nutrients for the microbes are the analytes and the detection method is based on either the amperometric monitoring of microbial respiration through use of a Clark-type oxygen electrode or on the mediated electron transfer from the microbes to an electrode. Hall (1991) describes this approach as tapping in to the normal electron transfer cycles of the living organisms.

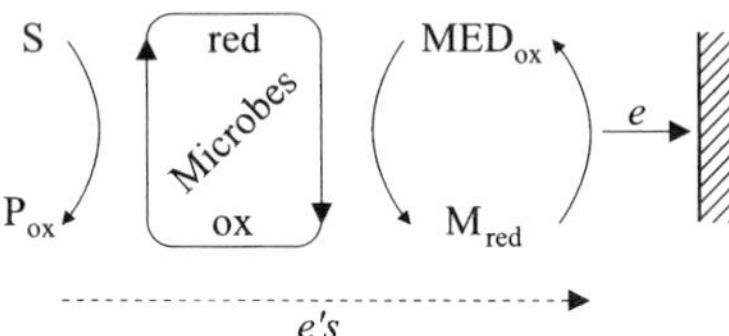

In the mediated microbial cell sensor, the biocatalytic efficiency is dependent on the interaction of the mediator with the microbes. This involves not only a reasonable choice of mediator but also the actual electron transfer process from microbe to mediator, permeation of the mediator into the membrane of the cells, lipophilicity of the mediator, and the relative rates of electron transfer in the cells and at the electrode. All of these factors involve the same kinds of phenomena that have been discussed for membrane electrodes, but the whole system becomes more complicated in this context of cellular interactions which other enzymes or cofactors may facilitate. A systematic quantitative assessment of biocatalytic efficiency could provide for microbial sensors the same type of informed design that has been made possible by various models of membrane electrodes.

Ikeda *et al.* (1996) have treated both the theoretical and experimental aspects of bacterial catalytic activity using $Fe(CN)_6^{3-}$, *p*-benzoquinone and dichloroindophenol as mediators, which interact with intact enzymes in the cellular membranes. Using an amperometric detection method they were able to devise a basic diffusion–reaction model for the mediated reactions with a mechanism akin to the ping-pong mechanism for isolated enzymes.

$$S_{cell} + E_{ox,cell} \underset{k_{-1}}{\overset{k_1}{\rightleftarrows}} ES_{cell} \overset{k_2}{\rightarrow} P_{cell} + E_{red,cell}$$

$$M_{ox,cell} + E_{red,cell} \underset{k_{-3}}{\overset{k_3}{\rightleftarrows}} EM_{cell} \overset{k_4}{\rightarrow} M_{red,cell} + E_{ox,cell}$$

A diffusion–reaction model based on the the mediator as the detectable species enabled them to use the membrane model to assess the biocatalytic efficiency of the cells in their bacterial system. Their model and experimental data indicated that the catalytic efficiency could be characterized by three parameters: a maximum reaction rate and a (Michaelis constant/partitioning constant) term for the substrate and for the mediator:

$$\nu_B = \frac{\nu_{B,max}}{1 + \frac{K_{s,cell}}{K_{s,p}}[S] + \frac{K_{m,cell}}{K_{m,p}}[M]}$$

$$\nu_{B,max} = \nu_{cell,max}[B] = \underbrace{k_{cat,cell}[E_{cell}]A}_{\text{Single cell}}[B]$$

Here [B] is the concentration of bacterial cells and A is the area of the membrane of a cell. Using their model they found that the rate determining step seemed to be k_4 and that benzoquinone appeared to be a better mediator than dichloroindophenol for their system.They observed some evidence for the separate effects of the partitioning and kinetic effects, but were unable to sort them with their model. Example calculations of predicted efficiencies were provided.They compared favorably to the rates of oxidoreductase-catalyzed reactions.

4.3.7. Low Levels of Detection

Amperometric methods are noted for good detection limits. The combination of improved instrumentation and some very creative sensing schemes have produced some fascinating limits of determination for a variety of methods. The enzyme recycling scheme has been discussed as a standard route to amplification of signals. The report of Bauer, *et al.* (1996) of zeptomolar limits for alkaline phosphatase attests to the potential for that approach. Scheller *et al.* (1995) have indicated that, in general, enzymatic recycling sensors should exhibit upward of 10^6 times the sensitivity of ordinary enzyme electrodes.

Coupling recycling with electrochemical immunoassay (Heineman and Halsall, 1985) just strengthens the effort. Heineman's group pioneered much of the ECIA work. Basically, typical enzyme-labeled antibodies and sandwich assays may be used, but the use of alkaline phosphatase and the substrate aminophenyl phosphate produces an electroactive product, aminophenol, which allows very low levels of determination when oxidized at ~0.2 V versus Ag/AgCl. Duan and Meyerhoff (1994) reported a variation on the assay which was designed to eliminate the usual washing steps which separate free reactants from those bound. By feeding substrate in through the back of a mesh electrode, thus keeping all reactants close to the electrode surface, they were able to achieve detection limits comparable to other assay methods without the washings.

Electrochemical immunoassay is definitely not the only evidence of the good detection limits offered by amperometric methods. Mikkelson (1996) has compared several analytical approaches to DNA determinations (Millan and Mikkelson, 1993; Millan *et al.*, 1994; Hashimoto *et al.*, 1994). The LODs reported by her group, and by others—femtomoles to attomoles—put amperometic methodology at the forefront, particularly considering that portable instrumentation may be used to accomplish the goals.

4.4. SENSORS BASED ON CONDUCTANCE AND IMPEDANCE

Conductimetric and impedimetric methods are fundamentally non-selective methods. Only with the advent of modified surfaces for selectivity and much improved instrumentation have these become more viable methods for designing biosensors (Cass, 1990 and references therein). There are some very practical considerations that make the methods attractive for inexpensive, simple measurement methods. No reference electrodes are needed. The sensing units themselves can be fabricated at relatively low cost. Improved instrumentation for the rapid determination of conductivity or impedance have facilitated the use of the methods. Both the Cass (1990) and Turner *et al.* (1989) books include chapters on conductimetric and impedimetric detection with reference especially to biosensing applications.

One of the most basic of all electrical measurements is the null-point detection of resistance or impedance with a Wheatstone or Wien bridge. Modern electronics have produced several other circuitries with phase sensitive detection, and designs that allow very accurate measurements with minimal data manipulation (Kissinger and Heineman, 1996; Cass, 1990). Integrated instruments also facilitate the use of the same instrumentation for both conductance and impedance measurements in either the frequency or time domain.

Any time an alternating voltage of frequency ω is imposed across electrodes, there will be both capacitive and resistive factors that will tend to impede the flow of current. The capacitive factor arises from charging the double layer at the electrode surface. The cell may be modeled by an equivalent circuit for which the total impedance or conductance of the assembly can be stated in terms of the components ($X_c = 1/\omega C$).

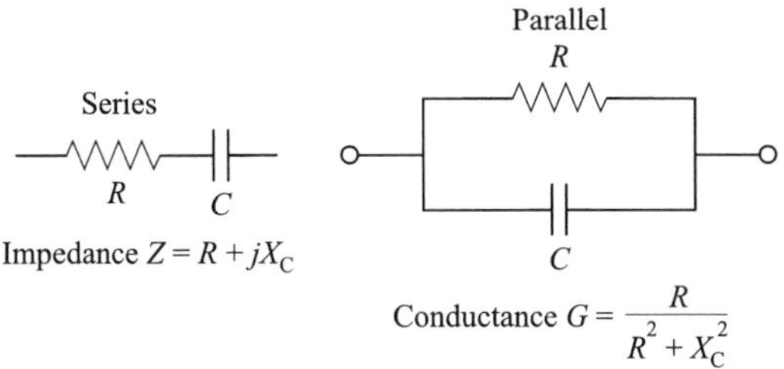

The capacitive reactance X_c of the cell is frequency dependent and decreases with increasing frequency. The common way to decrease the capacitive factor for conductimetric measurements is to increase the surface area of the electrode, usually by coating the surface with a classical platinizing preparation (Cass, 1990). Commonly, conductimetric enzymatic biosensors utilize IDEs, and are paired with reference and sample units for differential measurements.

When a sinusoidal voltage is applied there is a phase difference between the voltage and current, and the impedance is a vector quantity of magnitude $|Z|$ and phase angle θ.

$$|Z| = \sqrt{X_c^2 + R^2} \qquad \tan\theta = \frac{X_c}{R} = \frac{1}{\omega RC}$$

Both solution and surface characteristics determine the total impedance of a device.

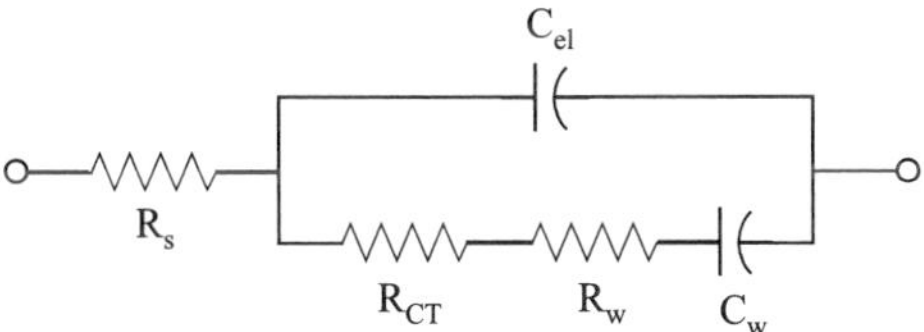

Here the subscripts are: S, solution; CT, charge transfer: w, mass transfer; and el, electrode charging. This is why it is basically so non-selective. By using biocomponents for molecular recognition, however, these conductance and impedance measurements may be converted to relatively selective methods. Several conductimetric sensors have been reported in recent years. The devices of Hendji *et al.* (1994) and Sheppard *et al.* (1995) illustrate some of the modern approaches in biosensing. The Hendji sensor is a micromachined device with interdigitated pairs of gold film electrodes. The method was designed for a differential measurement between an enzyme-modified surface and a non-active protein modified surface. Urease was used for urea determinations, glucose oxidase for glucose as analyte and cholinesterases for acetylcholine and butyrylcholine. Performance was comparable to standard methods for each analyte. The acetylcholinesterase (AchE) proved to have a better senitivity than the butyryl enzyme. Therefore, the authors concluded that the AchE would be better for pesticide analyses.

The Sheppard device was designed as a pH-sensitive conductimetric unit. It, too, was fabricated with the IDEs, platinum in this case, but the surface was coated with a pH-sensitive hydrogel. As pH changed, polymeric swelling occurred, thus changing the ionic mobilities and the resistance of the membrane. Sensor response changes as large as 45% per pH unit change were observed.

The impedimetric sensor for viruses was used as an example in Chapter 2 (Knichel *et al.*, 1995). That sensor illustrates well the selectivity conveyed by effective molecular recognition schemes. Another very interesting impedimetric sensor was reported in 1995 by McNeil *et al.* They spray-coated the surface of their screen-printed gold ink electrodes with enteric polymer which dissolves at pH's higher than 7. Urease was immobilized on a precast membrane which was laid on the polymer-coated surface. In the presence of urea as analyte, the enteric polymer dissolves and the impedance of the electrode changes. For 2–100 mM urea a response change of 4 orders of magnitude was observed. The method was also adapted with very good results to immunoassay by using urease-labeled antibodies in a sandwich format.

4.5. LIGHT ADDRESSABLE POTENTIOMETRIC SENSOR

When using amperometric sensors, immobilized microbial preparations can be used to produce measurands detected by the current measurement. In that case the microbial component is employed as the molecular recognition function to provide selectivity in a conventional amperometric mode. In 1988 Hafeman, Parce and McConnell introduced a completely different type of sensing sytem based on the physiological phenomena of living cells, a device called the microphysiometer, and it is a light addressable potentiometric sensor (LAPS). The actual response arises from biochemical effects related to the energetics and physiological function of cells. This device actually monitors physiological response as a function of time or concentration of a substance that alters the cellular physiology. The response depends on pH changes manifested at a solution–insulator interface. Information about specific catalytic or binding reactions may be extracted, but the direct response is an indicator of the integrated functions within the cellular preparation.

For practical purposes the microphysiometer may be viewed as a miniaturized, instrumentalized version of classical bioassays. A good summary of some of the early experiments with the microphysiometer appeared in McConnell *et al.*, 1992. Figure 4-21 illustrates the basic design of the microphysiometer. The practical devices are miniaturized with eight microvolume flow chambers on one 23 mm × 23 mm microchip (Baxter *et al.*, 1994). Cells are immobilized between two polycarbonate membranes.

The microphysiometer is designed for detection of net changes in proton concentration as a result of cellular processes such as phosphorylation, diffusion and ion pumping. The basic operational scheme is as follows.

(a) A potentiostat controls the potential between the solution and the silicon substrate. This surface potential is a function of proton concentration at the solution/Si_3N_4 interface, which is pH sensitive.

(b) Light from the LED produces hole–electron pairs in the silicon. As a result of the voltage gradient between the silicon and solution a charge asymmetry develops in the silicon substrate.

(c) A photocurrent induced by the charge asymmetry flows through the external circuit. The E_{appl} necessary for effecting the photocurrent reflects the changes in surface potential at the solution/Si_3N_4 interface.

(d) Using a stop-flow arrangement for the sensing cycle, the cellular sample may be interrogated once per second. The voltage signal is linearly related to the pH.

McConnell's group has demonstrated the effectiveness of the LAPS for monitoring cellular physiology affected by changes in carbon source for glycolysis, ligand receptor–agonist interactions, drugs, enzymatic inhibition and viral infection, including the effects of AZT administration (McConnell *et al.*, 1992). Rogers *et al.* (1992) utilized the LAPS for studying effects of toxins on nicotinic receptors. They found well-defined signals related to the receptor/ligand binding and attained a sensitivity comparable to radiolabeling techniques.

An Italian group introduced a variant of the LAPS device (Adami *et al.*, 1995a). Their system is called the potentiometric alternating biosensor (PAB), also shown in Figure 4-21. Sample chambers are interchangeable in the PAB, which has been characterized with cells and enzymatic catalysis involving pH changes (Adami *et al.*, 1994). The PAB has been extended to

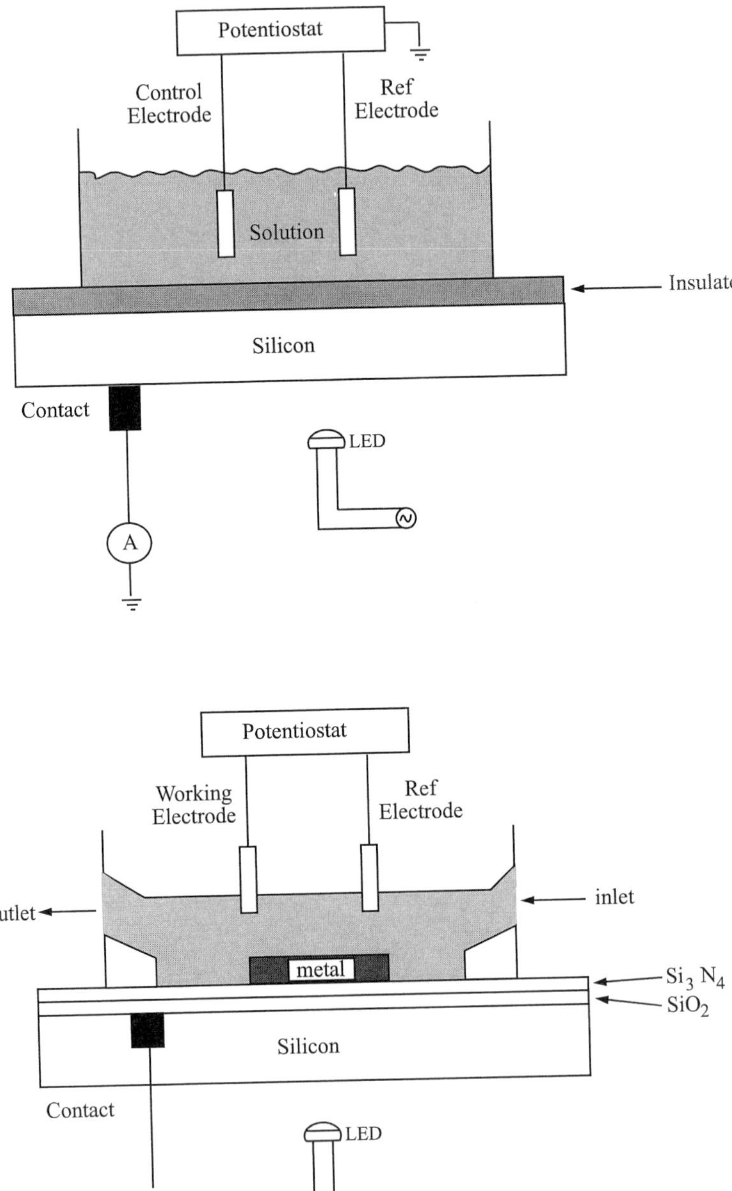

FIG. 4-21. Physiological sensing systems showing (top) the light-addressable potentiometric sensor (LAPS), and (bottom) the potentiometric alternating biosensor (PAB), shown in the redox mode with metal layer (see Parce *et al.*, 1989 and Adami *et al.*, (1995a,b, respectively).

redox measurements by incorporation of a metal vapor-deposited on the Si_3N_4 layer (Adami *et al.*, 1995b). Biphasic curves dependent on the bias potential are explained in terms of a pH-indicating segment and a redox segment.

Using the PAB in its pH mode Nicolini *et al.* (1995) performed real-time toxicity tests. They compared effects of mitoxantrone, an RNA and DNA inhibitor, on normal and malignant liver cells. The second part of the study examined the effects of Ara-C, a DNA antimetabolite, on mouse fibroblasts. Extracellular acidification rates were determined and found to correlate fairly well with conventional staining tests. The authors suggest that two of the advantages of the PAB are the real-time monitoring capability and the rapidity of the assay.

5. Optically Based Energy Transduction

Photometric methods have been a mainstay of bioanalytical procedures for many years. Many of the original enzyme assays and kinetic studies of biochemical reactions were based on the fundamental relationships for the interactions between electromagnetic radiation and matter. Aromatic amino acids in proteins, enzymatic cofactors and optical labels or tags have been ideal probes for absorbance, fluorescence and fluorescence quenching measurements in analytical biochemistry. Light scattering and polarization effects have also been utilized, though early on they were used primarily in structural studies rather than bioanalytical quantitation.

As options for measurement based on the interaction of light and matter are expanded, it should be emphasized that all of the fundamental characteristics of light are open to experimental examination. Intensity, frequency, phase and polarization are the basic phenomena which form the basis of analytical protocols. Intensity variation is the phenomenon most often encountered in routine methods. There are also established procedures for studying surface species by several reflection or refraction methods, some of which are often coupled with electrochemical variants. More recently, lifetime-based fluorescence has received considerable attention, particularly as a non-invasive method for biological systems. While undergraduates may have had some experience in optical measurements involving many of these phenomena, intensity variations have probably been the basis of most laboratory experience.

Options for optical sensing strategies have mushroomed as a result of newer optical materials, optoelectronic devices, and advantages that accrue from designed immobilization of molecules on surfaces. Although the same fundamental principles of matter–radiation interactions lie at the heart of the new developments, translation of those principles into the realm of newer technologies has been a real challenge. Much has been gained, however, in that translation. The expansion from traditional optical assay techniques to the enhanced information content from modern bioanalytical methods is truly one of the major accomplishments of twentieth-century science.

Several intoductory reviews allow one to trace the development of modern optical methods in biosensing. The Seitz (1988) and Arnold (1992) reviews provide good background for understanding the principles and early applications of fiber-optic sensors. Lübbers (1992) discussed fluorescence-based sensors. Later he dealt specifically with optical sensors in clinical monitoring (Lübbers, 1995). The 1994 Tsien review provides a good introduction to cellular dynamics using primarily fluorescent probes for monitoring calcium variations. Giuliano *et al.* (1995) reviewed fluorescent protein biosensors with an emphasis on measurement of cellular dynamics. The Wolfbeis (1991) volumes, and also Wise and Wingard (1991), are widely used and include much of the foundational material on which many modern sensors are based.

Brecht and Gauglitz (1995) have presented an updated review of optical probes and transducers. Their review gives special attention to biosensing applications, with considerations of sensitivity and spatial resolution obtainable by different surface-sensitive methods. The Tan and Kopelman (1996) chapter on near-field methodologies is replete with references and applications pertinent to biosensing problems. Emphasis is placed on spatial resolution and intracellular phenomena.

Consider the usual aspects of an optically based measurement protocol, usually introduced in an introductory course emphasizing instrumental analysis. There is a source of electromagnetic radiation, a sample that may absorb, emit, rotate, refract or reflect radiation, and a detector that will measure the exiting radiation. When choosing a source and detector for intensity-based measurements the options are restricted to those appropriate for a particular spectral region for the measurement of interest (see Figure 5-1). For example, tungsten and deuterium sources have different spectral outputs, even in the visible region, and they both differ from the usual infrared sources. Detectors that are effective in the ultraviolet/visible regions are not the same as those for infrared measurements. Spectral efficiencies

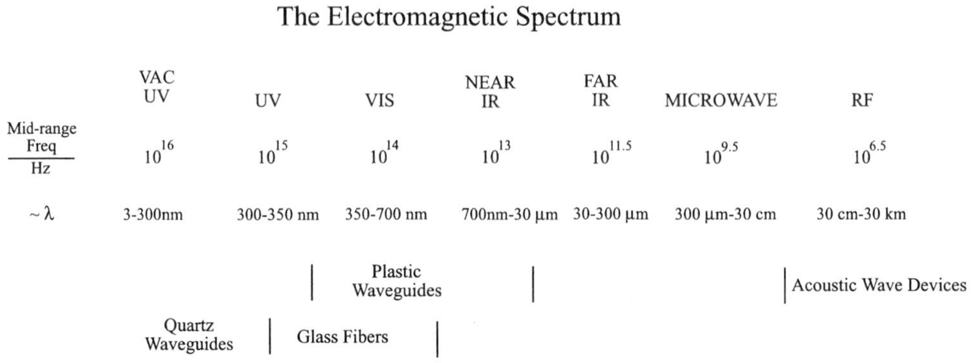

The Electromagnetic Spectrum

	VAC UV	UV	VIS	NEAR IR	FAR IR	MICROWAVE	RF
Mid-range Freq / Hz	10^{16}	10^{15}	10^{14}	10^{13}	$10^{11.5}$	$10^{9.5}$	$10^{6.5}$
$\sim\lambda$	3-300nm	300-350 nm	350-700 nm	700nm-30 μm	30-300 μm	300 μm-30 cm	30 cm-30 km

Plastic Waveguides

Acoustic Wave Devices

Quartz Waveguides

Glass Fibers

FIG. 5-1. Electromagnetic spectrum showing the ranges in which types of optical biosensors are useful.

vary even among sources or detectors for a particular wavelength range. In addition to the usual source illumination factors and detector capture characteristics, the full optical design might also include polarizers necessary at either the source or detector for signal manipulation.

The considerations above drive all experimentation involving the interaction of electromagnetic radiation with matter. Newer methods that have been developed for bioanalytical sensors are really no different. They are merely couched in the language and experimental advantages of newer tools such as fiber optics, guided waves, evanescent waves, surface plasmons, light-emitting diodes (LEDs), interferometers, diode arrays and charge-coupled devices (CCDs). The enhanced tools enable more effective control and monitoring of optically based interactions, even those involving low level signals arising from phenomena relatively unexplored until recent years. The new terminologies and tools do not represent new discoveries about fundamental optics or radiation–matter relationships. The basic game has not been changed. Analysts have merely refined the markers on the field so they can follow and control the game more intricately.

There are essentially four major developments that have fostered the quantum leap in measurement through use of optical methods:

(1) use of fiber optics in a variety of formats, including remote sensing;
(2) enhanced utilization of the evanescent wave associated with total internal reflection in waveguides;
(3) miniaturized components for optical measurements; and
(4) development of near-field optics for microscopy and spectroscopy.

New concepts of sample or reagent containment for radiative stimulation should also be considered fundamental to modern biosensing techniques.

Immobilization methods such as those discussed in Chapter 3 have been essential to many of the advances, particularly those that involve penetration of an evanescent wave into a thin layer. Those developments, however, cut across all types of transduction schemes, not just optically based methods.

Before embarking on a discussion of devices that utilize guided waves and the evanescent wave, three biosensing approaches more akin to traditional optical methods should be introduced. None requires the background or application of concepts more advanced than basic detection of absorbance, fluorescence, or rays that are refracted or reflected. Yet, two of the devices, capillary sensors and reflectometric layers, have many advantages in that they are inexpensive, disposable and very effective in monitoring trace concentrations. The third approach explored, near-IR absorption measurements, relies on the chemometric treatment of data to determine glucose by a non-invasive method.

Capillary optical sensors. In 1994 Weigl and Wolfbeis described an effective optical sensing strategy based on the use of capillaries chemically modified on the inner surface. They introduced into the capillary a carbon dioxide-sensitive indicator which was immobilized by a "reverse pumping" technique. Using fiber-optic couplers at the exterior walls of the capillary, they measured by absorbance CO_2 in the physiologically significant range of 0–8% in air. The t_{90} was 100–300 ms. Weigl and Wolfbeis pointed out that one advantage of the capillary sensor was that it could be utilized as an integrated sensor, considered as a sampling device as well as a sensing device.

In earlier discusssions of microsampling, the Cosford and Kuhr (1996) capillary sensor for glutamate has been mentioned as a forerunner to a CE biosensing detector. In this case they immobilized glutamate dehydrogenase on the inner wall of a capillary, then monitored via a fluorescence microscope the NADH fluorescence produced in the enzymatic reaction. The extension from the Weigl and Wolfbeis approach is obvious and provides a good example of the application envisioned by them.

More recently, Narang *et al.* (1997) have adopted the same capillary principle for their immunosensing determination of TNT in water using a spectrofluorometer. Earlier they had introduced a sensitive TNT sensor based on a tapered fiber-optic device operated in an evanescent wave mode (Anderson *et al.*, 1993, 1994; Shriver-Lake *et al.*, 1995). The capillary sensor is based on a competitive immunoassay with antibody bound to a fluorescent TNT antigen-analog immobilized on the inner wall of the capillary. As a sample containing TNT as analyte is passed through the capillary a displacement of the immobilized antigen-analog occurs, as indicated by a decrease in fluorescence from the immobilized layer. Their entire assay could be completed in less than 3 min with an LOD of 15 pg per ml (7 fmol) of TNT.

Reflectometric Interference Spectroscopy (RIfS). One of the basic properties of light at interfaces is demonstrated by this straightforward method based on interference effects created by the interfaces of thin films (Piehler *et al.*, 1996). Light passing through interfaces of differing indices of refraction will be refracted, but there is also partial reflection at each of the boundaries of a thin film deposited on a transparent substrate.

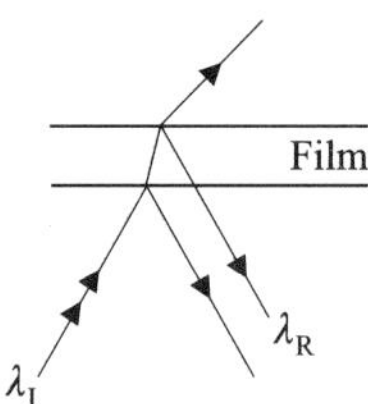

Because the reflected beams travel different distances, there will be interference resulting in a phase difference between the two reflected rays. The reflectance spectrum is a function of the index of refraction of the film (n), the thickness of the film (d), the wavelength (λ) and the phase angle (ϕ) between the reflected rays. A thin film of silica is used as the optical substrate for a biosensing layer. Chemical changes at the sensing surface result in a detectable film thickness change that can be monitored by acquisition of the reflectance spectrum. Spectral data are then converted to thicknesses which indicate the amount of analyte effecting the thickness changes. Using a streptavidin sensing layer on the silica, Piehler, Brecht and Gauglitz (1996) have applied this method to the determination of biotin with an approximate LOD of 40 nM and measurable thickness changes of > 2 pm. One of the major points made by this team is that RIfS can be used to detect low molecular weight analytes on an absolute basis, whereas many other mass-sensitive methods require higher molecular weight materials for adequately sensitive measurements.

In their review of optical methods, Brecht and Gauglitz (1995) have compared RIfS to other techniques on the basis of both depth and lateral spatial resolutions. They discuss the fact that RIfS provides much better depth resolution (< 10 pm) than do evanescent-wave methods ($\sim$100 nm). In that report the authors also summarize their work involving an RIfS immunosensor for the determination of atrazine at the 0.1 ppb level.

Near-IR of Glucose in Complex Matrices. In a series of reports from the early 1990s Arnold and coworkers presented sequential development of near-IR absorption methods for clinically relevant concentrations of glucose in complex matrices. The progression of studies provides a sound lesson in the application of basic chemometric methods for extracting information from complex data sets. Having demonstrated previously that the 5000–4000 cm^{-1} range provided adequate near-IR information for the determination of glucose, two 1993 reports dealt with multivariate calibration using digital filtering of signals and partial least squares regression (PLS) for calibration and prediction of glucose in protein matrices and in plasma (Marquardt *et al.*, 1993; Small *et al.*, 1993). A response surface (grid search) method was used for optimization of the filtering parameters and spectral range. The calibration model based on 12–14 PLS factors was sufficient for standard errors of prediction (SEP) of 0.4–0.5 mM in the 2.5–25.5 mM concentration range. This translated to ~4% mean percentage error.

Application of the near-IR measurements and multivariate calibration was later extended to glucose in protein and triglyceride matrices which exhibit significant spectral overlap with the glucose spectrum (Pan *et al.*, 1996). The calibration model resulted in prediction errors of 0.5 mM and 0.2 mM glucose for triacetin and protein solutions, respectively. In this study, however, the digital filtering method for pre-processing data was less effective due to similar spectral bandwidths for glucose, protein and triacetin.

In the Bangalore *et al.* (1996) and Shaffer *et al.* (1996) studies, genetic algorithms (GA) were introduced for optimization of five essential variables identified in the previous work: position and bandwidth of the digital filter, the two limits of the spectral range used for PLS regression, and the number of terms in the calibration model. Genetic algorithms are optimization tools based on the concept of genetics and natural selection. Basically, the "fittest" set of variables emerges from the evolutionary process of applying several transformational operators, which are analogs to natural genetic selection processes. Several complex matrices were subjected to analysis, including glucose in a human serum matrix. The authors concluded that the use of genetic algorithms provided a calibration model with fewer PLS factors and similar or better errors of prediction of unknown concentrations. They suggested that the advantages of the GA approach outweigh the greater complexity of the optimization procedure.

5.1. FIBER OPTIC DEVICES

Fiber-optic sensors, frequently referred to as optodes, have extended the opportunities for utilizing spectrophotometric methods in bioanalytical chemistry. There are three significant advantages in using optical fibers. Very small sample domains can be probed, remote sensing is facilitated, and also optical fibers are very simple, convenient devices for utilizing measurements based on interactions with the evanescent waves which result from total internal reflections.

Sources and detectors used with optical fibers are essentially the same as those used in most spectrophotometric instruments. Tables 5-1 and 5-2 show some comparative information about optical fibers, ordinary sources and detectors. The size and transmission qualities of optical fibers make them particularly suitable for using miniaturized sources and detectors such as LEDs, laser diodes and photodiodes. That provides an opportunity for integrated, miniaturized optical systems which are essential for many biosensing applications. Tran and Gao (1996) have recently introduced an erbium-doped fiber amplifier and near-infrared spectrophotometer that the authors describe as all solid-state, compact and characterized by light throughput sufficient for *in vivo* studies. The new integrated fluorescence device for multianalyte analysis reported by Bruno *et al.* (1997) represents another state-of-the-art instrument utilizing innovative optics. High-intensity LEDs and photodiodes comprise the source–detector units of the array. Assemblage of this device involves "pigtailing" fabrication, which refers to joining components with index-matching adhesives, thus minimizing or eliminating deleterious optical interface effects (Bruno *et al.*, 1994).

Optical fibers are available in various lengths and sizes, usually ranging from a few micrometers up to 100–200 μm in diameter (Seitz, 1988; Uiga,1995). The fibers may be tapered by HF etching for enhanced sensitivity (Shriver-Lake *et al.*, 1992; Anderson *et al.*, 1993, 1994). For near-field measurements (Section 5.3) fibers may be precision-pulled to tips less than 1

TABLE 5-1 Types of Fiber Optic

Material	Approximate Diameter (μm)	Numerical Aperture, NA	Comments
All plastic	600	0.5–0.6	Multimode, visible region, low cost
Plastic-clad silica	150	0.2-0.3	Multimode, visible,NIR, low cost
Silica	100	0.16-0.5	Multimode, UV-vis, low cost
Silica	50	0.08-0.15	Single mode, UV-vis, necessary for interferometric measurements
Imaging bundle	350	—	3000 individual fibers, used for imaging/ sensing arrays

TABLE 5-2 Sources and Detectors

	IR/NIR	Visible	Ultraviolet
Sources			
Tungsten lamp	×	×	
Deuterium lamp			×
Xenon		×	×
LEDs			
GaN, In/GaN		470 nm	
GaP		570 nm	
GaP/GaAsP		580–650nm	
AlGaAs	650–900 nm		
GaAs	900 nm		
InGaAs	1000–1300 nm		
Lasers[a]			
N_2			337 nm
Ar+		488–568 nm	
He–Ne		633 nm	
Ruby	694.3 nm		
Diodes	800–904 nm		
Nd/YAG	1060 nm		
CO_2	9200–10800 nm		
Detectors[b]			
Photomultipliers			
Photodiodes			
Photodiode arrays			
Phototransistors			
Charge-coupled devices			

[a] Tunable dye lasers range 400–800 nm.
[b] Response curves depend on material and doping.

μm in diameter (Tan and Kopelman, 1996). Fiber optic bundles used for imaging and optical arrays have outer diameters of about 350 μm, as shown in Table 5-1, and contain about 6000 individual fibers in the bundle. Walt's group has developed immobilization schemes for these bundles which can be used to create chemical arrays on the fibers (Healy *et al.*, 1995; Li, L. and Walt, 1995; Ferguson and Walt, 1996; Healy and Walt, 1997). Thus, the bundles can be designed as optical arrays which exhibit extremely fast response times, can be used for multicomponent analysis and can provide optical measurements with spatial resolution. In the 1997 Healy and Walt report, the authors presented pH and O_2 sensors that exhibited response times of 200–300 ms attributed to radial diffusion rather than planar diffusion to the microarray. A possibility of 7 μm spatial resolution was proposed. These sensors were presented as an alternative to Kopelman's nanoscale optodes (Tan and Kopelman, 1996) because microscopic observation of the signal is not required, thus enabling remote sensing. The spatial resolution, however, is less than that obtainable with the nanoptodes and the overall size of the bundle precludes it from intracellular monitoring.

The amount of signal loss with increasing length of an optical fiber depends on both the material of which the fiber is made and the wavelength being propagated. To minimize loss the optical material is covered with a cladding and usually encapsulated in a protective jacket (Figure 5-2). The cladding has an index of refraction n which is less than that of the core. The significance of the relative indices of refraction will be examined later. Sections of the jacket and cladding may be removed without damage to the core of the fiber. That, too, will be an important factor later in the discussion. The fibers may be plastic, glass or quartz. As is true of conventional spectrophotometer cells, plastic fibers are the choice for work in the visible and near-infrared regions of the spectum, whereas quartz is the preferred material at the other spectral extreme in the ultraviolet.

5.1.1. Guided Waves

In order to understand fundamentals of optical fibers applicable to bioanalytical sensing, the phenomenon of guided waves must be introduced. More complete treatments of the details can be found in Snyder and Love (1983), Seitz (1988), Uiga (1995), Wolfbeis (1991), Zhu and Yappert (1994), Benoit and Yappert (1996a), Wise and Wingard (1991) and Turner *et al.* (1989). The basic properties of waveguides are related to refraction and reflection phenomena at interfaces. At an interface between two dielectric materials of differing refractive indices, refraction occcurs due to the difference in speed of the light in the two media. According to Snell's law (Figure 5-3), light moving from an optically denser medium to an optically less dense medium

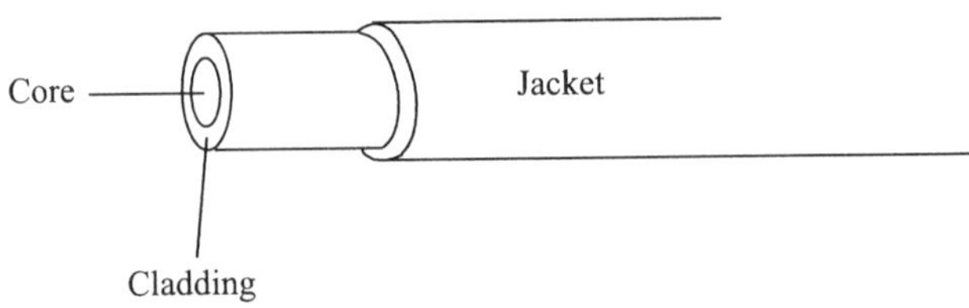

FIG. 5-2. Basic optical fiber.

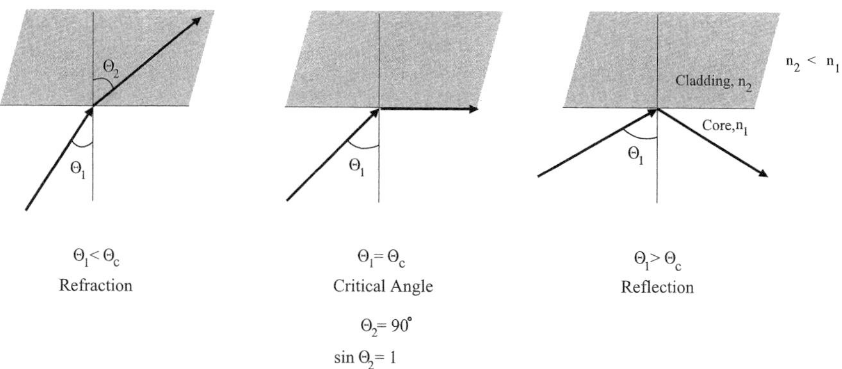

FIG. 5-3. Refracted and reflected light and Snell's law, $n_1 \sin \Theta_1 = n_2 \sin \Theta_2$. The critical angle Θ_c is illustrated in the center part.

can be refracted at an interface unless a critical angle Θ_c is exceeded. Once the critical angle is exceeded, there is total internal reflection of the light from the interface between the two media.

To be exact, in a derivation of the origin of guided waves, it would be necessary to specify whether planar or circular waveguides, straight or bent fibers, meridional or skew rays, etc., are to be discussed (Seitz, 1988; Benoit and Yappert, 1996a). For purposes of conceptualization it is possible to simplify to one particular situation, a generic optical fiber. As indicated in Figure 5-4, an optical fiber in air is characterized by light interactions at both the air–core and the core–cladding interfaces. For a fiber constructed with a cladding that has an index of refraction less than the core material, the situation described above as necessary for total internal reflection exists. That is, at the core–cladding interface light moves from an optically denser medium to an optically less dense medium.

As shown in Figure 5-4, multiple reflections occur through the core of the fiber if the incident light is such that total internal reflection at the core/cladding interface is achieved. Based on Snell's law, for reflection the light striking the core–cladding interface must strike at an angle exceeding the critical angle at that interface. Logically, the angles of incidence at the air–core interface which will produce that effect must be determined. By using Snell's law of refraction for the air-core interface and some fairly simple geometric relationships it can be determined that there is a set of incident angles which will induce the multiple reflections. This set of acceptable incident angles is included in the *cone of acceptance*, which becomes a qualifying property of a particular fiber. More commonly the fiber is described in terms of its numerical aperture (NA), a parameter also stated in terms of the

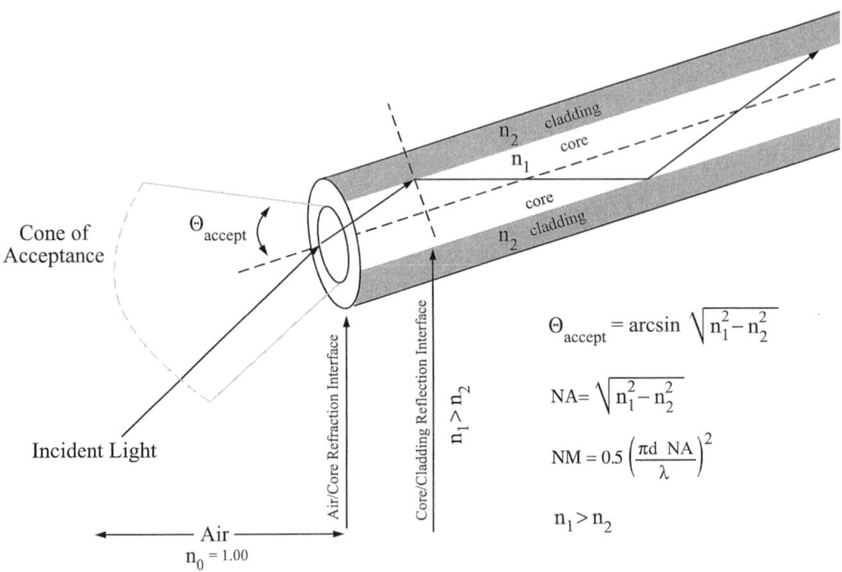

FIG. 5-4. Total internal reflection in an optical fiber: the *n*'s are refractive indices, λ is the wavelength of incident light, NA is the numerical aperture, and NM is the number of modes.

relative indices of refraction, and given in Figure 5-4. A larger NA implies a wider cone of acceptance of effective light into the core. There is a comparable cone of illumination at the exit end of a fiber, though the exact nature of that zone depends on the matrix through which the exiting light must pass.

Maxwell's equations for light propagation describe the behavior to be expected—they are wave equations with wavefunctions as solutions. Conceptually this is quite akin to the mathematical approaches for determining the atomic orbitals. Application of Maxwell's equations results in only certain allowed solutions which satisfy appropriate boundary conditions for the circular fiber. Practically speaking, this means that for light to be propagated constructively down the fiber, only certain combinations of wavelength and reflection angle are allowed. These allowed solutions are referred to as the *modes* of the fiber.

In Figure 5-4 it is shown that the number of modes, NM, is dependent on the NA, the diameter of the core and the wavelength of light. For a given wavelength of the light, the number of modes decreases with decreasing diameter of the fiber. For a planar waveguide rather than a circular fiber, the analogous dimension would be the thickness of the layer through which the wave is guided. For practical purposes the diameter of a single mode fiber is about 1–5 μm, whereas multimode fibers commonly have diameters of 100–200 μm. For most, but not all, applications relative to bioanalytical sensors the less expensive multimode fibers are sufficient. The notable exception is that of interferometric sensors, which require a single mode.

Yappert's group has analyzed the fluorescence signal enhancement factor observed when the distal end of an optical double-fiber is placed in a capillary containing sample (Zhu and Yappert, 1994; Benoit and Yappert, 1996a). This enhancement may increase sensitivity by as much as 2 orders of magnitude and is dependent on the length of the capillary relative to the depth at which the fiber tip is placed, the relative diameters of the fiber and capillary, and the sensor configuration. Benoit and Yappert (1996a) provided a model based on the capillary/sample cell acting as a partially reflective waveguide in which the fibers' excitation and collection cones overlap to make light collection more efficient.

5.1.2. The Evanescent Wave

When total internal reflection occurs at an interface, there is interference between the incident and reflected rays of light. This results in a standing wave within the fiber, as sketched in Figure 5-5(a). There would be one of these standing waves at each point of internal reflection. The standing waves are associated with a very important quality that is the basis for many sensing applications. Vector analysis of the waves at the points of reflection shows that an intensity component extends beyond the edge boundary of the waveguide, the core in this case. This extension of light intensity is referred to as the *evanescent wave*, depicted in Figure 5-5(b).

Taking advantage of the evanescent wave is not really new in spectral designs. For example, some of the earliest work in optically transparent electrodes used internal reflection spectroscopy to monitor changes near an electrode surface (see Heineman *et al.*, 1984). The evanescent wave's profile indicates that its intensity falls off in an exponential decay from the waveguide boundary. It penetrates into the surrounding space for a depth d_p (Figure 5-5(c)). Practical depths of penetration using ordinary optical materials and visible light are usually in the ~100–200 nm range, sufficient for probing thin films of immobilized components on the fiber's surface.

As the number of reflections increases for a given length of fiber, the distance between reflection points is shortened. This has the practical effect of overlapping evanescent waves, and the overlap provides approximately a continuous evanescent region extending along the external surface of the waveguide. This region, circumscribed by linear extension along the boundary and the depth of penetration, creates an interaction space that is ideally suited to the dimensions of typical immobilized chemical layers.

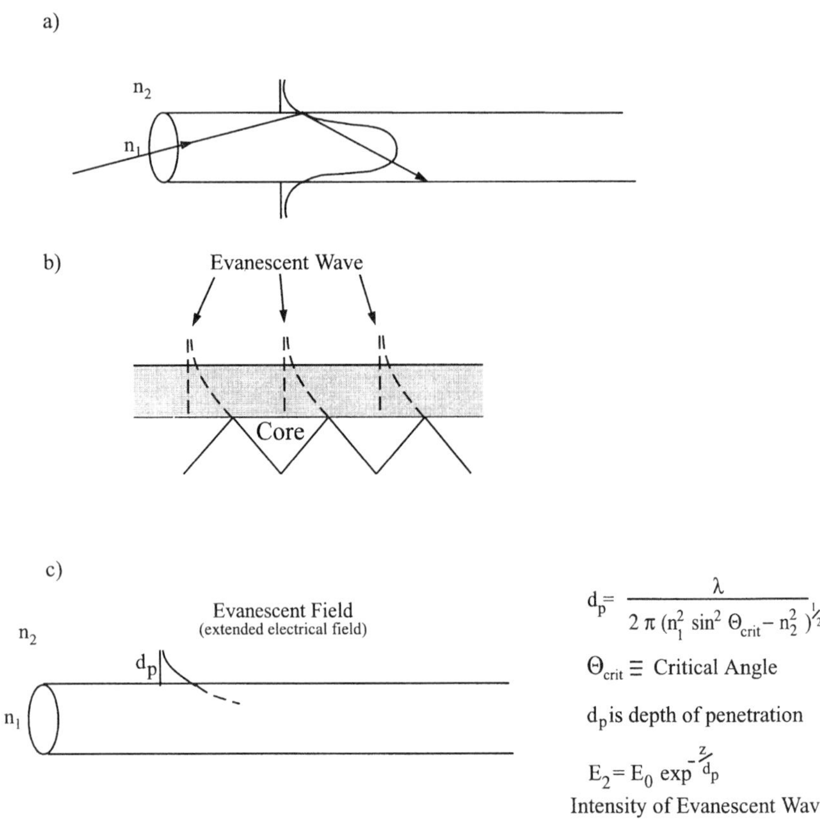

FIG. 5-5. Schematic of the evanescent wave. The energy field extends beyond the core to a penetration depth d_p.

The reason the evanescent wave is important is that it provides additional options for excitatory interactions between light and matter, and it is especially applicable to sensor designs for probing thin layers. In some excitation configurations for fiber-optic sensors a reaction matrix is applied on the end of the core at the point of exit for the propagated wave—i.e., a distal-end optode. Intensity variations in aborbance or fluorescence are most commonly used for detection using distal-end sensors. Alternatively, if cladding is stripped from a length of fiber core, immobilized reagents may be layered directly onto the longitudinal portion of the core surface, thus promoting interaction with the evanescent wave. Interactions between the light and the immobilized layer can be monitored in several different ways, depending on the information sought. Combined with the repertoire of immobilization strategies presented in Chapter 3, these options enhance enormously the optical measurement schemes that can be utilized.

5.1.3. Fiber-optic Sensor Designs

Conventional fiber-optic sensors emphasized in this section may be configured and modified in various ways. Nanoscale optodes used for near-field methodologies are discussed in Section 5.3. When used simply as a convenient conduit to direct light from a source onto a sample, return non-absorbed light or capture light emitted from a sample, the fiber is operating in an *extrinsic* mode. The *intrinsic* light-propagating characteristics of the fiber are unchanged. For example, absorption, fluorescence or quenching of emission by the sample may be observed with guided radiation traveling from source to sample to detector via the fiber-optic paths, even if a sample is remote from the instrumentation for control and signal processing.

As is true in conventional spectrophotometric examples, light paths differ depending on the experimental goals and optimization considerations. Three different fiber-optic configurations are depicted in Figure 5-6. In (a) light is piped into the fiber, the propagated ray interacts in the confined area at the end of the fiber and the altered signal may exit in a straight-line geometry to the detector. The thickness of the reaction layer and scattering effects through the immobilized matrix can affect results for sensitivity (SEN) and the limit of determination (LOD). A typical example might involve a pH sensitive dye immobilized on the fiber. This provides a sensing mechanism as pH variations in a sample alter the ratio of immobilized protonated to unprotonated dye. The absorbance changes can then be monitored at the λ_{max} of either form or both. Ratioing signals at two wavelengths can improve the figures of merit for the procedure (Seitz, 1988). Tan *et al.* (1992b) have found ratioing with background subtraction particularly effective in the calibration of nanoptodes.

A slight variation on the example of (a) would be the cylindrical sensor geometry of Kar and Arnold (1994). They used a cylindrical gas-permeable membrane to contain a sample of Bromthymol blue. Optical fibers transmitted the light to and from the reaction cylinder. When the gas-permeable membrane was exposed to ammonia as analyte, the non-protonated form of Bromthymol blue, which absorbs at 610 nm, was monitored. Intensity changes were monitored at a photomultiplier tube for the characterization studies. Although conventional optical components were used in the original work by Kar and Arnold, a very important set of variants was implied by the sensor design. Because the Bromthymol blue absorbs in the visible region of the spectum, relatively inexpensive plastic optical fibers, an LED source and a photodiode detector could be used. Thus, an inexpensive, easily constructed sensor is implied. The authors suggest that chromophores

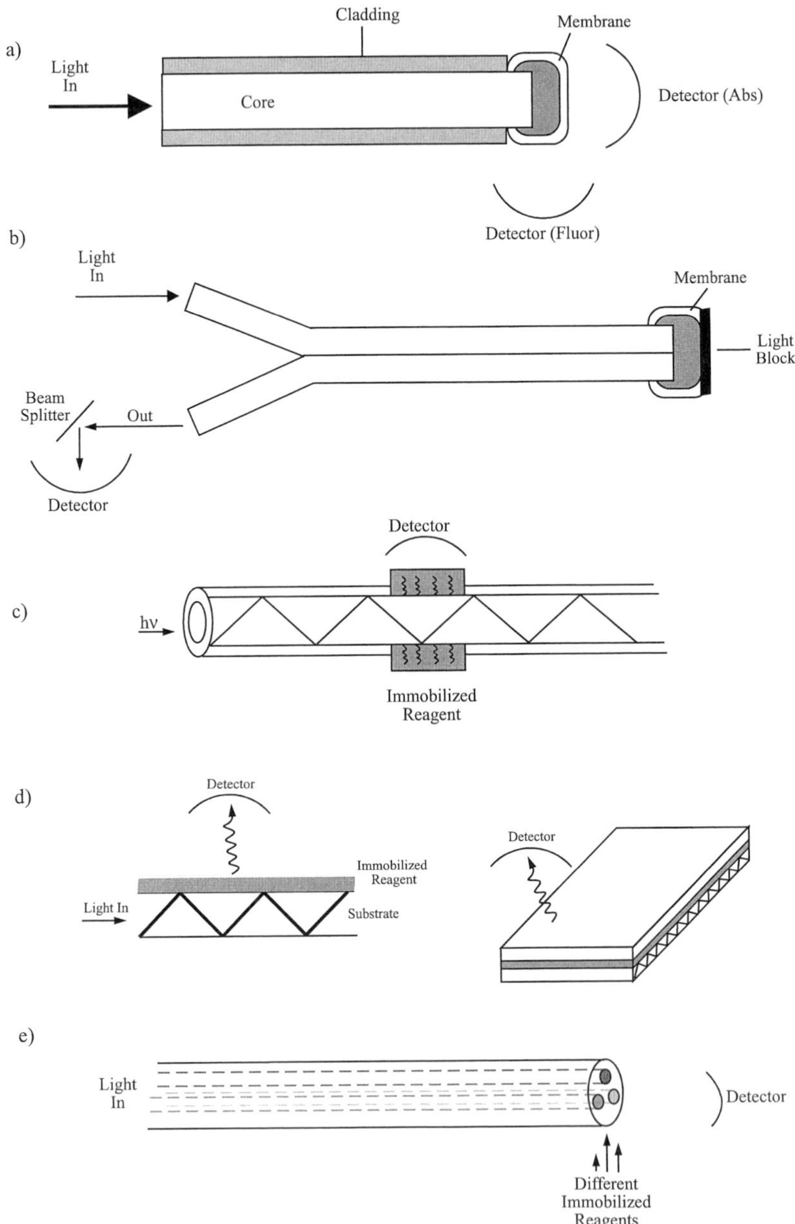

FIG. 5-6. Typical configurations of optically based biosensors: (a) distal-end immobilization; (b) bifurcated fiber; (c) immobilized biocomponent layered to interact with the evanescent field (cladding of the fiber is stripped where the immobilization layer is applied); (d) planar wave-guide; and (e) fiber bundle for optical array.

are frequently better than fluorophores in a detection scheme because most fluorophores often require UV wavelengths for excitation, thus eliminating LEDs as sources. However, Bruno *et al.* (1997) have reported a fluorescence-based pH sensor using excitation from newly developed GaN blue LEDs. The LED emission maximum at 470 nm is appropriate for the pH-sensitive fluorescein indicators immobilized by a variation on Walt's procedure (Pantano and Walt, 1995).

The development of near-infrared fluorogenic agents has also provided important alternatives for fluorescence monitoring. Patonay and Antoine (1991) discussed the instrumental and chemical advantages of using near-infrared fluorogenic agents. Two of the basic advantages lie in the simplification of instrumentation, particularly miniaturized versions, and the lessened possibility of radiative damage to living systems. The most widely used agents are of the carbocyanine and phthalocyanine dye families. The former group is comprised of those highly conjugated dyes so often used for particle-in-the-box calculations in introductory physical chemistry. Patonay and coworkers have applied the near-infrared dyes for immunoassays and labeling DNA oligomers. They also have proposed instrumentation, using a laser-diode source and silicon-photodiode detector, for solid-phase immunoassay based on near-infrared excitation and emission (Casay *et al.*, 1994; Williams *et al.*, 1994; Shealy *et al.*, 1995).

In Figure 5-6(b) a so-called bifurcated fiber is used to excite through one fiber and monitor the altered signal via a different return fiber. This geometry requires a blocking cap at the reagent end of the fiber. The advantage of using the twinned fiber is that it is simpler to separate the excitation light from the return light. This is particularly helpful if the returning light is emitted from a fluorescent species in the reaction matrix.

Lin, J. *et al.* (1996) have proposed an improved two-fiber probe (penetrometer) that can be used for *in situ* environmental monitoring, as well as other applications. They used fibers, polished at predetermined angles, arranged with a geometry that provides optimal recovery of emitted light. An enhancement factor of 14 was observed when measuring organics in water, even when the penetrometer was used in sands and clay.

In Figure 5-6(c) the evanescent wave has excited a fluorescent species immobilized on the longitudinal surface of the fiber, and the resulting emitted light is monitored. As an alternative approach, the fluorescent reaction of (c) could have occurred in a confined region at the end of the fiber with the detector placed perpendicularly to the propagated ray, as shown in Figure 5-6(a).

Several types of evanescent-wave biosensors are included in Table 5-3. The basic approach of utilizing the evanescent wave can be applied for both fibers and planar waveguides and to a variety of chemical transduction

TABLE 5-3 Examples of Fiber-optic Evanescent-wave Sensors

Reference	Type of sensor	Comments
Abel *et al.*, 1996	Oligonucleotides	Biotinylated oligonucleotides were bound to an avidin-treated quartz fiber. Fluorescein was used as the fluorophore. 24 fmol LOD for complementary oligonucleotide.
Devine *et al.*, 1995	Immunosensor for cocaine	Fluorophore-labeled benzoylecgonine is displaced from monoclonal antibodies by cocaine, thus decreasing fluorescence. LOD 5 ng/ml. Cross-reactivities with cocaine metabolites reported.
Piunno *et al.*, 1995	DNA sensor	Tethered ssDNA (dT_{20}) synthesized on surfaces of quartz fiber. Ethidium bromide used as fluorophore. LOD 86 ng/ml cDNA.
Shriver-Lake *et al.*, 1995	Immunosensor for TNT in water	Tapered optical fibers. Fluorophore-labelled TNT analog was prepared. Monoclonal Ab's raised against the analog were bound to the fiber using Bhatia *et al.*'s method (1993). TNT in sample then displaced analog, thus decreasing fluorescence. LOD 8 ppb TNT.

modes. The examples of Table 5-3 illustrate the versatility of the methodology.

Pantano and Walt (1995) have summarized the developmental research and various applications of their imaging optical fiber bundle, mentioned in Section 5-1. The bundle is comprised of thousands of individual fiber units, yet is only 350 μm in diameter. Using a site-specific photopolymerization method developed in Walt's laboratory, individual optical units of the fiber bundle can be modified with the same immobilized species or with a variety of reagents (Healy *et al.*, 1995; Ferguson and Walt, 1996; Healy and Walt, 1997). This creates an optical array that has proven very successful for multianalyte determinations and imaging using a CCD camera as detector. The fibers modified in a variety of ways have been used for pH, glucose, dissolved gases, acetylcholine, pencillin and oligonucleotides (Bronk and Walt, 1994; Li, L. and Walt, 1995; Bronk *et al.*, 1995; Ferguson and Walt, 1996; Healy and Walt, 1997). Using the fiber in the imaging mode they have imaged samples a few tens of micrometers in diameter (Bronk *et al.*, 1995). As mentioned earlier, the Healy and Walt (1997) report emphasized the fast response time of a pH-sensing array, but the authors also suggested that 7 μm spatial resolution might be attainable with the array. The ability to perform multianalyte determinations in spatially resolved spaces is a potentially powerful analytical technique.

5.2. PLANAR WAVEGUIDES

Discussion to this point has centered on the optical interactions employing an optical fiber as the waveguide. There are many other designs of devices in the general area of optical sensors. The principle of total internal reflection still applies for many of these sensors, as does the involvement of the evanescent wave. A generalized planar sensor is depicted in Figure 5-6(d). In this configuration the optical substrate is used as the waveguide. In this figure, perpendicular detection of fluorescence is indicated for illustration, but other detection schemes and geometries are possible. For example, end-on detection can be utilized with filtering of the excitation light. Attenuated total reflection and various phase-selective modes are also widely used. Sutherland and Dähne (1989) have summarized the basic principles of internal reflection devices including examples of planar waveguides. The Dessy (1989) review also discusses basic principles with some of the earliest examples drawn from reported applications to that date. Snyder and Love's (1983) monograph is foundational for designing optical waveguides.

The recent review by Ingersoll and Bright (1997) describes a series of studies in which that group has investigated interfacial dynamics of biosensors using time-resolved total internal reflection fluorescence (TIRF). Results related to binding affinities, effects of oriented immobilizations of antibodies and Fab fragments, and the mobilities of immobilized biocomponents in sol–gels are summarized in Bright *et al.* (1990), Lundgren *et al.* (1994), Narang *et al.* (1994); Jordan *et al.* (1995), Wang, R. *et al.* (1995), Ingersoll *et al.* (1996).

Figure 5-7 depicts another variation for guiding waves—i.e., using thin films as the waveguide. The device illustrated is for variable angle operation, but fixed angle configurations with optical array detectors are often simpler to implement. The film's refractive index in a sensor of this type would be higher than that of the supporting substrate. Light may be coupled to the

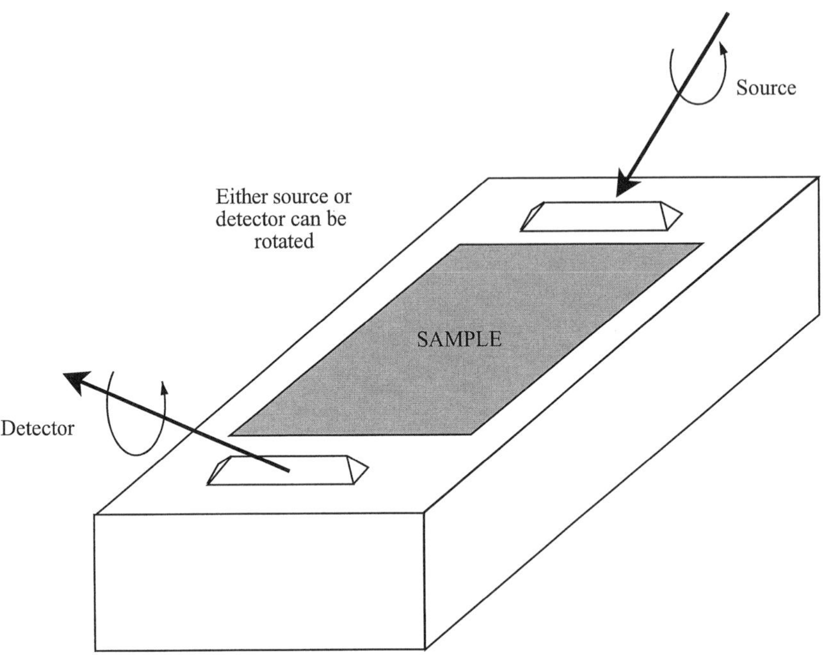

FIG. 5-7. Planar waveguide with prism coupling. The film acts as the waveguide in this example.

film waveguide through the use of coupling prisms, as indicated in Figure 5-7, or a grating, either integrated into the device or imprinted on the film may be used. Any chemical transformations of the film—e.g., analyte binding—are manifested in changes related to intensity, frequency, phase or polarization.

A series of reports from Saavedra's group provides an instructional progression of development of integrated optical waveguides using sol–gel layers for both waveguiding and sensing (Yang, Lin *et al.*, 1994; Yang, Lin and Saavedra, 1995; Yang, Lin *et al.*, 1996). In a recent report the same group demonstrated the use of waveguides in an attenuated total reflectance (ATR) mode and in total internal reflection fluorescence (TIRF) for studying the oriented immobilization of yeast cytochrome *c* on supported lipid bilayers (Edmiston *et al.*, 1997).

Schmid *et al.* (1997) have also studied reversible binding of a protein with TIRF. By immobilizing histidine-tagged proteins (His-Prot) at a nickel-chelated surface on quartz they were able to utilize TIRF to determine association constants and kinetic parameters for the binding of His-tagged green fluorescent protein. The nitrilotriacetic acid–Ni surface anchor is adapted from affinity chromatography and the His-tagged proteins are genetically engineered. Fluorescence stimulated by the evanescent wave was used to determine surface coverages and to test the immobilization method for reversibility on the basis of imidazole competition for the binding sites. The authors suggest that this immobilization–TIRF combination is generally applicable to studying biochemical ligand-binding at surfaces with oxygen functionalities.

Heideman, Kooyman and Greve (1993) developed and characterized a sensitive immunosensor based on a planar waveguide with coupling gratings and interferometric detection. The incoming laser beam was split between a sample path and reference path of the interferometer. An antigen–antibody binding event at the surface resulted in a change in refractive index in the sample arm, thus changing the propagation velocity of the light wave. This refractive index variation induces a phase change for the propagated ray that is monitored interferometrically. The sensor exhibited high sensitivity (pM) for a determination of human chorionic gonadotropin (hCG), but the total analysis and alignment times were long. Lechuga *et al.* (1995) attempted to apply the same approach as an immunosensor sufficiently sensitive for low molecular weight pesticides. The waveguide was incorporated into a flow system which reduced t_{90} from about 20 min to 2–8 min, dependent upon the concentration. They were unable, however, to get an LOD low enough for the required pesticide limits.

As a followup to the Heideman report, Schipper *et al.* (1996) tried to overcome some of the problems with the interferometric sensor discussed above. They introduced a novel immunosensing design in which the critical angle at an interface changes, but the detection of the event can be measured via relative intensity changes of two beams affected by the interface. A portion of a waveguide's sensing surface was shielded from the eventual immunoreaction, thus providing one area of constant index of refraction n_c. In the unshielded area the index of refraction n_1 is different from n_c and changes due to an immunoreaction at the surface. Therefore, the critical angle Θ_c for reflection at the interface of the two areas changes:

$$\Theta_c = \arcsin\left(\frac{n_1}{n_c}\right)$$

Before analyte is introduced the device is tuned so that the amount of reflected (R) and transmitted (T) light at the boundary of the two areas is balanced. After adsorption or the immunoreaction occurs, this balance is disturbed. On-chip photodiodes are used to detect the changes in intensity of

R and *T*. The difference (*R* − *T*), normalized to the original value for R + T, may be used to monitor either concentration effects or time-dependences of surface-film alterations. The Schipper sensor was not as sensitive as the Heideman interferometric design, but it was easier to fabricate and use routinely. The authors concluded that there are numerous applications where it should be sufficiently sensitive, particularly competitive binding immunoreactions as opposed to direct binding studies.

The planar waveguide of Figure 5-8 has been used for an interesting application for oligonucleotide sequencing by hybridization. Stimpson *et al.*(1995) designed a very simple planar waveguide with an array of DNA sequences immobilized on the waveguide surface. Then biotinylated complementary sequences were introduced through a capillary-action channel for binding. Finally, a light–scattering colloidal selenium–antibiotin conjugate was introduced into the channel. The evanescent wave of the device was used to scatter light from the bound selenium sites at the DNA capture sites. The light scattering was monitored with a CCD camera.

5.3. NEAR-FIELD OPTICAL SENSING

The fiber-optic and waveguide devices discussed in the previous sections utilize conventional optical techniques for observing the interactions of light and matter. In recent years there has been perhaps no greater revolution in optical methodolgy as that enabled by near-field optics (NFO), near-field scanning optical microscopy (NSOM) and near-field scanning optical spectroscopy (NSOS). Near-field methods have allowed scientists to make optical measurements and to image spatial domains heretofore inaccessible without the use of sometimes damaging high-energy electron or ion beams. The foundational reports by Betzig and coworkers (Betzig *et al.*, 1991; Betzig and Trautman 1992) and the reviews by Lewis and Lieberman (1991), Harris *et al.* (1994) and Tan and Kopelman (Kopelman and Tan, 1994; Tan and Kopelman, 1996) introduce the revolutionary breakthroughs of modern near-field optics, plus some biosensing applications of the near-field devices. The Tan and Kopelman (1996) review includes specifics about fabrication of the optical tips and scaling considerations that affect response time, resolution and *in vivo* measurements.

Optical techniques designated above as "conventional" utilize elements such as lenses and slits as focusing devices with sources and detectors at a relatively large distance from the sample. With the advent of NFO the more traditional type of arrangement is now routinely distinguished as far-field optics. According to the physical laws of optics far-field instruments exhibit a variety of interference phenomena and diffraction limitations to spatial resolution. For example, in light microscopy these effects lead to a diffrac-

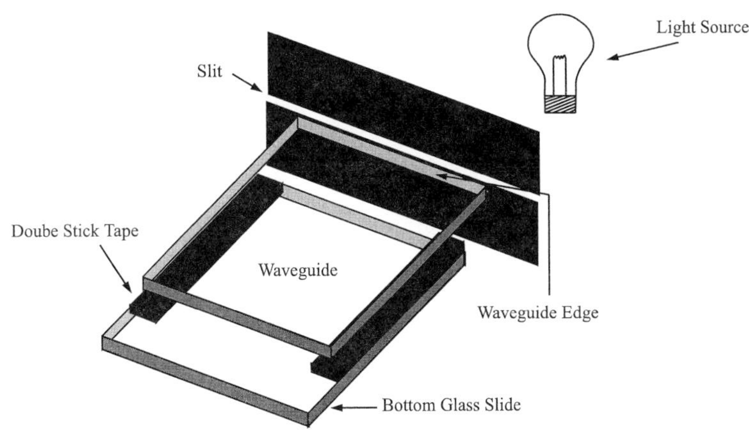

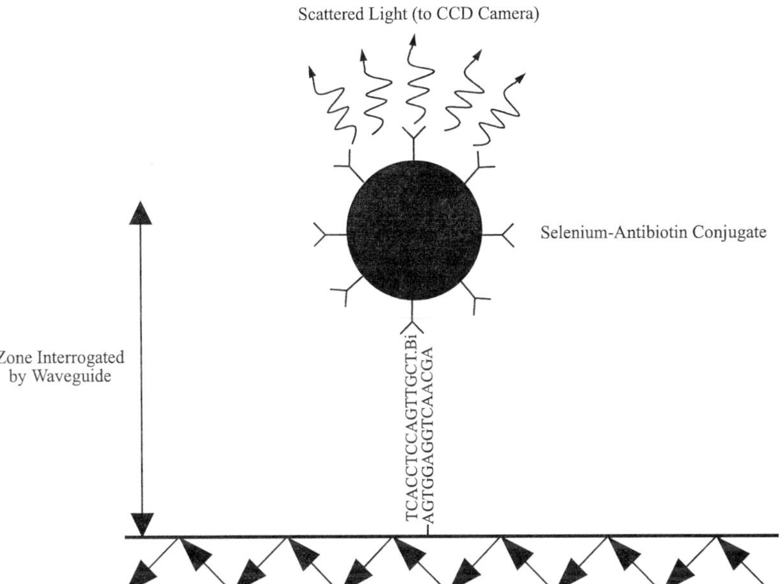

FIG. 5-8. Waveguide for studying sequencing by hybridization. The waveguide assembly is shown at the top, and oligonucleotides are immobilized on the waveguide. A second slide is added to form a channel for the introduction of reactants. Adapted, with permission, from Stimpson, D. I. *et al.*, *Proc. Natl. Acad. Sci. USA* **92**: 6379–6383. Copyright 1995, National Academy of Sciences, USA.

tion-limited spatial resolution of $\sim \lambda/2$ where λ is the wavelength of the light source. If violet light of ~ 400 nm is used, this means that resolution can be no better than ~ 200 nm or 0.2 μm. Referring back to the relative dimensions given in Figure 1-6, it will be obvious that image resolution is limited to objects slightly smaller than bacteria or mitochrondria, which would be at ~0.5 μm on the scale.

If light is directed to a sample through a narrow aperture, rather than being focused by far-field optics, the wavelength-dependent limits to resolution disappear. Resolution may be, theoretically, any value limited only by the size of the aperture. There are, of course, practical limits imposed by fabrication considerations, but current technologies suggest practical limits of $\sim \lambda/50$ (Tan and Kopelman, 1996). For the violet light cited above, this should provide an approximate resolution of 10 nm or less.

Near-field optical tips may be fabricated from precision-pulled micropipets or from optical fibers. To create the aperture the outer walls are metal coated, usually with aluminum. The optical tips may act as a conduit for light from laser sources to the sample (passive tips) or they may be converted into active tips by inclusion of a fluorescing crystal which becomes a radiative light source at the very tip of a micropipet (Tan and Kopelman, 1996). An example of this is represented by the device at the top of Figure 5-9.

While much of the early work in NFO emphasized the imaging aspects of the approach, NSOM, there have been accompanying developments in NSOS that have a direct bearing on quantitative and time-dependent phenomena pertinent to bioanalytical sensing. The use of NSOS allows spectroscopic investigation on a nanoscale basis. Thus, spectroanalytical methodologies can be utilized, for example, for intracellular studies.

A schematic of the near-field optical design for biosensing is shown in Figure 5-9. In the lower sketch the near-field is indicated as that region bounded in width by the diameter of the light aperture and in depth by the extension of the collimated source of radiation emerging from that aperture. The near-field sampling region falls within those boundaries. The included volume is on the order of femto- or attoliters. This drastically reduced volume is one of the primary factors in improving detection and resolution through the use of near-field methods (see Harris *et al.*, 1994).

Micropositioning must be used to locate the radiative source above the sample, and scanning techniques can be utilized to move the apertured tip around a sample surface with a resolution of a few nanometers. The analytical power of the method is illustrated well by the work of Birnbaum *et al.* (1993). Using an apertured tip 100 nm in diameter, they characterized spectroscopically nanoscale heterogeneity in adjoining microcystallites of α- and β-perylene.

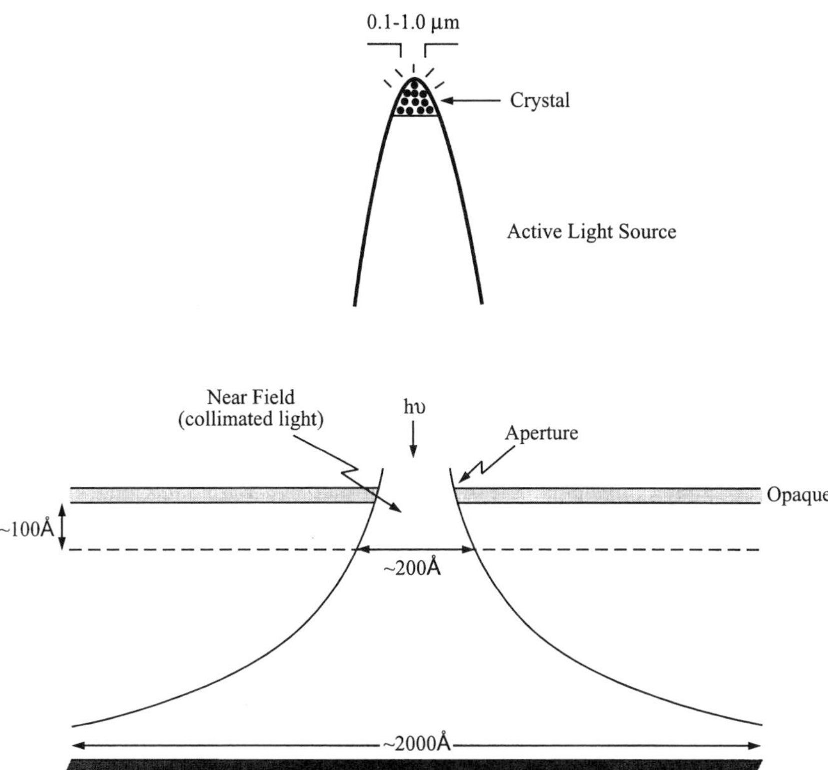

FIG. 5-9. (Top) "Active" light source with emitting crystal (e.g., anthracene or perylene) in tip of a pulled pipet (inner diameter < 1 μm). An outer metal coating on the tip is not shown. (Bottom) Schematic of near-field optics (NFO) using apertured light. Passive or active light sources can be used. The sampling depth is designated by the dotted line. Adapted, with permission, from Tan, W. and Kopelman, R., In Wang, X. F. and Herman, B. (eds.) *Fluorescence Imaging Spectroscopy and Microscopy*, p. 410. Copyright 1996. John Wiley and Sons, New York.

The usual near-field experiment involves excitation of a potentially fluorescing sample, with collection of the emission by an inverted fluorescence microscope objective. The emission is then passed to a spectrometer/photomultiplier (PMT) or optical multichannel analyzer (OMA) with computerized data acquistion. Due to the minute sizes of the optical tips and the method of data acquisition the response times of these sensors are on the order of milliseconds compared with conventional optical sensors that may have t_R's in the seconds range (Tan and Kopelman, 1996).

Kopelman's group has been quite active in applying NSOS to biosensing problems. Some of their earliest work involved intracellular pH-monitoring in embryos (Tan *et al.*, 1992b). Their more recent pH-sensors, 2 μm and 125 μm in diameter, are based on a fluorescein derivative and they use a ratiometric method with single excitation wavelength and dual emission detection (Song *et al.*, 1997). Precision of ± 0.03 pH units was reported, with response times of 1 ms and 400 ms for the smaller and larger sizes, respectively. That same group has developed several chemically modified near-field optical sensors that can be used for calcium and sodium ions (Shortreed *et al.*, 1996a,b), oxygen (Rosenzweig and Kopelman, 1995), glucose (Rosenzweig and Kopelman, 1996) and clinically relevant anions (Barker *et al.*, 1997). Some of these sensors have been discussed in prior chapters within the context of ion-selective sensing and oxygen determinations by quenching of Ru-complex fluorescence. In all cases low limits of determination, sensitivity, and stability of the optical probes have been demonstrated.

5.4. SURFACE PLASMON RESONANCE

Surface plasmon resonance (SPR) is another widely used optical method for investigating thin-film phenomena, including binding effects at surfaces. This technique has been adapted for a variety of problems, but has been especially valuable in elucidating biospecific interaction analysis (BIA). As an indication of the growing importance of SPR for BIA, the June 1994 issue of the *Journal of Immunological Methods* was devoted entirely to applications of SPR in immunology. The primary impact of SPR in this area is the ability to monitor in real-time the binding interactions of immuno-components. Thus, one is able to conduct studies based on not only binding equilibria data, but also the kinetics of the on–off binding processes. The mechanistic as well as equilibria information is essential to understanding the intricacies of biochemical events. The Schmid *et al.* (1997) report on binding studies, mentioned in Section 5-2, dealt with a TIRF method that the authors proposed as a possible alternative or complementary method to SPR for obtaining this kind of information.

As introduced in first physics courses, electromagnetic radiation is characterized by electrical and magnetic components which are orthogonal. When Maxwell's equations are solved for the propagation of light, the single-mode solution is doubly degenerate, with one solution corresponding to the tranverse electric mode (TE) and other to the transverse magnetic (TM) mode. These are the TE and TM polarization states. This means that states can be oriented by passing light through polarizers. For example, when using light to impinge on a surface, a TM polarized beam would have the magnetic component perpendicular to the plane of the surface normal while the TE component would be parallel to the plane of incidence. This TM polarized light is essential to the phenomenon of SPR since it is the TE component that is responsible for creating surface plasmons in the metal at the prism interface. (see Raether, 1977, 1980; Bender *et al.*, 1994; VanderNoot and Lai, 1991).

Metals have often been described as seas of electrons. That concept is useful when describing surface plasmons. One can envision a surface wave created by an oscillating electrical charge on the surface of the metal. This oscillating charge density is called a *plasmon*. Discussions of plasmons and their origins can be found in Raether (1977, 1980), Sadowski *et al.* (1991), VanderNoot and Lai (1991), Liedberg *et al.* (1993), and Jordan *et al.* (1994).

For the purposes of this discussion it is sufficient to say that the surface plasmons of the metal layer cannot be excited directly by incident light due to restrictions dealing with the laws for conservation of energy and momentum. The plasmons can be induced, however, by the evanescent wave from total internal reflection using TM polarized light. The surface plasmon effect is extremely sensitive to the refractive index and thickness of films on the opposite surface of the metal layer.

The usual Kretschmann configuration for SPR is that of Figure 5-10. Light undergoes total internal reflection so that the evanescent wave interacts with a thin metal film on the surface of the prism. Silver and gold have the highest optical constants for this purpose (see Sadowski *et al.*, 1991). The incidence angle of coherent, TM polarized light is varied. At some particular incidence angle the electric vector of the incident light induces a resonance with the surface plasmons.

If a thin film of differing dielectric is immobilized on the opposite metal surface, the evanescent wave of the surface plasmon couples to that layer. Chemical changes in that film then modulate the reflected light from the prism. The reflected light is monitored and the resonance condition is indicated by a decreasing intensity of the reflected light as a function of the incidence angle (Figure 5-10). The angle at which the minimum occurs is referred to as the SPR angle. For example, a resonance valley would occur at one angle of incidence for an antibody immobilized on the metal surface, whereas the SPR angle would shift when the antibody binds to antigen in a sample. Shifts in the SPR angle may be converted into thicknesses of the thin film on the gold. Coupled with ellipsometric data for optical properties of the thin films, thicknesses below monolayer coverage can be determined. The reader is referred to Liedberg *et al.*(1993), Frey *et al.* (1995), Löfas (1995a,b), Löfas *et al.* (1995) and Sigal *et al.* (1996), and references therein, for examples of practical units and conversions to thicknesses. This set of reports includes applications using avidin–biotin protein immobilizations, and also carboxymethyldextran and alkanethiolates (SAMs) as coupling

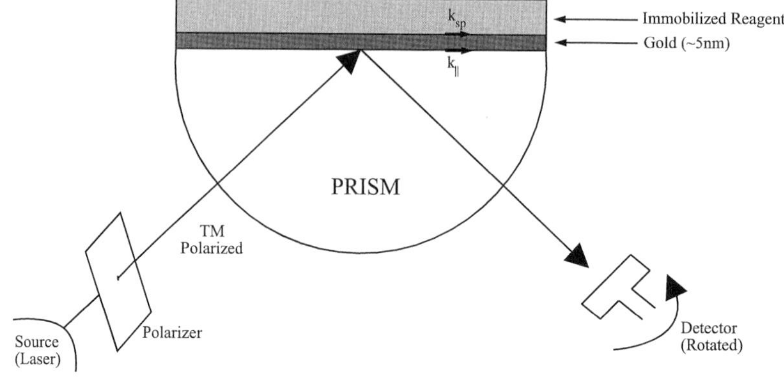

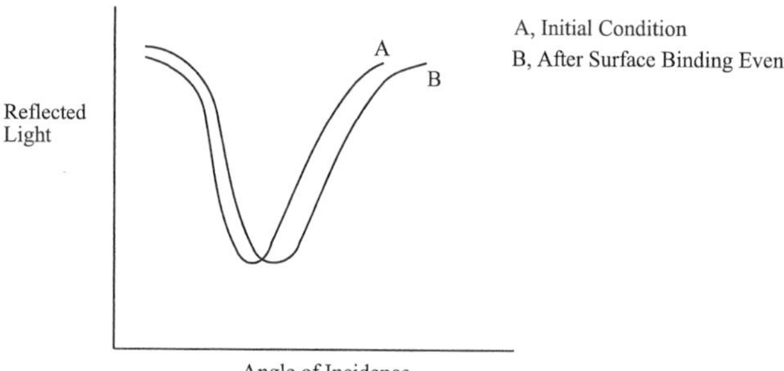

FIG. 5-10. Surface plasmon resonance showing (top) the Kretschman configuration. The resonance condition results when the evanescent wave of TM polarized light couples to produce surface plasmons in the metal, and changes as a function of angle of incident light in the presence of the immobilized layer on the metal. Underneath is shown the shift resulting from the effect of the layer on the signal.

matrices. Sigal *et al.* (1996) have presented a comparison of the latter two approaches as chemical modifications for studying proteins.

A typical experimental setup for SPR consists of a laser source, a carefully controlled polarizer, the prism/metal film assembly, a rotation mechanism for changing the angle of incident light (thousandths of a degree per step), wide angle optics and a sensitive detector for monitoring the reflected light (see Figure 5-11). The changes in reflectivity are small and one is confronted with the usual dilemma of trying to measure a small signal on top of a much larger signal, but resolution with modern instruments is remarkable.

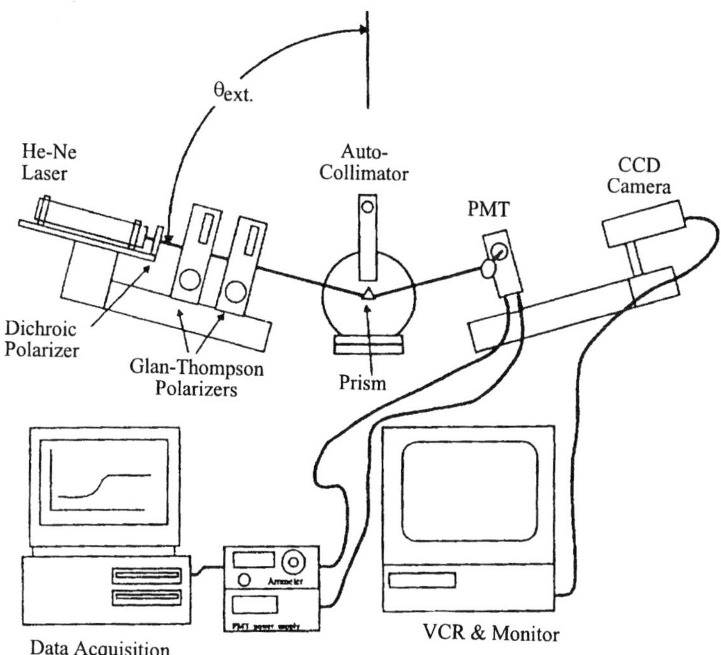

FIG. 5-11. Assembly for surface plasmon microscopy used to study gradients of refractive index or thickness with spatial resolution of about 5 micrometers. Reprinted with permission from Brennan *et al.* In Pecsek, J. J. and Leigh, I. E. (eds) *Chemically Modified Surfaces*, p. 75. Copyright 1994, The Royal Society of Chemistry, Cambridge, UK.

Variations on the configuration described above have been developed. Corn's group (Frey *et al.*, 1995; Jordan and Corn, 1997) used fixed-angle SPR with a CCD camera to image patterned protein-modified surfaces. VanderNoot and Lai (1991) used photothermal deflection of a laser beam immediately above the sensing layer for detection. Based on the fact that, for polychromatic light, different wavelengths are reflected at different angles, Eigen and Rigler(1994) employed a polychromatic source with a linear diode array as a spatially resolving detector. Karlsen *et al.* (1995) have proposed an instrumental design using multiple angles and multiple wavelengths, thus enabling acquisition of both refractive index and absorbance data. These authors presented this as an approach to potential multicomponent analyses using multivariate techniques. Lawrence *et al.* (1996) have proposed the use of disposable diffraction gratings in place of the prism configuration. The gratings were fabricated as Perspex (Mooney Plastics, Ltd) replicas of a holographic diffraction grating, then they were sputter-coated with gold on the surface. The eventual goal for use of the gratings is in a portable SPR biosensor.

A significant modification to the conventional SPR configuration was presented by Cush and others (Cush *et al.*, 1993; Buckle *et al.*, 1993). They produced a device called a resonant mirror (Figure 5-12) that was designed to capitalize on the sensitivity of waveguide devices while using the basic construction principles of an SPR device. The major difference lies in replacing the SPR's thin metal layer with a dielectric resonant layer of high refractive index titania. The evanescent waves from reflections in the prism are coupled to the resonant layer through a low refractive index silica layer. A guided wave is produced in the resonant layer and the evanescent wave from that layer interacts with the topmost sensing layer containing reagent or sample. Both TE and TM polarizations produce resonance, as is true of waveguides. The device can be used with a focused beam and an optical detecting array or the beam and detector can be rotated through angles to detect the intensity changes upon resonance.

Even with the variations for detection and enhanced information retrieval, prism-based SPR and the resonant mirror are limited in applications. The incorporation of the prism and associated optics tends to complicate portability and miniaturization. Thus, they are effective biosensing tools, but difficult to translate into a practical biosensor to be applied outside the realm of research laboratories. The disposable gratings proposed by the

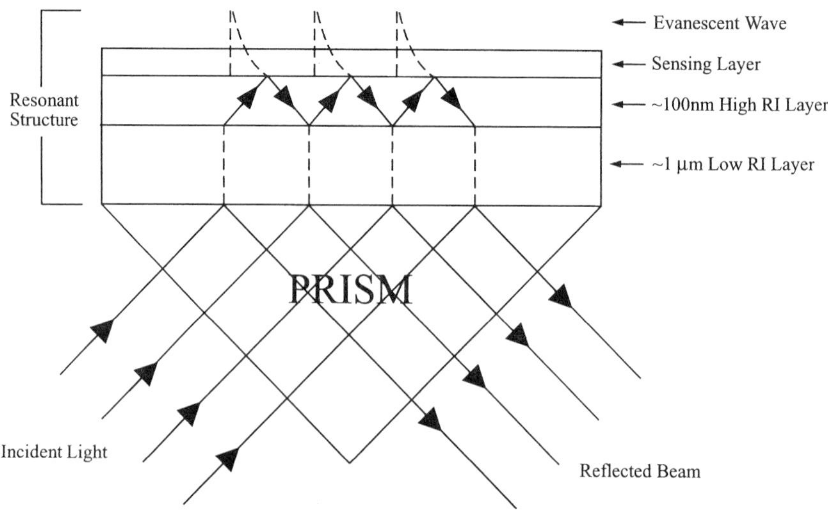

FIG. 5-12. Schematic of the resonant mirror (see text).

Lawrence group is one approach to this problem of practical portability. Another alternative to the prism-based system, a fiber-optic SPR device, was proposed by Bender, Dessy, Miller and Claus (1994). The device is schematized in Figure 5-13. The authors contend that the fiber-optic format lends itself to miniaturization and responsiveness in the refractive index ranges encountered in biochemical analyses, though the initial characterization did not include behavior in biosensing environments. They also propose that the sensitivity should be approximately the same as a prism-based instrument if the materials and processes of fabrication are optimized.

Bender's sensor is referrred to as a "four-layer device", though part of the cladding comprises one of the layers . The cladding is removed except for a thin layer at the core surface. The surface is polished and a thin metal layer deposited. A silicon monoxide overlay is deposited for refractive index "tuning" to the range of sample refractive indices to be encountered. A diode laser source and photodiode detector were used. The authors calculate that the active surface area is only 10^{10} nm^2 and a minimal sample volume for use is 2 nL.

Sigal *et al.* (1996) reported on an SPR study of histidine-tagged proteins bound to a gold surface via a mixed self-assembled monolayer (SAM). Essentially the same nitrilotriacetic acid–Ni chemistry described in the Schmid *et al.* (1997) TIRF study was utilized. The authors compared the efficiency of oriented protein binding to the SAM with protein immobiliztion by covalent binding to a thin dextran gel, as proposed by Löfas (1995) and Löfas *et al.* (1995). The conclusion from the Sigal SPR data was that the

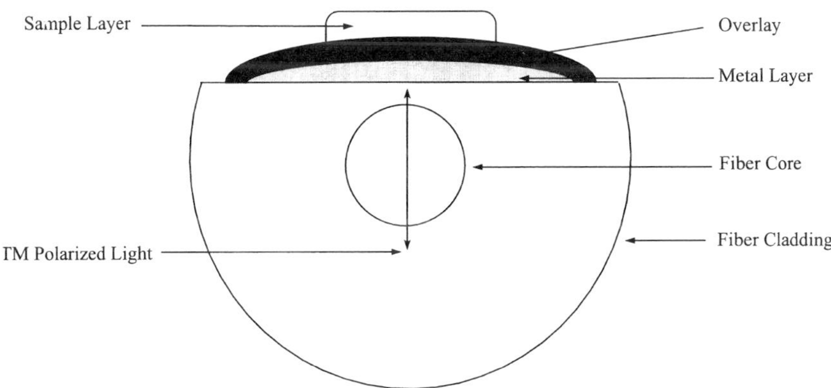

FIG. 5-13. Optical fiber adapted for SPR: this is a schematic of a design by Bender, *et al.* (1994).

NTA–Ni affinity binding is specific for His-tagged proteins and that the binding of the SAM to the NTA–Ni was more efficient than that for the dextran gel.

5.5. OPTICAL SOL–GEL SENSORS

Optically transparent, porous sol–gel matrices containing immobilized biochemicals can be produced in a variety of sizes and geometries that are amenable to both direct spectrophotometric measurements and the various methods involving guided waves. The sol–gel products can be formed into monoliths (single blocks), films or fibers (MacCraith,1993; MacCraith *et al.*, 1993; Avnir *et al.*, 1994; Dave *et al.*, 1994; Lev *et al.*, 1995; Ingersoll and Bright 1997). Of all the optical sensing approaches the use of sol–gel matrices is, perhaps, most closely related to some of the original spectrophotometric bioanalytical protocols.

Instead of having reagents in a conventional photometric cell, the biomolecules are immobilized in the sol–gel. Smaller analyte molecules diffuse into the monolith or sol–gel film where the reaction of interest may be monitored via the spectral response. The monoliths are sometimes referred to as biofunctional glasses, and may be thought of as surrogates for the usual photometric cell.

Absorbance or luminescence behavior may be followed quite easily using any of the spectral methods which would be applicable for ordinary excitation/emission spectrophotometry. While monoliths are quite amenable to ease of handling and optical measurements, response times are diffusion dependent, and thus they are controlled by the thickness of the device.

Monoliths have been convenient platforms for extensive investigations of physicochemical phenomena associated with analyte diffusion and the effects of sol–gel entrapment on biomolecular species (Wang, R. *et al.*, 1993; Dave, *et al.* 1994; Jordan *et al.*, 1995; Ingersoll *et al.*, 1996). Several generalized conclusions emerge from the combined studies: (a) low molecular weight analytes do diffuse into the sol–gels, thus having acceptable access to the entrapped biomolecules; (b) the biomolecular function of the immobilized species appears to be essentially the same as that observed in aqueous solutions, though some differences are observed; and (c) immobilization in the sol–gel matrix does not restrict the internal and global motions of the entrapped proteins. The implications of these findings for optical measurements are very important. (1) Biomolecules may be altered less by sol–gel entrapment than by other immobilization methods. Thus, the optical measurements involving the species should mimic very closely the natural systems. (2) Having the unencumbered biomolecule trapped, yet free to undergo some of the natural motions, means that valuable binding–reaction information can be obtained from spectral studies of the sol–gel sensors.

Although monoliths are extremely convenient, sol–gel films promote faster response times and are more amenable to miniaturization. Films may be used on planar substrates, optical fibers or waveguides. Many of the same measurement strategies discussed in previous sections apply. MacCraith (1993) and MacCraith *et al.* (1993) utilized sol–gel coatings on fiber-optic evanescent-wave sensors for detecting oxygen by fluorescence quenching of Ru complexes. Bright's group has used films for phase-resolved evanescent-wave fluorescence to study interfaces (Lundgren *et al.*, 1994; Ingersoll and Bright, 1997). Dunuwila *et al.* (1994) proposed a titanium carboxylate sol–gel film for optical test strips. In characterizing their devices they used immobilized metalloporphyrins reacting with cyanide, and conventional absorbance methods for measurements.

As discussed in Section 5.2, Yang, Lin and Saavedra (1995) prepared a planar waveguide (integrated optical waveguide) for use with attentuated total reflectance as the monitoring mode. The device was tested for pH and Pb^{2+} determinations. They found that the device was sensitive with rapid response, thus showing the response advantages of thin-film sol–gel approaches. Narang *et al.* (1994b) have tested a sol–gel/glucose oxidase/sol–gel sandwich device as a biosensing platform. The devices were tested optically and amperometrically as fabrication conditions were varied and compared. The device was judged to be stable, rapid in response and capable of high enzyme loading.

5.6. LUMINESCENCE AND LIFETIMES

Throughout this chapter numerous examples have been provided for sensors based on fluorescence intensity. Different configurations for generating and detecting fluorescence have been introduced. In previous discussions of oxygen sensing the use of luminescence quenching of ruthenium and rhenium complexes has been introduced. What has not been examined is the elucidation of the quenching process in the microenvironment of polymeric matrices. This becomes an important consideration for understanding experimental observations for quenching-based sensors, and it is a topic still under investigation.

For solution measurements, the classical Stern–Volmer relationship is used to describe bimolecular quenching kinetics:

$$A + h\nu \rightarrow A^*$$

$$A^* \xrightarrow{k_1} A + h\nu$$

or

$$A^* + Q \xrightarrow{k_2} A + Q^*$$

The Stern–Volmer equations for this scheme are:

$$\frac{I_0}{I} = 1 + K_{SV}[Q]$$

and

$$K_{SV} = k_2\tau_0$$

I_0, and I are emission intensities in the absence and presence of quencher, K_{SV} is the Stern–Volmer quenching constant and $\tau_0(= 1/k_1)$ is the luminescence lifetime. Plots of I_0/I versus [Q] should be linear, and luminescence decay curves can be represented by single exponentials. Either the concentration-based or lifetime-based modes can be used for quantification. Most of the developmental work has been done with O_2 as the quencher, and thus [Q] would be either the concentration or the partial pressure of oxygen, with appropriate changes in the units. If there are additional quenchers, then additional terms would be added, with a K_{SV} for each quenching species.

In the usual polymeric microenvironment of an immobilized luminophore, the physical conditions of free solution species do not pertain. There may be microdomains of the polymeric matrix that exert different influences on the luminescence quenching process and the matrix itself may act a quencher. This environmental heterogeneity and its effect on the response of quenching-based sensors has been the topic of much discussion. Linear Stern–Volmer plots are no longer observed and the decay curves are no longer single-exponential in nature. If quenching-based sensors are to be designed for optimal performance, then it is necessary to model accurately the observed behavior of the devices.

In a series of reports Demas and coworkers developed a model based on studies in different types of polymeric matrices (Xu *et al.*, 1994,1995; Demas *et al.*, 1995). They postulated domains differing in their polarities, and proposed what they referred to as a two-site model. Using curve-fitting routines with four free-floating parameters they were able to model their data for quantification of oxygen. Unfortunately, when they compared their model with one based on nonlinear solubility (adsorption-type), the curve-fitting worked equally well for either case. They suggested that while quantitative information might be obtained using either model, their complementary data characterizing the microenvironment tended to favor the physical interpretation of the two-site model.

Primarily interested in lifetime-based sensing, Draxler and Lippitsch (1996) proposed a different theoretically based model for explaining the behavior of luminophores in non-uniform environments. They formulated their model on the time-dependent probability functions for radiative and non-radiative de-excitation of the luminophores in a heterogeneous medium. The resulting integrated decay function for the de-excitation process in a 3-dimensional matrix, including the presence of a quencher, has only three model parameters, and contains in the exponential expression both a linear and a nonlinear term:

$$N(t) = N(\mathrm{o}) \exp\left[-(1 + \mathrm{c})\frac{\mathrm{t}}{\tau} - \mathrm{a}\left(\frac{\mathrm{t}}{\tau}\right)^{1/2}\right]$$

N(t) is the number of luminophores in the excited state at time t; a is a parameter proportional to the density of quenching sites of the matrix, and c is a quenching parameter proportional to the concentration of the quencher, oxygen in this case. The fit to experimental data was excellent for the two systems studied, one of which was an oxygen sensor.

Lakowicz and coworkers have applied lifetime-based sensing in the frequency domain through the use of phase-modulation (Bambot *et al.*, 1995a,b; Szmacinski and Lakowicz, 1993). Excitation light is sinusoidally modulated. Emission must oscillate at the same frequency but, due to the finite luminescence lifetime, the emission is phase shifted:

$$\tan\phi = \omega\tau \qquad m = [1 + \omega^2\tau^2]^{1/2}$$

where ϕ is the phase shift, ω is the frequency of modulation and m is the demodulation factor. The method has been applied to pH, CO_2 and oxygen sensing. In developing a non-invasive method for sensing oxygen through skin, the authors noted that while emission intensity was attenuated by a factor of 50, the phase-based response was unaffected by the presence of the skin (Bambot *et al.*, 1995b).

5.7. ELECTROGENERATED CHEMILUMINESCENCE

Chemiluminescence and bioluminescence have been utilized in several types of sensors, and the potential uses of such sensors have been discussed in Chapter 31 of the Turner, Karube and Wilson (1989) book. The chemiluminescent (CL) sensors for determining NO, introduced in Chapter 2, provide good examples of the former type (Kikuchi *et al.*, 1993; Zhou and Arnold, 1996). Collins and Rose-Pehrsson (1995) have also introduced potentially portable luminol-based CL sensors for oxygen and nitrogen dioxide. They were able to achieve some degree of selectivity by varying the immobilization matrix. The use of arrays with variable matrices was suggested as a possible strategy for better selectivity. Wood and Gruber (1996) have recently used bioluminescence as a physiological probe for possible environmental monitoring. Holt *et al.* (1996) developed a very sensitive heroin sensor based on the bioluminescence of bacterial luciferease in a reaction coupled to an oxidoreductase action on the heroin metabolites. They reported detection of 250 pmol of heroin.

One of the most versatile applications of this family of sensing modes, however, has been developed using electrogenerated chemiluminescence (ECL). As indicated during the 1995 Pittsburgh Conference symposium, "Advances in Electrogenerated Chemiluminescence," ECL methodology was developed more than thirty years ago (see Bard and Faulkner, 1980; Faulkner and Bard, 1977). Yet, practical biosensing applications have emerged only in recent years.

Yang, H. *et al.* (1994) and Jamieson *et al.* (1996) have described applications for immunoassays, DNA probe assays and several clinical analytes. The chemical sequences of the ECL process are not simple and Mikkelson (1996) has noted that the proposed mechanisms are a bit perplexing for the DNA probe analyses. The method provides, however, sub-picomolar LODs for PCR products.

A basic ECL instrument usually consists of a flow system, a potentiostatically controlled electrochemical cell and a luminometer. A common ECL reporter molecule is the excited state ruthenium complex $Ru(bpy)_3^{2+*}$. A promoter of an excited state, changes depending on the application. For immunoassays and DNA probe analysis tripropylamine has been used (Yang, H. *et al.*, 1994). The 1996 Jameison *et al.* report involves reactions in which either NADH or oxalate ion is used as the promoter for determinations of glucose, ethanol, CO_2, cholesterol and glucose-6-phosphate dehydrogenase activity. The reader should consult the original reports for the complete reaction cycles.

Both promoter and reporter may be oxidized at the electrode, then react in an electron-transfer sequence to form the chemiluminescent species for detection:

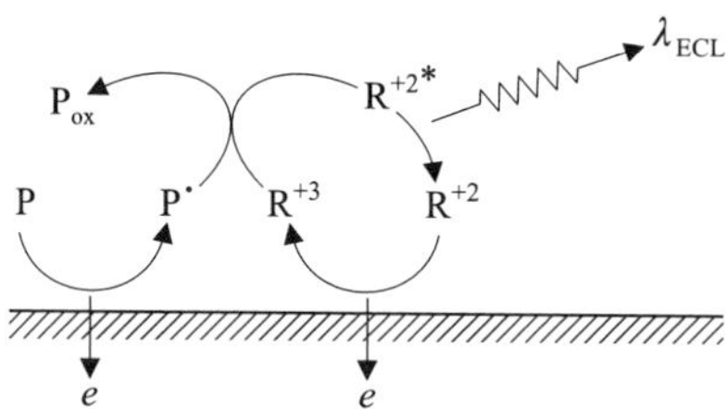

Bioreaction components, tagged with the ECL reagents, may be immobilized on magnetic beads with steptavidin/biotin chemistry, then captured at the electrode by the magnet. This localization at the detector provides significant ECL signal enhancement (Yang, H., *et al.* 1994).

A much simpler kind of ECL device has been proposed by Preston and Nieman (1996). It is designed to be used as a dip-probe with a diameter of only a few centimeters. No flowing stream is required for supply of reagents and the cell is designed so that protection from ambient light is unnecessary. The working electrode is incorporated into the probe. A fiber-optic emission collector is positioned 3 mm above the working electrode. The probe was tested with several standard ECL chemistries, including the H_2O_2/luminol reaction where the H_2O_2 was produced in an enzymatically catalyzed glucose oxidation. The LODs were reported as comparable to those of corresponding HPLC or flow-injection ECL analyses.

In summary, the preceding sections indicate that optically based sensors have been utilized in various measurement modes ranging from conventional absorbance and luminescence techniques to a wide variety of indirect interactions via the evanescent wave resulting from total internal reflection. Applications have run the gamut from straighforward quantitative determination of analytes to nanoscale spatial resolution to measurement of dynamic phenomena related to biospecific interactions and interfacial phenomena. The newer technologies and methodologies for optical sensing suggest a bright future in bioanalytical applications.

6. Thermal and Acoustic-Wave Transduction

Although electrochemical and optical sensors predominate in practice, there are several other modes of energy transduction, such as thermal and acoustic-wave detection, that can be quite effective in biosensing applications. Thermal devices have the longer history in biosensing (Danielsson and Mosbach, 1989). Acoustic-wave devices have taken on new importance with greater understanding of physicochemical contributors to the overall responses, including their behavior in liquid environments (Auld, 1990; Andle and Vetelino, 1994; Jossé, 1994 and references therein; Martin *et al.*, 1997) and their versatility with regard to the informational output (Yang, M. and Thompson, 1993 and references therein; Yang, M. *et al.*, 1993 and references therein; Thompson and Stone, 1997).

Lack of inherent selectivity is a common trait of the energy-transduction modes mentioned above. Yet, some very selective quantitative measurements are possible by well-designed chemical modification strategies using thermal and mass-sensitive acoustic-wave devices. Also, both thermal and acoustic-wave devices can be miniaturized for array formats as well as any of the energy transducers (Xie, B. *et al.*, 1995; Zellers *et al.*, 1995; Grate *et al.*, 1993c; Mandelis and Christofides, 1993).

6.1. THERMAL SENSORS

Thermal biosensors are based on principles of enthalpimetry for which $\Delta H = C_p \Delta T$ and $\Delta H_{tot} = \Sigma \Delta H_i$ according to Hess' law when multiple (*i*) reactions are in equilibrium. The calorimetric devices are designed for either adiabatic measurements using thermistors or for a heat conduction mode using thermopiles. Configurations range from sensing systems with miniaturized, well-insulated immobilized enzyme reactors (IMER) to microflow cells with micromachined reaction chambers. Various types of thermal sensors, most of them based on enthalpic changes related to enzymatic reactions, have been used in clinical and industrial applications (Danielsson and Mosbach, 1989; Danielsson and Winquist, 1990). In addition to Danielsson, Bataillard (1993) has provided a brief overview of calorimetric sensing in bioanalytical chemistry.

Thermal responses depend on reaction enthalpies which may range from tens of kJ mol^{-1} to several hundred kJ mol^{-1} for many biochemical reactions:

	ΔH (kJ mol^{-1})
$\text{Glucose} + O_2 \xrightarrow{\text{GOx}} \text{Gluconolactone} + H_2O_2$	–50
$\text{Uric acid} + O_2 \xrightarrow{\text{UOx}} \text{Allantoin} + CO_2 + H_2O$	–50
$\text{Oxalic acid} + O_2 \xrightarrow{\text{OOx}} 2CO_2 + H_2O_2$	–143
$H_2O_2 \xrightarrow{\text{Catalase}} H_2O + \frac{1}{2}O_2$	–100
$\text{Urea} + H_2O \xrightarrow{\text{Urease}} 2NH_3 + CO_2$	–61

In some applications enzymatic reactions are coupled by design, thus providing enhanced output due to a thermochemical amplification factor as a result of Hess' law (Chapter 1). Danielsson and Winquist (1990) use the example of a typical response: 1 ml of a 1 mM solution introduced at a flow rate of 1 ml min^{-1} will produce ~0.01 °C. From this example the need for reduction of thermal noise and in-built amplification factors become fairly obvious.

Coupling enzymatic reactions is not the only useful strategy for amplification. Danielsson and Winquist (1990) also reported efforts toward using organic solvents with lower C_p's than aqueous solutions, thus providing larger ΔT's for a given ΔH. When they determined triglycerides in cyclohexane the response was ~2.5 times that obtained in an aqueous solution. With enhanced data on the behavior of enzymes in organic media, as mentioned in Chapter 1, this could prove to be a significant tactic for thermal sensing.

Another approach to amplification of signal is based on total enthalpic output, not necessarily all of which is derived from the coupled enzyme reactions. A thermal sensor's response is a result of all contributing ΔH's, including hydration, protonations, etc. For many of the enzymatically based sensing schemes a buffer solution with a high enthalpy of protonation can be utilized for improving S/N. The common bioanalytical buffer, Trishydroxy-methylaminomethane (Tris), fits this profile well with its ΔH of ~ -50 kJ mol^{-1}. Interestingly, Tris also has a good fairly high temperature coefficient of its pK_a, $-0.027°C^{-1}$. Straub and Seitz (1993) used this property of Tris to design a fiber-optic thermal probe based on the temperature dependence of the HIn/In^- ratio of Phenol red in Tris buffer.

As indicated in Chapter 1, thermal enzyme probes (TEP) based on thermistor responses were among the first enthalpimetric devices adopted for widespread usage. A thermistor is a mixed-oxides semiconductor, and the resistance is a function of the semiconductor's bandgap energy. A thermistor is usually characterized by its actual resistance at room temperature and by its temperature coefficent α, which may be either positive or negative. Figure 6-1 includes a sketch of typical resistance versus temperature behavior. Resistances at room temperature are usually several thousand ohms and temperature coefficients are on the order of 3–4% per ^{0}C. As shown in Figure 6-1(a), the response in nonlinear, but normally the actual devices are used only at temperatures that lie in the linear range. High-quality Wheatstone bridges, multimeters or operational amplifier circuits for accurate measurment of resistance are used for detection of the sensor's response. Response times for thermal sensors vary, but can range from fractions of a second to minutes. Older types of thermistors with glass encasements were notorious for long response times, but the thermistor ultramicrobeads of the Shimohigoshi and Karube (1996) study cited in Chapter 2 had a response time of about one second.

The original TEP devices had the biocomponent immobilized directly on the glass encasement of the device (Danielsson and Mosbach, 1989). A protective membrane was added over the immobilized component. This is one of the simplest designs, but a probe of this type exhibits long response times and inefficient heat transfer to the sensing surface, which result in low sensitivity (Bataillard, 1993; Carr and Bowers, 1980).

A more effective approach in thermal sensing was demonstrated using the thermistor in conjuction with an enzyme reactor column incorporated into a flow system (Figure 6-1(b)). In this configuration the reactor/thermistor combination is commonly referred to as an enzyme thermistor, or ET (Danielsson and Mosbach, 1989). In the modern parlance of analytical scientists, it probably would have been designated as an IMER with thermal detection, or an IMER-TD. This would be consistent with the other IMERs used, for example that of Niwa *et al.* (1996) with electrochemical detection

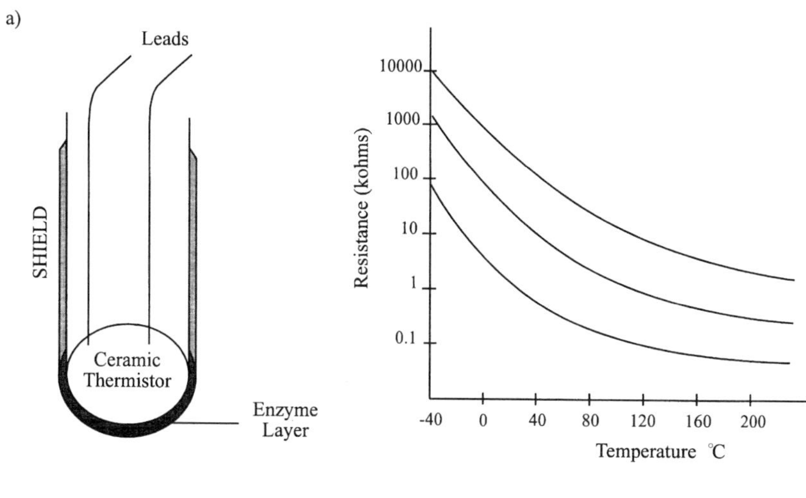

FIG. 6-1. Schematics of thermal sensing components showing: (a) a thermistor and its typical resistance versus temperature plot with approximate linear response over small temperature ranges; and (b) the general arrangement for a flow-through enzyme reactor column with thermal sensing (an enzyme thermistor, ET).

which could be designated an IMER-EC. The transition in design from a discrete device (TEP) to the ET is one of the early examples of the distinction between biosensors and biosensing systems, as indicated in Chapter 1.

Using the IMER-TD configuration a small reactor column, μl to ml in volume, normally contains one or more enzymes that interact with the analyte and, perhaps, with enzymatic products involved in sequential reactions:

$$S + E_1 \longrightarrow P_1 + E_1 \xrightarrow{E_2} P_2 + E_2$$

The enzyme(s) is (are) usually immobilized on a stationary-phase substrate such as Sephadex, Sepharose or derivatized controlled pore glass (CPG). The use of a packed-column reactor enables high enzyme loadings which lead to excellent sensitivities. As is true of most flow-through columns, the CPG-packed column is subject to fewer complications arising from compression under flow conditions. For example, Rank *et al.* (1993) used 80–120 mesh CPG for their miniaturized IMER. Their system was designed for split flow to incorporate a reference reactor with inert protein immobilized. Penicillin, glucose and ethanol were determined in an industrial application.

Temperature resolution for these units can be very good if proper attention is given to insulation sufficient for the adiabatic conditions required. Performance also improves considerably with inclusion of a reference thermistor for differential measurements, but this requires well-matched thermistors, as emphasized by Bataillard (1993). Temperature resolution on the order of 10^{-4} to 10^{-5} K and micromolar detection limits for analytes are not uncommon. Danielsson and Mosbach (1989) and Bataillard (1993) have presented comparative data from numerous studies including applications in clinical measurements, bioprocess control, environmental monitoring, and immunoassays.

Flow rates for some of the earlier macroscale reactors were on the order of 1 ml min^{-1}, but with miniaturized systems that value has dropped to a few μl min^{-1} (Danielsson and Winquist, 1990; Rank *et al.*, 1993). In dealing with flow systems for biochemical samples from various sources, clogging and contamination of the columns is always a problem. Some sample clean-up—e.g., pre-column filtering or microdialysis—may be accomplished using the same strategies developed for chromatographic procedures and *in vivo* sampling. Those variations are usually adopted, however, at the expense of response time. Amine *et al.* (1995) have developed a microdialysis probe for continuous subcutaneous monitoring of glucose using a miniaturized thermal sensor with an FIA system. They obtained an 85 s response time (42 samples per hour). The enzyme reactor tube in this case was only 1.5 mm × 15 mm, while the whole sampling assembly was only 25 mm × 54 mm.

One of the problems associated with the earlier thermal devices was the difficulty with miniaturization. In order to move toward miniaturized devices that would be portable and capable of handling multiple analytes, the entire system needed to be scaled down. With the advent of thin-film thermistor arrays (Urban *et al.*, 1991), micromachined flow-cells (Xie, B. *et al.*, 1992, 1995) and thermopiles (Muehlbauer *et al.*, 1989; Xie, B. *et al.*,

1994b; Bataillard *et al.*, 1993) thermal sensors could be scaled for multi-analyte analysis, microvolume samples, portability and FIA detection. An example of this type of cell design is illustrated in Figure 6-2.

Thermopiles are based on detection by heat conduction. Heat transfer is detected by the voltage that develops across thermocouple pairs connected in series. The total response is given by the Seebeck effect, which predicts a voltage proportional to the number of thermocouples in the pile (Figure 6-2). A significant advantage in using thermopiles is that no zero-concentration offset occurs because there is no signal until heat is produced from the reaction of interest.

Muehlbauer *et al.* (1989) developed a thermopile-based glucose sensor that could be used for *in vitro* and *in vivo* measurements. It was an Sb/Bi thermopile prepared with 50+ thermocouple pairs as a thin film on Mylar sheets. Temperature resolution was about 0.1 mK but the LOD was high, primarily due to thermal noise. In a later report that group suggested a vibrating probe thermopile to reduce thermal gradients, improve flux of analyte and improve the S/N ratio (Towe and Guilbeau, 1996). The vibrational source is a piezoelectrically driven bender which induces a 3 mm excursion of the probe in solution. The enzyme was immobilized in a hollow dialysis membrane on the surface of the probe. Noise-limited temperature measurement was found to be 40 μ°C, but LOD problems arose again, attributed by the authors to the method of immobilization.

Bataillard *et al.* (1993) utilized a silicon thermopile for a microsensor for glucose, urea and penicillin. These devices exhibited excellent performance with a time constant of 16 ms, an intrinsic sensitivity of 70 mV K^{-1}, temperature resolution of 0.01 mK and glucose sensitivity of 35 μV mM^{-1}. Using a quartz substrate with micromachined channels (17.5 mm × 3.6 mm × 0.32 mm) for enzyme immobilization, Xie, B. *et al.* (1994b) utilized a thermopile design (similar to Figure 6-2) for which the authors reported $\sim 10^{-4}$–10^{-3} °C changes for the glucose oxidase–catalase coupled reaction with 1 μl samples of glucose. The linear range was 0–20 mM glucose.

The Danielsson group has reported two other new designs that have proven effective for totally different types of application. In Xie, B. *et al.* (1995) they described an integrated multianalyte design using 5 thin-film thermistors on a quartz chip. In an interesting arrangement they alternated thermistors and immobilized enzyme sites with a dead zone between to prevent thermal carryover. By immobilizing different enzymes and using pairs of thermistors for an individual reaction site, the array was used for glucose/urea and urea/penicillin V mixtures. The fifth thermistor was for assessment of the carryover effect. Flow rate became a major factor to optimize for sensitivity and linear range, but the performance was satisfactory in FIA analyses.

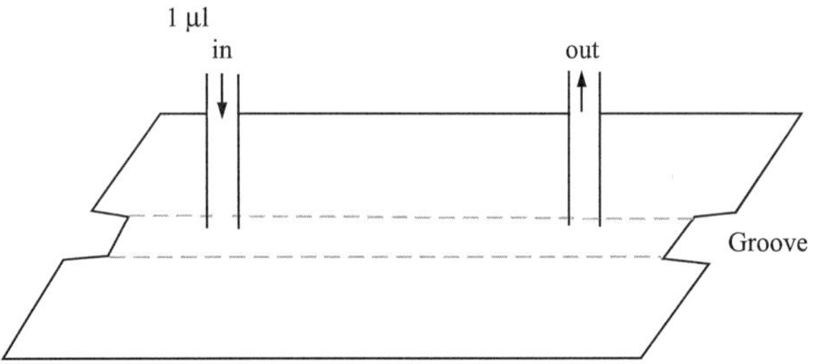

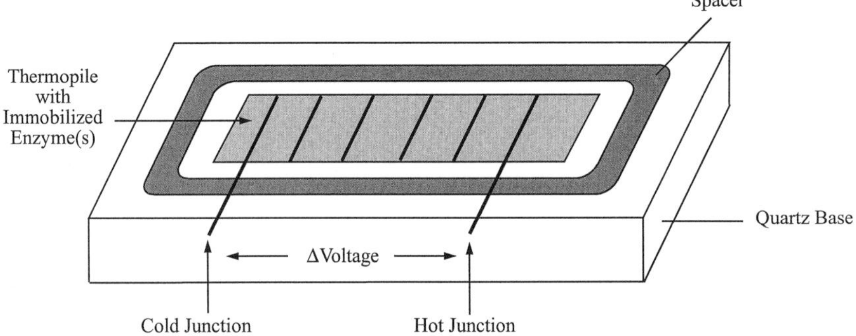

FIG. 6-2. Miniaturized thermopile sensor. The voltage change is proportional to the number of thermocouples N, the temperature difference Δ T, and the Seebeck constant *a*, which is the voltage response per °C—e.g. μV/°C. $\Delta V = N\Delta T\, a$). Adapted from *Anal. Chim. Acta*, Vol. **299**, Microsensor based on an integrated thermopile, pp. 165–170 (1994) with kind permission of Elsevier Science B.V., Amsterdam, The Netherlands.

In a clever design for the thermal detection of a ferrocene-mediated glucose oxidase catalysis, Xie, B. *et al.* (1993) immobilized the enzyme on crushed reticulated vitreous carbon (RVC) and packed that into a platinum foil column. Using a simple dc power source across the ends of the platinum column as the applied potential, they measured instead of Faradaic current the temperature differential between a reference thermistor and a measuring thermistor, located at opposite ends of the column.

Thermistors and thermopiles are not the only thermally sensitive modes of transduction. Materials with a permanent internal charge polarization and no center of symmetry are usually pyroelectric—i.e., they produce an electrical signal upon thermal stress. Polyvinylidene fluoride films, and also some other related polymeric materials, can be manufactured to have pyroelectric characteristics. Dessy *et al.* (1989) reported a pyroelectric biosensor for use in a flow system. Two metal-coated polyvinylidene fluoride films were placed back-to-back in what they termed a "bimorph" configuration. The mated surface(s) were grounded and the exposed sides served as sample and reference sides. Catalase or urease was immobilized on the underside of a cover used to form a channel for flow. A linear calibration was achieved for hydrogen peroxide concentrations of 0.004–1.0 M.

In the same report Dessy's group described the use of an enthalpimeter based on thermal changes affecting the intrinsic properties of an optical fiber. Two fibers, which formed the arms of an interferometer, were coupled so their evanescent waves interacted. Enzyme was immobilized on one fiber and the assembly was placed in an FIA cell. As substrate flowed over the enzyme, the heat generated changed the light propagation in the sample fiber. A linear optical array was used as the detector. The phase angle difference depended on the concentration of substrate.

6.2. ACOUSTIC-WAVE DEVICES

Acoustic waves (AW) are designated as that part of the frequency spectrum corresponding to ultrasound (see Figure 5-1). These are the frequencies related to the elastic vibrations of materials, usually discussed in terms of those materials that are piezoelectric. However, the propagation of acoustic waves is based entirely on mechanical phenomena. For example, Li, P.C.H. *et al.* (1993) and Li, P.C.H. and Thompson (1996) have developed a thin-rod AW device consisting of an AW transmitter and receiver with the thin-rod (radius $<$ acoustic wavelength) acting as the delay line between them. For the purposes of this introduction only piezoelectric AW devices will be considered. Much of the discussion will center on α-quartz as the piezoelectric material of interest. The reader interested in other types of piezoelectric materials should peruse, for example, O'Toole *et al.* (1992) for a description of a miniaturized thin aluminum nitride (AlN) device of very high sensitivity.

Crystals that are non-centrosymmetric have a polar axis due to the orientation of dipoles in the lattice. When a crystal of this type is placed under stress the dipoles are realigned, and therefore the surface charges on the crystal. Conversely, if an electrical field is imposed across the crystal the dipoles are realigned and the crystal becomes strained, thus altering the vibrational energy states of the crystal. When an alternating electrical field is applied to electrodes mounted on the surface of the crystal there will be a resonant frequency at which the vibrating crystal behaves as a harmonic oscillator. The acoustic waves generated by these transitions are launched and propagated through the bulk of the crystal, guided on a surface as a surface wave or reflected from the surfaces as a plate mode (Lu and

Czanderna, 1984; Auld, 1990; Kittel, 1996; Thompson and Stone, 1997). The mass sensitivity S_m of the devices arises from the fact that any additional mass on the surface alters the wave propagation through the crystal.

The usual abbreviations for different types of AW devices, some of which are illustrated in Figures 6-3 and 6-4, are based on the different propagation modes (Grate *et al.*, 1993a,b):

TSM, thickness shear mode;
SAW, surface acoustic wave;
STW, surface transverse wave;
FPW, flexural plate wave; and
SH-APM, shear horizontal acoustic plate mode.

An example of a TSM device, often referred to as the quartz crystal microbalance (QCM), with electrodes on each side of the crystal is shown in Figure 6-3. For all examples above except the TSM the acoustic waves are launched on the surface by an interdigital transducer (IDT) that acts as a transmitter with a second IDT as receiver (Figure 6-4). This is referred to as a delay-line configuration.

The basic relationships for mass sensitivity of the TSM oscillator–resonator behavior of these devices were derived (Section 6.2.1) with the assumption of air or vacuum in contact with the surface of the device. The resonant frequency of oscillation is a function of the mass of the sensor,

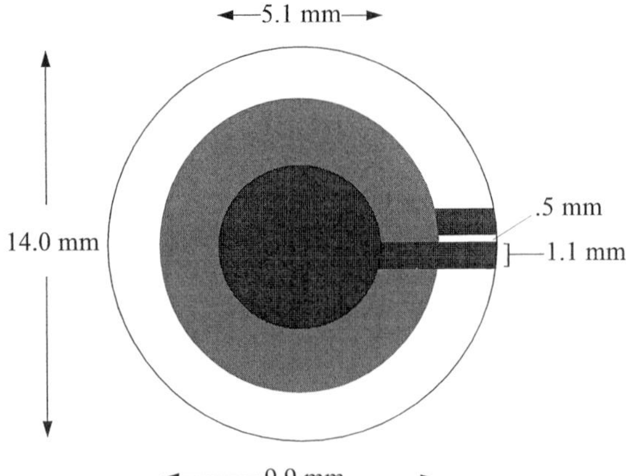

FIG. 6-3. Typical thickness shear mode (TSM) quartz crystal with electrodes mounted on both sides of the crystal. Reprinted, with permission, from Dunham, G. C. *et al.*, *Anal. Chem.* **67**: 267–272. Copyright 1995, American Chemical Society, Washington, DC.

a) Top View - SAW

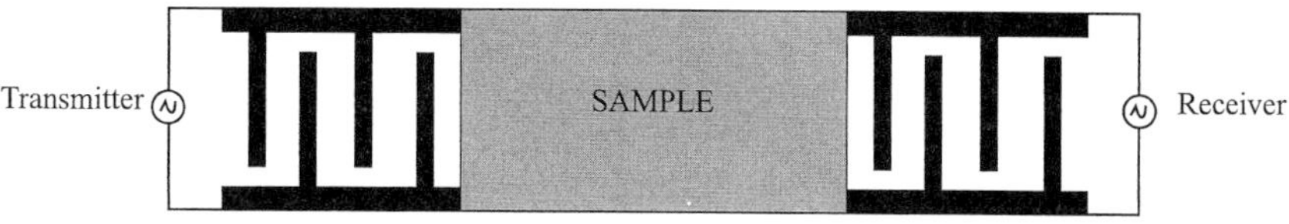

b) Side View - SAW

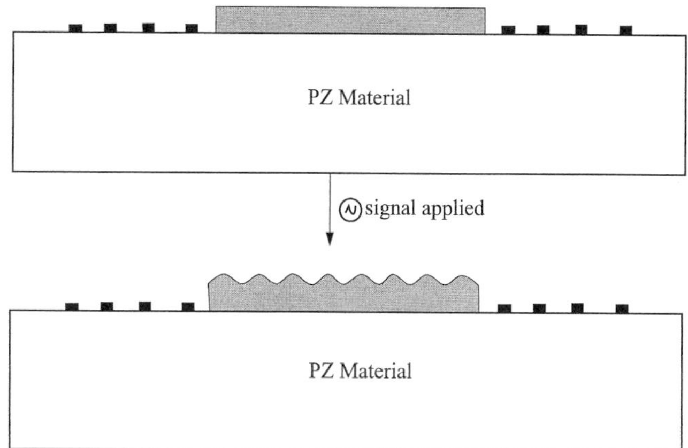

c) Side View - APM

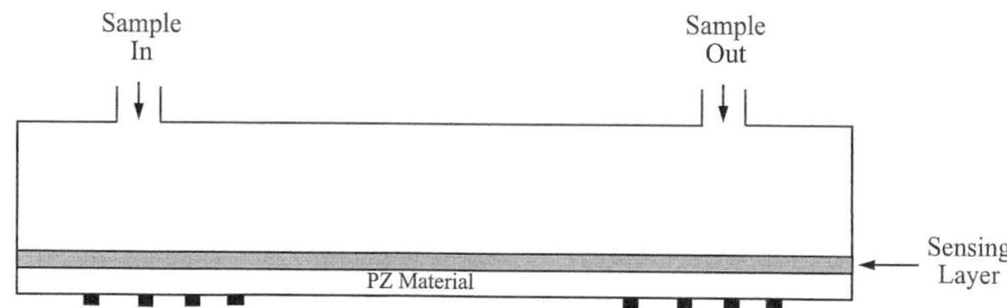

FIG. 6-4. Examples of acoustic-wave devices (immobilized layer shaded): (a) top view of a SAW device showing the interdigitated electrodes serving as transmitter and receiver; (b) propagation of acoustic wave through SAW layer; and (c) an APM device in flow-cell configuration.

including modifying layers on the surface. This wave perturbation is the source of the sensing function. Changes in the resonant frequency for a resonator–oscillator operating in air or vacuum are linearly related to changes in mass.

As illustrated in Figure 6-5, acoustic-wave devices may be used in media other than air or vacuum. Any other type of interface or changes of the interfacial region at the surface of the device—e.g., a liquid phase in contact with the surface or polymeric swelling—will also alter the characteristics of wave propagation, that is, the frequency, amplitude or wave velocity. For example, the viscosity, density and conductivity of liquid phases in contact with the oscillator surface can create viscoelastic and acoustoelectric effects which introduce additional frequency-dependent factors (Grate *et al.*, 1993b; Yang, M. and Thompson, 1993; Yang, M. *et al.*, 1993; Bruckenstein *et al.*, 1994; Jossé, 1994; Tessier *et al.*, 1994; Martin *et al.*, 1997 and references therein). This means that mass is no longer the sole determinant of response, and data interpretation must be adjusted accordingly. For example, a shift in the resonant frequency of a TSM under liquid load is a function of $(\rho\ \eta)^{1/2}$ where ρ is the density of the liquid and η is its viscosity (Martin, 1997; Kipling and Thompson, 1990; Martin *et al.*, 1991).

The nature of the crystalline material, the geometric arrangement of the excitation electrodes on the surface and the thickness of the device determine the wave characteristics. Kittel (1996) covers in good detail the theory of the piezoelectric effect, including the thermodynamic and quantum mechanical treatments. The two-part review by Grate, Martin and White (1993a,b) compares types of AW devices and summarizes the physical behavior of each type (Table 6-1). Andle and Vetelino (1994) have reviewed the field of acoustic-wave devices with particular emphasis on biosensors. Guilbault and Jordan (1988) surveyed analytical applications. Minnuni *et al.* (1995) dealt specifically with applications using the quartz crystal microbalance. Jossé (1994) has discussed considerations of AW devices pertaining to liquid-phase measurements. Buttry (1991) and Buttry and Ward (1992) focused on the use of AW devices coupled to electrochemical measurements. Thompson and Stone (1997) have provided a thorough introduction to devices based on surface-launched (IDTs) acoustic waves. They also explain fully the various quartz crystal cuts — AT, BT, etc. — and notational standards adopted in acoustic engineering.

In the Grate, Martin and White (1993b) report there is a succinct comment on the nomenclature, actually misnomers, regarding acoustic wave devices: "Most reviews of chemical microsensors group acoustic wave sensors under mass sensors or piezoelectric sensors. Neither of these classifications directly addresses the transduction mechanisms involved: Sensing occurs because of the perturbation of acoustic waves at ultrasonic frequencies. It should be evident that the term 'mass sensor' is limiting and, in many

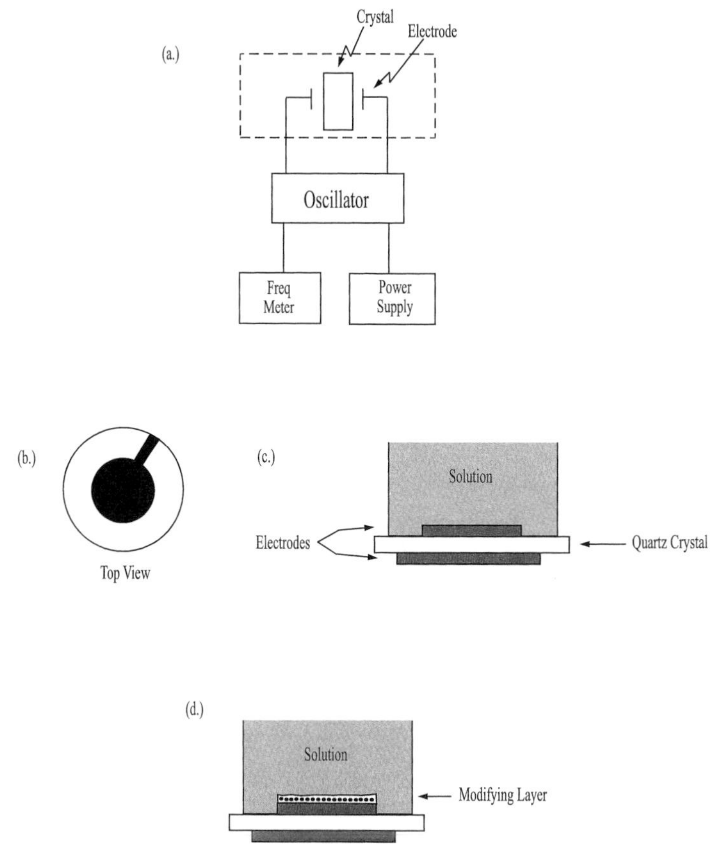

FIG. 6-5. Basic quartz crystal microbalance (QCM) or thickness shear mode (TSM) acoustic-wave device: (a) instrumental setup for simple resonant frequency monitoring; (b) one side only of the crystal (another electrode of the "flag" type would be on the opposite side of the crystal); (c) and (d) side-views of a QCM (TSM) cell with liquid sample where (d) includes the chemically selective layer that would be applied for biosensing, thus altering the resonant frequency.

cases, inaccurate. Mass detection is not the only, nor always the dominant, mechanism by which acoustic wave devices function as sensors, and piezoelectricity is simply a material property that is useful, but not necessary, for generating acoustic waves in small devices. The recognition of additional mechanisms by which an acoustic wave device can act as a sensor creates many new opportunities for the design of physical and chemical sensors and provides the means by which sensitivities can be much greater than those that might be achieved by mass detection alone." In spite of the clear

TABLE 6-1 Some Characteristics of Acoustic-wave Devices[a]

Device	Type of Wave	Primary Media	Typical Frequency (MHz)
TSM	Bulk	Vapor, liquid	5–10
SAW	Surface	Vapor	30–300
STW	Surface	Liquid	30–300
SH-APM	Plate	Liquid	25–200

[a] S_m (mass sensitivity) varies and is a function of several variables in addition to frequency of operation.

distinctions made by these authors, the usage of deeply entrenched nomenclature still appears in the literature.

As indicated above, there are several types of AW devices which may be used in biosensing. A thickness shear mode (TSM) device is classified as a bulk-wave device. Exact characteristics of wave propagation depend on the type of quartz crystal cut, with the AT-cut having one of the lowest temperature coefficients (Buttry, 1991). The acoustic waves of a TSM sensor (~5–10 MHz) travel through the bulk of the device in a transverse mode perpendicular to the surfaces. Particle displacement in the crystal is parallel to the surface. Thus, vapor sorption, immobilized materials or changes in a modifying layer on the surface modulate the signal. Studies of the distribution of vibrational amplitudes indicate that the amplitude is maximum at the center of the disk and decays with distance from the center, approaching zero at the edge of the electrodes (Buttry, 1991).

As mentioned above, the frequency response of a liquid-loaded TSM involves more than just the mass dependence. One approach to compensating for this is to use both a sample and reference crystal. The TSM component illustrated in Figure 6-3 has been proposed by Janata's group (Dunham *et al.*, 1995). The whole device is designed as a dual crystal dip-probe designed to compensate for complex effects of viscosity, temperature and conductivity when using the crystals in liquid media. Bruckenstein *et al.* (1994) had suggested approaches to the same problems for a laboratory-based system for electrochemical measurements. Whereas the simplest response of a TSM sensor is detected as the oscillator–resonator frequency change proportional to mass changes, Janata's dual crystal arrangement, as would any design with sample and reference crystals, requires a differential frequency measurement.

Several types of surface waves generated by ac signals applied to interdigitated transducers (IDTs) are utilized (Thompson and Stone, 1997). In Figure 6-4 two of these types of devices—the SAW (Wohltjen, *et al.* 1989;

Kepley *et al.*, 1992; Grate *et al.*, 1993c; Zellers, 1995; Dickert *et al.*, 1996, 1997) and the SH-APM (Ricco *et al.*, 1989; Andle and Vetelino, 1994; Dahint *et al.*, 1994; Renken *et al.*, 1996)—are illustrated. Because the Rayleigh wave propagated along the surface of the SAW has a significant transverse component normal to the surface, it is attenuated by liquid-loading and there is high propagational loss. Thus, SAWs have been used essentially as vapor-phase sensors. It should be noted in Figure 6-4 that the SH-APM, characterized by a surface-parallel component, is a device depicted with the IDTs on the surface opposite the sensing surface. This arrangement facilitates sensing in liquid environments, since the electrodes are not exposed to the sampling environment.

Tom-Moy *et al.* (1995) chose to use an STW device for their atrazine sensor. The STW device with IDTs uses an ST-cut quartz and exhibits shear horizontal waves parallel to the surface. An STW device also includes a waveguiding grating on the surface. The sensing system includes a sample and reference configuration with interferometric detection of signal variations between the two lines. The authors' description of the logic in choosing the STW configuration is instructive with regard to practical aspects of device comparisons.

Because chemical selectivity of an acoustic-wave device is conveyed entirely by the nature of the modifying layer(s) on the surface, numerous applications in biosensing become obvious through molecular recognition schemes discussed in earlier chapters. Various types of discrete AW sensors have been adapted for bioanalytical problems, including important applications in immunosensing and nucleic acid interactions. Unfortunately, the complex response functions of AW devices in liquid media hindered data interpretation and practical applications until fairly recently.

Acoustic-wave devices are by their very nature non-selective in that the response is a function of mass changes, regardless of the source of the mass. The use of sorptive-selective or molecularly selective modifying layers can convert the devices into sensors that are, at least, partially selective. Some of the most successful applications early on, however, derived selectivity from the use of sensor arrays (Murray *et al.*, 1989; Mandelis and Christofides, 1993; Grate *et al.*, 1993c) functioning as higher order analytical instruments. The use of AWs in arrayed formats of partially selective sensors was introduced in earlier chapters in the context of chemometric analysis of data (Booksh and Kowalski, 1994; Carey, W.P., 1994). Because AW devices are small, relatively inexpensive and amenable to mass production, they are ideally suited to array formats. In 1994 Vellekoop *et al.* introduced a new type of integrated circuit-compatible AW device that could be used for biosensing applications. The design and fabrication of single-chip AWs was discussed. With increasing use of sensing arrays this type of design should be adaptable to various formats for biosensing.

Grate *et al.* (1993a,b) have explained in full detail that the mass sensitivity S_m of an acoustic-wave device is defined as "the incremental signal change occurring in response to an incremental change in mass per unit area on one surface of the device."

$$S_m = \text{limit}(\Delta f / F_o)/\Delta m$$

Here $\Delta f/F_o$ is the frequency change relative to the fundamental frequency. In this formulation, S_m has the units of (Hz/MHz)/(ng/cm^2). As pointed out by Grate, a common unit reported in the literature is cm^2 g^{-1}. The two are related by:

$$(\text{Hz/MHz})/(\text{ng/cm}^2) \times 1000 = \text{cm}^2\text{g}^{-1}$$

The authors emphasize that this mass sensitivity of the device itself is not the same as analyte sensitivity, which is the incremental signal change per unit concentration change of the analyte. Further, they point out that S_m is a function of more than just the frequency factor, which is often quoted in a simpified relationship. The S_m dependence on the crystal's characteristics and thickness is different for each type of AW device. Grate *et al.* (1993a,b) or Thompson and Stone (1997) should be consulted for the appropriate equations.

6.2.1. TSM (QCM) Devices

The basic TSM sensor consists of a piezoelectric crystal with vapor-deposited gold electrodes, a broad-band oscillator circuit for inducing the crystal's resonant oscillation (1–10 MHz) and, most commonly, a frequency detector (Figure 6-5). Typical oscillator circuit designs may be found in Buttry (1991) and Buttry and Ward (1992) or any basic electronics text. The gold electrodes may be sputtered with some other metal or carbon to achieve a different binding surface. For example, Tatsuma and Buttry (1997) used sputtered carbon for developing a heme-sensitive device. The diameters of the crystal are about 1.5 cm while the electrode diameters are about 0.5 cm. Thicknesses are on the order of 0.03 cm. The thin-film electrodes on the crystal are about 300 nm thick (Buttry, 1991). Selectivity of the device is achieved by the choice of the chemically modified surface, which often involves structuring with SAMs and immobilized biocomponents on the electrode surface. Teuscher and Garrell's (1995) note includes some very practical guidance for mounting and stabilizing responses using commercially available components.

The basic TSM instrumental design is a relatively simple configuration that can be easily packaged and designed for routine use as a portable unit. For the simplest interpretation of mass-sensitive frequency behavior, the oscillator configuration indicated in Figure 6-5 is typical. For full evaluation of the contributors to frequency dependence, however, a more detailed examination of the responses must be made, particularly when the device is used in liquid media, thus requiring more sophisticated circuitry for impedance analysis.

The basic governing relationship between mass and frequency variations, known as the Sauerbrey (1959) equation, were derived for the TSM device in air or a vacuum, assuming a rigid mass on the surface and low relative frequency changes (in Hz):

$$\Delta f = -C_{\mathrm{f}} \Delta \mathrm{m}_{\mathrm{f}},$$

C_{f} is a calibration constant, $(2F_{\mathrm{o}}^{2}/(\mu_{\mathrm{q}}\rho_{\mathrm{q}})^{1/2})$ that depends on the crystal density ρ_{q} and shear-wave modulus (stiffness) μ_{q}, and on F_0, which is the fundamental resonant frequency of the crystal. Δm_{f} is the mass change producing the frequency variation. The negative sign indicates that a decrease in frequency will result from an increase in mass of the crystal. For AT-cut quartz the values of density and shear-wave velocity are 2.648 g cm^{-3} and 2.947×10^{11} g $\mathrm{cm}^{-1}\mathrm{s}^{-2}$, respectively (Buttry, 1991). This corresponds to a calibration constant C_{f} of 56.6 Hz $\mu\mathrm{g}^{-1}$ cm^{2} for quartz with a 5 MHz fundamental frequency mode. The equation may be stated in its practical form for quartz as

$$\Delta f = -2.26 \times 10^{-6} F_0^2 \Delta m / A$$

where A is the area of the electrode surface on which immobilized material is deposited. Typically that area will be < 1 cm^2. Calculations will show that nanograms of mass change per unit of area will produce frequency changes of a few Hz. Detection limits in the picogram range are not uncommon. Janata (1989), Buttry (1991), Buttry and Ward (1992), Mandelis and Christofides (1993) and Guilbault and Jordan (1988) should be perused for a more thorough treatment of theory and experimental details.

For a more complete analysis of response factors a detection mode other than the simple resonator–oscillator circuitry is used. The whole assembly—crystal/electrode/modifying layer—may be modeled by a resonant frequency circuit characterized by both the resonant frequency and a complex impedance.

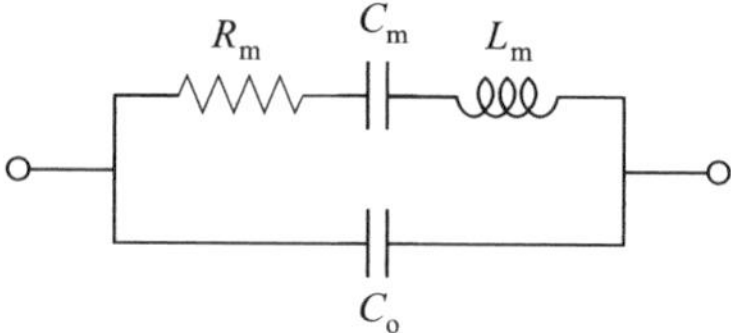

The components of the equivalent circuit may be related to the mechanical/electrical phenomena of the assembly. The resistance R_m is related to the energy dissipation into a modifying film. Energy stored during oscillation of the crystal, analogous to the compliance of a spring, is represented by C_m; L_m relates to motional inertia with respect to displacement of the film. C_o is an electrical capacitance due to the electrodes on the crystal surface. It may be measured at any frequency away from the resonant frequency band. When the oscillator is operated in gas phase measurements, the resonant frequency f_R is not a function of the energy dissipation factor R_m but is the frequency of maximum conductance in the circuit, and can be represented as

$$f_R = \frac{1}{2\pi}\left(\frac{1}{L_m C_m}\right)^{1/2}$$

If, however, the device is operated under liquid load (in solution), the equivalent circuit must be altered to include additional factors (R_L, L_L) related to viscoelastic and acoustoelectrical coupling, which affect responses. There are two resonant frequencies at which maximum conductance occurs and the phase angle is zero (series and parallel):

$$f_S = f_R\left(1 + \frac{r}{2Q^2}\right)$$

$$f_P = f_R\left(1 + \frac{1}{2r} + \frac{r}{2Q^2}\right)$$

where $r = C_o/C_m$ and $Q = 2\pi f_R L_m/R_m$ (Buttry, 1991; Yang, M. and Thompson, 1993; Martin *et al.*, 1997). Q is an indicator of the sharpness of the resonance band and is also referred to as the quality factor of the resonator. The change in the characteristic frequency upon liquid loading is a function of the density and viscosity of the liquid (Martin *et al.*, 1997).

$$\Delta f_S = -\frac{2nf_S^2}{N\sqrt{\mu_q \rho_q}}\left(\frac{\rho\eta}{4\pi f_S}\right)^{1/2}$$

Here n is the number of sides of the crystal in contact with the solution; ρ and η are the liquid's density and viscosity, respectively. N is the harmonic number.

Network analyzers may be used to extract the information for calculation of the variables in the circuit. These are sophisticated electronic devices with high-quality output of data in various formats (Kipling and Thompson, 1990; Yang, M. and Thompson, 1993; Yang, M. *et al.*, 1993). Alternating voltages are applied to the crystal over a frequency range. The amplitude and phase relationships are measured and the output provides data regarding conductance, reactance, resistance and phase angles. Frequencies at maximum conductance and zero phase angles may be determined and the behavior characterized in terms of the contributors to the equivalent circuit model.

The relatively simple resonator/oscillator circuit design is easier to implement but, due to the multiple AW coupling mechanisms in liquid media, the response interpretations may not produce valid mass–frequency correlations (Kipling and Thompson, 1990; Martin *et al.*, 1991). Consequently, experienced personnel often prefer the network analyzer output for full equivalent circuit analysis (Yamaguchi *et al.*, 1993; Jossé, 1994; Martin *et al.*, 1994; Shana and Jossé, 1994; Su *et al.*, 1994, 1995; Tessier, 1994; Dunham *et al.*, 1995; Tom-Moy *et al.*, 1995; Welsch *et al.*, 1996). On the other hand, a simpler instrument that could be reliable for direct mass-frequency interpretation would be desirable.

Martin *et al.* (1997) have addressed this problem and proposed a rather simple electronic alteration to resonator–oscillator circuitry which might bridge this theory–application gap. The proposed circuitry is simple and is based on an electronic tuning mechanism that minimizes the net capacitance, thus allowing the response to track to "true" resonant frequency. This electronic correction removes the disparity of responses between network analyzer and resonator–oscillator circuitries, thus allowing the use of the simpler instrumentation for routine measurements.

6.2.2. TSM Applications

Although the mechanism giving rise to the TSM signal in liquid media has not always been well understood, the device has been used for numerous biosensing applications. Some basic work in understanding the behavior of macromolecules is exemplified by the Yang, M. *et al.* (1993b) study of protein adsorption on TSM sensors. The behavior was characterized with network analysis and interpreted in terms of conformational changes of the protein after electrostatic adsorption at the interface.

Yamaguchi *et al.* (1993) examined DNA adsorption and hybridization using the TSM sensor. They also used impedance analysis, and were able to separate adsorption effects from the mass-sensitive response of DNA hybridizations. The frequency at maximum conductance was utilized to distinguish viscoelastic effects in the adsorbed layer from actual mass changes due to both ssDNA and dsDNA, and also hybrids formed with known sequences. Both the Yang and Yamaguchi studies provide good instruction in the use of network analysis for interpreting complex interfacial interactions.

Caruso *et al.* (1997) took a slightly different approach to obtaining true mass-sensitive data. They followed the adsoprtion with time of avidin from solution, then they removed the AW crystals from the solution and measured in air via the TSM the actual amount of bound avidin and the increment from binding biotinlyated DNA at the electrode. Monolayer coverage of avidin was confirmed. As discussed in Chapter 3, the Caruso study was intended to demonstrate that the use of multilayer films improves the sensitivity for DNA hybridization studies. Avidin denatured upon exposure to the air, and thus freshly deposited avidin had to be used for the DNA-binding experiments.

Su *et al.* (1994) have used AW methodology to study nucleic acid hybrids. Su *et al.* (1995) also investigated the interactions of DNA with anticancer platinum drugs. In addition to attaining an LOD of 10^{-7} M for the drugs, they were also able to perform a kinetic analysis of the data, and suggested that the DNA binding is with hydrolysis products of the drugs.

Wang, Juan and coworkers (1993) reported a pH-sensitive TSM sensor using amphoteric polymer films. The authors suggested that it could be used as a "threshold" sensor or could be adapted for enzyme-linked immunoassays. Lasky and Buttry (1990) used the TSM approach for a glucose sensor based on a hexokinase catalyzed reaction. They reported higher than predicted responses and spurious effects which they suggested at that time were due to the non-mass-dependent factors in the detected response.

Reversing the usual direction of mass dependence, Másson *et al.* (1995) developed an assay for biotin that results in an increase of frequency with loss of mass from the QCM. The measurement of small molecules as analytes via TSM experiments is difficult because the mass changes due to simple adsorption or binding to the surfaces are small. The Másson approach was based on an indirect method whereby BSA was derivatized with desthiobiotin and adsorbed onto the sensor surface. Using either stopped-flow or continuous-flow, avidin was introduced for binding to the immobilized component, then the analyte, biotin, was added. Because the binding constant for biotin–avidin is greater than that for desthiobiotin–

avidin, the analyte "removes" avidin from the sensor surface, thus providing a mass-dependent *increase* in frequency rather than the usual negative response predicted for adsorption. In the stopped-flow mode, the linear range of response was 0.05–100 μg per ml biotin.

Guilbault and coworkers have reported several procedures for determining immunocomponents and microorgansims (Guilbault and Jordan, 1988; Jacobs *et al.*, 1995; Minnuni *et al.*, 1995). For the *Listeria monocytogenes* determination (Jacobs *et al.*, 1995), both dry and flow-cell measurements were made with the flow-cell arrangement providing detection of 5×10^5 cells. The antibody to the microorganism was immobilized via a Protein A anchor on the electrodes. In a totally different application involving detection of microorganisms, Chambers' group used the QCM for long-term monitoring of microbial accumulations (Nivens *et al.*, 1993).

A significant body of work in biosensing with TSM devices has focused on various immunochemical strategies. The selectivity conveyed by antibody–antigen interactions has been an effective complement to the sensitivity of the TSM sensing methodology. Carter *et al.* (1995) have determined ricin and anti-ricin antibody. Suri *et al.* (1995) have reported an immunoassay for insulin. The König and Gratzel (1994) sensor for discriminating among various herpes viruses was used as an example in Chapter 2. They describe their method as ". . . cheap, rapid (about 90 min), sensitive (LOD for human herpes viruses is about 5×10^3 cells), and selective."

The Nakanishi *et al.* (1996) report is related directly to using the TSM device as a monitor for studying oriented binding of antibodies at the surface, and was used in Chapter 3 to illustrate random versus oriented binding. Using ethylenediamine plasma-polymerized films on gold, they examined several approaches to random immobilization and Protein A anchoring of antibodies and Fab fragments. The oriented ProtA/antibody definitely gave better immunoactivity than the random case, though more antibody bound randomly. Their results for Fab fragments did not agree with Lu, B. *et al.* (1995), who had found that oriented Fab immobilization enhanced immunoactivity by a factor of ~3 times that of random immobilization. The authors suggested that the amount–orientation argument applied to the full antibodies may not apply for Fab fragments.

Tatsuma and Buttry (1997) have proposed a dual-response sensor based on amperometric measurements and TSM responses, thus providing a higher order instrument than that possible with either mode separately. The gold electrodes of the TSM device were coated with carbon paint, then heme-peptide (microperoxidase) was immobilized. In the presence of H_2O_2 the heme-peptide goes through a series of Fe(III/IV) redox conversions resulting in the uptake of two electrons per mole of heme-protein, thus producing a Faradaic current proportional to the heme concentration. A ligand-bound heme will not undergo the original H_2O_2 oxidation step.

Therefore, the Faradaic current is diminished in the presence of ligands. At the same time, the binding of ligand to heme should produce a positive mass change that can be detected by the TSM device. With histidine and imidazole as analytes (ligands for heme), Tatsuma and Buttry demonstrated that the dual-responses enabled simultaneous qualitative–quantitative or quantitative–quantitative determinations, if only two ligands were present and, in the latter case, if the identities of the ligands were known *a priori*. These are rather severe limitations on applicability. The authors indicate that the dual-response principle may be extended to other chemical systems using the electrochemical/piezoelectric combination or to applications involving multiple repsonses. This suggests the use of first- or second-order instrumentation for improved selectivity and array formats.

6.2.3. Surface and Plate Wave Modes

In Figure 6-4 both an acoustic plate mode (APM) and a surface acoustic wave (SAW) device are depicted. It has been indicated that SAW devices are utilized primarily for vapor-phase sensing. For an excellent adaptation of the SAW for sensing 100 ppb of the nerve gas DIMP the Kepley *et al.* (1992) design should be examined. They used SAMs with terminal $-COO^-$ groups to complex Cu^{2+}, which then reversibly binds to the DIMP, generating a signal with 35 s response times. The authors described the sensor as selective, sensitive, reversible and durable.

The APM waves travel either in a transverse mode through the bulk of the material or along the surface with an inherent interaction with the liquid or film adjacent to the surface (SH-APM). The thicker plate of the SH-APM devices and higher operating frequencies are distinguishing factors. The analytical value of each lies in the fact that the interaction with a liquid layer or immobilized species is enhanced relative to other modes based on the surface-generated waves.

The second feature to note about the surface-generated wave devices is that wave propagation is initiated and detected by the specifically designed transmitter for the acoustic-wave frequencies and a corresponding receiver (IDTs). Operable frequencies, and thus wavelengths, are determined either by the spacing between the fingers of the IDTs or that spacing and plate thickness. The frequency for flexural-plate wave devices is on the same order as that of the TSM, but the STW and SH-APM devices may operate at much higher frequencies, even into the hundreds of MHz. The essential factor for biosensing is that transmission along the surface between the two IDT sets is affected by the immobilized material placed on the surface between the IDTs.

For SAWs the IDTs are located on the active surface of the device. In the FPM and APM devices, however, the IDTs are placed on the bottom of the device separated from the sensing layer. This arrangement facilitates liquid sensing (Dahint *et al.*, 1994; Renken *et al.*, 1996). Welsch *et al.* (1996) have reported an SH-SAW overlaid with silicone rubber for use in immunosensing. The sensing system is designed for reference and sample paths. With the higher frequencies and mass sensitivity of the SH-SAW, they were able to approximate an LOD of 33 pg for IgG antigens. They observed a 12 × signal enhancement using 10 nm Au colloid-labeled antigens.

Numerous applications utilizing surface-generated acoustic wave devices for biosensing have been reported. The Andle-Vetelino group has reported work using the acoustic plate mode devices as immunosensors and as sensors for DNA hybridization (Andle *et al.*, 1992, 1993). They reported LODs in the ng per ml range for DNA. Using an improved design they also determined 200 ng per ml of PCR-amplified DNA from cytomegalovirus.

In a well-designed study of SH-APM devices for immunosensors, Dahint *et al.* (1994) have reported a 50 MHz APM device (ZX–$LiNbO_3$) operated at its third harmonic, 150 Mhz, as a biosensor for immunochemical reactions. They used immunoglobulin G (IgG) immobilized on the surface via a conventional aminosilane procedure. A reference/sample dual delay line configuration was used with injection of antigen and no stirring. The authors indicated that the sample throughput could be improved with a convective system, but that was not the purpose of their developmental study. By using the higher frequencies they were able to report a highly sensitive biosensor. The mass-detection limit was ~200 pg mm^{-2}, which corresponded to 4×10^{10} IgG molecules on the surface.

In a follow-up study of the Dahint results, Renken *et al.* (1996) extended the investigation of response as a function of several immobilization methods, including conventional aminosilane coupling, CM-dextran coupling and an immunosorbent, poly(etherurethane) (XP-5). As expected, the latter two hydrogel preparations resulted in higher loading of antibody, but the viscoelastic effects of the polymeric materials on AW signals negated advantages of that loading by decreasing the sensitivity of the method. Because the XP-5 had been shown previously to minimize non-specific adsorption from natural samples of blood and body fluids, there would be real analytical advantage in using it as an immobilization matrix, particuarly since the multilayer antibody-binding could improve detection limits. The conclusion was that optimization of the device would require minimization of viscoelastic effects, but that requires further data about the the polymer's viscoelastic properties.

The STW immunosensor developed by Tom-Moy *et al.* (1995) was used to determine atrazine in the 0.06–10 ppm range. The sensor operates at 250 Mz and was designed for measurements in liquids. A reference-sample dual-line design with a pumping system was used with interferometric detection. Their device has a mass sensitivity about 10 times that of either a TSM or a

SH-APM sensor. One disadvantage, as pointed out by the authors, is that their sensing scheme involves immobilized antigen (atrazine analog) with antibody in the flow stream. This means that to adapt the system to other analytes would require individualized immobilization strategies for each small organic. The article includes a good discussion of the cost effectiveness of their system, with the conclusion that probably it will be applied best in those situations that require the high throughput of repetitive measurements.

The examples given in the preceding sections indicate that there is great momentum in the application of acoustic-wave devices to biosensing. Many of the originally complicating factors are now better understood and are being addressed by new designs. Because the acoustic-wave phenomena are complex in the interfacial characteristics, and accurate data interpretation is difficult, the conversion to more reliable, simplified instruments and thorough data analysis is a welcome trend in the area.

Appendix A-1

ANALYTES REFERENCED IN THE BIBLIOGRAPHY
(See Appendix A-2 for Glucose)

Analyte	Author(s)	Year
Acetaldehyde	Pariente *et al.*	1995
	Rank *et al.*	1995
Acetaminophen	Gilmartin and Hart	1994a
(paracetamol)	Moore *et al.*	1994
Acetylchlorides	Hendji *et al.*	1994
Acetylcholine	Kawagoe *et al.*	1991
(see also neurotransmitters)	Garguilo *et al.*	1993
	Bronk *et al.*	1995
	Nikolelis and Siontoru	1995
	Riklin and Willner	1995
	Shear *et al.*	1995
	Yang, Liu *et al.*	1995
Acetylthiocholine	Alegret *et al.*	1996
Alcohol(s)	Borziex *et al.*	1995
(see also individual alcohols)	Wang, J. *et al.*	1995c
Alcohol dehydrogenase	Adami *et al.*	1995b
Aldose	Smolander *et al.*	1995

Analyte	Author(s)	Year
Alkaline phosphatase	Avnir *et al.*	1994
	Duan and Meyerhoff	1994
	LaGal LaSalle *et al.*	1995
	Bauer, C.G. *et al.*	1996
	Limoges and Degrand	1996
AMD (Aleutian mink disease) (see also Viruses)	Babkina *et al.*	1996
Amino Acids	Chen, Q. *et al.*	1993
(see also individual amino acids)	Jacobson *et al.*	1994e
Aminophenol	Niwa *et al.*	1993
	Duan and Meyerhoff	1994
Ammonium ions	Goldberg *et al.*	1994
	Wei *et al.*	1994
	Nikolelis *et al.*	1996
Ammonia	Kar and Arnold	1992, 1994
cAMP	Tsien	1994
Anesthetics (tetracaine, lidocaine)	Babb *et al.*	1995
Antimonite	Scott *et al.*	1997
anti-Ricin	Carter *et al.*	1995
Aromatic compounds	Kobatake *et al.*	1995
	Pandey and Weetall	1995a
	Wang, J.	1996d
Arsenite	Scott *et al.*	1997
Ascorbate/ascorbic acid	Korrell and Lennox	1992
	Wang, J. and Angnes	1992
	Cahill and Wightman	1995
	Janda *et al.*	1996
Atrazine	Oroszlan *et al.*	1993a
	Matsui *et al.*	1995
	Preuss and Hall	1995
	Tom-Moy *et al.*	1995
	Muldoon and Stanker	1997
Avidin	Pritchard *et al.*	1995a,b
	Sugawara *et al.*	1995
	Vreeke *et al.*	1995a
	Limoges and Degrand	1996
Bacteria	Ogert *et al.*	1992
	Kumar *et al.*	1994
	Watts *et al.*	1994
	Gatto-Menking *et al.*	1995
	Jacobs *et al.*	1995
	Strachan and Gray	1995
	Brewster *et al.*	1996
	Ikeda *et al.*	1996
	Wood and Gruber	1996
Bicarbonate ions	Pace and Hamerslag	1992
Bilirubin	Vaezi and Richter	1995

Analyte	Author(s)	Year
Biotin	Másson *et al.*	1995
	Pritchard *et al.*	1995a,b
	Sugawara *et al.*	1995
	Vreeke *et al.*	1995a
	Wright *et al.*	1995
	Piehler *et al.*	1996
Blister agents (see also Organic vapors)	Noble	1994
Blood electrolytes (see also individual electrolytes)	Yim *et al.*	1993
Blood Gases (see also Oxygen, Carbon dioxide)	Yim *et al.*	1993
Biotoxoids	Gatto-Menking *et al.*	1995
Botulinum toxin	Ogert *et al.*	1992
	Kumar *et al.*	1994
Bradykinin	Shear *et al.*	1995
	Fishman *et al.*	1996
Butyrylcholine	Hendji *et al.*	1994
Cadaverine	Gasparini *et al.*	1994
Cadium Ions	Turyan and Mandler	1994
Calcium ions	Moody *et al.*	1970
	Grynkiewicz *et al.*	1984
	Otto and Thomas	1985
	Lemké *et al.*	1992
	Blair *et al.*	1994
	Goldberg *et al.*	1994
	Knoll *et al.*	1994
	Tsien	1994
	Vo-Dinh *et al.*	1994b
	Li, C. *et al.*	1995
	Shear *et al.*	1995
	Suzuki *et al.*	1995
	Shortreed *et al.*	1996b
	Takahashi *et al.*	1996
Cancer/liver cells	Nicolini *et al.*	1995
Carbohydrates	Bilitewski *et al.*	1993
	Buttler *et al.*	1993
	Chen, Q. *et al.*	1993
Carbon dioxide	Severinghaus and Bradley	1958
	Lee, K.S. *et al.*	1993
	Arquint *et al.*	1994
	Mills and Chang	1994
	Telting-Diaz *et al.*	1994
	Jameison *et al.*	1996
Carcinoembryonic antigen (CEA)	Shiku *et al.*	1996

Analyte	Author(s)	Year
Catecholamines (see also Neurotransmitters)	Cooper *et al.*	1992
	Schroeder *et al.*	1992
	Chen, T.K. *et al.*	1994
	Ciolkowski *et al.*	1994
	Tabei *et al.*	1994
	Cahill and Wightman	1995
	Cahill *et al.*	1996
	Chang and Yeung	1995
	Pihel *et al.*	1996
Catechols	Chen, T.K. *et al.*	1992
	Yang, Liu and Murray	1994
	Burestedt *et al.*	1996
Chemical warfare agents (see also Organic vapors)	Noble	1994
Chloride ions	Pace and Hamerslag	1992
	Ozawa *et al.*	1996
Chlorinated hydrocarbons	Krska *et al.*	1993
	Conzen *et al.*	1994
	Goebel *et al.*	1994
	Henshaw *et al.*	1994
	Rosenberg *et al.*	1994
	Smilde *et al.*	1994
	Tauler *et al.*	1994
	Roberts and Durst	1995
Chlorpromazine	Wang, J. *et al.*	1993
	Moore *et al.*	1994
Cholesterol	Gilmartin and Hart	1994b
	Valencia-Gonzales and Diaz-Garcia	1994
	Whitcomb *et al.*	1995
	Jameison *et al.*	1996
Choline (see also Neurotransmitters)	Garguillo *et al.*	1993
	Garguillo and Michael	1994, 1995
	Leca *et al.*	1995b
	Yang, Liu *et al.*	1995
Cobalt ions	Satoh and Iijima	1995
Cocaine	Newton and Justice	1994
	Devine *et al.*	1995
Copper ions	Satoh and Iijima	1995
Corticosteroids	Ramström *et al.*	1996
Creatinine	Tsuchida and Yoda	1983
	Osborne and Girault	1995
	Yamato *et al.*	1995
	Mădăraş and Buck	1996
Cyanide	Dunuwila *et al.*	1994
	Licht *et al.*	1996
	Tatsuma and Oyama	1996
Cytochrome *c*	Lion-Dagan *et al.*	1994b
	Scott and Bowden	1994

Analyte	Author(s)	Year
Cytochrome c_3	Zhang, D. *et al.*	1994
Cytomegalovirus (CMV)	Andle *et al.*	1993
Diuron	Preuss and Hall	1995
	Rouillon *et al.*	1995
DNA fragments	Swerdlow *et al.*	1997
	Tan, H. and Yeung	1997
DNA hybrids	Andle *et al.*	1993
(see also Oligonucleotides,	Millan and Mikkelson	1993
Gene probes, Nucleic acids)	Yamaguchi	1993
	Millan *et al.*	1994
	Kynclova *et al.*	1995
	Stimpson *et al.*	1995
	Blanchard *et al.*	1996
	Ferguson and Walt	1996
Dopamine	Malem and Mandler	1993
(see also Neurotransmitters)	Strein and Ewing	1993
	Wang, J. *et al.*	1993
	Newton and Justice	1994
	Wang, J.	1994
	Wang, J. and Liu	1994
	Pihel *et al.*	1996
	Sreenivas *et al.*	1996
	Swanek *et al.*	1996
	Zhang, X. *et al.*	1996
	Cosnier *et al.*	1997
Drugs	Hogan *et al.*	1994
	Pidgeon *et al.*	1994
Dyes	Poscio *et al.*	1994
Epinephrine	Pihel *et al.*	1994
(see also Neurotransmitters)	Wang, J. and Liu	1994
Ethanol	Buttler *et al.*	1993
	Mitsubayashi *et al.*	1994
	Rank *et al.*	1995
	Jameison *et al.*	1996
Ethelene	Bodenhöfer *et al.*	1996
Foot and mouth disease virus (FMDV)	Knichel *et al.*	1995
	Rickert *et al.*	1996
Formaldehyde	Wang, J. *et al.*	1995e
Fructose	Khan *et al.*	1992
Galactose	Manowitz *et al.*	1995
Gene probes	Graham *et al.*	1992
(see also DNA hybrids,	Hashimoto *et al.*	1994
Olignucleotides, Nucleic acids)	Vo-Dinh *et al.*	1994a
	Chee *et al.*	1996

Analyte	Author(s)	Year
Glucose, See Appendix A-2		
Glucuronide	Wang, A-J. and Rechnitz	1993
Glutamate	Kar and Arnold	1992
(see also Neurotransmitters)	Pantano and Kuhr	1993
	Berners *et al.*	1994
	Pandey and Weetall	1995b
	White *et al.*	1995
	Cosford and Kuhr	1996
	Niwa *et al.*	1996
	Cosnier *et al.*	1997
Glutamine	White *et al.*	1995
Glutathione	Wring *et al.*	1991, 1992
Glycerol	Rank *et al.*	1995
Glycerophosphate	Katakis and Heller	1992
Heavy metals	Booksh *et al.*	1994
(see also individual metal ions)	Lin, Z. and Burgess	1994
	Lin, Z. *et al.*	1994
	Turyan and Mandler	1997
Hemoglobin	Chen, H-Y. *et al.*	1994
Heparin	Fu *et al.*	1994
	Van-Kerkhof *et al.*	1995
	Wang, E. *et al.*	1995
	Meyerhoff *et al.*	1996
Herbicides	Brewster *et al.*	1995
(see also individual herbicides)	Preuss and Hall	1995
	Rouillon *et al.*	1995
Heroin	Holt *et al.*	1996
Herpes virus	König and Grätzel	1994
Histamine	Pihel *et al.*	1995
HIV	Uda	1995
	Wang, J. *et al.*	1996a
Hormones		
(see also individual hormones)		
Horseradish peroxidase	Adami	1995b
	Vreeke	1995a
Human chorionic gonadotropin (hCG)	Duan and Meyerhoff	1995
Human growth hormones	Shimura and Karger	1994
Hydrazines	Wang, J. and Chen	1995
	Wang, J. and Pamidi	1995
	Wang, J. *et al.*	1996b

Analyte	Author(s)	Year
Hydrogen peroxide	Tatsuma *et al.*	1992a
(see also Peroxides)	Vreeke *et al.*	1992
	Fernández-Romero and Luque de Castro	1993
	Garguilo *et al.*	1993
	Horrocks *et al.*	1993b
	Kerner *et al.*	1993
	Palmisano and Zambonin	1993
	Csöregi *et al.*	1994a
	Dong and Guo	1994
	Hermes *et al.*	1994
	Korell and Spichiger	1994
	Lötzbeyer *et al.*	1994
	Tatsuma *et al.*	1994a, 1995
	Yang, Liu and Murray	1994
	Johnston *et al.*	1995
	Meyer *et al.*	1995
	Tsai and Cass	1995
	Vreeke *et al.*	1995b
	Wang, J. *et al.*	1995b
	Yang, Liu *et al.*	1995
	Zhou, X.Z. and Arnold	1995
	Alegret *et al.*	1996
	Lei and Deng	1996
	Towe and Guilbeau	1996
	Wang, J. and Chen	1996
Hydrogen sulfite	Hutchins *et al.*	1994
Hydroperoxides	Wang, J. *et al.*	1991
	Yang, Liu and Murray	1994
	Tsai and Cass	1995
	Wang, C-L. and Mulchandani	1995
	Postlethwaite *et al.*	1996
Hydroquinone	McAleron and Slater	1994
7-Hydroxycoumarin	Deasy *et al.*	1994
5-Hydroxytryptamine (see Serotonin)		
Imidazoles	Tatsuma and Watanabe	1992a
Immunocomponents	Danielsson and Mosbach	1989
(see also individual antigens, antibodies)	Bright *et al.*	1990
	Eenink *et al.*	1990
	Attridge *et al.*	1991
	Bowyer *et al.*	1991
	VanderNoot and Lai	1991
	Morgan and Taylor	1992
	Walczak *et al.*	1992
	Buckle *et al.*	1993
	Heideman *et al.*	1993
	Liedberg *et al.*	1993
	Shriver-Lake *et al.*	1993
	Zhou, Y. *et al.*	1993

Analyte	Author(s)	Year
Immunocomponents (contd)	Andle and Vetelino	1994
	Dahint *et al.*	1994
	Duan and Meyerhoff	1994
	Hayes *et al.*	1994
	Song *et al.*	1994
	Sutherland *et al.*	1994a,b
	Williams and Blanch	1994
	Bernard and Bosshard	1995
	Cao *et al.*	1995
	Gao, H. *et al.*	1995
	Knichel *et al.*	1995
	McNeil *et al.*	1995
	Owaku *et al.*	1995
	Ruegg *et al.*	1995
	Hale *et al.*	1996
	Invitski and Rishpon	1996
	Lawrence *et al.*	1996
	Nakanishi *et al.*	1996
	Polzius *et al.*	1996
	Reiken *et al.*	1996
	Renken *et al.*	1996
	Sadik and Van Emon	1996
	Schipper *et al.*	1996
	Welsch *et al.*	1996
	Chiem and Harrison	1997
	Rabbany *et al.*	1997
Insulin	Kennedy *et al.*	1993, 1995
	Schultz *et al.*	1995
	Suri *et al.*	1995
	Tao and Kennedy	1996
Interleukin 2 receptor	Wu *et al.*	1995
Interleukin 5	Bennett *et al.*	1995
Interleukin 6	Ward *et al.*	1995
Lactate	Katakis and Heller	1992
	Ohara *et al.*	1994
	Wang, J. and Chen	1994a
	Baker and Gough	1995
	deLumley-Woodyear *et al.*	1995
	Pandey and Weetall	1995b
	Sprules *et al.*	1995
	Wang, J. *et al.*	1995b
	Yang, Liu *et al.*	1995
	Sampath and Lev	1996
	Silber *et al.*	1996
Lactose	Wood and Gruber	1996

Analyte	Author(s)	Year
Lead ions	Lerchi *et al.*	1992
	Wang, J. and Tian	1992 , 1993a
	Feldman *et al.*	1994
	Yang, Lin and Saavedra	1995
Linamarin	Tatsuma *et al.*	1996
Listeria monocytogens	Jacobs *et al.*	1995
	Strachan and Gray	1995
Lithium ions	Silber *et al.*	1996
Luteinizing hormones (hLH)	Invitski and Rishpon	1996
Magnesium ions	Otto and Thomas	1985
	Suzuki *et al.*	1995
Maltose	Gilardi *et al.*	1994
Mangnese ions	Otto and Thomas	1985
	Smit and Rechnitz	1992
Melanocyte stimulating hormone (MSH)	Paras and Kennedy	1995
Metal inhibitors	Wang, J. and Chen	1993
Mercury ions	Wang, J. and Tian	1993b
	Lerchi *et al.*	1994
Microbial cells/films	Nivens *et al.*	1993
	Watts *et al.*	1994
Microperoxidase	Lötzbeyer *et al.*	1994
	Mabrouk	1996
Morphines	Kriz *et al.*	1995b
Multi-ions	Lemké *et al.*	1992
	Pace and Hamerslag	1992
	Cosofret *et al.*	1994, 1995a
	Goldberg *et al.*	1994
	Knoll *et al.*	1994
	Satoh and Iijima	1995
NADH	Gorton *et al.*	1992
	Vreeke *et al.*	1992
	Wang, J. *et al.*	1992
	Pantano and Kuhr	1993
	Sprules *et al.*	1994
	Nowall and Kuhr	1995
	Pariente *et al.*	1994,1995,1996
Nerve agents (see also organic vapors)	Noble	1994
Neu differentiation factor	Lu, H.S. *et al.*	1995
Neurotransmitters	Pantano and Kuhr	1993
(see also individual neurotransmitters)	Pihel *et al.*	1994,1995,1996
	Boulton *et al.*	1995
	Hayes and Kuhr	1995
	Tan, W. *et al.*	1995a
	Zhou, S.Y. *et al.*	1995
	Stamford and Justice	1996

Analyte	Author(s)	Year
Nitrate	Cosnier *et al.*	1994
	Knoll *et al.*	1994
	Hutchins and Bachas	1995
Nitrite	Schaller *et al.*	1994
	Badr *et al.*	1995
	Strehlitz *et al.*	1996
Nitric oxide	Ichimori *et al.*	1994
	Etches *et al.*	1995
	Lantoine *et al.*	1995
	Friedemann *et al.*	1996
	Zhou, X.Z. and Arnold	1996
Nitrogen dioxide	Collins and Rose-Pehrsson	1995
Nitrophenol	Benoit and Yappert	1996a
Nitroxide radicals	Bauer, J.E. *et al.*	1996
Norepinephrine	Moore *et al.*	1994
(see also Neurotransmitters)	Pihel *et al.*	1994
	Burnette *et al.*	1996
Nucleic acids	Andle *et al.*	1993
(see also DNA hybrids, Gene probes,	deVries *et al.*	1994
Oligonucleotides, DNA, RNA)	Palaček and Fajita	1994
	Su, H. *et al.*	1994
	Ueno and Yeung	1994
	Piunno *et al.*	1995
	Wang, J. and Chen	1995a
	Kolakowski *et al.*	1996
	Wooley *et al.*	1996
Octapeptides	Witkowski *et al.*	1993
Oligonucleotides	Effenhauser *et al.*	1994
(see also DNA hybrids, Gene probes,	Pease *et al.*	1994
Nucleic acid, DNA,RNA)	Stimpson *et al.*	1995
	Abel *et al.*	1996
	Blanchard *et al.*	1996
	Ferguson and Walt	1996
	Kawazoe *et al.*	1996
	Wang, J. *et al.*	1996a
Opiate	Holt *et al.*	1995
Organic compounds	Conzen *et al.*	1993
(see also individual compounds)	Buerck *et al.*	1994
Organic vapors	Gardner *et al.*	1987
	Kepley *et al.*	1992
	Grate *et al.*	1993c, 1996
	Schierbaum *et al.*	1994
	Zellers *et al.*	1995
	Bodenhöfer *et al.*	1996
	Dickert *et al.*	1996
	White *et al.*	1996
	Sutter and Jurs	1997
Organonitriles	Liu, Z. T. *et al.*	1993, 1995

Analyte	Author(s)	Year
Organophosphorus compounds (see Organic vapors)		
Organosulfur compounds (see Organic vapors)		
Oxalic acid	Shimohigoshi and Karube	1996
Oxygen	Clark, L.C. *et al.*	1958
	Severinghaus and Bradley	1958
	Clark, L. C. and Lyons	1962
	Clark, L. C.	1971
	Lau *et al.*	1992
	MacCraith *et al.*	1993
	Arquint *et al.*	1994
	Harsanyi *et al.*	1994
	Hermes *et al.*	1994
	Oglesby *et al.*	1994
	Preininger *et al.*	1994
	Suzuki	1994
	Bambot *et al.*	1995b
	Collins and Rose-Pehrsson	1995
	Hartmann *et al.*	1995
	Klimant and Wolfbeis	1995
	Meyer *et al.*	1995
	Rosenzwieg and Kopelman	1995
	Tarnowsky *et al.*	1995
	Tsionsky and Lev	1995
	Xu *et al.*	1995
	Bowyer *et al.*	1996
	Gooding and Hall	1996
	Hartmann and Trettnak	1996
	Hartmann and Ziegler	1996
	Postlethwaite *et al.*	1996
Paracetamol (see Acetaminophen)		
Pectin	Horie and Rechnitz	1995
Penicillin	Nishizawa *et al.*	1992
	Bataillard *et al.*	1993
	Csöregi *et al.*	1994a
	Li, L. and Walt	1995
	Nikolelis and Siontoru	1995
Peroxides	Korell and Spichiger	1994
	Wang, C-L. and Mulchandani	1995
	Mulchandani *et al.*	1995

Analyte	Author(s)	Year
Pesticides	Bier *et al.*	1992
	Hart and Hartley	1994
	Hartley and Hart	1994
	Mionetto *et al.*	1994
	Cagnini *et al.*	1995
	Lechuga *et al.*	1995
	Schipper *et al.*	1995
	Bauer, C.G. *et al.*	1996
	Sadik and Van Emon	1996
pH-Protons	Kirkbright *et al.*	1984
	van der Schoot and Bergveld	1988
	Buck *et al.*	1992
	Lemké *et al.*	1992
	Pace and Hamerslag	1992
	Tan, W. *et al.*	1992b
	Horrocks *et al.*	1993a
	Szmacinski and Lakowicz	1993
	Wang, Juan *et al.*	1993
	Adami *et al.*	1994
	Arquint *et al.*	1994
	Bronk and Walt	1994
	Goldberg *et al.*	1994
	Greenway *et al.*	1994
	Hitzmann and Kullick	1994
	Shakhsher *et al.*	1994
	Telting-Diaz *et al.*	1994
	Tsionsky *et al.*	1994
	Bronk *et al.*	1995
	Cosofret *et al.*	1995a
	Fiedler *et al.*	1995
	Sheppard *et al.*	1995
	Yang, Lin and Saavedra	1995
	Zhang, S. *et al.*	1995
	Cheng and Brajter-Toth	1996
	Espadas-Torre *et al.*	1996
	Bruno *et al.*	1997
	Nomura *et al.*	1997
	Song *et al.*	1997
pH-Protons (for cellular dynamics)	Hafeman *et al.*	1988
	McConnell *et al.*	1992
	Owicki and Parce	1992
	Rogers *et al.*	1992
	Baxter *et al.*	1994
	Nicolini *et al.*	1995a
	Yoshinobu	1996
Pharmaceuticals, oxidizable (see also individual drugs)	Moore *et al.*	1994

Analyte	Author(s)	Year
Phenols	Wang, J. *et al.*	1991
	Wang, J. *et al.*	1994b
	Deng, Q. and Dong	1995
	Kotte *et al.*	1995
	Wang, J. and Chen	1996
L-Phenylalanine	Kriz *et al.*	1995
	Sampath and Lev	1996
Phenytoin	Astles and Miller	1994
	La Gal La Salle *et al.*	1995
Phosphate	Cary and Riggan	1994
	Su, Y.S. and Mascini	1995
	Meruva and Meyerhoff	1996
cis/trans-Platins	Su, H. *et al.*	1995
Polychlorinated biphenyls	Roberts and Durst	1995
Polyions	Fu *et al.*	1994
	Meyerhoff *et al.*	1996
Potassium ions	Otto and Thomas	1985
	Buck *et al.*	1992
	Lemké *et al.*	1992
	Pace and Hamerslag	1992
	Cheung and Yeung	1994
	Goldberg *et al.*	1994
	Knoll *et al.*	1994
	Wei *et al.*	1994
	Cosofret *et al.*	1995a
	Brooks *et al.*	1996
	Silber *et al.*	1996
s-Propranolol	Reinhoudt *et al.*	1994
	Anderssen	1996
	Silber *et al.*	1996
Protamine	Meyerhoff *et al.*	1996
Proteins	Witkowski *et al.*	1993
	Lillard *et al.*	1996
	Ramsden *et al.*	1996
	Schneiderheinze and Hogan	1996
Protons (see pH-Protons and pH-Protons, cellular dynamics)		
Ricin (see also anti-Ricin)	Carter *et al.*	1995
RNA	Wang, J. *et al.*	1995a
Salicylate	Frew *et al.*	1989
Salmonella	Brewster *et al.*	1996
Serotonin	Jackson *et al.*	1995
	Pihel *et al.*	1995
	Rivot *et al.*	1995
	Lillard *et al.*	1996
	Zhang, X. *et al.*	1996

Analyte	Author(s)	Year
Serum albumins	Buckle *et al.*	1993
	Hayes *et al.*	1994
	Jordan *et al.*	1995
	Welsch *et al.*	1996
Sex hormone binding globulin (SHBG)	Morgan and Taylor	1992
Sialic acid	Kugimiya *et al.*	1995
Silver ions	Lerchi *et al.*	1994, 1996
Sodium ions	Otto and Thomas	1985
	Cadogan *et al.*	1992
	Lemké *et al.*	1992
	Pace and Hamerslag	1992
	Careri *et al.*	1993
	Cheung and Yeung	1994
	Knoll *et al.*	1994
	Tsujimura *et al.*	1995
	Shortreed *et al.*	1996a
Spermine	Gasparini *et al.*	1994
Sucrose	Tobias-Katona and Pecs	1995
	Filho *et al.*	1996
Sulfamethazines	Sternesjo *et al.*	1995
Superoxide	Lvovich and Scheeline	1997
Surfactants	Lundgren and Bright	1996
Tetrachloroethene	Bodenhöfer *et al.*	1996
Theophylline	Foulds *et al.*	1990
	Buckle *et al.*	1993
	Chiem and Harrison	1997
Thiocholine	Hart and Hartley	1994
Thyrotrophin	Athey *et al.*	1993
TNT	Shriver-Lake *et al.*	1993, 1995
Triolein	Sangodkar *et al.*	1996
Trypsin	Buckle *et al.*	1993
	Avnir *et al.*	1994
Tyrosine inhibitors	Stancil *et al.*	1995
	Wang, J. and Chen	1995
Urea	Bataillard *et al.*	1993
	Kallury *et al.*	1993
	Wang, Juan *et al.*	1993
	Hendji *et al.*	1994
	Narang *et al.*	1994a
	Andres and Narayanaswamy	1995
	Liu, D.H. *et al.*	1995
	McNeil *et al.*	1995
	Nikolelis and Siontoru	1995
	Sangodkar *et al.*	1996
	Silber *et al.*	1996
Urease	Andres and Narayanaswamy	1995

Analyte	Author(s)	Year
Uric acid	Gilmartin *et al.*	1992, 1994
	Gilmartin and Hart	1994c
	Lisdat *et al.*	1996
	Shimohigoshi and Karube	1996
Uranium	Wang, J. *et al.*	1994b
Vapors (see Organic vapors)		
Viruses	Andle *et al.*	1993
(see also Herpes, HIV, FMDV, CMV, AMD)	Rickert *et al.*	1996
Zinc ions	Wei *et al.*	1994
	Satoh and Iijima	1995
	Thompson and Patchan	1995

Appendix A-2

GLUCOSE AS ANALYTE

Authors	Year	Authors	Year
Clark, L.C.	1971	Ohara *et al*	1993
Armour *et al.*	1990	Palmisano and Zambonin	1993
Lasky and Buttry	1990	Palmisano *et al.*	1993
Kawagoe *et al.*	1991	Small *et al.*	1993
Pishko *et al.*	1991	Willner *et al.*	1993
Abe *et al.*	1992	Ye, L. *et al.*	1993
Ballarin *et al.*	1992	Avnir *et al.*	1994
Bartlett *et al.*	1992	Csöregi *et al.*	1994b
Fan and Harrison	1992	Furbee *et al.*	1994
Hoa *et al.*	1992	Gasparini *et al.*	1994
Maidan and Heller	1992	Hendji *et al.*	1994
Reach and Wilson	1992	Kaku *et al.*	1994
Wang, J. and Angnes	1992	Kulys and Hansen	1994
Wang, J. *et al.*	1992	Kuwabata and Martin	1994
Wolowacz *et al.*	1992	Lowry *et al.*	1994
Bartlett and Birkin	1993, 1994	Moussy *et al.*	1994
Bataillard *et al.*	1993	Narang *et al.*	1994b
Cardosi and Birch	1993	Rhodes *et al.*	1994
Fernández-Romero and Luque de Castro	1993	Shul'ga *et al.*	1994
Kerner *et al.*	1993	Tsionsky *et al.*	1994
Marquardt *et al.*.	1993	Wang, J.	1994
		Wang, J. and Chen	1994a,b

Authors	Year	Authors	Year
Wang, J. *et al.*	1994a	Tarnowski *et al.*	1995
Ward *et al.*	1994	Wang, J. *et al.*	1995c
Zhang, Y. *et al.*	1994	White *et al.*	1995
Amine *et al.*	1995	Wilkins *et al.*	1995
Ammon *et al.*	1995	Yabuki and Mizutani	1995
Chen, Q. *et al.*	1995	Yang, Lin and Saavedra	1995
Chi and Dong	1995	Zhang, H. *et al.*	1995
Clark, A. D. *et al.*	1995	Zhou, X.Z. and Arnold	1995
Csöregi *et al.*	1995	Alegret *et al.*	1996
deLumley-Woodyear *et al.*	1995	Bangalore *et al.*	1996
Dong and Guo	1995	Bu *et al.*	1996
Gamburzev *et al.*	1995	Calvo *et al.*	1996
Hoshi *et al.*	1995	Du *et al.*	1996
Ito *et al.*	1995	Ikeda *et al.*	1996
Jaffari and Turner	1995	Jameison *et al.*	1996
Jimenez *et al.*	1995	Jung and Wilson	1996
Jin *et al.*	1995	Khan *et al.*	1996
Kuwabata *et al.*	1995	Kikuchi *et al.*	1996
Liu, H. Y. and Deng	1995	Lisdat *et al.*	1996
Meyer *et al.*	1995	Pan *et al.*	1996
Nagata *et al.*	1995a,b	Preston and Nieman	1996
Nishida *et al.*	1995	Rickert *et al.*	1996
Ohashi and Karube	1995	Rosenzweig and Kopelman	1996
Palmisano *et al.*	1995c	Sampath and Lev	1996
Pandey and Weetall	1995b	Sangodkar *et al.*	1996
Pravda *et al.*	1995a,b	Schmidtke *et al.*	1996
Quinn *et al.*	1995b	Shaffer *et al.*	1996
Ramanathan *et al.*	1995	Siber *et al.*	1996
Riklin and Willner	1995	Thomé-Duret *et al.*	1996
Saby *et al.*	1995	Towe and Guilbeau	1996
Shimohigoshi *et al.*	1995	Wang, J. *et al.*	1996c
Shin and Kim	1995		

Appendix B

SELECTED TOPICAL REVIEWS

Primary Topic	Authors	Year
Acoustic-wave devices	Grate *et al.*	1993a,b
(see also Piezoelectric devices)	Andle and Vetelino	1994
	Jossé	1994
Antibodies	Killard *et al.*	1995
	Lu, B. *et al.*	1996
Arrays	Diamond	1993
	Carey	1994
	Niwa	1995
Automated systems	Trojanowicz	1996
Biochemical aspects	Byfield and Abuknesha	1994
Bioelectronic devices	Nicolini	1995
Biomolecular interactions	Yeung *et al.*	1995
Bioprocess control	Mulchandani and Bassi	1995
Biosensors—general	Arnold and Myerhoff	1988
	Connolly	1995
Chemometrics	Brown and Bear	1993
	Booksh and Kowalski	1994a
	Ferrus and Egea	1994
	Thomas	1994
Clinical diagnostics	Connolly	1995

Primary Topic	Authors	Year
DNA arrays	Hunkapiller *et al.*	1991
	Blanchard *et al.*	1996
	Borman	1996
	Gette and Kreiner	1997
Electrochemical DNA sensors	Mikkelson	1996
Environmental sensors	Rawson *et al.*	1989
	Van Emon and Lopez-Avila	1992
	Damgaard *et al.*	1995
	Karube *et al.*	1995b
	Rogers	1995
	Rogers and Williams	1995
	Sadik and Van Emon	1996
Enzyme–modified electrodes	Heller	1990
	Pantano and Kuhr	1995
Enzymes in non-aqueous media	Kamat *et al.*	1995
Fiber-optic sensors	Arnold	1992
Fluorescence imaging	Tsien	1994
Fluorescent protein biosensors	Giuliano *et al.*	1995
General	Arnold and Meyerhoff	1988
	Wolfbeis	1990
	Scheller *et al.*	1991
	Sethi	1994
Genosensors	Beattie *et al.*	1995
Glucose oxidase	Wilson and Turner	1992
Immobilization/avidin–biotin	Anzai *et al.*	1994
Immobilization/proteins	Bartlett and Cooper	1993
	Weetall	1993
	Williams and Blanch	1994
Immobilization/supported membranes	Sackmann	1996
Intracellular/single cells	Ewing *et al.*	1992
	Weber	1995
In vivo monitoring	Meyerhoff	1993
	Jaffari and Turner	1995
	Stamford and Justice	1996
Ion-selective sensors	Yim *et al.*	1993
	Beer	1996
Micromachining	Karube *et al.*	1995a
	Kovacs *et al.*	1996
Microscale sensors	Damgaard *et al.*	1995
Modeling—electrochemical sensors	Bartlett and Pratt	1993
	Speiser	1996
Modified electrodes	Heller	1990
	Scheller *et al.*	1991
	Murray	1992
	Gilmartin and Hart	1995
	Gorton	1995
	Kalcher *et al.*	1995
	Pantano and Kuhr	1995

Primary Topic	Authors	Year
Molecular imprinting	Mosbach	1994
Multivariate calibration	Thomas	1994
Near-field spectroscopy	Lewis and Lieberman	1991
	Betzig and Trautman	1992
	Harris *et al.*	1994
	Kopelman and Tan	1994
	Tan and Kopelman	1996
Near-infrared labels	Patonay and Antoine	1991
Optical imaging fibers	Pantano and Walt	1995
Optical methods	Angel	1987
	Seitz	1988
	Brecht and Gauglitz	1995
	Lübbers	1995
Oxidoreductases	Scheller *et al.*	1991
	Wilson and Turner	1992
	Schmidt and Schuhmann	1996
Piezoelectric detection (see also Acoustic-wave devices)	Guilbault and Jordan	1988
	Buttry and Ward	1992
	Mandelis and Christofides	1993
	Minunni *et al.*	1995
Potentiometric sensors	Yim *et al.*	1993
Self-assembled monolayers (SAMS)	Creager and Olsen	1995
	Mitler-Neher *et al.*	1995
	Zhong and Porter	1995
	Finklea	1996
Screen-printing	Goldberg *et al.*	1994
Sol–gels and sensors	Dave *et al.*	1994
	Lev *et al.*	1995
Surface characterization—sensors	Perry and Somorjai	1994
	Göpel and Heiduschka	1995
Thermal sensing	Bataillard	1993
Thick or thin films	Swalen *et al.*	1987
	Galan-Vidal *et al.*	1995
	Zhong and Porter	1995
	Frank *et al.*	1996
Ultramicroelectrodes	Wightman and Wipf	1989
	Heinze	1993
	Zoski	1996
Waveguides	Dessy	1989
"Wired" redox enzymes	Heller	1990

Bibliography

Abbott, N. L., Kumar, A. and Whitesides, G. M. (1994) Using micromachining, molecular self-assembly and wet etching to fabricate 0.1–1μm-scale structures of gold and silicon. *Chem. Mater.* **6**: 596–602.

Abe, T., Lau, Y. Y. and Ewing, A. G. (1992) Characterization of glucose microsensors for intracellular measurements. *Anal. Chem.* **64**: 2160–2163.

Abel, A. P., Weller, M. G., Dureneck, G. L., Ehrat, M. and Widmer, H. M. (1996) Fiber-optic evanescent wave biosensor for the detection of oligonucleotides. *Anal. Chem.* **68**: 2905–2912.

Abeles, R. H., Frey, P. A. and Jencks, W. P. (1992) *Biochemistry*. Jones and Bartlett, Boston, MA.

Adami, M., Piras, L., Lanzi, M., Fanigliulo, A., Vakula, S. and Nicolini, C. (1994) Monitoring of enzymatic activity and quantitative measurements of substrates by means of a newly designed silicon-based potentiometric sensor. *Sens. Actuators* (B) **18**: 178–182.

Adami, M., Sartore, M. and Nicolini, C. (1995a) PAB: a newly designed potentiometric alternating biosensor system. *Biosens. Bioelec.* **10**: 155–167.

Adami, M., Martini, M. and Piras, L. (1995b) Characterization and enzymatic application of a redox potential biosensor based on a silicon transducer. *Biosens Bioelec.* **10**: 633–638.

Adams, R. E., Betso, S. R. and Carr, P. W. (1976) Electrochemical pH-stat and controlled current coulometric acid-base analyser. *Anal. Chem.* **48**: 1989–1996.

Adamson, A. W. (1990) *Physical Chemistry of Surfaces*, 5th Edition. John Wiley and Sons, New York.

Ahluwalia, A., DeRossi, D., Ristori, C., Shirone, A. and Sierra, G. (1991) A comparative study of immobilization techniques for optical immunosensors. *Biosens. Bioelec.* **7**: 207–214.

Aizawa, M., Nishiguchi, K., Imamura, M., Kobatake, E., Haruyama, T. and Ikariyami, Y. (1995) Integrated molecular systems for biosensors. *Sens. Actuators* (B) **24**: 1–5.

Alberts, B., Bray, D., Lewis, J., Raff, M., Roberts, K. and Watson, J. D. (1989) *Molecular Biology of the Cell*, 2nd Edition. Garland Publishing, New York.

Albery, W. J. and Craston, D. H. (1989) Amperometric enzyme electrodes: theory and experiment. In Turner, A. P. F., Karube, I. and Wilson, G. S. (eds) *Biosensors: Fundamentals and Applications*. Oxford University Press, New York.

Alegret, S., Céspedes, F., Martínez-Fábregas, E., Nartirekk, D., Nirakes, A., Centelles, E. and Muñoz, J. (1996) Carbon-polymer biocomposites for amperometric sensing. *Biosens. Bioelec.* **11**: 35–44.

Allara, D. L. (1995) Critical issues in applications of self-assembled monolayers. *Biosens. Bioelec.* **10**: 771–783.

Alvarez-Icaza, M. and Bilitewski, U. (1993) Mass production of biosensors. *Anal. Chem.* **65**: 525A–533A.

Amankawa, L. N. and Kuhr, W. G. (1993) Online peptide mapping by capillary zone electrophoresis. *Anal. Chem.* **65**: 2693–2697.

Amatore, C., Savéant, J. M. and Tessier, D. (1983) Charge transfer at partially blocked surfaces. A model for the case of microscopic active and inactive sites. *J. Electroanal. Chem.* **147**: 39–51.

Amine, A., Deni, J. and Kauffmann, J–M. (1994) Preparation and characterization of octadecylamine-containing carbon paste electrodes. *Anal. Chem.* **66**: 1595–1599.

Amine, A., Dugna, K., Xie, B. and Danielsson, B. (1995) A microdialysis probe coupled with a miniaturized thermal glucose sensor for *in vivo* monitoring. *Anal. Lett.* **28**: 2275–2286.

Amman, D. (1986) *Ion-Selective Microelectrodes*. Springer-Verlag, Heidelberg.

Ammon, H. P., Ege, W., Oppermann, M., Göpel, W. and Eisels, S. (1995) Improvement in the long-term stability of an amperometric glucose sensor system by introducing a cellulose membrane of bacterial origin. *Anal. Chem.* **67**: 466–471.

Anderson, G. P., Golden, J. P. and Ligler, F. S. (1993) Fibre-optic biosensor: combination tapered fibres designed for improved signal acquisition. *Biosens. Bioelec.* **8**: 249–256.

Anderson, G. P., Golden, J. P. and Ligler, F. S. (1994) An evanescent wave biosensor—Part I: Fluorescent signal acquisition from step-etched fiber optic probes. *IEEE Trans. Biomed. Eng.* **41**: 578–584.

Anderson, J. L. and Winograd, N. (1996) Film electrodes. In Kissinger, P. Y. and Heineman, W. R. (eds) *Laboratory Techniques in Electroanalytical Chemistry*, 2nd Edition. Marcel Dekker, New York.

Anderssen, L. I. (1996) Application of molecular imprinting to the development of aqueous buffer and organic solvent based radioligand binding assays for (S)-propranolol. *Anal. Chem.* **68**: 111–117.

Andle, J. C., Vetelino, J. F., Lade, M. W. and McAllister, D. J. (1992) An acoustic wave biosensor. *Sens. Actuators* (B) **8**: 191–198.

Andle, J. C., Weaver, J. T., McAllister, D. J., Jossé, F. and Vetelino, J. F. (1993) An improved acoustic plate mode biosensor. *Sens. Actuators* (B) **13**: 437–442.

Andle, J. C. and Vetelino, J. F. (1994) Acoustic wave biosensors. *Sens. Actuators* (A) **44**: 167–176.

Andres, R. T. and Narayanaswamy, R. (1995) Effect of the coupling reagent on the metal inhibition of immobilized urease in an optical biosensor. *Analyst* **120**: 1549–1554.

Andrieux, C. P., Audebert, P., Bacchi, P. and Divisia-Blohorn, B. (1995) Kinetic behaviour of an amperometric biosensor made of an enzymatic carbon paste and a Nafion gel investigated by rotating electrode studies. *J. Electroanal. Chem.* **394**: 141–148.

Angel, S. M. (1987) Optrodes: chemically selective fibre-optic sensors. *Spectroscopy* **2**: 38–48.

Ansell, R. J., Ramström, O. and Mosbach, K. (1996) Towards artificial antibodies prepared by molecular imprinting. *Clin. Chem.* **42**: 1506–1512.

Anzai, J., Hoshi, T. and Osa, T. (1994) Avidin–biotin complexation for enzyme sensor applications. *Trends Anal. Chem.* **13**: 205–210.

Arbault, S., Pantano, P., Jankowski, J. A., Vuillaume, M. and Amatore, C. (1995) Monitoring an oxidative stress mechanism at a single human fibroblast. *Anal. Chem.* **67**: 3382–3390.

Armour, J. C., Lucisano, J. Y., McKean, B. D. and Gough, D. A. (1990) Application of a chronic intravascular blood glucose sensor in dogs. *Diabetes* **39**: 1519–1526.

Arnold, M. A. and Meyerhoff, M. E. (1988) Recent advances in the development and analytical applications of biosensing probes. *CRC Crit. Revs. Anal. Chem.* **20**: 149–196.

Arnold, M. A. and Rechnitz, G. A. (1989) Biosensors based on plant and animal tissue. In Turner, A. P. F., Karube, I. and Wilson, G. S. (eds), *Biosensors: Fundamentals and Applications*. Oxford University Press, New York.

Arnold, M. A. (1992) Fiber-optic chemical sensors. *Anal. Chem.* **64**: 1015A–1025A.

Arquint, P., Koudelka-Hep, M., van der Schoot, B. H., van der Wal, P. and de Rooij, N. F. (1994) Micromachined analyzers on a silicon chip. *Clin. Chem.* **40**: 1805–1809.

Aschauer, E., Fasching, R., Urban, G., Nicolussi, G. and Husinsky, W. (1995) Surface characterization of thin-film platinum electrodes for biosensors by means of cyclic voltammetry and laser SNMS. *J. Electroanal. Chem.* **381**: 143–150.

Astles, J. R. and Miller, W. G. (1994) Measurement of free phenytoin in blood with a self-contained fiber-optic immunosensor. *Anal. Chem.* **66**: 1675–1682.

Athey, D., Ball, M. and McNeil, C. J. (1993) Avidin–Biotin based electrochemical immunoassay for thyrotrophin. *Ann. Clin. Biochem.* **30**: 570–577.

Atkins, P. W. (1990) *Physical Chemistry*, 4th Edition. Oxford University Press, New York.

Attridge, J. W., Daniels, P. B., Deacon, J. K., Robinson, G. A. and Davidson, G. P. (1991) Sensitivity enhancement of optical immunosensors by the use of a surface plasmon resonance fluoroimmunoassay. *Biosens. Bioelec.* **6:** 201–214.

Auld, B. A. (1990) *Acoustic Fields and Waves in Solids*. Krieger Publishing, Malibar, FL.

Avnir, D., Braun, S., Lev, O. and Ottolenghi, M. (1994) Enzymes and other proteins entrapped in sol–gel materials. *Chem. Mater.* **6**: 1605–1614.

Babb, C. W., Coon, D. R. and Rechnitz, G. A. (1995) Biomagnetic neurosensors. 3. Noninvasive sensors using magnetic stimulation and biomagnetic detection. *Anal. Chem.* **67**: 763–769.

Babkina, S. S., Medyantseva, E. P., Budnikov, H. C. and Tyshiek, M. P. (1996) New variants of enzyme immunoassay of antibodies to DNA. *Anal. Chem.* **68**: 3827–3831.

Bacha, S., Bergel, A. and Comtat, M. (1995) Transient response of multilayer electroenzymatic biosensors. *Anal. Chem.* **67**: 1669–1678.

Badr, I. H., Meyerhoff, M. E. and Hassan, S. S. (1995) Potentiometric anion selectivity of polymer membranes doped with palladium organophosphine complex. *Anal. Chem.* **67:** 2613–2618.

Baker, D. A. and Gough, D. A. (1993) Dynamic concentration challenges for biosensor construction. *Biosens. Bioelec.* **8**: 433–441.

Baker, D. A. and Gough, D. A. (1995) A continuous, implantable lactate sensor. *Anal. Chem.* **67**: 1536–1540.

Baker, D. A. and Gough, D. A. (1996) Dynamic delay and maximal dynamic error in continuous biosensors. *Anal. Chem.* **68**: 1292–1297.

Baker, D. R. (1996) *Capillary Electrophoresis.* John Wiley and Sons, New York.

Bakker, E. and Simon, W. (1992) Selectivity of ion-sensitive bulk optodes. *Anal. Chem.* **64**: 1805–1812.

Bakker, E., Meruva, R. K., Pretsch, E. and Meyerhoff, M. E. (1994) Selectivity of polymer membrane-based ion-selective electrodes: self-consistent model describing the potentiometric response in mixed ion solutions of different charge. *Anal. Chem.* **66**: 3021–3030.

Bakker, E. (1997) Determination of unbiased selectivity coefficients of neutral carrier-based cation-selective electrodes. *Anal. Chem.* **69**: 1061–1069.

Ballarin, B., Brumlik, C. J., Lawson, D. R., Liang, W., Van Dyke, L. S. and Martin, C. R. (1992) Chemical sensors based on ultrathin-film composite membranes — a new concept in sensor design. *Anal. Chem.* **64**: 2647–2651.

Bambot, S. B., Lakowicz, J. R., Sipior, J., Carter, G. and Rao, G. (1995a) Optical measurement of bioprocess and clinical analytes using lifetime-based phase fluorimetry. In Rogers, K. R., Mulchandani, A. and Zhou, W. (eds), *Biosensor and Chemical Sensor Technology.* ACS Symposium Series #613, American Chemical Society, Washington, DC.

Bambot, S. B., Rao, G., Romauld, M., Carter, G. M., Sipior, J., Terpetchnig, E. and Lakowicz, J. R. (1995b) Sensing oxygen through skin using a red diode laser and fluorescence lifetimes. *Biosens. Bioelec.* **10**: 643–652.

Bangalore, A. S., Shaffer, R. E., Small, G. W. and Arnold, M. A. (1996) Genetic algorithm-based method for selecting wavelengths and model size for use with partial least-squares regression: application to near-infrared spectroscopy. *Anal. Chem.* **68**: 4200–4212.

Baptista, M. S., Tran, C. D. and Gao, G. H. (1996) Near-infrared detection of flow injection analysis by acoustooptic tunable filter-based spectrophotometry. *Anal. Chem.* **68**: 971–976.

Bard, A. J. and Faulkner, L. R. (1980) *Electrochemical Methods: Fundamentals and Applications.* John Wiley and Sons, New York.

Bard, A. J., Fan, F-R.F., Pierce, D. T., Unwin, P. R., Wipf, D.O. and Zhou, F. (1991) Chemical imaging of surfaces with the scanning electrochemical microscope. *Science* **254**: 68–74.

Bard, A. J. (1994) *Integrated Chemical Systems: A Chemical Approach to Nanotechnology.* John Wiley and Sons, New York.

Bard, A. J., Fan, F-R.F. and Mirkin, M. V. (1994) Scanning electrochemical microscopy. In Bard, A. J. (ed.) *Electroanalytical Chemistry*, Vol. 18. Marcel Dekker, New York.

Barker, S. L. R., Shortreed, M. R. and Kopelman, R. (1997) Utilization of lipophilic ionic additives in liquid polymer film optodes for selective anion activity measurements. *Anal. Chem.* **69**: 990–995.

Bartlett, P. N. (1990) Conducting organic salt electrodes. In Cass, A. E. G. (ed.) *Biosensors: A Practical Approach.* Oxford University Press, New York.

Bartlett, P. N., Tebbutt, P. and Whitaker, R. G. (1991) Kinetic aspects of the use of modified electrodes and mediators in bioelectrochemistry. *Prog. Reaction Kinetics* **16**: 55–155.

Bartlett, P. N., Tebbutt, P. and Tyrell, C. H. (1992) Electrochemical immobilization of enzymes. 3. Immobilization of glucose oxidase in thin films of electrochemically polymerized phenols. *Anal. Chem.* **64**: 138–142 (Correction: p. 1635).

Bartlett, P. N. and Birkin, P. R. (1993) Enzyme switch responsive to glucose. *Anal. Chem.* **65**: 1118–1119.

Bartlett, P. N. and Cooper, J. M. (1993) A review of the immobilization of enzymes in electropolymerized films. *J. Electroanal. Chem.* **362**: 1–12.

Bartlett, P. N. and Pratt, K. F. E. (1993) Modelling of processes in enzyme electrodes. *Biosens. Bioelec.* **8**: 451–462.

Bartlett, P. N. and Birkin, P. R. (1994) Micro-electrochemical enzyme transistor responsive to glucose. *Anal. Chem.* **66**: 1552–1559.

Bataillard, P. (1993) Calorimetric sensing in bioanalytical chemistry: principles, applications and trends. *Trends Anal. Chem.* **12**: 387–394.

Bataillard, P., Steffgen. S., Haemmerli, S., Manz, A. and Widmer, H. M. (1993) Integrated silicon thermopile as biosensor for the thermal monitoring of glucose, urea and penicillin. *Biosens. Bioelec.* **8**: 89–98.

Bätz, P., Schmeisser, D. and Göpel, W. (1991) Electronic structure of polypyrrole film. *Phys. Rev. B.*, **43**: 9178–9189.

Bauer, C. G., Eremenko, A. V., Ehrentreich-Förster, E., Bier, F. F., Makower, A., Halsall, H. B., Heineman, W. R. and Scheller, F. W. (1996) Zeptomole-detecting biosensor for alkaline phosphatase in an electrochemical immunoassay for 2,4-dichlorophenoxyacetic acid. *Anal. Chem.* **68**: 2453–2458.

Bauer, J. E., Wang, S. and Brandt, M. C. (1996) Fast-scan voltammetry of cyclic nitroxide free radicals. *Anal. Chem.* **68**: 3815–3821.

Baxter, G. T., Bousse, L. J., Dawes, T. D., Libby, J. M., Modlin, D. N., Owicki, J. C. and Parce, J. W. (1994) Microfabrication in silicon microphysiometry. *Clin. Chem.* **40**: 1800–1804.

Beattie, K. L., Beattie, W. G., Meng, L., Turner, S. L., Coral-Vazques, R., Smith, D. D., McIntyre, R. M. and Dao, D. D. (1994) Advances in genosensor research. *Clin. Chem.* **41**: 700–706.

Beer, P. D. (1996) Anion selective recognition and optical/electrochemical sensing by novel transition-metal receptor systems. *J. Chem. Soc., Chem. Commun.* 689–691.

Bender, W. J. H., Dessy, R. E., Miller, M. S. and Claus, R. O. (1994) Feasibility of a chemical microsensor based on surface plasmon resonance on fiber optics modified by multilayer vapor deposition. *Anal. Chem.* **66**: 963–970.

Bennett, D., Morton, T., Breen, A., Hertzberg, R., Cusimano, D., Appelbaum, E., McDonnell, P., Young, P., Matico, R. and Chaiken, I. (1995) Kinetic characterization of the interaction of biotinylated human interleukin 5 with an Fc chimera of its receptor alpha subunit and development of an ELISA screening assay using real-time interaction biosensor analysis. *J. Mol. Recognit.* **8**: 52–58.

Benoit, V. and Yappert, M. C. (1996a) Effect of capillary properties on the sensitivity enhancement in capillary/fiber optical sensors. *Anal. Chem.* **68**: 183–188.

Benoit, V. and Yappert, M. C., (1996b) Characterization of a simple Raman/fiber optical sensor. *Anal. Chem.* **68**: 2255–2258.

Berggren, K. K., Bard, A., Wilbur, J. L., Gillaspy, J. D., Helg, A. G., McClelland, J. J., Rolston, S. L., Phillips, W. D., Prentiss, M. and Whitesides, G. M. (1995) Microlithography by using neutral metastable atoms and self-assembled monolayers. *Science* **269**: 1255–1257.

Bernard, A. and Bosshard, H. R. (1995) Real-time monitoring of antigen–antibody recognition on a metal oxide surface by an optical grating coupler sensor. *Eur. J. Biochem.* **230**: 416–423.

Berners, M. O., Boutelle, M. G. and Fillenz, M. (1994) On-line measurement of brain glutamate with an enzyme/polymer-coated tubular electrode. *Anal. Chem.* **66**: 2017–2021.

Betzig, E., Trautman, J. K., Harris, T. D., Weiner, J. S. and Kostelak, R. L. (1991) Breaking the diffraction barrier: optical microscopy on a nanometric scale. *Science* **251**: 1468–1470.

Betzig, E. and Trautman, J. K. (1992) Near-field optics: microscopy, spectroscopy, and surface modification beyond the diffraction limit. *Science* **257**: 189–195.

Beyer, P. J., Lee, R., Gilman, S. D. and Ewing, A. G. (March 1995). Picoliter beakers for sample holders in voltammetry and capillary electrophoresis. Abstract #796, Pittsburgh Conference, New Orleans.

Bhatia, S., Teixeira, J., Anderson, M., Shriver-Lake, L., Calvert, J., Georger, J., Hickman, J., Dulcey, C., Schoen, P. and Ligler, F. (1993) Fabrication of surfaces resistant to protein adsorption and application of two-dimensional protein patterning. *Anal. Biochem.* **208**: 197–205.

Bier, F. F., Stoecklein, W., Boecher, M., Bilitewski, U. and Schmid, R. D. (1992) Use of a fibre optic immunosensor for the detection of pesticides. *Sens. Actuators.* (B) **7**: 509–512.

Bilitewski, U., Jaeger, A., Rueger, R. and Weise, W. (1993) Enzyme electrodes for the determination of carbohydrates in food. *Sens. Actuators* (B) **15**: 113–118.

Birnbaum, D., Kook, S-K. and Kopelman, R. (1993) Near-field scanning optical spectroscopy: spatially resolved spectra of microcrystals and nanoaggregates in dyed polymers. *J. Phys. Chem.* **97**: 3091–3094.

Blackburn, G. F. (1989) Chemically sensitive field-effect transistors. In Turner, A. P. F., Karube, I. and Wilson, G. S. (eds), *Biosensors: Fundamentals and Applications*. Oxford University Press, New York.

Blair, T., Yang, S-T., Smith-Palmer, T. and Bachas, L. G. (1994) Fiber optic sensor for Ca^{+2} based on an induced change in the conformation of the protein calmodulin. *Anal. Chem.* **66**: 300–302.

Blanchard, A. P., Kaiser, R. J. and Hood, L. E. (1996) High-density oligonucleotide arrays. *Biosens. Bioelec.* **11**: 687–690.

Blank, T. B. and Brown, S. D. (1993) Nonlinear multivariate mapping of chemical data using feed-foward neural networks. *Anal. Chem.* **65**: 3081–3089.

Blank, T. B., Sum, S. T., Brown, S. D. and Monfre, S. L. (1996) Transfer of near-infrared multivariate calibrations without standards. *Anal. Chem.* **68**: 2987–2995.

Blonder, R., Katz, E., Cohen, Y., Itzhak, N., Riklin, A. and Willner, I. (1996) Application of redox enzymes for probing the antigen–antibody association at monolayer interfaces: development of amperometric immunosensor electrodes. *Anal. Chem.* **68**: 3151–3157.

Blum, L. J. and Coulet, P. R. (eds) (1991) *Biosensor Principles and Applications*. Marcel Dekker, New York.

Bodenhöfer, K., Hierlemann, A., Noetzel, G., Weimar, U. and Göpel, W. (1996) Performance of mass-sensitive devices for gas sensing: thickness shear mode and surface acoustic wave transducers: *Anal. Chem.* **68**: 2210–2218.

Boguslavsky, L., Kalash, H., Xu, Z., Beckles, D., Geng, L., Skotheim, T., Laurinavicius, V. and Lee, H. S. (1995) Thin film bienzyme amperometric biosensors based on polymeric redox mediators with electrostatic bipolar protecting layer. *Anal. Chem. Acta* **311**: 15-21.

Boitieux, J. L., Desmet, G., Wilson, G. and Thomas, D. (1990) The specific immobilization of antibody fragments on membranes for the development of multifunctional biosensors. *Ann. N.Y. Acad. Sci.* **613**: 390–395.

Booksh, K. S. and Kowalski, B. R. (1994) Theory of analytical chemistry. *Anal. Chem.* **66**: 782A–791A.

Booksh, K. S., Lin, Z., Wang, Z. and Kowalski, B. R. (1994) Extension of trilinear decomposition method with an application to the flow probe sensor. *Anal. Chem.* **66**: 2561–2569.

Borman, S. (1996) DNA chips come of age. *Chem Eng. News*, Dec. 9, 42–43.

Borzeix, F., Monot, F. and Vandercasteele, J-P. (1995) Bi-enzymatic reaction for alcohol oxidation in organic media: from purified enzymes to cellular systems. *Enz. Microb. Technol.* **17**: 615–622.

Boulton, A. A., Baker, G. B. and Adams, R. N. (eds). (1995) *Voltammetric Methods in Brain Systems*. The Humana Press, Totowa, NJ.

Boumans, P. W. J. M. (1994) Detection limits and spectral interferences in atomic emission spectrometry. *Anal. Chem.* **66**: 459A-467A.

Bourdillon, C., Demaille, C., Moiroux, J. and Savéant, J. M. (1994) Step-by-step immunological construction of a fully active multilayer enzyme electrode. *J. Amer. Chem. Soc.* **116**: 10328–10329.

Bowyer, J. R., Alaire, J. P., Sepaniak, M. J., Vo-dinh, T. and Thompson, R. Q. (1991) Construction and evaluation of a regenerable fluoroimmunochemical-based fibre-optic biosensor. *Analyst* **116**: 117–122.

Bowyer, W. J., Clark, M. E. and Ingram, J. L. (1992) Electrochemical measurements in submicroliter volumes. *Anal. Chem.* **64**: 459–462.

Bowyer, W. J., Xie, J. and Engstrom. R. C. (1996) Fluorescence imaging of the heterogeneous reduction of oxygen. *Anal. Chem.* **68**: 2005–2009.

Bratten, C. D. T., Cobbold, P. H. and Cooper, J. M. (1997) Micromachining sensors for electrochemical measurement in subnanoliter volumes. *Anal. Chem.* **69**: 253–258.

Brecht, A. and Gauglitz, G. (1995) Optical probes and transducers. *Biosens. Bioelec.* **10**: 923–936.

Brewster, J. D., Lightfield, A. R. and Bermel, P. L. (1995) Storage and immobilization of photosystem II reaction centers used in an assay for herbicides. *Anal. Chem.* **67**: 1296–1299.

Brewster, J. D. Gehring, A-G., Mazenko, R. S., Van Houten, L. J. and Crawford, C. J. (1996) Immunoelectrochemical assays for bacteria: use of epifluorescence microscopy and rapid-scan electrochemical techniques in development of an assay for *Salmonella*. *Anal. Chem.* **68**: 4153–4159.

Bright, F. V., Betts, T. A. and Litwiler, K. S. (1990) Regenerable fibre-optic-based immunosensor. *Anal. Chem.* **62**: 1065–1069.

Bronk, K. S. and Walt, D. R. (1994) Fabrication of patterned sensor arrays with aryl azides on a polymer-coated imaging optical fiber bundle. *Anal. Chem.* **66**: 3519–3520.

Bronk, K. S., Michael, K. L., Pantano, P. and Walt, D. R. (1995) Combined imaging and chemical sensing using a single optical imaging fiber. *Anal. Chem.* **67**: 2750–2757.

Brooks, K. A., Allen, J. R., Feldhoff, P. W. and Bachas, L. G. (1996) Effect of surface-attached heparin on the response of potassium-selective electrodes. *Anal. Chem.* **68**: 1439–1443.

Brown, S. D. and Bear Jr, R. S. (1993) Chemometric techniques in electrochemistry: a critical review. *Crit. Revs. Anal. Chem.* **24**: 99–131.

Bruckenstein, S., Michalski, M., Fensor, A., Li., Z. and Hillman, A. R. (1994) Dual quartz crystal microbalance oscillator circuit. Minimizing effects due to liquid viscosity, density, and temperature. *Anal. Chem.* **66**: 1847–1852.

Brumlik, C. J. and Martin, C. R. (1992) Microhole array electrodes based on microporus alumina membranes. *Anal. Chem.* **64**: 1201–1203.

Bruno, A. E., Maystre, F., Kralliger, B., Nussbaum, P. and Grassmann, E. (1994) The pigtailing approach to optical detection in capillary electrophoresis. *Trends Anal. Chem.* **13**: 190–198.

Bruno, A. E., Barnard, S., Rouilly, M., Waldne, A., Berger, J. and Ehrat, M. (1997) All-solid-state miniaturized fluorescence sensor array for the determination of critical gases and electrolytes in the blood. *Anal. Chem.* **67**: 507–513.

Bu, H., Mikkelsen, S. R. and English, A. M. (1995) Characterization of a ferrocene-containing polyacrylamide-based redox gel for biosensors use. *Anal. Chem.* **67**: 4071–4076.

Bu, H., English, A. M. and Mikkelsen, S. R. (1996) Modification of ferrocene-containing redox gel sensor performance by copolymerization of charged monomers. *Anal. Chem.* **68**: 3951–3957.

Buck, R. P., Cosofret, V. V., Nahir, T. M., Johnson, T. A., Kusy, R. P., Reinbold, K. A., Simon, M. A., Neuman, M. R., Ash, R. B. and Nagle, H. T. (1992) Macro- to microelectrodes for *in vivo* cardiovascular measurements. In Edelman, P. G. and Wang, J. (eds) *Biosensors and Chemical Sensors: Optimizing Performance Through Polymeric Materials.* ACS Symposium Series #487, American Chemical Society, Washington, DC.

Buck, R. P., Hatfield, W. E., Umana, M. and Bowden, E. F. (eds). (1990) *Biosensor Technology—Fundamentals and Applications.* Marcel Dekker, New York.

Buckle, P. E., Davies, R. J., Kinning, T., Yeung, D., Edwards, P. R., Pollard-Knight, D. and Lowe, C. R. (1993) The resonant mirror: a novel optical sensor for direct sensing of biomolecular interactions. Part II: Applications. *Biosens. Bioelec.* **8**: 355–363.

Buerck, J., Conzen, J. P., Beckhaus, B. and Ache, H. J. (1994) Fibre-optic evanescent-wave sensor for *in situ* determination of non-polar organic compounds in water. *Sens. Actuators* (B) **18**: 291–295.

Burestedt, E., Navaez, A., Ruzas, T., Gorton, L., Emnéus, J., Domínguez, E. and Marko-Varga, G. (1996) Rate-limiting steps of tyrosinase-modified electrodes for the detection of catechol. *Anal. Chem.* **68**: 1605–1611.

Burnette, W. B., Bailey, M. D., Kukoyi, S., Blakely, R. D., Trowbridge, C. G. and Justice Jr., J. B. (1996) Human norepinephrine transporter kinetics using rotating disk electrode voltammetry. *Anal. Chem.* **68**: 2932–2938.

Buttler, T. A., Johansson, K. A. J., Gorton, L. G. O. and Marko-Varga, G. A. (1993) On-line fermentation process monitoring of carbohydrates and ethanol using tangential flow filtration and column liquid chromatography. *Anal. Chem.* **65**: 2628–2636.

Buttry, D. A. (1991) Application of the quartz crystal microbalance to electrochemistry. In Bard, A. J. (ed.), *Electroanalytical Chemistry*, Vol. 17. Marcel Dekker, New York.

Buttry, D. A. and Ward, M. D. (1992) Measurements of interfacial processes at electrode surfaces with the ECQC microbalance. *Chem. Revs.* **92**: 1355–1379.

Byfield, M. P. and Abuknesha, R. A. (1994) Biochemical aspects of biosensors. *Biosens. Bioelec.* **9**: 373–400.

Cabral, J. M. and Kennedy, J. F. (1991) Covalent and coordination immobilization of proteins. *Bioprocess Technology* **14**: 73–138.

Cadogan, A., Gao, Z., Lewenstam, A., Iraska, A. and Diamond, D. (1992) All-solid-state sodium-selective electrode based on a calixarene ionophore in a poly(vinyl choloride) membrane with a polypyrrole solid contact. *Anal. Chem.* **64**: 2496–2501.

Cagnini, A., Palchetti, I., Lionti, I., Mascini, M. and Turner, A. P. F. (1995) Disposable ruthenized screen-printed biosensors for pesticides monitoring. *Sens. Actuators* (B) **24**: 85–89.

Cahill, P. S. and Wightman, R. M. (1995) Simultaneous amperometric measurement of ascorbate and catecholamine secretion from individual bovine adrenal medullary cells. *Anal. Chem.* **67**: 2599–2605.

Cahill, P. S., Walker, Q. D., Finnegan, J. M., Mickelson, G. E., Travis, R. R. and Wightman, R. M. (1996) Microelectrodes for the measurement of catecholamines in biological systems. *Anal. Chem.* **68**: 3180-3186.

Calvert, J. M. (1993) Lithographic patterning of self-assembled films. *J. Vac. Sci. Technol.* B **11**: 2155–2163.

Calvo, E. J., Etchenique, R., Danilowicz, C. and Diaz, L. (1996) Electrical communications between electrodes and enzymes mediated by redox hydrogels. *Anal. Chem.* **68**: 4186–4193.

Campbell, A. M. (1989) Monoclonal antibodies and immunosensor technology. In VanderVliet, P. C. (ed.), *Laboratory Techniques in Biochemistry*, Elsevier, New York.

Cao, L. K., Anderson, G. P., Ligler, F. S. and Ezzell, J. (1995) Detection of *Yersinia pestis* fraction 1 antigen with a fiber optic biosensor. *J. Clin. Microbiol.* **33**: 336–341.

Cardosi, M. F. and Birch, S. W. (1993) Screen-printed glucose electrodes based on platinized carbon particles and glucose oxidase. *Anal. Chim. Acta* **276**: 69–74.

Careri, M., Casnati, A., Guarinoni, A., Mangia, R., Mori, G., Pochini, A., and Ungaro, R. (1993) Study of the behavior of calix[4]arene-based sodium-selective electrodes by means of ANOVA. *Anal. Chem.* **65**: 3156–3160.

Carey, C. M. and Riggan, Jr., W. B. (1994) Cyclic polyamine ionophore for use in a dibasic phosphate selective electrode. *Anal. Chem.* **66**: 3587–3591.

Carey, W. P. (1994) Multivariate sensor arrays as industrial and environmental monitoring systems. *Trends Anal. Chem.* **13**: 210–218.

Carlyon, E. E., Lowe, C. R., Reid, D. and Bennion, I. (1992) Single-mode fibre-optic evanescent wave biosensor. *Biosens. Bioelec.* **7**: 141–146.

Carr, P. W. and Bowers, L. D. (1980) *Immobilized Enzymes in Analytical and Clinical Chemistry*. John Wiley and Sons, New York.

Carter, R. M., Jacobs, M. B., Lubrano, G. J. and Guilbault, G. G. (1995) Piezoelectric detection of ricin and affinity-purified goat anti-ricin antibody. *Anal. Lett.* **28**: 1379–1386.

Caruso, F., Rodda, E., Furlong, D. N., Nikura, K. and Okahata, Y. (1997) Quartz crystal microbalance study of DNA immobilization and hybridization for nucleic acid sensor development. *Anal. Chem.* **69**: 2043–2049.

Casay, G. A., Daneshvar, M. I. and Patonay, G. (1994) Development of a fiber optic biomolecular probe instrument using near infrared dyes and semiconductor laser diodes. *Instrum. Sci. Tech.* **22**: 323–341.

Cass, A. E. G. (ed) (1990) *Biosensors: A Practical Approach*. Oxford University Press, New York.

Cassis, L. A. and Lodder, R. A. (1993) Near-IR imaging of atheromas in living arterial tissue. *Anal. Chem.* **65**: 1247–1256.

Chance, J. J. and Purdy, W. C. (1996) Bile acid measurements using a cholestyramine-coated TSM acoustic wave sensor. *Anal. Chem.* **68**: 3104-3111.

Chang, H. T. and Yeung, E. S. (1995) Determination of catecholamines in single adrenal medullary cells by capillary electrophoresis and laser-induced native fluorescence. *Anal. Chem.* **67**: 1079–1083.

Chang, W., Bard. A. J., Nagy, G. and Toth, K. (1995) Scanning electrochemical microscopy. 28. Ion-selective neutral carrier-based microelectrode potentiometry. *Anal. Chem.* **67**: 1346–1356.

Charpentier, L. and El Murr, N. (1995) Amperometric determination of cholesterol in serum with use of a renewable surface peroxidase electrode. *Anal. Chim. Acta* **318**: 89–93.

Chaudhury, M. K., (1995) Self-assembled monolayers on polymer surfaces. *Biosens. Bioelec.* **10**: 785–788.

Chee, M., Yang, R., Hubbell, E., Berno, A., Huang, X. C., Stern, D., Winkler, J., Lockhart, D. J., Morris, M. S. and Fodor, S. P. A. (1996) Accessing genetic information with high-density DNA arrays. *Science* **274**: 610–614.

Chen, H.-Y., Ju, H.-X. and Xun, Y.-G. (1994) Methylene blue/perfluorosulfonated ionomer modified microcylinder carbon fiber electrode and its application for the determination of hemoglobin. *Anal. Chem.* **66**: 4538–4542.

Chen, P. and McCreery, R. L. (1996) Control of electron transfer kinetics at glassy carbon electrodes by specific surface modification. *Anal. Chem.* **68**: 3958–3965.

Chen, Q., Wang, J., Rayson, G., Tian, B. and Lin, Y. (1993) Sensor array for carbohydrates and amino acids based on electrocatalytic modified electrodes. *Anal. Chem.* **65**: 251–254.

Chen, Q., Pamidi, P. V. A., Wang, J. and Kutner, W. (1995) beta-Cyclodextrin cation-exchange polymer membrane for improved second-generation glucose biosensors. *Anal. Chim. Acta* **306**: 201–208.

Chen, T. K., Lau, Y. Y., Wong, D. K. Y. and Ewing, A. G. (1992) Pulse voltammetry in single cells using platinum microelectrodes. *Anal. Chem.* **64**: 1264–1268.

Chen, T. K., Luo, G. and Ewing, A. G. (1994) Amperometric monitoring of stimulated catecholamine release from rat pheochromocytoma (PC12) cells at the zeptomole level. *Anal. Chem.* **66**: 3031–3035.

Chen, X., Davies, M. C., Roberts, C. J., Shakesheff, K. M., Tendler, S. J. B. and Williams, P. M. (1996) Dynamic surface events measured by simultaneous probe microscopy and surface plasmon detection. *Anal. Chem.* **68**: 1451–1455.

Cheng, Q. and Brajter-Toth, A. (1995) Permselectivity and high sensitivity at ultrathin monolayers. Effect of film hydrophobicity. *Anal. Chem.* **67**: 2767–2775.

Cheng, Q. and Brajter-Toth, A. (1996) Permselectivity, sensitivity and amperometric pH sensing at thioctic acid monolayer microelectrodes. *Anal. Chem.* **68**: 4180–4185.

Cheung, N. and Yeung, E. S. (1994) Distribution of sodium and potassium within individual human erythrocytes by pulsed-laser vaporization in a sheath flow. *Anal Chem.* **66**: 929–936.

Chi, Q. J. and Dong, S. J. (1995) Amperometric biosensors based on the immobilization of oxidases in a Prussian blue film by electrochemical codeposition. *Anal. Chim. Acta* **310**: 429–436.

Chidsey, C. E. D. and Loiacono, D. N. (1990) Chemical functionality in self-assembled monolayers: structural and electrochemical properties. *Langmuir* **6**: 682–691.

Chiem, N. and Harrison, D. J. (1997) Microchip-based capillary electrophoresis for immunoassays: analysis of monoclonal antibodies and theophylline. *Anal. Chem.* **69**: 373–378.

Ciolkowski, E. I., Maness, K. M., Cahill, P. S., Wightman, R. M., Evans, D. H., Fosset, B. and Amatore, C. (1994) Disproportionation during electrooxidation of catecholamines at carbon-fiber microelectrodes. *Anal. Chem.* **66**: 3611–3617.

Claremont, D. J. and Pickup, J. C. (1989) *in vivo* Chemical sensors and biosensors in clinical medicine. In Turner, A. P. F., Karube, I., Wilson, G. S. (eds), *Biosensors: Fundamentals and Applications*. Oxford University Press, New York.

Clark, A. D., Chapli, M. F., Rousston, S. A. and Dunne, L. J. (1995) Development of a thin-film glucose biosensor using semiconducting amorphous carbons. *Biosens. Bioelec.* **10**: 237–241.

Clark Jr, L. C., Mishray, G. and Fox, R. P. (1958) Chronically implanted polarographic electrodes. *J. Appl. Physiol.* **13**: 85–91.

Clark Jr, L. C., and Lyons, C. (1962) Electrode systems for continuous monitoring in cardiovascular surgery. *Ann. N.Y. Acad. Sci.* **102**: 29–45.

Clark Jr, L. C., (1971) Oxygen transport and glucose metabolism in septic shock. In Hershet, G. S. *et al.* (eds), *Septic Shock in Man*. Brown & Co., Boston, MA.

Clark Jr, L. C. (1989) The enzyme electrode. In Turner, A. P. F., Karube, I. and Wilson, G. S. (eds), *Biosensors: Fundamentals and Applications*. Oxford University Press, New York.

Clark Jr, L. C., (1993) Guest editorial. *Biosens. Bioelec.* **8**: iii–vii.

Clark, R. A., Hietpas, P. B. and Ewing, A. G. (1997) Electrochemical analysis in picoliter microvials. *Anal. Chem.* **69**: 259–263.

Coche–Guerente, L., Cosnier, S. and Innocent, C. (1995a) Poly(amphiphilic pyrrole) PPO electrodes for organic-phase-enzymatic assay. *Anal. Lett.* **28**: 1005–1016.

Coche-Guerente, L., Cosnier, S., Innocent, C. and Mailley, P. (1995b) Development of amperometric biosensors based on the immobilization of enzymes in polymer films electrogenerated from a series of amphiphilic pyrrole derivatives. *Anal. Chim. Acta* **311**: 23–30.

Collins, G. E., and Rose-Pehrsson, S. L. (1995) Chemiluminescent chemical sensors for oxygen and nitrogen dioxide. *Anal. Chem.* **67**: 2224–2230.

Colon, L. A., Dadoo, R. and Zare, R. N. (1993) Determination of carbohydrates by capillary zone electrophoresis with amperometric detection at a copper electrode. *Anal. Chem.* **65**: 476–481.

Connolly, P. (1995) Clinical diagnostics: opportunities for biosensors and bioelectronics. *Biosens. Bioelec.* **10**: 1–6.

Conzen, J. P., Buerck, J. and Ache, H. J. (1993) Characterization of a fibre-optic evanescent wave absorbance sensor for non-polar organic compounds. *Appl. Spectros.* **47**: 753–763.

Conzen, J. P., Buerck, J. and Ache, H. J. (1994) Determination of chlorinated hydrocarbons in water by fibre-optic evanescent wave spectroscopy and partial least-squares regression. *Fresenius' J. Anal. Chem.* **348**: 501–505.

Coon, D. R., Babb, C. W. and Rechnitz, G. A. (1994) Biomagnetic neurosensors. 2. Magnetically stimulated sensors. *Anal. Chem.* **66**: 3193–3197.

Cooper, B. R., Jankowski, J. A., Leszczyszyn, D. J., Wightman, R. M. and Jorgenson, J. W. (1992) Quantitative determination of catecholamines in individual bovine adrenomedullary cells by reversed-phase microcolumn liquid chromatography with electrochemical detection. *Anal. Chem.* **64**: 691–694.

Córdova, E., Gao, J. and Whitesides, G. M. (1997) Noncovalent polycationic coatings for capillaries in capillary electrophoresis of proteins. *Anal. Chem.* **69**: 1370–1379.

Cosford, R. J. O. and Kuhr, W. (1996) Capillary biosensor for glutamate. *Anal. Chem.* **68**: 2164–2169.

Cosnier, S., Innocent, C. and Jouanneau, Y. (1994) Amperometric detection of nitrate via a nitrate reductase immobilized and electrically wired at the electrode surface. *Anal. Chem.* **66**: 3198–3201.

Cosnier, S., Innocent, C., Allien, L., Poitry, S. and Tsacopoulos, M. (1997) An electrochemical method for making enzyme microsensors. Application to the detection of dopamine and glutamate. *Anal. Chem.* **69**: 968–971.

Cosofret, V. V., Buck, R. P. and Erdősy, M. (1994) Carboxylated poly(vinyl chloride) as a substrate for ion sensors: effects of native ion exchange on response. *Anal. Chem.* **66**: 3592–3599.

Cosofret, V. V., Erdősy, M., Johnson, T. A., Buck, R. P., Ash, R. B. and Neuman, M. R. (1995a) Microfabricated sensor arrays sensitive to pH and K^+ for ionic distribution measurements in the beating heart. *Anal. Chem.* **67**: 1647–1653.

Cosofret, V. V., Erdősy, M., Johnson, T. A., Bellinger, D. A., Buck, R. P., Ash, R. B. and Neuman, M. R. (1995b) Electroanalytical and surface characterization of encapsulated implanted planar microsensors. *Anal. Chim. Acta* **314**: 1–11.

Coulet, P. (1993) What are the criteria for belonging to the "biosensor family?" *Biosens. Bioelec.* **8**; xxvi–xxvii.

Creager, S. E. and Olsen, K. G. (1995) Self-assembled monolayers and enzyme electrodes: progress, problems and prospects. *Anal. Chim. Acta* **307**: 277–289.

Cremisini, C., Di-Sario, S., Mela, J., Pilloton, R. and Palleschi, G. (1997) Evaluation of the use of free and immobilized acetylcholinesterase for paraoxon detection with an amperometric choline oxidase-based biosensor. *Anal. Chim. Acta* **311**: 273–280.

Crespi, F., England, T., Ratti, E. and Trist, D. G. (1995) Carbon fibre micro-electrodes for concomitant *in vivo* electrophysiological and voltammetric measurements: no reciprocal influences. *Neurosci. Lett.* **188**: 33–36.

Csöregi, E., Gorton, L., Marko-Varga, G., Tüdös, A. J. and Kok, W. T. (1994a) Peroxidase-modified carbon fiber microelectrodes in flow-through detection of hydrogen peroxide and organic peroxides. *Anal. Chem.* **66**: 3604–3610.

Csöregi, E., Quinn, C. P., Schmidtke, D. W., Lindquist, S-E., Pishko, M. V., Ye, L., Katakis, I., Hubbell, J. A. and Heller, A. (1994b) Design, characterization, and one-point *in vivo* calibration of a subcutaneously implanted glucose electrode. *Anal. Chem.* **66**: 3131–3138.

Csöregi, E., Schmidtke, D. W. and Heller, A. (1995) Design and optimization of a selective subcutaneously implantable glucose electrode based on "wired" glucose oxidase. *Anal. Chem.* **67**: 1240–1244.

Cush, R., Cronin, J. M., Stewart, W. J., Maule, C. H., Molloy, J. and Goddard, N. J. (1993) The resonant mirror: a novel optical biosensor for direct sensing of biomolecular interactions. Part I: Principle of operation and associated instrumentation. *Biosens. Bioelec.* **8**: 347–353.

Czarnik, A. W. (ed.) (1993) *Fluorescent Chemosensors for Ion and Molecule Recognition.* ACS Symposium Series #538, American Chemical Society, Washington, DC.

Dahint, R., Grunze, M., Jossé, F. and Renken, J. (1994) Acoustic plate mode sensor for immunochemical reactions. *Anal. Chem.* **66**: 2888–2892.

Damgaard, L. R., Larsen, L. H. and Revsbech, N. P. (1995) Microscale biosensors for environmental monitoring. *Trends Anal. Chem.* **7**: 300–303.

Danielsson, B. and Mosbach, K. (1989) Theory and application of calorimetric sensors. In Turner, A. P. F., Karube I, and Wilson, G. S. (eds) *Biosensors: Fundamentals and Applications.* Oxford University Press, New York.

Danielsson, B. and Winquist, F. (1990) Thermometric sensors. In Cass, A. E. G. (ed.) *Biosensors: A Practical Approach.* Oxford University Press, New York.

Danzer, T. and Schwedt, G. (1996) Chemometric methods for the development of a biosensor system and the evaluation of inhibition studies with solutions and mixtures of pesticides and heavy metals. Part 1. Development of an enzyme electrode system for pesticide and heavy metal screening using selected chemometric methods. *Anal. Chim. Acta* **318**: 275–286.

Dave, B. C., Dunn, B., Valentine, J. S. and Zink, J. I. (1994) Sol–gel encapsulation methods for biosensors. *Anal. Chem.* **66**: 1121A–1127A.

Davidson, V. L. (ed.) (1993) *Principles and Applications of Quinoproteins.* Marcel Dekker, New York.

Davis, J., Vaughan, D. H. and Cardosi, M. F. (1995) Elements of biosensor construction. *Enz. Microb. Tech.* **17**: 1030–1035.

D'Costa, E. J., Higgins, I. J. and Turner, A. P. F. (1986) Quinoprotein glucose dehydrogenase and its application in an amperometric glucose sensor. *Biosens.* **2**: 71–89.

de Alwis, U. and Wilson, G. S. (1989) Strategies for the reversible immobilization of enzymes by use of biotin-bound anti-enzymes antibodies. *Talanta* **36**: 249–253.

Deasy, B., Demsey, E., Smyth, M. R., Egan, D., Bogan, D. and O'Kennedy, R. (1994) Development of an antibody-based biosensor for determination of 7-hydroxycoumarin (umbelliferone) using horseradish peroxidase labelled anti-7-hydroxycoumarin antibody. *Anal. Chim. Acta* **294**: 291–297.

Decher, G., Lehr, B., Lowack, K., Lvov, Y. and Schmitt, J. (1994) New nanocomposite films for biosensors: layer-by-layer absorbed films of polyelectrolytes, proteins or DNA. *Biosens. Bioelec.* **9**: 677–684.

deLumley-Woodyear, T., Rocca, R., Lindsay, J., Dror, Y., Freeman, A. and Heller, A. (1995) Polyacrylamide-based redox polymer for connecting redox centers of enzymes to electrodes. *Anal. Chem.* **67**: 1332-1338.

Demas, J. N., DeGraff, B. A. and Xu, W. (1995) Modeling of luminescence quenching-based sensors: comparison of multisite and nonlinear gas solubility models. *Anal. Chem.* **67**: 1377–1380.

Deng, Q. and Dong, S. J. (1995) Construction of a tyrosinase-based biosensor in pure organic phase. *Anal. Chem.* **67**: 1357–1360.

deRooij, N. F. and van den Vlekkert, H. H. (1991) Microstructured ISFETs. In Yamazoe, K. (ed.) *Chemical Sensor Technology*, Vol. 3, pp 213–231. Elsevier, New York.

Dessy, R. E. (1989) Waveguides as chemical sensors. *Anal. Chem.* **61**: 233–239.

Dessy, R. E., Burgess, L., Arney, L. and Peterson, J. (1989) Fiber-optic and polymer film-based enthalpimeters for biosensor applications. In Murray, R. W., Dessy, R. E., Heineman, W. R.., Janata, J. and Seitz, W. R. (eds.) *Chemical Sensors and Microinstrumentation*. ACS Symposium Series #403, American Chemical Society, Washington, DC.

Devine, P. J., Anis, N. A., Wright, J., Kim, S., Eldefrawi, A. T. and Eldefrawi, M. E. (1995) A fibre-optic cocaine biosensor. *Anal. Biochem.* **227**: 216–224.

deVries, E. F. A., Schasfoort, R. B. M., van der Plas, J. and Greve, J. (1994) Nucleic acid detection with surface plasmon resonance using cationic latex. *Biosens. Bioelec.* **9**: 509–514.

Diamond, D. (1993) Progress in sensor array research. *Electroanalysis* **5**: 795–802.

Dickert, F. L., Haunschild, A., Kuschow, V., Reif, M. and Stathopulos, H. (1996) Mass-sensitive detection of solvent vapors. Mechanistic studies on host–guest sensor principles by FT-IR spectroscopy and BET adsorption analysis. *Anal. Chem.* **68**: 1058–1061.

Dickert, F. L., Bäumler, U.P.A. and Stathopulos, H. (1997) Mass-sensitive solvent vapor detection with calix[4]resorcinarenes: tuning sensitivity and predicting sensor effects. *Anal. Chem.* **69**: 1000–1005.

Dong, S. and Guo, Y. (1994) Organic phase enzyme electrode operated in water-free solvents. *Anal. Chem.* **66**: 3895–3899.

Dong, X. D., Lu, J. T. and Cha, C.S. (1995) Kinetic analysis of the enzyme electrode composed of glucose oxidase adsorbed on an alkanethiol SAM-modified platinum surface. *J. Electroanal. Chem.* **381**: 195–201.

Dontha, N., Nowall, W. B. and Kuhr, W. G. (1997) Generation of biotin/avidin/enzyme nanostructures with maskless photolithography. *Anal. Chem.* **69**: 2619–2625.

Dou, X., Takama, T., Yamaguchi, Y., Yamamoto, H. and Ozaki, Y. (1997) Enzyme immunoassay utilizing surface-enhanced Raman scattering of the enzyme reaction product. *Anal. Chem.* **69**: 1492–1495.

Draxler, S. and Lippitsch, M. E. (1996) Life-time based sensing: influence of the microenvironment. *Anal. Chem.* **68**: 753–757.

Du, G., Lin, C. and Bocarsly, A. B. (1996) Electroanalytical detection of glucose using a cyanometalate-modified electrode: requirements for the oxidation of buried redox sites in glucose oxidase. *Anal. Chem.* **68**: 796–806.

Duan, C. and Meyerhoff, M. E. (1994) Separation-free sandwich enzyme immunoassays using microporous gold electrodes and self-assembled monolayer/immobilized capture antibodies. *Anal. Chem.* **66**: 1369–1377.

Duan, C. and Meyerhoff, M. E. (1995) Immobilization of proteins on gold coated porous membranes via an activated self-assembled monolayer of thioctic acid. *Mikrochim. Acta* **117**: 195–206.

Duevel, R. V. and Corn, R. M. (1992) Amide and ester surface attachment reactions for alkanethiol monolayers at gold surfaces as studied by polarization modulation Fourier transform spectroscopy. *Anal. Chem.* **64**: 337–342.

Dunbar, R. A., Jordan, J. D. and Bright, F. V. (1996) Development of chemical sensing platforms based on sol–gel–derived thin films: origin of film age vs performance trade-offs. *Anal. Chem.* **68**: 604–610.

Dunham, G. C., Benson, N. H., Petelenz, D. and Janata, J. (1995) Dual quartz crystal microbalance. *Anal. Chem.* **67**: 267–272.

Dunuwila, D. D., Torgerson, B. A., Chang, C. K. and Berglund, K. A. (1994) Sol–gel derived titanium carboxylate thin films for optical detection of analytes. *Anal. Chem.* **66**: 2739–2744.

Eddowes, M. J. (1990) Theoretical methods for analysing biosensor performance. In Cass, A. E. G. (ed.) *Biosensors: A Practical Approach*. Oxford University Press, New York.

Edelman, P. G. and Wang, J. (eds) (1992) *Biosensors and Chemical Sensors: Optimizing Performance Through Polymeric Materials*. ACS Symposium Series #487, American Chemical Society, Washington, DC.

Edmiston, P. L., Wambolt, C. L., Smith, M. K. and Saavedra, S. S. (1994) Spectroscopic characterization of albumin and myoglobin entrapped in bulk sol-gel glasses. *J. Colloid Interface Sci.* **163**: 395–406.

Edmiston, P. L., Lee, J. E., Wood, L. L. and Saavedra, S. S. (1997) Molecular orientation distributions in heme protein films: site-specific immobilization of yeast cytochrome *c* to supported planar lipid bilayers. Abstract #013, Division of Analytical Chemistry, 213th National Meeting, American Chemical Society, San Francisco, CA. April 13–17.

Eenink, R. G., De-Bruijn, H. E., Kooyman, R. P. H. and Greve, J. (1990) Fibre-fluorescence immunosensor based on evanescent wave detection. *Anal. Chim. Acta* **238**: 317–321.

Effenhauser, C. S., Paulus, A., Manz, A. and Widmer, H. M. (1994) Glass chips for high-speed capillary electrophoresis separations with submicrometer plate heights. *Anal. Chem.* **65**: 2637–2642.

Effenhauser, C. S., Paulus, A., Manz, A. and Widmer, H. M. (1994) High-speed separation of antisense oligonucleotides on a micromachined capillary electrophoresis device. *Anal. Chem.* **66**: 2949–2953.

Ege, S. N. (1989) *Organic Chemistry*, 2nd Edition. D. C. Heath, Lexington, MA.

Eggert, A. A. (1983) *Electronics and Instrumentation for the Clinical Laboratory*. John Wiley and Sons, New York.

Eggins, B. R. (1996) *Biosensors: an Introduction*. Wiley-Teubner, Chichester-Stuttgart.

Eigen, M. and Rigler, R. (1994) Sorting single molecules: application to diagnostics and evolutionary biotechnology. *Proc. Natl Acad. Sci.* **91**: 5740–5747.

Eisele, S., Ammon, P. T., Kindervater, R., Grobe, A. and Göpel, W. (1994) Optimized biosensor for whole blood measurements using a new cellulose based membrane. *Biosens. Bioelec.* **9**: 119–124.

Eldefrawi, M., Eldefrawi, A., Wright, J., Emanuel, P., Valdes, J. and Rogers, K. R. (1995) Immunosensors for detection of chemical mixtures. In Rogers, K. R., Mulchandani, A. and Zhou, W. (eds), *Biosensor and Chemical Technology*. ASC Symposium Series #613, American Chemical Society, Washington, DC.

Enfors, S-O. (1989) Compensated enzymes-electrodes for *in situ* process control. In Turner, A. P. F., Karube, I. and Wilson, G. S. (eds), *Biosensors: Fundamentals and Applications*. Oxford University Press, New York.

Engstrom, R. C., Ghaffari, S. and Qu, H. (1992) Fluorescence imaging of electrode–solution interfacial processes. *Anal. Chem.* **64**: 2525–2529.

Espadas-Torre, C. and Meyerhoff, M. E. (1995) Thrombogenic properties of untreated and poly(ethylene oxide)-modified polymeric matrices useful for preparing intraarterial ion selective electrodes. *Anal. Chem.* **67**: 3108–3114.

Espadas-Torre, C., Bakker, E., Barker, S. and Meyerhoff, M. E. (1996) Influence of nonionic surfactants on the potentiometric response of hydrogen ion-selective polymeric membrane electrodes. *Anal. Chem.* **68**: 1623–1631.

Etches, P. C., Harris, M. L., McKinley, R. and Finer, N. N. (1995) Clinical monitoring of inhaled nitric oxide: comparison of chemiluminescent and electrochemical sensors. *Biomed. Instrum. Technol.* **29**: 134–140.

Ewing, A. G., Strein, T. G. and Lau, Y. Y. (1992) Analytical chemistry in microenvironments: single nerve cell. *Accts Chem. Res.* **25**: 440–447.

Fairbank, R. W. P., Xiang, Y. and Wirth, M. J. (1995) Use of methyl spacers in a mixed horizontally polymerized stationary phase. *Anal. Chem.* **67**: 3879–3885.

Fan, Z. H. and Harrison, D. J. (1992) Permeability of glucose and other neutral species through recast perfluorosulfonated ionomer films. *Anal. Chem.* **64**: 1304–1311.

Fan, Z. H. and Harrison, D. J. (1994) Micromachining of capillary electrophoresis injectors and separators on glass chips and evaluation of flow at capillary intersections. *Anal. Chem.* **66**: 177–184.

Faulkner, L. R. and Bard, A. J. (1977) Techniques of electrogenerated chemiluminescence. In Bard, A. J. (ed.) *Electroanalytical Chemistry*, Vol 10. Marcel Dekker, New York.

Feldman, B. J., Osterloh, J. D., Hata, B. H. and D'Alessandro, A. (1994) Determination of lead in blood by square wave anodic stripping voltammetry at a carbon disk ultramicroelectrode. *Anal. Chem.* **66**: 1983–1987.

Ferguson, J. A. and Walt, D. R. (1996) A fiberoptic DNA biosensor microarray for the analysis of gene expression. *Nature Biotechnology* **14**: 1681–1684.

Fernández-Romero, J. M. and Luque de Castro, M. D. (1993) Flow-through optical biosensor based on a permanent immobilization of an enzyme and transient retention of a reaction product. *Anal. Chem.* **65**: 3048–3052.

Ferré, J. and Rius, F. X. (1996) Selection of the best calibration sample subset for multivariate regression. *Anal. Chem.* **68**: 1565–1571.

Ferrus, R. and Egea, M. R. (1994) Limit of discrimination, limit of detection and sensitivity in analytical systems. *Anal. Chim. Acta* **287**: 119–145.

Fiedler, S., Hagedorn, R., Schnelle, T., Richter, E., Wagner, B. and Fuhr, G. (1995) Diffusion electrotitrations: generation of pH gradients over arrays of ultramicroelectrodes detected by fluorescence. *Anal. Chem.* **67**: 820–828.

Filho, J. L. L., Pandey, P. C. and Weetall, H. H. (1996) An amperometric flow injection analysis enzyme sensor for sucrose using a tetracyanoquinodimethane modified graphite paste electrode. *Biosens. Bioelec.* **11**: 719–723.

Finklea, H. O., Avery, S., Lynch, M. and Furtsch, T. (1987) Blocking oriented monolayers of alkyl mercaptans on gold electrodes. *Langmuir* **3**: 409–413.

Finklea, H. O. (1996) Electrochemistry of organized monolayers of thiols and related molecules on electrodes. In Bard, A. J. and Rubinstein, I. (eds), *Electroanalytical Chemistry*, Vol. 19. Marcel Dekker, New York.

Fishman, H. A., Orwar, O., Scheller, R. H. and Zare, R. N. (1995) Identification of receptor ligands and receptor subtypes using antagonists in a capillary electrophoresis single-cell biosensor separation system. *Proc. Natl Acad. Sci. USA* **92**: 7877–7881.

Fishman, H. A., Orwar, O., Allbritton, N. L., Modi, B. P., Shear, J. B., Scheller, R. H. and Zare, R. N. (1996) Cell-to-cell scanning in capillary electrophoresis. *Anal. Chem.* **68**: 1181–1186.

Fluri, K., Fitzpatrick, G., Chiem, N. and Harrison, D. J. (1996) Integrated capillary electrophoresis devices with an efficient postcolumn reactor in planar quartz and glass chips. *Anal. Chem.* **68**: 4285–4290.

Fodor, S. P. A., Read, J. L., Pirrung, C., Stryer, L., Lu, A. T. and Solas, D. (1991) Light-directed spatially addressable parallel chemical synthesis. *Science* **251**: 767–773.

Fogg, A. G., Pirzad, R., Moreira, J. C. and Davies, A. E. (1995) Improving the performance of screen printed carbon and polished platinum and glassy carbon voltammetric electrodes by modification with poly-L-lysine. *Anal. Proc.* **32**: 209–212.

Foulds, N. C., Wilshere, J. M. and Green, M. J. (1990) Rapid electrochemical assay for theophylline in whole blood based on the inhibition of bovine liver alkaline phosphatase. *Anal. Chim. Acta* **229**: 57–62.

Fraden, J. (1993) *AIP Handbook of Modern Sensors: Physics, Designs and Applications.* American Institute of Physics, New York.

Frank, C. W., Rao, V., Despotopoulou, M. M., Pease, R. F. W., Hinsberg, W. D., Miller, R. D. and Rabolt, J. F. (1996) Structure in thin and ultrathin spin-cast polymer films. *Science* **273**: 912–915.

Frew, J. E., Bayliff, S. W., Gibbs, P. N. B. and Green, M. J. (1989) Amperometric biosensor for the rapid determination of salicylate in whole blood. *Anal. Chim. Acta* **224**: 39–46.

Frey, B. L., Jordan, C. E., Kornguth, S. and Corn, R. M. (1995) Control of the specific adsorption of proteins onto gold surfaces with poly(L-lysine) monolayers. *Anal. Chem.* **67**: 4452–4457.

Frey, B. L. and Corn, R. M. (1996) Covalent attachment and derivatization of poly(L-lysine) monolayers on gold surfaces as characterized by polarization-modulation FT-IR spectroscopy. *Anal. Chem.* **68**: 3187–3193.

Friedemann, M. N., Robinson, S. W. and Gerhardt, G. A. (1996) *o*-Phenylenediamine-modified carbon fiber electrodes for the detection of nitric oxide. *Anal. Chem.* **68**: 2621–2628.

Fritch-Faules, I. and Faulkner, L. R. (1992a) Use of microelectrode arrays to determine concentration profiles of redox centers in polymer films. *Anal. Chem.* **64**: 1118–1127.

Fritch-Faules, I. and Faulkner, L. R. (1992b) Relationships between measured potential and concentrations of redox centers in polymer networks. *Anal. Chem.* **64**: 1127–1131.

Fu, B., Bakker, E., Yun, J. H., Yang, V. C. and Meyerhoff, M. E. (1994) Response mechanism of polymer membrane-based potentiometric polyion sensors. *Anal. Chem.* **66**: 2250–2259.

Furbee Jr, J. W., Kuwana, T. and Kelly, R. S. (1994) Fractured carbon fiber-based biosensor for glucose. *Anal. Chem.* **66**: 1575–1577.

Galan-Vidal, C. A., Muñoz, J., Domínguez, C. and Alegret, S. (1995) Chemical sensors, biosensors and thick-film technology. *Trends Anal. Chem.* **14**: 225–231.

Gamburzev, S., Atanasov, P. and Wilkins, E. (1995) Glucose biosensor based on oxygen electrode. III. Long-term performance of the glucose biosensor in blood plasma at body temperature. *Anal. Chim. Acta* **28**: 1143–1157.

Gao, H., Sanger, M., Luginbuhl, R. and Sigrist, H. (1995) Immunosensing with photo-immobilized immunoreagents on planar optical wave guides. *Biosens. Bioelec.* **10**: 317–328.

Gao, X., Le, J. and White, H. S. (1995) Natural convection at microelectrodes. *Anal. Chem.* **67**: 1541–1545.

Gardner, J. W., Hines, E. L. and Tang, H. C. (1987) Detection of vapors and odors from a multisensor array using pattern recognition techniques. *Sens. Actuators* (B) **9**: 9–15.

Gardner-Medwin, A. R. (1980) Fundamental biophysical issues. In Nicholson, C. (ed.) *Dynamics of the Brain Cell Microenvironment*. Neurosci. Res. Prog. Bull. Vol. 18. MIT Press, Cambridge, MA.

Gargallo, R., Sánchez, F. C., Izquierdo-Ridorsa, A. and Massart, D. L. (1996) Application of eigenstructure tracking analysis and SIMPLISMA to the study of the protonation equilibria of cCMP and several polynucleotides. *Anal. Chem.* **68**: 2241–2247.

Garguilo, M. G., Huynh, N., Proctor, A. and Michael, A. C. (1993) Amperometric sensors for peroxide, choline and acetylcholine based on electron transfer between horseradish peroxidase and a redox polymer. *Anal. Chem.* **65**: 525–528.

Garguilo, M. G. and Michael, A. C. (1994) Quantitation of choline in the extracellular fluid of brain tissue with amperometric microsensors. *Anal. Chem.* **66**: 2621–2629.

Garguilo, M. G. and Michael, A. C. (1995) Optimization of amperometric microsensors for monitoring choline in the extracellular fluid of brain tissue. *Anal. Chim. Acta* **307**: 291–299.

Garnier, F., Hajlaoui, R., Yassar, A. and Srivastava, P. (1994) All polymer field-effect transistor realized by printing techniques. *Science* **265**: 1684–1686.

Garten, R. P. H. and Werner, H. W. (1994) Trends in applications and strategies in the analysis of thin films, interfaces and surfaces. *Anal. Chim. Acta* **297**: 3–14.

Gasparini, R., Scarpa, M., Vianello, F., Mondovi, B. and Rigo, A. (1994) Renewable miniature enzyme-based sensing devices. *Anal. Chim. Acta* **294**: 299–304.

Gatto-Menking, D. L., Yu, H., Bruno, J. G., Goode, M. T., Miller, M. and Zulich, A. W. (1995) Sensitive detection of biotoxoids and bacterial spores using an immunomagnetic electrochemiluminescence sensor. *Biosens. Bioelec.* **10**: 501–507.

Gavin, P. F. and Ewing, A. G. (1996) A microfabricated electrochemical array detector for channel electrophoresis. *J. Amer. Chem. Soc.* **118**: 8932–8936.

Gebert, M. S. and Pekala, R. W. (1994) Fluorescence and light-scattering studies of sol–gel reactions. *Chem. Mater.* **6**: 220–226.

Gemperline, P. E., Cho, J. H., Aldridge, P. K. and Sekulic, S. S. (1996) Appearance of discontinuities in spectra transformed by the piecewise direct instrument standardization procedure. *Anal. Chem.* **68**: 2913–2915.

Gette, W. and Kreiner, T. (1997) Precision scanning technology for complex genetic analysis. *Amer. Lab.* March: 15–17.

Ghindilis, A. L., Makower, A. and Scheller, F. W. (1995) Nanomolar determination of the ferrocene derivatives using a recycling enzyme electrode. Development of a redox label immunoassay. *Anal. Lett.* **28**: 1–11.

Gilaman, S. (1967) The anodic film on platinum electrodes. In Bard, A. J. (ed.), *Electroanalytical Chemistry*, Vol. 2. Marcel Dekker, New York.

Gilardi, G., Zhou, L. Q., Hibbert, L. and Cass, A. E. G. (1994) Engineering the maltose binding protein for reagentless fluorescence sensing. *Anal. Chem.* **66**: 3840–3847.

Gilman, S. D. and Ewing, A. G. (1995) Analysis of single cells by capillary electrophoresis with on-column derivatization and laser-induced fluorescence detection. *Anal. Chem.* **67**: 58–64.

Gilmartin, M. A. T., Hart, J. P. and Birch, B. (1992) Voltammetric and amperometric behaviour of uric acid at bare and surface-modified screen-printed electrodes: studies towards a disposable uric acid sensor. *Analyst* **117**: 1299–1303.

Gilmartin, M. A. T. and Hart, J. P. (1994a) Rapid detection of paracetamol using a disposable, surface-modified screen printed carbon electrode. *Analyst* **119**: 2431–2437.

Gilmartin, M. A. T. and Hart, J. P. (1994b) Fabrication and characterization of a screen-printed, disposable, amperometric cholesterol biosensor. *Analyst* **119**: 2331–2336.

Gilmartin, M. A. T., and Hart, J. P. (1994c) Novel, reagentless, amperometric biosensor for uric acid based on a chemically-modified screen-printed carbon electrode coated with cellulose acetate and uricase. *Analyst* **119**: 833–840.

Gilmartin, M. A. T., Hart, J. P. and Birch, B. J. (1994) Development of amperometric sensors for uric acid based on chemically modified graphite-epoxy resin and screen-printed electrodes containing cobalt phthalocyanine. *Analyst* **119**: 243–252.

Gilmartin, M. A. T. and Hart, J. P. (1995) Sensing with chemically and biologically modified carbon electrodes. *Analyst* **120**: 1029–1045.

Giuliano, K. A., Post, P. L., Hahn, K. M. and Taylor, D. L. (1995) Fluorescent protein biosensors: measurement of molecular dynamics in living cells. *Ann. Rev. Biophys. Biomol. Struct.* **24**: 405–434.

Goebel, R., Krska, R., Kellner, R., and Katzir, A. (1994) Development of protective polymer coatings for silver halide fibres and their application as threshold level sensors for chlorinated hydrocarbons in sea-water. *Fresenius' J. Anal. Chem.* **348**: 780–781.

Goldberg, H. D., Brown, R. B., Liu, D. P. and Meyerhoff, M. E. (1994) Screen printing: a technology for the batch fabrication of integrated chemical-sensor arrays. *Sens. Actuators* (B) **21**: 171–183.

Golden, J. P., Anderson, G. P., Rabbany, S. Y. and Ligler, F. S. (1994) An evanescent wave biosensor—Part II: Fluorescent signal acquisition from tapered fiber optic probes. *IEEE Trans. Biomed. Eng.* **41**: 585–591.

Gooding, J. J. and Hall, E. A. H. (1996) Parameters in the design of oxygen detecting oxidase enzyme electrodes. *Electroanalysis* **8**: 407–413.

Göpel, W., Hesse, J. and Zemel, J. N. (eds) (1989) *Sensors: A Comprehensive Survey*, Vol. I, *Fundamentals and General Aspects*. VCH, Weinheim.

Göpel, W., Hesse, J. and Zemel, J. N. (eds) (1991) *Sensors: A Comprehensive Survey*, Vols 2 and 3, *Chemical and Biochemical Sensors*. VCH, Weinheim.

Göpel, W. and Heiduschka, P. (1995) Interface analysis in biosensor design. *Biosens. Bioelec.* **10**: 853–883.

Gorton, L., Persson, B., Hale, P. D., Boguslavsky, L. I., Karan, H. I., Lee, H. S., Skotheim, T. A., Lan, H. L. and Okamoto, Y. (1992) Electrocatalytic oxidation of nicotinamide adenine dinucleotide cofactor on chemically modified electrodes. In Edelman, P. G. and Wang, J.

(eds), *Biosensors and Chemical Sensors*. ACS Symposium Series #487, American Chemical Society, Washington, DC.

Gorton, L. (1995) Carbon paste electrodes modified with enzymes, tissues and cells. *Electroanalysis* **7**: 23–45.

Gough, D. A. and Leypoldt, J. K. (1979) Membrane-covered rotated disk electrode. *Anal. Chem.* **51**: 439–444.

Grabar, K. C., Freeman, R. G., Hommer, M. B. and Natan, M. J. (1995) Preparation and characterization of Au colloid monolayers. *Anal. Chem.* **67**: 735–743.

Grabar, K. C., Brown, K. R., Keating, C. D., Stranick, S. J., Tang, S-L. and Natan, M. J. (1997) Nanoscale characterization of gold colloid monolayers. A comparison of four techniques. *Anal. Chem.* **69**: 471–477.

Graham, C. R., Leslie, D. and Squirrell, D. J. (1992) Gene-probe assays on a fibre-optic evanescent-wave biosensor. *Biosens. Bioelec.* **7**: 487–493.

Grate, J. W., Martin, S. J. and White, R. M. (1993a) Acoustic wave microsensors (Part I). *Anal. Chem.* **65**: 940A–948A.

Grate, J. W., Martin, S. J. and White, R. M. (1993b) Acoustic wave microsensors (Part II). *Anal. Chem.* **65**: 987A–996A.

Grate, J. W., Rose-Pherson, S. L., Venezky, D. L., Klusty, M., and Wohltjen, H. (1993c) Smart sensor system for trace organophosphorous and organosulfur vapor detection employing a temperature-controlled array of surface acoustic wave sensors, automated sample preconcentration, and pattern recognition. *Anal. Chem.* **65**: 1868–1881.

Grate, J. W., Patrash, S. J., Abraham, M. H. and Du, C. M. (1996) Selective vapor sorption by polymers and cavitands on acoustic wave sensors: is this molecular recognition? *Anal. Chem.* **68**: 913–917.

Greenway, R. M., Gardiner, S. J., Hart, A. L. and Dunlop, J. (1994) Stainless steel electrodes in enzyme-based potentiometric and "pH-stat" biosensors. *Biosens. Bioelec.* **9**: 457–461.

Gross, G. W., Rhoades, B. K., Azzazy, H. M. E. and Wu, M-C. (1995) The use of neuronal networks on multielectrode arrays as biosensors. *Biosens. Bioelec.* **10**: 553–567.

Groves, J. T., Ulman, N. and Boxer, S. G. (1997) Micropatterning fluid lipid bilayers on solid supports. *Science* **275**: 651-653.

Grynkiewicz, G., Ponie, M. and Tsien, R. T. (1984) A new generation of Ca^{2+} indicators with greatly improved fluorescence properties. *J. Biol. Chem.* **260**: 3440–3450.

Guilbault, G. G. (1984) *Analytical Uses of Immobilized Enzymes*. Marcel Dekker, New York.

Guilbault, G. G. and Jordan, J. M. (1988) Analytical uses of piezoelectric crystals: a review. *CRC Crit. Revs Anal. Chem.* **19**:1-28.

Guilbault, G. G. and Suleiman, A. (1990) Piezoelectric crystal biosensors. *Amer. Biotech. Lab.*, March: 28–32.

Guilbault, G. G. and Mascini, M. (eds) (1993) *Uses of Immobilized Biological Compounds*. Kluwer, Dordrecht, The Netherlands.

Guoy, Y. and Dong, S. (1997) Organic phase enzyme electrodes based on organohydrogel. *Anal. Chem.* **69**: 1904–1908.

Hafeman, D., Parce, W. and McConnell, H. M. (1988) Light addressable potentiometric sensor for biochemical systems. *Science* **240**: 1182–1185.

Hale, P. D., Boguslavsky, L. I., Skotheim, T. A., Liu, L. F., Lee, H. S., Karan, H. I., Lan, H. L. and Okamoto, Y. (1992) Electrical wiring of flavoenzymes with flexible redox polymers. In Edelman, P. G. and Wang, J. (eds), *Biosensors and Chemical Sensors*. ACS Symposium #487, American Chemical Society, Washington, DC.

Hale, Z. M., Payne, F. P., Marks, R. S., Lowe, C. R. and Levine, M. M. (1996) The single mode tapered optical fibre loop immunosensor. *Biosens. Bioelec.* **11**: 137–148.

Hall, E. A. H. (1991) *Biosensors.* Prentice Hall, Englewood Cliffs, NJ.

Hall, E. A. H. (1992) Overview of biosensors. In Edelman, P. G. and Wang, J. (eds), *Biosensors and Chemical Sensors.* ACS Symposium Series #487, American Chemical Society, Washington, DC.

Harris, T. D., Grober, R. D., Trautman, J. K. and Betzig, E. (1994) Super-resolution imaging spectroscopy. *Appl. Spectros.* **48**: 14a–21a.

Harrison, D. J., Fan, Z., Seiler, K., Manz, A. and Widmer, H. M. (1993) Rapid separation of fluorescein derivatives using a micromachined capillary electrophoresis system. *Anal. Chim. Acta* **283**: 361–366.

Harsanyi, G., Peteri, I. and Deak, I. (1994) Low-cost ceramic sensors for biomedical use: a revolution in transcutaneous blood oxygen monitoring? *Sens. Actuators* (B) **18**: 171–174.

Hart, J. P. and Wring, S. A. (1991) Carbon-based electrodes and their application as electrochemical sensors for selected biomolecules. *Anal. Proc.* **28**: 4–7.

Hart, J. P. and Wring, S. A. (1994) Screen-printed voltammetric and amperometric electrochemical sensors for decentralized testing. *Electroanalysis* **6**: 617–624.

Hart, J. P. and Hartley, I. C. (1994) Voltammetric and amperometric studies of thiocholine at a screen-printed carbon electrode chemically modified with cobalt phthalocyanine: studies toward a pesticide sensor. *Analyst* **119**: 259–263.

Hartgerink, J. D., Granja, J. R., Milligan, R. A. and Ghadiri, M. R. (1996) Self-assembling peptide nanotubes. *J. Amer. Chem. Soc.* **118**: 43–50.

Hartley, I. C. and Hart, J. P. (1994) Amperometric measurement of organophosphate pesticides using a screen-printed disposable sensor and biosensor based on cobalt phthalocyanine. *Anal. Proc.* **31**: 333–337.

Hartmann, P., Leiner, J. P. and Lippitsch, M. E. (1995) Luminescence quenching behavior of an oxygen sensor based on a Ru(II) complex dissolved in polystyrene. *Anal. Chem.* **67**: 88–93.

Hartmann, P. and Trettnak, W. (1996) Effects of polymer matrices on calibration functions of luminescent oxygen sensors based on porphyrin ketone complexes. *Anal. Chem.* **68**: 2615–2620.

Hartmann, P. and Ziegler, W. (1996) Lifetime imaging of luminescent oxygen sensors based on all-solid state technology. *Anal. Chem.* **68**: 4512–4514.

Hashimoto, K., Ito, K. and Ishimori, Y. (1994) Sequence-specific gene detection with a gold electrode modified with DNA probes and an electrochemically active dye. *Anal. Chem.* **66**: 3830–3833.

Haswell, S. J. (1992) *Practical Guide to Chemometrics.* Marcel Dekker, New York.

Hayes, F. J., Halsall, H. B. and Heineman, W. R. (1994) Simultaneous immunoassay using electrochemical detection of metal ion labels. *Anal. Chem.* **66**: 1860–1865.

Hayes, M. and Kuhr, W. G. (1995) Development of small scale, fast response chemical probes for neuroactive compounds. Abstracts (#1085), Pittsburgh Conference, New Orleans, LA.

Healy, B. G., Foran, S. E. and Walt, D. R. (1995) Photodeposition of micrometer-scale polymer patterns on optical imaging fibers. *Science* **269**: 1078–1080.

Healy, B. G. and Walt, D. R. (1997) Fast temporal response fiber-optic chemical sensors based on the photodeposition of micrometer-scale polymer arrays. *Anal. Chem.* **69**: 2213–2216.

Heideman, R. G., Kooyman, R. P. H. and Greve, J. (1993) Performance of a highly sensitive optical waveguide Mach–Zehnder interferometer immunosensor. *Sens. Actuators* (B) **10**: 209–217.

Heineman, W. R., Hawkridge, F. M. and Blount, H. N. (1984) Spectroelectrochemistry at optically transparent electrodes. In Bard, A. J. (ed.), *Electroanalytical Chemistry*, Vol. 13. Marcel-Dekker, New York.

Heineman, W. R. and Halsall, H. B. (1985) Electrochemical immunoassay. *Anal. Chem.* **57**: 1321A–1331A.

Heinze, J. (1993) Ultramicroelectrodes in electrochemistry. *Angew. Chem. (Int. Ed. Engl.)* **32**: 1268–1288.

Heller, A. (1990) Electrically wired enzymes. *Accts Chem. Res.* **23**: 128–148.

Heller, A. (1992) Electrical connection of enzyme redox centers to electrodes. *J. Phys. Chem.* **96**: 3579–3587.

Hendji, A. M. N., Jaffrezic-Renault, N., Martelet, C., Shul'ga, A. A., Dzydevich, S. V., Sodatkin, A. P. and El'skaya, A. V. (1994) Enzyme biosensor based on a micromachined interdigitated conductometric transducer: application to the detection of urea, glucose, acetyl- and butyrylcholine chlorides. *Sens. Actuators* (B) **21**: 123–129.

Henke, C., Steinem, C., Janshoff, A., Steffan, G., Luftmann, H., Sieber, M. and Galla, H-J. (1996) Self-assembled monolayers of monofunctionalized cyclodextrins onto gold: a mass spectrometric characterization and impedance analysis of host–guest interaction. *Anal. Chem.* **68**: 3158–3165.

Henshaw, J. M., Burgess, L. W., Booksh, K. S. and Kowalski, B. R. (1994) Multicomponent determination of chlorinated hydrocarbons using a reaction-based chemical sensor. 1. Multivariate calibration of Fujiwara reaction products. *Anal. Chem.* **66**: 3328–3336.

Hermes, T., Bühner, M., Bücher, S., Sundermeier, C., Dumschat, C., Börchardt, M., Cammann, K. and Knoll, M. (1994) An amperometric microsensor array with 1024 individually addressable elements for two-dimensional concentration mapping. *Sens. Actuators* (B) **21**: 33–37.

Higson, S. P. J. and Vadgama, P. M. (1995a) Diamond-like carbon coated films for enzyme electrodes: characterization of biocompatibility and substrate diffusion limiting properties. *Anal. Chim. Acta* **300**: 77–83.

Higson, S. P. J. and Vadgama, P. M. (1995b) Diamond-like carbon films for enzyme electrodes: Characterization of novel overlying permselective barriers. *Anal. Chim. Acta* **300**: 85–90.

Hisamoto, H., Tsubuku, M., Enomoto, Y., Watanabe, K., Kawaguchi, H., Koike, Y. and Suzuki, K. (1993) Theory and practice of rapid flow-through analysis based on optode detection and its application to pH measurement as a model case. *Anal. Chem.* **68**: 3871–3878.

Hitzmann, B. and Kullick, T. (1994) Evaluation of pH field effect transistor measurement signals by neural networks. *Anal. Chim. Acta* **294**: 243–249.

Hitzmann, B., Gomersall, R., Brandt, J. and vanPutten, A. (1995) An expert system for the supervision of a multichannel flow injection analysis system. In Rogers, K. R., Mulchandani, A. and Zhou, W. (eds), *Biosensor and Chemical Sensor Technology*. ACS Symposium Series #613, American Chemical Society, Washington, DC.

Hoa, D. T., Kumar, T. N. S., Punekar, N. S., Srinivasa, R. S., Lal, R. and Contractor, A. B. (1992) Biosensor based on conducting polymers. *Anal. Chem.* **64**: 2645–2646.

Hofmeister, F. (1888) Zur Lehre von der Wirking der Salze. *Arch. Exp. Pathol. Pharmakol.* **24**: 247.

Hogan, B. L. and Yeung, E. S. (1992) Determination of intracellular species at the level of a single erythrocyte via capillary electrophoresis with direct and indirect fluorescence detection. *Anal. Chem.* **64**: 2841–2845.

Hogan, B. L., Lunte, S. M., Stobaugh, J. F. and Lunte, C. E. (1994) On-line coupling of *in vivo* microdialysis sampling with capillary electrophoresis. *Anal. Chem.* **66**: 596–602.

Holt, P.-J., Stephens, L. D., Bruce, N. C. and Lowe, C. R. (1995) An amperometric opiate assay. *Biosens. Bioelec.* **10**: 517–526.

Holt, P.-J., Bruce, N. C. and Lowe, C. R. (1996) Bioluminescent assay for heroin and its metabolites. *Anal. Chem.* **68**: 1877–1882.

Home, P. D. and Alberti, K. G. M. M. (1989) Biosensors in medicine: the clinician's requirements. In Turner, A. P. F., Karube, I. and Wilson, G. S. (eds), *Biosensors: Fundamentals and Applications*. Oxford University Press, New York.

Hong, H. G., Bohn, P. W. and Sligar, S. G. (1993) Optical determination of surface density in oriented metalloprotein nano-structures. *Anal. Chem.* **65**: 1635–1638.

Hong, H. G., Jiang, M., Sligar, S. G. and Bohn, P. W. (1994) Cysteine-specific surface tethering of genetically engineered cytochromes for fabrication of metalloprotein nanostructures. *Langmuir* **10**: 153–158.

Horie, H. and Rechnitz, G. A., (1995) Hybrid tissue/enzyme biosensor for pectin. *Anal. Chim. Acta* **306**: 123–127.

Horrocks, B. R., Mirkin, M. V., Pierce, D. T., Bard, A. J., Nagy, G. and Toth, K. (1993a). Scanning electrochemical microscopy. 19. Ion-selective potentiometric microscopy. *Anal. Chem.* **65**: 1213–1224.

Horrocks, B. R., Schmidtke, D., Heller, A. and Bard, A. J. (1993b) Scanning electrochemical microscopy. 24. Enzyme ultramicroelectrodes for the measurement of hydrogen peroxide at surfaces. *Anal. Chem.* **65**: 3605–3614.

Hoshi, T., Anzai, J. and Osa, T. (1995) Controlled deposition of glucose oxidase on platinum electrode based on an avidin/biotin system for the regulation of output current of glucose sensors. *Anal. Chem.* **67**: 770–774.

Hoyer, B., Sørensen, G., Jensen, N., Nielsen, D. B. and Larsen, B. (1996) Electrostatic spraying: a novel technique for preparation of polymer coatings on electrodes. *Anal. Chem.* **68**: 3840–3844.

Hseueh, C. and Brajter-Toth, A. (1994) Electrochemical preparation and analytical applications of ultrathin overoxidized polypyrrole films. *Anal. Chem.* **66**: 2458–2464.

Hubbard, A. T. (ed.) (1995) *Handbook of Surface Imaging and Visualization*. CRC Press, Boca Raton, FL.

Hughes, R. C., Ricco, A. J., Butler, M. A. and Martin, S. J. (1991) Chemical microsensors. *Science* **254**: 74–80.

Hunkapiller, T., Kaiser, R. J., Koop, B. F., and Hood, L. (1991) Large-scale and automated DNA sequence determination. *Science* **254**: 59–66.

Hulanicki, A., Glab, S. and Ingman, F. (1991) Chemical sensors: definitions and classification. *Pure Appl. Chem.* **63**: 1248–1250.

Hutchins, R. S., Molina, P., Alajarin, M., Vidal, A. and Bachas, L. G. (1994) Use of a guanidinium ionophore in a hydrogen sulfite-selective electrode. *Anal. Chem.* **66**: 3188–3192.

Hutchins, R. S. and Bachas, L. G. (1995) Nitrate-selective electrode developed by electrochemically mediated imprinting/doping of polypyrrole. *Anal. Chem.* **67**: 1654–1660.

Hutchins, R. S., Bansal, P., Molina, P., Alajarín, M., Vidal, A. and Bachas, L. G. (1997) Salicylate-selective electrode based on a biomimetic guanidinium ionophore. *Anal. Chem.* **69**: 1273–1278.

Hutchinson, A. M. (1995) Evanescent wave biosensors. Real-time analysis of biomolecular interactions. *Mol. Biotechnol.* **3**: 47–54.

Hutton, R. S. and Williams, D. E. (1995) Scanning laser photoelectrochemical microscopy of reaction dynamics at a microelectrode array. *Anal. Chem.* **67**: 280–282.

Ichimori, K. Ishida, H., Fukahori, M., Nakazawa, H. and Murakami, E. (1994) Practical nitric oxide measurement employing a nitric oxide-selective electrode. *Rev. Sci. Instrum.* **65**: 1–5.

Ikeda, T., Kurosaki, T., Takayama, K. and Kano, K. (1996) Measurement of oxidoreductase-like activity of intact bacterial cells by an amperometric method using a membrane-coated electrode. *Anal. Chem.* **68**: 192–198.

Ingersoll, C. M., Jordan, J. D. and Bright, F. V. (1996) Accessibility of the fluorescent reporter group in native, silica-adsorbed, and covalently attached acrylodan-labeled serum albumins. *Anal. Chem.* **68**: 3194–3198.

Ingersoll, C. M. and Bright, F. V. (1997) Using fluorescence to probe biosensor interfacial dynamics. *Anal. Chem.* **69**: 403A–408A.

Invitski, D. and Rishpon, J. (1996) A one-step, separation-free amperometric enzyme immunosensor. *Biosens. Bioelec.* **11**: 409–417.

Inzelt, G. (1994) Mechanism of charge transport in polymer- modified electrodes. In Bard, A. J. (ed.), *Electroanalytical Chemistry*, Vol. 18. Marcel Dekker, New York.

Ippolito, J. A., Baird, T. T., McGee, S. A., Christianson, D. W. and Fierke, C. A. (1995) Structure–assisted redesign of a protein-zinc-binding site with femtomolar affinity. *Proc. Natl Acad. Sci.* USA **92**: 5017–5021.

Ito, N., Saito, A., Kayashima, S., Kimura, J., Kuriyama, T., Nagata, N., Arai, T. and Kikuchi, M. (1995) Transcutaneous blood glucose monitoring system based on an ISFET glucose sensor and studies on diabetic patients. *Front. Med. Biol. Eng.* **6**: 269–280.

IUPAC (1991) Chemical sensors: definitions and classification. *Pure Appl. Chem.* **63**: 1248–1250 (see Hulanicki *et al.*, 1991).

IUPAC (1995) Selectivity coefficients for ion-selective electrodes: Recommended methods for reporting $K^{pot}_{A,B}$ values. *Pure Appl. Chem.* **67**: 507–518 (see Umezawa *et al.*, 1995).

IUPAC (1996) Electrochemical biosensors: Proposed definitions and classification. Synopsis of Report, as presented in *Sens. Actuators* (B) **30**: 81 (see Thevenot *et al.*, 1996).

Iwuoha, E. I., Adeyoju, O., Dempsey, E., Smyth, M. R., Liu, J. and Wang, J. (1995) Investigation of the effects of polar organic solvents on the activity of tyrosinase entrapped in a poly(ester-sulphonic acid) polymer. *Biosens. Bioelec.* **10**: 662–667.

Iwuoha, E. I., Leister, I., Miland, E., Smyth, M. R. and Fágáin, C.Ó. (1997) Reactivities of organic-phase biosensors. 1. Enhancement of the sensitivity and stability of amperometric peroxidase biosensors using chemically modified enzymes. *Anal. Chem.* **69**: 1674–1681.

Izquierdo, A. and Luque de Castro, M. D. (1995) Ion-sensitive field-effect transistors and ion-selective electrodes as sensors in dynamic systems. *Electroanalysis* **7**: 505–519.

Jackson, B. P., Dietz, S. M. and Wightman, R. M. (1995) Fast-scan cyclic voltammetry of 5-hydroxytryptamine. *Anal. Chem.* **67**: 1115–1120.

Jacobs, M. B., Carter, R. M., Lubrano, G. L. and Guilbault, G. G. (1995) A piezoelectric biosensor for *Listeria monocytogens. Amer. Laboratory* July: 26–28.

Jacobson, S. C., Hergenroder, R., Koutny, L. B. and Ramsey, J. M. (1994a) High-speed separations on a microchip. *Anal. Chem.* **66**: 1114–1118.

Jacobson, S. C., Hergenroder, R., Koutny, L. B., Warmack, R. J. and Ramsey, J. M. (1994b) Effects of injection schemes and column geometry on the performance of microchip electrophoresis devices. *Anal. Chem.* **66**: 1107–1113.

Jacobson, S. C., Hergenroder, R., Koutny, L. B., Warmack, R. J. and Ramsey, J. M. (1994c) Open channel electrochromatography on a microchip. *Anal. Chem.* **66**: 2369–2373.

Jacobson, S. C., Hergenroder, R., Moore Jr, A. W. and Ramsey, J. M. (1994d) Precolumn reactions with electrophoretic analysis integrated on a microchip. *Anal. Chem.* **66**: 4127–4132.

Jacobson, S. C., Koutny, L. B., Hergenroder, R., Moore Jr, A. W., Ramsey, J. M. (1994e) Microchip capillary electrophoresis with an integrated postcolumn reactor. *Anal. Chem.* **66**: 3472–3476.

Jacobson, S. C. and Ramsey, J. M. (1996) Integrated microdevice for DNA restriction fragment analysis. *Anal. Chem.* **68**: 720–723.

Jaffari, S. A. and Turner, A. P. (1995) Recent advances in amperometric glucose biosensors for *in vivo* monitoring. *Physiol. Meas.* **16**: 1–15.

Jameison, F., Sanchez, R. I., Dong, L., Leland, J. K., Yost, D. and Martin, M. T. (1996) Electrochemiluminescence-based quantitation of classical clinical chemistry analytes. *Anal. Chem.* **68**: 1298–1302.

Janata, J. (1989) *Principles of Chemical Sensors.* Plenum Press, New York.

Janda, P., Weber, J., Dunsch, L. and Lever, A. B. P. (1996) Detection of ascorbic acid using a carbon fiber microelectrode coated with cobalt tetramethylpyridoporphyrazine. *Anal. Chem.* **68**: 960–965.

Jimenez, C., Bartroli, J., de Rooij, N. F. and Koudelka-Hep, M. (1995) Glucose sensor based on an amperometric microelectrode with a photopolymerizable enzyme membrane. *Sens. Actuators* (B) **27**: 421–424.

Jin, W., Wollenberger, U., Bier, F. and Scheller, F. (1995) Construction and characterization of multilayer enzyme electrode. Covalent binding of quinoprotein glucose dehydrogenase onto gold electrodes. *Biosens. Bioelec.* **10**: 823–829.

Jin, Z., Chen, R. and Colón, L. A. (1997) Determination of glucose in submicroliter samples by CE-LIF using precolumn or on-column enzymatic reactions. *Anal. Chem.* **69**: 1326–1331.

Johnston, D. A., Cardosi, M. F. and Vaughn, D. H. (1995) Electrochemistry of hydrogen peroxide on evaporated gold/palladium composite electrodes. *Electroanalysis* **7**: 520–526.

Jordan, C. E., Frey, B. L., Kornguth, S. and Corn, R. M. (1994) Characterization of poly-L-lysine adsorption onto alkanethiol-modified gold surfaces with polarization-modulation Fourier transform infrared spectroscopy and surface plasmon resonance measurements. *Langmuir* **10**: 3642–3648.

Jordan, C. E. and Corn, R. M. (1997) Surface plasmon resonance imaging measurements of electrostatic biopolymer adsorption onto chemically modified gold surfaces. *Anal. Chem.* **69**: 1449–1456.

Jordan, J. D., Dunbar, R. A. and Bright, F. V. (1995) Dynamics of acrylodan-labeled bovine and human serum albumin entrapped in a sol–gel derived biogel. *Anal. Chem.* **67**: 2436–2443.

Josowicz, M. (1995) Applications of conducting polymers in potentiometric sensors. *Analyst* **120**: 1019–1024.

Jossé, F. (1994) Acoustic wave liquid-phase-based microsensors. *Sens. Actuators* (A) **44**: 199–208.

Jung, S–K. and Wilson, G. S. (1996) Polymeric mercaptosilane-modified platinum electrodes for elimination of interferants in glucose biosensors. *Anal. Chem.* **68**: 591–596.

Justice Jr, J. B. (1987) Introduction to *in vivo* voltammetry. In Justice Jr, J. B. (ed.), *Voltammetry in the Neurosciences.* Humana Press, Clifton, NJ.

Justice Jr, J. B. and Parsons, L. H. (1991) Microvoltammetry and brain dialysis. In Yamazoe, N. (ed.), *Chemical Sensor Technology*, Vol. 3. Kodansha, Tokyo and Elsevier, Amsterdam. pp. 249–262.

Kaku, T., Karan, H. I., and Okamoto, Y. (1994) Amperometric glucose sensors based on immobilized glucose oxidase–polyquinone system. *Anal. Chem.* **66**: 1231–1235.

Kalcher, K., Kauffman, J. M., Wang, J., Švancara, I., Vytřas, K., Neuhold, C. and Yang, Z. (1995) Sensors based on carbon paste in electrochemical analysis: a review with particular emphasis on the period 1990–1993. *Electroanalysis* **7**: 5–22.

Kallury, K. M. R., Lee, W. E. and Thompson, M. (1992) Enhancement of the thermal and storage stability of urease by covalent attachment to phospholipid-bound silica. *Anal. Chem.* **64**: 1062–1068.

Kallury, K. M. R., Lee, W. E. and Thompson, M. (1993) Enhanced stability of urease immobilized onto phospholipid covalently bound to silica, tungsten, and fluoropolymer surfaces. *Anal. Chem* **65**: 2459–2467.

Kallury, K. M. R. and Thompson, M. (1994) Extraordinary thermal stabilization of enzymes through surface attachment by covalently bound phospholipids. In Pecsek, J. J. and Leigh, I. E. (eds), *Chemically Modified Surfaces*. The Royal Society of Chemistry, Cambridge, UK.

Kamat, S. V., Beckman, E. J. and Russell, A. J. (1995) Enzyme activity in supercritical fluids. *Crit. Revs Biotechnol.* **15**: 41–71.

Kar, S. and Arnold, M. A. (1992) Fiber-optic ammonia sensor for measuring synaptic glutamate and extracellular ammonia. *Anal. Chem.* **64**: 2438–2443.

Kar, S. and Arnold, M. A. (1994) Cylindrical sensor geometry for absorbance-based fiber-optic ammonia sensors. *Talanta* **41**: 1051–1058.

Karlsen, S. R., Johnston, K. S., Jorgenson, R. C. and Yee, S. S. (1995) Simultaneous determination of refractive index and absorbance spectra of chemical samples using surface plasmon resonance. *Sens. Actuators* (B) **24–25**: 747–749.

Karube, I. (1989a) Micro-organism based sensors. In Turner, A. P. F., Karube, I, and Wilson, G. S. (eds), *Biosensors: Fundamentals and Applications*. Oxford University Press, New York.

Karube, I. (1989b) Micro-biosensors based on silicon fabrication technology. In Turner, A. P. F., Karube, I. and Wilson, G. S. (eds), *Biosensors: Fundamentals and Applications*. Oxford University Press, New York.

Karube, I., Ikebukuro, K., Murakami, Y. and Yokoyama, K. (1995a) Micromachining technology and biosensors. *Ann. N.Y. Acad. Sci.* **750**: 101–108.

Karube, I., Nomura, Y. and Azikawa, Y. (1995b) Biosensors for environmental control. *Trends Anal. Chem.* **14**: 295–299.

Karyakin, A. A., Karyakina, E. E., Gorton, L., Bobrova, O. A. Luckachora, L. V., Gladilin, A. K. and Levashov, A. V. (1996) Improvement of electrochemical biosensors using enzymes immobilization from water–organic mixture with a high content of organic solvent. *Anal. Chem.* **68**: 4335–4341.

Katakis, I. and Heller, A. (1992) L-α-Glycerophosphate and L-lactate electrodes based on the electrochemical "wiring" of oxidases. *Anal. Chem.* **64**: 1008–1013.

Katakis, I. and Dominguez, E. (1995) Characterization and stabilization of enzyme biosensors. *Trends Anal. Chem.* **14**: 310–319.

Kats, M., Richberg, P. C. and Hughes, D. M. (1997) pH-Dependent isoform transitions of a monoclonal antibody monitored by micellar electrokinetic capillary chromatography. *Anal. Chem.* **69**: 338–343.

Katz, E., Lötzbeyer, T., Schlereth, D. D. and Schuhmann, W. (1994a) Electrocatalytic oxidation of reduced nicotinamide coenzymes at gold and platinum electrode surfaces modified with a monolayer of pyrroloquinoline quinone—effect of Ca^{2+} cations. *J. Electroanal. Chem.* **373**: 189–200.

Katz, E., Schlereth, D. D. and Schmidt, H-L. (1994b) Electrochemical study of pyrroloquinoline quinone covalently immobilized as a monolayer onto a cysteamine-modified gold electrode. *J. Electroanal. Chem.* **367**: 59–70.

Kawagoe, J. L., Niehaus, D. E. and Wightman, R. M. (1991) Enzyme-modified organic conducting salt microelectrode. *Anal. Chem.* **63**: 2961–2965.

Kawazoe, N., Ito, Y. and Imonishi, Y. (1996) Patterned staining by fluorescein-labeled oligonucleotides obtained by *in vitro* selection. *Anal. Chem.* **68**: 4309–4311.

Kempe, M. (1996) Antibody-mimicking polymers as chiral stationary phases in HPLC. *Anal. Chem.* **68**: 1948–1953.

Kenausis, G., Chen, Q. and Heller, A. (1997) Electrochemical glucose and lactate sensors based on "wired" thermostable soybean peroxidase operating continuously and stably at 37 °C. *Anal. Chem.* **69**: 1054–1060.

Kennedy, R. T., Huang, L., Atkinson, M. A. and Dush, P. (1993) Amperometric monitoring of chemical secretions from individual pancreatic β-cells. *Anal. Chem.* **65**: 1882-1887.

Kennedy, R. T., Huang, L., Paras, R., Gorski, W. and Shen, H. (1995) High resolution monitoring of insulin secretion from single pancreatic β-cells and single islets. Abstracts (#625). Pittsburgh Conference, New Orleans, LA.

Kepley, L. T., Crooks, R. M. and Ricco, A. J. (1992) Selective surface acoustic wave-based organophosphonate chemical sensor employing a self-based composite monolayer: a new paradigm for sensor design. *Anal. Chem.* **64**: 3191–3193.

Kerner, W., Kiwit, M., Linke, B., Keck, F. S., Zier, H. and Pfeiffer, E. F. (1993) The function of a hydrogen peroxide-detecting electroenzymatic glucose electrode is markedly impaired in human sub-cutaneous tissue and plasma. *Biosens. Bioelec.* **8**: 473–482.

Khan, G. F., Kobataka, E., Shinohara, H., Ikariyama, Y. and Aizawa, M. (1992) Molecular interface for an activity controlled enzyme electrode and its application for the determination of fructose. *Anal. Chem.* **64**: 1254–1258.

Khan, G. F., Ohwa, M. and Wernet, W. (1996) Design of a stable charge transfer complex electrode for a third-generation amperometric glucose sensor. *Anal. Chem.* **68**: 2939–2945.

Kikuchi, A., Suzuki, K, Okabayashi, O., Hoshino, H., Kataoka, K., Sakutia, Y. and Okano, T. (1996) Glucose-sensing electrode coated with polymer complex gel containing phenylboronic acid. *Anal. Chem.* **68**: 823–828.

Kikuchi, K., Nagano, T., Hayakawa, H., Hirata, Y. and Hirobe, M. (1993) Detection of nitric oxide production from a perfused organ by a luminol–H_2O_2 system. *Anal. Chem.* **65**: 1794–1799.

Killard, A. J., Deasy, B., O'Kennedy, R. and Smyth, M. R. (1995) Antibodies: production, function, and applications in biosensors. *Trends Anal. Chem.* **14**: 257–266.

Kim, W., Chung, S., Park, S. B., Lee, S. C., Kim, C. and Sung, D. D. (1997) Sol–gel method for the matrix of chloride-selective membranes doped with tridodecylmethylammonium chloride. *Anal. Chem.* **69**: 95–98.

Kimura, J., Murakami, T., Kuriyama, T. and Karube, I. (1988) An integrated multibiosensor for simultaneous amperometric and potentiometric measurement. *Sens. Actuators* **15**: 435–443.

Kipling, A. L. and Thompson, M. (1990) Network analysis applied to liquid-phase acoustic wave sensors. *Anal. Chem.* **62**: 1514–1519.

Kirkbright, G. F., Narayanaswamy, R. and Welti, N. A. (1984) Fiber optic pH probe based on the use of an immobilized colorimetric indicator. *Analyst* **109**: 1025–1028.

Kissinger, P. Y. and Heineman, W. R. (eds) (1996) *Laboratory Techniques in Electroanalytical Chemistry*, 2nd Edition. Marcel Dekker, New York.

Kittel, C. (1996) *Introduction to Solid State Physics*. 7th Edition. John Wiley and Sons, New York.

Klein, L. C. (ed.) (1994) *Sol–gel Optics: Processing and Applications*. Kluwer, Norwell, MA.

Klimant, I. and Wolfbeis, O. S. (1995) Oxygen-sensitive luminescent materials based on silicone-soluble ruthenium diimine complexes. *Anal. Chem.* **67**: 3160–3166.

Klunder, G. L., Buerck, J., Ache, H. J., Silva, R. J. and Russo, R. E. (1994) Temperature effect on a fibre-optic evanescent wave absorption sensor. *Appl. Spectrosc.* **48**: 387–393.

Knichel, M., Heiduschka, P., Beck, W., Jung, G. and Göpel, W. (1995) Utilization of a self-assembled peptide monolayer for an impedimetric immunosensor. *Sens. Actuators* (B) **28**: 85–94.

Knoll, M., Cammann, K., Dumschat, C., Eshold, J. and Sundermeier, C. (1994) Micromachined ion-selective electrodes with polymer matrix membranes. *Sens. Actuators* (B) **21**: 71–76.

Kobatake, E., Niimi, T., Haruyama, T., Ikariyama, Y. and Aizawa, M. (1995) Biosensing of benzene derivatives in the environment by luminescent *Escherichia coli*. *Biosens. Bioelec.* **10**: 601–605.

Kolakowski B., Battaglini, F., Lee, Y. S. Klironomos, G. and Mikkelsen, S. R. (1996) Comparison of an intercalating dye and an intercalant–enzyme conjugate for DNA detection in a microtiter-based assay. *Anal. Chem.* **68**: 1197–1200.

Kolb, D. M., Ullmann, R. and Will, T. (1997) Nanofabrication of small copper clusters on gold (111) electrodes by a scanning tunneling microscope. *Science* **275**: 1097–1102.

Koncki, R. and Glab, S. (1995) Kinetic model of pH-based potentiometric enzymic sensors. Part 4. Enzyme loading and lifetime factors. *Analyst* **120**: 489–493.

König, B. and Grätzel, M. (1994) A novel immunosensor for Herpes viruses. *Anal. Chem.* **66**: 341–344.

Kopelman, R. and Tan, W. (1993) Near-field optics: imaging single molecules. *Science* **262**: 778–781.

Kopelman, R. and Tan, W. (1994) Near-field optical microscopy, spectroscopy, and chemical sensors. *Appl. Spectrosc. Rev.* **29**: 39–66.

Korell, U. and Lennox, R. B. (1992) Determination of ascorbic acid using an organic conducting salt electrode. *Anal. Chem.* **64**: 147–151.

Korell, U. and Spichiger, U. E. (1994) Novel membraneless amperometric peroxide biosensor based on a tetrathiofulvalene–*p*-tetracyanoquinodimethane electrode. *Anal. Chem.* **66**: 510–515.

Koryta, J. (1992) *Ions, Electrodes and Membranes*, 2nd Edition. John Wiley and Sons, Chichester, UK.

Kotte, H., Gründig, B., Vorlop, K. D., Strehlitz, B. and Stottmeister, U. (1995) Methylphenazonium-modified enzyme sensor based on polymer thick films for sub-nanomolar detection of phenols. *Anal. Chem.* **67**: 65–70.

Kounaves, S. P., Deng, W., Hallock, P., Kovacs, G. T. A. and Storment, C. W. (1994) Iridium-based ultramicroelectrode array fabricated by microlithography. *Anal. Chem.* **66**: 418–423.

Kovacs, G. T. A., Petersen, K. and Albin, M. (1996) Silicon micromachining: sensors to systems. *Anal. Chem.* **68**: 407A–412A.

Kriz, C. B., Rådevik, K. and Kriz, D. (1996) Magnetic permeability measurements in bioanalysis and biosensors. *Anal. Chem.* **68**: 1966–1970.

Kriz, D. and Mosbach, K. (1995) Competitive amperometric morphine sensors based on an agarose immobilized molecularly imprinted polymer. *Anal. Chim. Acta* **300**: 71–75.

Kriz, D., Ramström, O., Svensson, A. and Mosbach, K. (1995) Introducing biomimetic sensors based on molecularly imprinted polymers as recognition elements. *Anal. Chem.* **67**: 2142–2144.

Kriz, D., Ramström, O. and Mosbach, K. (1997) Molecular imprinting: new possibilities for sensor technology. *Anal. Chem.* **69**: 345A–349A.

Krohn, D. A. (1988) *Fiber Optics Sensors. Fundamentals and Applications.* Instrument Society of America, Research Triangle Park, NC.

Krska, R., Taga, K. and Kellner, R. (1993) New IR fibre-optic chemical sensor for *in situ* measurements of chlorinated hydrocarbons in water. *Appl. Spectrosc.* **47**: 1484–1487.

Kugimiya, A., Matsui, J., Takeuchi, T., Yano, K., Muguruma, H., Elgersma, A. V. and Karube, I. (1995) Recognition of sialic acid using molecularly imprinted polymer. *Anal. Lett.* **28**: 2317–2323.

Kuhn, K. J. and Burgess, L. W. (1993) Chemometric-evaluation of the multiple mode response of an ion-diffused planar optical waveguide to liquid-phase analytes. *Anal. Chem.* **65**: 1390–1398.

Kulys, J. and Hansen, H. E. (1994) Carbon-paste biosensors array for long-term glucose measurement. *Biosens. Bioelec.* **9**: 491–500.

Kumar, P., Colston, J. T., Chambers, J. P., Rael, E. D. and Valdes, J. J. (1994) Detection of botulinum toxin using an evanescent wave immunosensor. *Biosens. Bioelec.* **9**: 57–63.

Kuwabata, S. and Martin, C. R. (1994) Mechanism of the amperometric response of a proposed glucose sensor based on a polypyrrole-tubule-impregnated membrane. *Anal. Chem.* **66**: 2757–2762.

Kuwabata, S., Okamoto, T., Kajiya, Y. and Yoncyama, H. (1995) Preparation and amperometric glucose sensitivity of covalently bound glucose oxidase to (2-aminoethyl)ferrocene on an Au electrode. *Anal. Chem.* **67**: 1684–1690.

Kynclova, E., Hartig, A. and Schalkhammer, T. (1995) Oligonucleotide labelled lipase as a new sensitive hybridization probe and its use in bio-assays and biosensors. *J. Mol. Recognit.* **8**: 139–145.

Kyrolainen, M., Rigsby, P., Eddy, S. and Vadgama, P. (1995) Bio-/haemocompatibility: implications and outcomes for sensors? *Acta Anaesthesiol Scand.* Suppl. **104**: 55–60.

La Gal La Salle, A., Limoges, B., Degrand, C. and Brossier, P. (1995) Enzyme immunoassays with an electrochemical detection method using alkaline phosphatase and a perfluorosulfonated ionomer-modified electrode. Application to phenytoin assay. *Anal. Chem.* **67**: 1245–1253.

Lagerkvist, A. C., Furebring, C. and Borrebaeck, C. A. (1995) Single antigen-specific B-cells used to generated Fab fragments using CD40-mediated amplification or direct PCR cloning. *Biotechniques* **18**: 862–869.

Lagraff, J. R. and Gewirth, A. A. (1995) Nanometer-scale mechanism for the constructive modification of Cu single-crystals and alkanethiol-passivated Au (111) with an atomic-force microscope. *J. Phys. Chem.* **99**: 10009–10018.

Lakowicz, J. R. (ed.) (1994) *Probe Design and Chemical Sensing.* Topics in Fluorescence Spectroscopy, Plenum Press, New York.

Lantoine, F., Trevin, S., Bedioui, F. and Devynck, J. (1995) Selective and sensitive electrochemical measurement of nitric oxide in aqueous solution: discussion and new results. *J. Electroanal. Chem.* **392**: 85–89.

Lasky, S. J. and Buttry, D. A. (1989) Sensor based on biomolecules immobilized on the piezoelectric quartz crystal microbalance: detection of glucose using hexokinase. In Murrary, R. W., Dessy, R. E., Heineman, W. R., Janata, J. and Seitz, W. R. (eds), *Chemical Sensors and Microinstrumentation.* ASC Symposium Series #403, American Chemical Society, Washington, DC.

Lasky, S. J. and Buttry, D. A. (1990) Development of a real-time-glucose biosensor by enzyme immobilization on the quartz crystal microbalance. *Amer. Biotech. Lab.* February: 8–16.

Lau, Y. Y., Abe, T., and Ewing, A. G. (1992) Voltammetric measurement of oxygen in single neurons using platinized carbon ring electrode. *Anal. Chem.* **64**: 1702–1705.

Laurell, T., Drott, J. and Rosengren, L. (1995) Silicon wafer integrated enzyme reactors. *Biosens. Bioelec.* **10**: 289–299.

Lawrence, C. R., Geddes. N. J. and Furlong, D. N. (1996) Surface plasmon resonance studies of immunoreactions utilizing disposable diffraction gratings. *Biosens. Bioelec.* **11**: 389–400.

Leca, B., Morelis, R. M. and Coulet, P. R. (1995a) A novel electropolymerized sensing layer for biosensors involving entrapped enzyme-phospholipid vesicles. *Ann. NY. Acad. Sci.* **750**: 109-111

Leca, B., Morelis, R. M. and Coulet, P. R. (1995b) Design of a choline sensor via direct coating of the transducer by photopolymerization of the sensing layer. *Sens. Actuators* (B) **27**: 436–439.

Lechuga, L. M., Lenfering, A. T. M., Kooyman, R. P. H. and Greve, J. (1995) Feasibility of evanescent wave interferometer immunosensors for pesticide detection: Chemical aspects. *Sens. Actuators* (B) **25**: 762–765.

Lee, K. S., Shin., J. H., Han., S. H., Cha, G. S., Shin, D. S. and Kim, H. D. (1993) Asymmetric carbonate ion-selective cellulose acetate membrane electrodes with reduced salicylate interference. *Anal. Chem.* **65**: 3151–3155.

Lee, W.-W., White, H. S. and Ward, M. D. (1993) Depletion layer effects on the response of the electrochemical quartz crystal microbalance. *Anal. Chem.* **65**: 3232–3237.

Leech, D. and Rechnitz, G. A. (1993) Biomagnetic neurosensors. *Anal. Chem.* **65**: 3262–3266.

Lehninger, A. L., Nelson, D. L. and Cox, M. M. (1993) *Principles of Biochemistry*, 2nd. Edition. Worth Publishers, New York.

Lei, C. and Deng, J. (1996) Hydrogen peroxide sensor based on coimmobilized methylene green and horseradish peroxidase in the same montmorillonite-modified bovine serum albumin–glutaraldehyde matrix on a glassy carbon electrode surface. *Anal. Chem.* **69**: 3344–3349.

Lemké, U., Cammann, K., Kötter, C., Sundermeir, C. and Knoll, M. (1992) Multisensor array for pH, K^+, Na^+, and Ca^{2+} measurements based on coated-film electrodes. *Sens. Actuators* (B) **7**: 488–491.

Lemmo, A. V., Fisher, J. T., Geysen, H. M. and Rose, D. J. (1997) Characterization of an inkjet chemical microdispenser for combinatorial library synthesis. *Anal. Chem.* **69**: 543–551.

Lennox, J. C. and Murray, R. W. (1978) Chemically modified electrodes. 10. Electron spectroscopy for chemical analysis and alternating current voltammetry of glassy carbon-bound tetra(aminophenyl)porphyrins. *J. Amer. Chem. Soc.* **100**: 3710–3714.

Lerchi, M., Bakker, E., Rusterholz, B. and Simon, W. (1992) Lead-selective bulk optodes based on neutral ionophores with subnanomolar detection limits. *Anal. Chem.* **64**: 1534–1540.

Lerchi, M., Reitter, E., Simon, W., Pretsch, E., Chowdhury, D. A. and Kamata, S. (1994) Bulk optodes based on neutral dithiocarbamate ionophores with high selectivity for silver and mercury cations. *Anal. Chem.* **66**: 1713–1717.

Lerchi, M., Orsini, T., Cimerman, Z., Pretsch, E., Chowdhury, D. A. and Kamata, S. (1996) Selective optical sensing of silver ions in drinking water. *Anal. Chem.* **68**: 3210–3214.

Lev, O., Tsionsky, M., Rabinovich, L., Glezer, V., Sampath, S., Pankratov, I. and Gun, J. (1995) Organically modified sol–gel sensors. *Anal. Chem.* **67**: 22A–30A.

Levi, R., McNiven, S., Piletsky, S. A., Cheong, S-H., Yano, K. and Karube, I. (1997) Optical detection of chloramphenicol using molecularly imprinted polymers. *Anal. Chem.* **69**: 2017–2021.

Lewis, A. and Lieberman, K. (1991) The optical near field and analytical chemistry. *Anal. Chem.* **63**: 625A–638A.

Li, C., Davletov, B. A. and Sudhof, T. C. (1995) Distinct Ca^{2+} and Sr^{2+} binding properties of synaptotagmins. Definition of candidate Ca^{2+} sensors for the fast and slow components of neutrotransmitter release. *J. Biol. Chem.* **270**: 24898–24902.

Li, D. Q., Buscher, C. T. and Swanson, B. I. (1994) Synthesis, characterization and properties of covalently bound, self-assembled porphyrin multilayer films. *Chem. Mater.* **6**: 803–810.

Li, L. and Walt, D. R. (1995) Dual-analyte fiber-optic sensor for the simultaneous and continuous measurement of glucose and oxygen. *Anal. Chem.* **67**: 3746–3752.

Li, P. C. H., Stone, D. C. and Thompson, M. (1993) Flexural thin-rod acoustic wave devices as chemical sensors. *Anal. Chem.* **65**: 2177–2180.

Li, P. C. H. and Thompson, M. (1996) Mass sensitivity of the thin-rod acoustic wave sensor operated in flexural and extensional modes. *Anal. Chem.* **68**: 2590–2597.

Liang, Z., Chiem, N., Ocvirk, G., Tang, T., Fluri, K. and Harrison, D. J. (1996) Microfabrication of a planar absorbance and fluorescence cell for integrated capillary electrophoresis devices. *Anal. Chem.* **68**: 1040–1046.

Licht, S., Myung, N. and Sun, Y. (1996) A light addressable photoelectrochemical cyanide sensor. *Anal. Chem.* **68**: 954–959.

Liedberg, B., Lundstrom, I. and Stenberg, E. (1993) Principles of biosensing with an extended coupling matrix and surface plasmon resonance. *Sens. Actuators* (B) **11**: 63–72.

Lillard, S. J., Yeung, E. S. and McCloskey, M. A. (1996) Monitoring exocytosis and release from individual mast cells by capillary electrophoresis with laser-induced native fluorescence detection. *Anal. Chem.* **68**: 1897–1904.

Limoges, B. and Degrand, C. (1996) Ferrocenylethyl phosphate: an improved substance for the detection of alkaline phosphatase by cathodic stripping ion-exchange voltammetry. Application to the electrochemical enzyme affinity of avidin. *Anal. Chem.* **68**: 4141–4148.

Lin, J., Hart, S-J. and Kenny, J. E. (1996) Improved two-fiber probe for *in situ* spectroscopic measurements. *Anal. Chem.* **68**: 3098–3103.

Lin, Z., Yip, C. M., Joseph, I. S. and Ward, M. D. (1993) Operation of an ultrasensitive 30-MHz quartz crystal microbalance in liquids. *Anal. Chem.* **65**: 1546–1551.

Lin, Z. and Burgess, L. W. (1994) Chemically facilitated Donnan dialysis and its application in a fiber optic heavy metal sensor. *Anal. Chem.* **66**: 2544–2551.

Lin, Z., Booksh, K. S., Burgess, L. W. and Kowalski, B. R. (1994) A second-order fiber optic heavy metal sensor employing second-order tensorial calibration. *Anal. Chem.* **66**: 2552–2560.

Lin, Z. and Ward, M. D. (1995) The role of longitudinal waves in quartz microbalance applications in liquids. *Anal. Chem.* **67**: 685–693.

Lion-Dagan, M., Katz, E. and Willner, I. (1994a) Amperometric transduction of optical signals recorded by organized monolayers of photoisomerizable biomaterials on Au electrodes. *J. Amer. Chem. Soc.* **116**: 7931–7941.

Lion-Dagan, M., Katz, E. and Willner, I. (1994b) A bifunctional monolayer electrode consisting of 4-pyridyl sulfide and photoisomerizable spiropyran: photoswitchable electrical communication between the electrode and cytochrome *c*. *J. Chem. Soc. Chem. Commun.* 2741–2742.

Lisdat, F., Miura, N. and Yamazoe, N. (1996) NASICON-based solid-electrolyte cell as transducer for enzyme sensors. *Sens. Actuators* (B) **30**: 195–200.

Liu, D. H., Ge, K., Chen, K., Nie, L. H. and Yao, S. H. (1995) Clinical analysis of urea in human blood by coupling a surface acoustic wave sensor with urease extracted from pumpkin seeds. *Anal. Chim. Acta* **307**: 61–69.

Liu, H. Y. and Deng, J. Q. (1995) Amperometric glucose sensor using tetrathiafulvalene in Nafion gel as electron shuttle. *Anal. Chim. Acta* **300**: 65–70.

Liu, Z. T., Wang, Y., Kounaves, S. P. and Brush, E. J. (1993) Determination of organonitriles using enzyme-based selectivity mechanisms. 1. An ammonia gas sensing electrode-based sensor for benzonitrile. *Anal. Chem.* **65**: 3134–3136.

Liu, Z. T., Wang, Y., Kounaves, S. P. and Brush, E. J. (1995) Determination of organonitriles using enzyme-based selectivity mechanisms. 2. A nitrilase-modified glassy carbon microelectrode sensor for benzonitrile. *Anal. Chem.* **67**: 1679–1683.

Lobmaier, C., Schalkhammer, T., Hawa, G., Ecker, B. and Pittner, F. (1995) Photostructurized electrochemical biosensors for bioreactor control and measurement in body fluids. *J. Mol. Recognit.* **8**: 146–150.

Löfas, S. and Johnsson, B. (1990) A novel hydrogel matrix on gold surface plasmon resonance sensors for fast and efficient covalent immobilization of ligands. *J. Chem. Soc. Chem. Commun.* **21**: 1526–1528.

Löfas, S. (1995) Dextran-modified self-assembled monolayer surfaces for use in biointeraction analysis with surface plasmon resonance. *Pure Appl. Chem.* **67**: 829–834.

Löfas, S., Johnsson, B., Edström, Å., Hansson, A., Lindquist, G., Müller-Hillgren, R-M. and Stigh, L. (1995) Methods for site controlled coupling to carboxymethyldextran surfaces in surface plasmon resonance sensors. *Biosens. Bioelec.* **10**: 813–822.

Lorber, A., Faber, K. and Kowalski, B. P. (1997) Net analyte signal calculation in multivariate calibration. *Anal. Chem.* **69**: 1620–1626.

Lötzbeyer, T., Schuhmann, W. and Schmidt, H-L. (1994) Direct electron transfer between covalently immobilized microperoxidase MP-11 and a cystamine-modified gold electrode. *J. Electroanal. Chem.* **377**: 291–294.

Lowry, J. P., McAteer, K., El Atrash, S. S., Duff, A. and O'Neill, R. D. (1994) Characterization of glucose oxidase-modified poly(phenylenediamine)-coated electrodes *in vitro* and *in vivo*: homogeneous interference by ascorbic acid in hydrogen peroxide detection. *Anal. Chem.* **66**: 1754–1761.

Lu, B., Xie, J. M., Lu, C. L., Wu, C. R. and Wei, Y. (1995) Oriented immobilization of Fab′ fragments on silica surfaces. *Anal. Chem.* **67**: 83–87.

Lu, B., Smyth, M. R. and O'Kennedy, R. (1996) Oriented immobilization of antibodies and its application in immunoassays and immunosensors. *Analyst* **121**: 29R–32R.

Lu, C. and Czanderna, A. W. (eds) (1984) *Applications of Piezoelectric Quartz Crystal Microbalances, Methods, and Phenomena*, Vol. 7, Elsevier, New York.

Lu, H. S., Chang, D., Brankow, D. and Wen, D. (1995) Detection of neu differentiation factor with a biospecific affinity sensor during chromatography. *J. Chrom.* A **705**: 163–169.

Lübbers, D. W. and Opitz, N. (1975) The pCO_2-/pO_2-optode. New probe for measurement of partial pressure of carbon dioxide or partial pressure of oxygen in fluids and gases. *Z. Naturforsch.* (*Biosci.*) **30c**: 532–533.

Lübbers, D. W. (1992) Fluorescence based chemical sensors. In Turner, A. P. F. (ed.), *Advances in Biosensors*, Vol. 2. JAI Press, Greenwich, CT.

Lübbers, D. W. (1995) Optical sensors for clinical monitoring. *Acta Anaesthesiol. Scand.* Suppl. 104: 37–54.

Lundgren, J. S., Bekos, E. J., Wang, R. and Bright, F. V. (1994) Phase-resolved evanescent wave induced fluorescence. An *in situ* tool for studying heterogeneous interfaces. *Anal. Chem.* **66**: 2433–2440.

Lundgren, J. S. and Bright, F. V. (1996) Biosensor for the nonspecific determination of ionic surfactants. *Anal. Chem.* **68**: 3377–3381.

Lvovich, V. and Scheeline, A. (1997) Amperometric sensors for simultaneous superoxide and hydrogen peroxide detection. *Anal. Chem.* **69**: 454–462.

Mabrouk, P. A. (1996) Direct electrochemistry for the imidazole complex of microperoxidase-11 in dimethyl sulfoxide solution at naked electrode substrates including glassy carbon, gold, and platinum. *Anal. Chem.* **68**: 189–191.

MacCraith, B. D. (1993) Enhanced evanescent wave sensors based on sol–gel-derived porous glass coatings. *Sens. Actuators* (B) **11**: 29–34.

MacCraith, B. D., McDonagh, C. M., O'Keefe, G., Keyes, E. T., Vos, J. G., O'Kelly, B. and McGilp, J. F. (1993) Fibre-optic oxygen sensor based on fluorescence quenching of evanescent-wave-excited ruthenium complexes in sol–gel-derived porous coatings. *Analyst* **118**: 385–388.

Mădăras, M. B. and Buck, R. P. (1996) Miniaturized biosensors employing electropolymerized permselective films and their use for creatinine assays in human serum. *Anal. Chem.* **68**: 3832–3839.

Madou, M. J. and Morrison, S. R. (1989) *Chemical Sensing with Solid State Devices.* Academic Press, San Diego, CA.

Maidan, R. and Heller, A. (1992) Elimination of electrooxidizable interferant-produced currents in amperometric biosensors. *Anal. Chem.* **64**: 2889–2896.

Malem, F. and Mandler, D. (1993) Self-assembled monolayers in electroanalytical chemistry: application of ω-mercapto carboxylic acid monolayers for the electrochemical detection of dopamine in the presence of a high concentration of ascorbic acid. *Anal. Chem.* **65**: 37–41.

Malinski, T. and Taha, Z. (1992) A porphyrinic microsensor for nitric oxide. *Nature* **353**: 676–678.

Mandelis, A. and Christofides, C. (1993) *Physics, Chemistry and Technology of Solid State Gas Sensor Devices.* John Wiley and Sons, New York.

Manowitz, P., Stoecker, P. W. and Yacynych, A. M. (1995) Galactose biosensors using composite polymers to prevent interferences. *Biosens. Bioelec.* **10**: 359–370.

Marquardt, L. A., Arnold, M. A. and Small, G. W. (1993) Near-infrared spectroscopic measurement of glucose in a protein matrix. *Anal. Chem.* **65**: 3271–3278.

Martens, H. and Naes, T. (1989) *Multivariate Calibration.* John Wiley and Sons, New York.

Martens, N. and Hall, E. A. H. (1994) Model for an immobilized oxidase enzyme electrode in the presence of two oxidants. *Anal. Chem.* **66**: 2763–2770.

Martens, N., Hindle, A. and Hall, E. A. H. (1995) An assessment of mediators as oxidants for glucose oxidase in the presence of oxygen. *Biosens. Bioelec.* **10**: 393–403.

Martin, C. R. (1994) Nanomaterials: a membrane-based synthetic approach. *Science* **266**: 1961–1966.

Martin, C. R. and Foss, C. A. (1996) Chemically modified electrodes. In Kissinger, P. Y. and Heineman, W. R. (eds) *Laboratory Techniques in Electroanalytical Chemistry*, 2nd Edition. Marcel Dekker, New York.

Martin, S. J., Granstaff, V. E. and Frye, G. C. (1991) Characterization of a quartz crystal microbalance with simultaneous mass and liquid loading. *Anal. Chem.* **63**: 2272–2281.

Martin, S. J., Frye, G. C., Ricco, A. J. and Sēnturia, S. D. (1993) Effect of surface roughness on the response of thickness-shear mode resonators in liquids. *Anal. Chem.* **65**: 2910–2922.

Martin, S. J., Frye, G. C. and Sēnturia, S. D. (1994) Dynamics and response of polymer-coated surface acoustic wave devices: effect of viscoelastic properties and film resonance. *Anal. Chem.* **66**: 2201–2219.

Martin, S. J., Spates, J. J., Wessendorf, K. O., Schneider, T. W. and Huber, R. J. (1997) Resonator/oscillator response to liquid loading. *Anal. Chem.* **69**: 2050–2054.

Masel, R. (1996) *Principles of Adsorption and Reaction on Solid Surfaces*. John Wiley and Sons, New York.

Maskus, M., Pariente, F., Wu, Q., Toffanin, A., Shapleigh, J. P. and Abruña, H. D. (1996) Electrocatalytic reduction of nitric oxide at electrodes modified with electropolymerized films of $[Cr(v\text{-}typ)_2]^{3+}$ and their application to cellular NO determinations. *Anal. Chem.* **68**: 3128–3134.

Másson, M., Yun, K., Haruyama, T. Kobatake, E. and Aizawa, M. (1995) Quartz crystal microbalance bioaffinity sensor for biotin. *Anal. Chem.* **67**: 2212–2215.

Mathews, C. K. and van Holde, K. E. (1990) *Biochemistry*. Benjamin Cummings, Redwood City, CA.

Matsue, T. and Uchida, I. (1991) Ultramicroelectrodes as sensing devices for biological molecules. In Yamazoe, N., (ed.), *Chemical Sensor Technology*, Vol. 3. Kodansha, Tokyo and Elsevier, Amsterdam.

Matsuhiko, N., Menon, V. P. and Martin, C. R. (1995) Metal nanotubule membrane with electrochemically switchable ion-transport selectivity. *Science* **268**: 700–702.

Matsui, J., Mitoshi, Y., Doblhoff-Dier, O. and Takeuchi, T. (1995) A molecularly imprinted synthetic polymer receptor selective for atrazine. *Anal. Chem.* **67**: 4404–4408.

Matsumoto, K., Baeza, J. J. B. and Mottola, H. A. (1993) Continuous-flow sensor strategy comprising a rotating bioreactor and a stationary ring amperometric detector. *Anal. Chem.* **65**: 636–639.

Mayes, A. G. and Mosbach, K. (1996) Molecularly imprinted polymer beads: suspension polymerization using a liquid perfluorocarbon as the dispersing phase. *Anal. Chem.* **68**: 3769–3774.

McAlernon, P. and Slater, J. M. (1994) Behaviour of hydroquinone at surface active screen-printed carbon electrodes. *Anal. Proc. (London)* **31**: 365–368.

McCann, J. (1989) Exploiting biosensors. In Turner, A. P. F., Karube, I. and Wilson, G. S. (eds), *Biosensors: Fundamentals and Application*. Oxford University Press, New York. pp. 737–746.

McCarley, T. D. and McCarley, R. L. (1997) Toward the analysis of electrochemically modified self-assembled monolayers. Electrospray ionization mass spectrometry of organothiolates. *Anal. Chem.* **69**: 130–136.

McConnell, H. M., Owicki, J. C., Parce, J. W., Miller, D. L., Baxter, G. T., Wada, H. G. and Pitchford, S. (1992) The cytosensor microphysiometer: biological applications of silicon technology. *Science* **257**: 1906–1912.

McCreery, R. L. (1991) Carbon electrodes: structural effects on electron transfer kinetics. In Bard, A. J. (ed.), *Electroanalytical Chemistry*, Vol. 17. Marcel Dekker, New York.

McCreery, R. L. and Cline, K. K. (1996) Carbon electrodes. In Kissinger, P. Y. and Heineman, W. R. (eds) *Laboratory Techniques in Electroanalytical Chemistry*, 2nd Edition. Marcel Dekker, New York.

McCurley, M. F. (1994) An optical sensor using a fluorescent, swelling sensing element. *Biosens. Bioelec.* **9**: 527–533.

McCurley, M. F. and Glazier, S. A. (1995) Optical control of enzymatic conversion of sucrose to glucose by bacteriorhodopsin incorporated into self-assembled phosphatidyl choline vesicles. *Anal. Chim. Acta* **311**: 211–215.

McGown, L. B., Joseph, M. J., Pitner, J. B., Vonk, G. P. and Linn, C. P. (1995) The nucleic acid ligand: a new tool for molecular recognition. *Anal. Chem.* **67**: 663A–668A.

McLean, M. A., Stayton, P. S. and Sligar, S. G. (1993) Engineering protein orientation at surfaces to control macromolecular recognition events. *Anal. Chem.* **65**: 2676–2678.

McNeil, C. J., Athey, D., Ball, M., Ho, W. O., Krause, S., Armstrong, R. D., Wright, J. D. and Rawson, K. (1995) Electrochemical sensors based on impedance measurements of enzyme-catalyzed polymer dissolution: theory and application. *Anal. Chem.* **67**: 3928–3935.

Meier, P. C. and Zund, R. E. (1993) *Statistical Methods in Analytical Chemistry*. John Wiley and Sons, New York.

Mell, L. D. and Maloy, J. T. (1975) A model for the amperometric enzyme electrode obtained through digital simulation and applied to the immobilized glucose oxidase system. *Anal. Chem.* **47**: 299–307.

Mell, L. D. and Maloy, J. T. (1976) Amperometric response enhancement of the immobilized glucose oxidase enzyme electrode. *Anal. Chem.* **48**: 1597–1601.

Menon, V., and Martin, C. R. (1995) Fabrication and evaluation of nanoelectrode ensembles. *Anal. Chem.* **67**: 1920–1928.

Meruva, R. K. and Meyerhoff, M. E. (1996) Mixed potential response mechanism of cobalt electrodes toward inorganic phosphate. *Anal. Chem.* **68**: 2022–2026.

Messick, N. J., Kalivas, J. and Lang, P. M. (1996) Selectivity and related measures for *n*th-order data. *Anal. Chem.* **68**: 1572–1579.

Meyer, H., Drewer, H., Gründig, B. and Cammann, K. (1995) Two-dimensional imaging of O_2, H_2O_2, and glucose distributions by an array of 400 individually addressable microelectrodes. *Anal. Chem.* **67**: 1164–1170.

Meyerhoff, M. E. (1993) *In vivo* blood-gas and electrolyte sensors: progress and challenges. *Trends Anal. Chem.* **12**: 257–266.

Meyerhoff, M. E., Fu, B., Bakker, E., Yun, J-H. and Yang, V. C. (1996) Polyion-sensitive membrane electrodes for biomedical analysis. *Anal. Chem.* **68**: 168A–175A.

Mikkelsen, S. R. (1996) Electrochemical biosensors for DNA sequence detection. *Electroanalysis* **8**: 15–19.

Miles, D. T., Knedlik, A. and Wipf, D. O. (1997) Construction of gold microbead ultramicroelectrodes. *Anal. Chem.* **69**: 1240–1243.

Millan, K. M. and Mikkelsen, S. R. (1993) Sequence-selective biosensor for DNA based on electroactive hybridization indicators. *Anal. Chem.* **65**: 2317–2323.

Millan, K. M., Saraullo, A. and Mikkelsen, S. R. (1994) Voltammetric DNA biosensor for cystic fibrosis based on a modified carbon paste electrode. *Anal. Chem.* **66**: 2943–2948.

Mills, A. and Chang, Q. (1994) Colorimetric polymer film sensors for dissolved carbon dioxide. *Sens. Actuators* (B) **21**: 83–89.

Minunni, M., Mascini, M., Guilbault, G. G. and Hock, B. (1995) The quartz crystal microbalance as biosensor. A status report on its future. *Anal. Lett.* **28**: 749–764.

Mionetto, N., Marty, J-L. and Karube, I. (1994) Acetylcholinesterase in organic solvents for the detection of pesticides: biosensor application. *Biosens. Bioelec.* **9**: 463–470.

Mitler-Neher, S., Spinke, J., Liley, M., Nelles, G., Weisser, M., Back, R., Wenz, G. and Knoll, W. (1995) Spectroscopic and surface-analytical characterization of self-assembled layers on Au. *Biosens. Bioelec.* **10**: 903–916.

Mitsubayashi, K., Yokoyama, K., Takeuchi, T. and Karube, I. (1994) Gas-phase biosensor for ethanol. *Anal. Chem.* **66**: 3297–3302.

Moody, G. J., Oke, R. B. and Thomas, J. D. R. (1970) A calcium-sensitive electrode based on a liquid ion exchanger in a poly(vinyl chloride) matrix. *Analyst* **95**: 910–918.

Moody, G. J. (1992) Role of polymeric materials in the fabrication of ion-selective electrodes and biosensors. In Edelman, P. G. and Wang, J. (eds), *Biosensors and Chemical Sensors.* ACS Symposium Series #487, American Chemical Society, Washington, DC.

Moore, T., Nam, G. G., Pipes, L. C. and Coury, L. A. Jr. (1994) Chemically amplified voltammetric enzyme electrodes for oxidizable pharmaceuticals. *Anal. Chem.* **66**: 3158–3163.

Morf, W. E. (1981) *Principles of Ion-Selective Electrodes and of Membrane Transport.* Elsevier, Amsterdam.

Morf, W. E., Seiler, K., Sorenson, P. R. and Simon, W. (1989) New sensors based on carrier membrane systems: theory and practice. In Pungor, E. (ed.) *Ion-selective Electrodes*, Vol. 5. Pergamon, Oxford, UK.

Morgan, H. and Taylor, D. M. (1992) A surface plasmon resonance immunosensor based on the streptavidin–biotin complex. *Biosens. Bioelec.* **7**: 405–410.

Morgan, H., Pritchard, D-J. and Cooper, J. M. (1995) Photo-patterning of sensor surfaces with biomolecular structures: characterization using AFM and fluorescence microscopy. *Biosens. Bioelec.* **10**: 841–846.

Morozov, V. N., Seeman, N. C. and Kallenback, N. R. (1993) New methods for depositing and imaging molecules in scanning-tunneling-microscopy. *Scanning Microscopy* **7**: 757–779.

Morton, T. A., Myszka, D. G. and Chaiken, I. M. (1995) Interpreting complex binding kinetics from optical biosensors: a comparison of analysis by linearization, the integrated rate equation, and numerical integration. *Anal. Biochem.* **227**: 176–185.

Mosbach, K. and Danielsson, B. (1974) An enzyme thermistor. *Biochim. Biophys. Acta* **364**: 140–145.

Mosbach, K., Borgerud, A. and Scott, M. (1975) Determination of the heat changes in the proximity of immobilized enzymes with an enzyme thermistor and its use for the assay of metabolites. *Biochim. Biophys. Acta* **403**: 256–265.

Mosbach, K. (Ed.) (1988) Immobilized enzymes and cells. In Colowick, S. P. and Kaplan, N. O. (eds.) *Methods in Enzymology*, Vol. 137, Part D. Academic Press, New York.

Mosbach, K. (1994) Molecularly imprinted polymers as recognition elements. *Trends Biochem. Science* **19**: 9–14.

Mosbach, K. (1996) The emerging technique of molecular imprinting and its future impact on biotechnology. *Bio-Technology* **14**: 163–170.

Motta, N. and Guadalupe, A. R. (1994) Activated carbon paste electrodes for biosensors. *Anal. Chem.* **66**: 566–571.

Mottola, H. A. and Steinmetz, J. R. (eds) (1992) *Chemically Modified Surfaces*. Elsevier, Amsterdam.

Moussy, F. and Harrison, D. J. (1994) Prevention of the rapid degradation of subcutaneously implanted Ag/AgCl reference electrodes using polymer coatings. *Anal. Chem.* **66**: 674–679.

Moussy, F., Jakeway, S., Harrison, D. J. and Rajotte, R. V. (1994) *In vitro* and *in vivo* performance and lifetime of perfluorinated ionomer-coated glucose sensors after high-temperature curing. *Anal. Chem.* **66**: 3882-3888.

Muehlbauer, M. J., Guilbeau, E. J. and Towe, B. C. (1989) Model for a thermoelectric enzyme glucose sensor. *Anal. Chem.* **61**: 77–83.

Muehlbauer, M. J., Guilbeau, E. J. and Towe, B. C. (1990) Applications and stability of a thermoelectric enzyme sensor. *Sens. Actuators* (B) **2**: 223-232.

Mulchandani, A. and Bassi, A. S. (1995) Principles and applications of biosensors for bioprocess. *Crit. Revs Biotechnol.* **15**: 105–124.

Mulchandani, A., Wang, C-L. and Weetall, H. H. (1995) Amperometric detection of peroxides with poly(anilinomethylferrocene)-modified enzyme electrodes. *Anal. Chem.* **67**: 94–100.

Muldoon, M. T. and Stanker, L. H. (1997) Molecularly imprinted solid phase extraction of atrazine from beef liver extracts. *Anal. Chem.* **69**: 803–808.

Murakami, Y., Takeuchi, T., Yokoyama, K., Tamiya, E., Karube, I. and Suda, M. (1993) Integration of enzyme-immobilized column with electrochemical flow cell using micromachining techniques for a glucose detection system. *Anal. Chem.* **65**: 2731–2735.

Muratsugu, M., Ohta, F., Miya, Y., Hosokawa, T., Kurosawa, S., Kamo, N. and Ikeda, H. (1993) Quartz crystal microbalance for the detection of microgram quantities of human serum albumin: relationship between the frequency change and the mass of protein adsorbed. *Anal. Chem.* **65**: 2933–2937.

Murray, R. W. (1984) Chemically modified electrodes. In Bard, A. J. (ed.), *Electroanalytical Chemistry*, Vol. 13. Marcel Dekker, New York.

Murray, R. W., Dessy, R. E., Heineman, W. R., Janata, J. and Seitz, W. R. (eds) (1989) *Chemical Sensors and Microinstrumentation*. ACS Symposium Series #403, American Chemical Society, Washington, DC.

Murray, R. W. (1992) Introduction to the chemistry of molecularly designed electrode surfaces. In Murray, R. W. (ed.), *Molecular Design of Electrode Surfaces*. John Wiley and Sons, New York.

Murray, R. W. (1994a) Chemical sensors and molecular selectivity (Editorial). *Anal. Chem.* **66**: 505A.

Murray, R. W. (1994b) Chemical sensors and fast separations (Editorial). *Anal. Chem.* **66**: 625A.

Nagata, R., Clark, S. A., Yokoyama, K., Tamiya, E. and Karube, I. (1995a) Amperometric glucose biosensor manufactured by a printing technique. *Anal. Chim. Acta* **304**: 157–164.

Nagata, R., Yokoyama, K., Clark, S. A. and Karube, I. (1995b) A glucose sensor fabricated by the screen printing technique. *Biosens. Bioelec.* **10**: 261–267.

Nakamura, R. M., Kasahara, Y. and Rechnitz, G. A. (eds) (1992) *Immunochemical Assays and Biosensor Technology for the 1990's*. American Society for Microbiology, Washington, DC.

Nakanishi, K., Muguruma, H. and Karube, I. (1996) A novel method of immobilizing antibodies on a quartz crystal microbalance using plasma-polymerized films for immunosensors. *Anal. Chem.* **68**: 1695–1700.

Nahir, T. M., Clark, R. A. and Bowden, E. F. (1994) Linear sweep voltammetry of irreversible electron transfer in surface-confined species using the Marcus theory. *Anal. Chem.* **66**: 2595–2598.

Narang, U., Prasad, P. N. and Bright, F. V. (1994a) A novel protocol to entrap active urease in a tetraethoxysilane-derived-sol–gel thin-film architecture. *Chem. Mater.* **6**: 1596–1598.

Narang, U., Prasad, P. N., Bright, F. V., Ramanathan, K., Kumar, N. D., Malhotra, B. D., Kamalasanan, M. N. and Chandra, S. (1994b) Glucose biosensor based on a sol–gel platform. *Anal. Chem.* **66**: 3139–3144.

Narang, U., Gauger, P. R. and Ligler, F. S. (1997) Capillary-based displacement flow immunosensor. *Anal. Chem.* **69**: 1961–1964.

Neikov, A. and Sokolov, S. (1995) Generalized model for enzyme amperometric biosensors. *Anal. Chim. Acta* **307**: 27–36.

Netchiporouk, L. I., Shul'ga, A. A., Jaffrezic-Renault, N., Martelet, C., Olier, R. and Cespuglio, R. (1995) Properties of carbon fibre microelectrodes as a basis for enzyme biosensors. *Anal. Chim. Acta* **303**: 275–283.

Newman, J. D., White, S. F., Tothill, I. E. and Turner, A. P. F. (1995) Catalytic materials, membranes, and fabrication technologies suitable for the construction of amperometric biosensors. *Anal. Chem.* **67**: 4594–4599.

Newton, A. P. and Justice, J. B. Jr. (1994) Temporal response of microdialysis probes to local perfusion of dopamine and cocaine followed with one-minute sampling. *Anal. Chem.* **66**: 1468–1472.

Nicholson, C. and Phillips, J. M. (1979) Diffusion of anions and cations in the extracellular microenvironment of the brain. *J. Physiol.* **296**: 66P.

Nicolini, C., Lanzi, M., Accossato, P., Fanigliulo, A., Mattioli, F. and Martelli, A. (1995) A silicon-based biosensor for real-time toxicity testing in normal versus cancer liver cells. *Biosens. Bioelec.* **10**: 723–733.

Nicolini, C. (1995) From neural chip and engineered biomolecules to bioelectronic devices: an overview. *Biosens. Bioelec.* **10**: 105–127.

Nicolsky, B. P., Schulz, M. M., Belijustin, A. A. and Lev, A. A. (1967) Recent developments in the ion-exchange theory of the glass electrode and its application in the chemistry of glass. In Eisenman, G. (ed.) *Glass Electrodes for Hydrogen and Other Cations*. Marcel Dekker, New York.

Nie, S., Chiu, D. T. and Zare, R. N. (1995) Real-time detection of single molecules in solution by confocal fluorescence microscopy. *Anal. Chem.* **67**: 2849–2857.

Nie, S. and Emory, S. R. (1997) Probing single molecules and single nanoparticles by surface-enhanced Raman scattering. *Science* **275**: 1102–1106.

Nikolelis, D. P. and Siontorou, C. G. (1995) Bilayer lipid membranes for flow-injection monitoring of acetylcholine, urea and penicillin. *Anal. Chem.* **67**: 936–944.

Nikolelis, D. P., Siontorou, C. G., Krull, U. J. and Katrivanos, P. L. (1996) Ammonium ion minisensors from self-assembled bilayer lipid membranes using gramicidin as an ionophore. Modulation of ammonium selectivity by platelet-activating factor. *Anal. Chem.* **68**: 1735–1741.

Nilsson, P., Persson, B., Uhlen, M. and Nygren, P. A. (1995) Real-time monitoring of DNA manipulations using biosensors technology. *Anal. Biochem.* **224**: 400–408.

Nishida, K., Sakakida, M., Ichinose, K., Uemura, T., Uehara, M., Kajiwara, K., Miyata, T., Shichiri, M., Ishihara, K. and Nakabayashi, N. (1995) Development of a ferrocene-needle-type glucose sensor covered with newly designed biocompatible membrane, 2-methacryloyloxethyl phosphorylcholine-co-n-butyl methacrylate. *Med. Prog. Technol.* **21**: 91–102.

Nishizawa, M., Matsue, T. and Uchida, I. (1992) Penicillin sensor based on a microarray electrode coated with pH-responsive polypyrrole. *Anal. Chem.* **64**: 2642–2644.

Nishizawa, M., Menon, V. P. and Martin, C. R. (1995) Metal nanotuble membranes with electrochemically switchable ion-transport selectivity. *Science* **268**: 700–702.

Nivens, D. E., Chambers, J. Q., Anderson, T. R. and White, D. C. (1993) Long-term, on line monitoring of microbial biofilms using a quartz crystal microbalance. *Anal. Chem.* **65**: 65–69.

Niwa, O., Xu, Y., Halsall, H. B. and Heineman, W. R. (1993) Small-volume voltammetric detection of 4-aminophenol with interdigitated array electrodes and its application to electrochemical enzyme immunoassay. *Anal. Chem.* **65**: 1559–1563.

Niwa, O. and Tabei, H. (1994) Voltammetric measurements of reversible and quasi-reversible redox species using carbon film based interdigitated array microelectrodes. *Anal. Chem.* **66**: 285–289.

Niwa, O. (1995) Electroanalysis with interdigitated array microelectrodes. *Electroanalysis* **7**: 606–613.

Niwa, O. and Morita, M. (1996) Carbon film-based interdigitated ring array electrodes as detectors in radial flow cells. *Anal. Chem.* **68**: 355–359.

Niwa, O., Torimitsu, K., Morita, M., Osborne, P. and Yamamoto, K. (1996) Concentration of extracellular L-glutamate released from cultured nerve cells measured with a small-volume online sensor. *Anal. Chem.* **68**: 1865–1870.

Noble, D. (1994) Back into the storm: reanalyzing health effects of the Gulf war. *Anal. Chem.* **66**: 805A–808A.

Nolan, M. A., Tan, S. H. and Kounaves, S. P. (1997) Fabrication and characterization of a solid state reference electrode for electroanalysis of natural waters with ultramicroelectrodes. *Anal. Chem.* **69**: 1244–1247.

Nomura, S., Nakao, M., Nakanishi, T., Takamatsu, S. and Tomita, K. (1997) Real-time imaging of microscopic pH distribution with a two-dimensional pH-imaging apparatus. *Anal. Chem.* **69**: 977–981.

Nowall, W. B. and Kuhr, W. G. (1995) Electrocatalytic surface for the oxidation of NADH and other anionic molecules of biological significance. *Anal. Chem.* **67**: 3583–3588.

O'Connor, K. D., Arrigan, D. W. M. and Svehla, G. (1995) Calixarenes in electroanalysis. *Electroanalysis* **7**: 205–215.

Ogert, R. A., Brown, J. E., Singh, B. R., Shriver-Lake, L. C. and Ligler, F. S. (1992) Detection of *Clostridium botulinum* toxin A using a fiber optic-based biosensor. *Anal. Biochem.* **205**: 1990–1994.

Oglesby, D. M., Upchurch, B. T., Leighty, B. D., Collman, J. P., Zhang, X. and Herrmann, P. C. (1994) Surface acoustic wave oxygen sensor. *Anal. Chem.* **66**: 2745–2751.

Ohara, T. J., Rajagopalan, R. and Heller, A. (1993) Glucose electrodes based on cross-linked bis(2,2′-bipyridine)chloroosmium (+/2+) complexed poly (1-vinylimidazole) films. *Anal. Chem.* **65**: 3512–3517.

Ohara, T. J., Rajagopalan, R. and Heller, A. (1994) "Wired" enzyme electrodes for amperometric determination of glucose or lactate in the presence of interfering substances. *Anal. Chem.* **66**: 2451–2457.

Ohashi, E. and Karube, I. (1995) Development of a thin membrane glucose sensor using beta-type crystalline chitin for implantable biosensor. *J. Biotechnol.* **40**: 3–19.

Okazaki, I., Hasegawa, Y., Shinohara, Y., Kamasaki, T., Bhikhabhai, R. (1995) Determination of the interactions between lectins and glycoproteins by surface plasmon resonance. *J. Mol. Recognit.* **8**: 95–99.

Oldham, K. B. and Myland, J. C. (1994) *Fundamentals of Electrochemical Science.* Academic Press, San Diego, CA.

Oldham, K. B. (1996) Effect of diffusion coefficient diversity on steady-state voltammetry when homogeneous equilibria and migration are encountered. *Anal. Chem.* **68**: 4173–4179.

Ong, S., Cal, S-J., Bernal, C., Rhee, D., Qiu, X. and Pidgeon, C. (1994) Phospholipid immobilization on solid surfaces. *Anal. Chem.* **66**: 782–792.

Oroszlan, P., Duveneck, G. L., Ehrat, M. and Widmer, H. M. (1993a) Fibre-optic atrazine immunosensor. *Sens. Actuators* (B)**11**: 301–305.

Oroszlan, P., Thommen, C., Wehrli, M., Duveneck, G. and Ehrat, M. (1993b) Automated optical sensing system for biochemical assays: a challenge for ELISA? *Anal. Methods Instrum.* **1**: 43–51.

Osborn, T. D. and Yager, P. (1995) Modeling success and failure of Langmuir–Blodgett transfer of phospholipid bilayers to silicon dioxide. *Biophys. J.* **68**: 1346–1373.

Osborne, M. D. and Girault, H. H. (1995) The micro water/1,2-dichloroethane interface as a transducer for creatinine assay. *Mikrochim. Acta* **17**: 175–185.

O'Toole, R. P., Burns, S. G., Bastiaans, G. J. and Porter, M. D. (1992) Thin aluminum nitride film resonators: miniaturized high sensitivity mass sensors. *Anal. Chem.* **64**: 1289–1294.

Otto, M. and Thomas, J. D. R. (1985) Model studies on multiple channel analysis of free magnesium, calcium, sodium, and potassium at physiological concentration levels with ion-selective electrodes. *Anal. Chem.* **57**: 2647–2651.

Owaku, K., Goto, M., Ikariyama, Y. and Aizawa, M. (1995) Protein A Langmuir–Blodgett film for antibody immobilization and its use in optical immunosensing. *Anal. Chem.* **67**: 1613–1616.

Owicki, J. C., Parce, J. W., Kereso, K. M., Sigal, G. B., Muir, V. C., Venter, J. C., Fraser, C. M. and McConnell, H. M. (1990) Continuous monitoring of receptor-mediated changes in the metabolic rates of living cells. *Proc. Natl Acad. Sci. USA* **87**: 4007–4011.

Owicki, J. C. and Parce, J. E. (1992) Biosensors based on the energy metabolism of living cells: the physical chemistry and cell biology of extracellular acidification. *Biosens. Bioelec.* **7**: 255–272.

Ozawa, S., Miyagi, H., Shibata, Y., Oki, N., Kunitake, T. and Keller, W. E. (1996) Anion-selective electrodes based on long-chain methyltrialkylammonium salts. *Anal. Chem.* **68**: 4149–4152.

Pace, S. J. and Hamerslag, J. D. (1992) Thick-film multilayer ion sensors for biomedical applications. In Edelman, P. G. and Wang, J. (eds), *Biosensors and Chemical Sensors: Optimizing Performance Through Polymeric Materials.* ACS Symposium Series #487, American Chemical Society, Washington, DC.

Paleček, E. and Fojita, M. (1994) Differential pulse voltammetric determination of RNA at the picomole level in the presence of DNA and nucleic acid components. *Anal. Chem.* **66**: 1566–1571.

Palet, C., Munoz, M., Daunert, S., Bachas, L. G. and Valiente, M. (1993) Vitamin B_{12} derivatives as anion carriers in transport through supported liquid membranes and correlation with their behavior in ion-selective electrodes. *Anal. Chem.* **65**: 1533–1536.

Palmisano, F. and Zambonin, P. G. (1993) Ascorbic acid interferences in hydrogen peroxide detecting biosensors based on electrochemically immobilized enzymes. *Anal. Chem.* **65**: 2690–2692.

Palmisano, F., Centonze, D., Guerrieri, A. and Zambonin, P. G. (1993) An interference-free biosensor based on glucose oxidase electrochemically immobilized in a non-conducting poly(pyrrole) film for continuous subcutaneous monitoring of glucose through microdialysis sampling. *Biosens. Bioelec.* **8**: 393–399.

Palmisano, F., Malitesta, C., Centonze, D. and Zambonin, P. G. (1995a) Correlation between permselectivity and chemical structure of overoxidized polypyrrole membranes used in electro-produced enzyme biosensors. *Anal. Chem.* **67**: 2207–2211.

Palmisano, F., Guerrieri, A., Quinto, M. and Zambonin, P. G. (1995b) Electrosynthesized bilayer of polymeric membrane for effective elimination of electroactive interferents in amperometric biosensors. *Anal. Chem.* **67**: 1005–1009.

Palmisano, F., Centonze, D., Malitesta, C. and Zambonin, P. G. (1995c) Electrochemical immobilization of enzymes on conducting organic salt electrodes: preparation of an oxygen-independent and interference-free glucose biosensor. *J. Electroanal. Chem.* **381**: 235–237.

Pan, S., Chung, H., Arnold, M. A. and Small, G. W. (1996) Near-infrared spectroscopic measurement of physiological glucose levels in variable matrices of protein and triglycerides. *Anal. Chem.* **68**: 1124–1135.

Pandey, P. C. and Weetall, H. H. (1994) Application of photochemical reaction in electrochemical detection of DNA intercalation. *Anal. Chem.* **66**: 1236–1241.

Pandey, P. C. and Weetall, H. H. (1995a) Detection of aromatic compounds based on DNA intercalation using an evanescent wave biosensor. *Anal. Chem.* **67**: 787–792.

Pandey, P. C. and Weetall, H. H. (1995b) Peroxidase- and tetracyanoquinodimethane-modified graphite paste electrodes for the measurement of glucose/lactate/glutamate using enzyme-packed bed reactor. *Anal. Biochem.* **224**: 428–433.

Pankratov, I. and Lev, O. (1995) Sol–gel-derived renewable-surface biosensors. *J. Electroanal. Chem.* **393**: 35–41.

Pantano, P., Morton, T. H. and Kuhr, W. G. (1991) Enzyme-modified carbon-fiber microelectrodes with millisecond response times. *J. Amer. Chem. Soc.* **113**: 1832–1833.

Pantano, P. and Kuhr, W. G. (1993) Dehydrogenase-modified carbon fiber microelectrodes for the measurement of neurotransmitter dynamics. 2. Covalent modification utilizing avidin–biotin technology. *Anal. Chem.* **65**: 623–630.

Pantano, P. and Kuhr, W. G. (1995) Enzyme-modified microelectrodes for *in vivo* neurochemical measurements. *Electroanalysis* **7**: 405–416.

Pantano, P. and Walt, D. R. (1995) Analytical applications of optical imaging fibers. *Anal. Chem.* **67**: 481A–487A.

Paras, C. D. and Kennedy, R. T. (1995) Electrochemical detection of exocytosis at single rat melanotrophs. *Anal. Chem.* **67**: 3633–3637.

Parce, J. W., Owicki, J. C., Kereso, K. M., Sigal, G. B., Wada, H. G., Muir, V. C., Bousse, L. J., Ross, K. L., Sikie, B. I. and McConnell, H. M. (1989) Detection of cell affecting agents with a silicon biosensor. *Science* **246**: 243–247.

Pariente, F., Lorenzo, E. and Abruña, H. D. (1994) Electrocatalysis of NADH oxidation with electropolymerized films of 3,4-dihydroxybenzaldehyde. *Anal. Chem.* **66**: 4337–4344.

Pariente, F., Lorenzo, E., Tobalina, F. and Abruña, H. D. (1995) Aldehyde biosensor based on the determination of NADH enzymatically generated by aldehyde dehydrogenase. *Anal. Chem.* **67**: 3936–3944.

Pariente, F., Tobalina, F., Darder, M., Lorenzo, E. and Abruña, H. D. (1996) Electrodeposition of redox-active films of dihydroxy benzaldehydes and related analogs and their electrocatalytic activity towards NADH oxidation. *Anal. Chem.* **68**: 3135–3142.

Patonay, G. and Antoine, M. D. (1991) Near-infrared fluorogenic labels: new approach to an old problem. *Anal. Chem.* **63**: 321A–327A.

Pease, A. C., Solas, D., Sullivan, E. J., Cronin, M. T., Holmes, C. P. and Fodor, S. P. (1994) Light-generated oligonucleotide arrays for rapid DNA sequence analysis. *Proc. Natl Acad. Sci. USA* **91**: 5022–5026.

Pecsek, J. J. and Leigh, I. E. (1994) *Chemically Modified Surfaces*. The Royal Society of Chemistry, Cambridge, UK.

Perry, A. A. and Somorjai, G. A. (1994) Characterization of organic surfaces. *Anal. Chem.* **66**: 403A-415A.

Petit-Dominguez, M. D., Shen, H., Heineman, W. R. and Seliskar, C. J. (1997) Electrochemical behavior of graphite electrodes modified by spin-coating with sol–gel-entrapped ionomers. *Anal. Chem.* **69**: 703–710.

Pettit, C. and Kauffmann, J. M. (1995) New carbon paste electrode for the development of biosensors. *Anal. Proc. (London)* **32**: 11–12.

Phillips, D. (1994) Luminescence lifetimes in biological systems. *Analyst* **119**: 543–550.

Pidgeon, C., Ong, S., Choi, H. and Liu, H. (1994) Preparation of mixed ligand immobilized artificial membranes for predicting drug binding to membranes. *Anal. Chem.* **66**: 2701–2709.

Piehler, J., Brecht, A. and Gauglitz, G. (1996) Affinity detection of low molecular weight analytes. *Anal. Chem.* **68**: 139–143.

Pierce, D. T., Unwin, P. R. and Bard, A. J. (1992) Scanning electrochemical microscopy. 17. Studies of enzyme-mediator kinetics for membrane- and surface-immobilized glucose oxidase. *Anal. Chem.* **64**: 1795–1804.

Pierce, D. T. and Bard, A. J. (1993) Scanning electrochemical microscopy. 23. Reaction localization of artificially patterned and tissue-bound enzymes. *Anal. Chem.* **65**: 3598–3604.

Pihel, K., Schroeder, T. J. and Wightman, R. M. (1994) Rapid and selective cyclic voltammetric measurements of epinephrine and norepinephrine as a method to measure secretions from single bovine adrenal medullary cells. *Anal. Chem.* **66**: 4532–4537.

Pihel, K., Hsiehs, S., Jorgenson, J. W. and Wightman, R. M. (1995) Electrochemical detection of histamine and 5-hydroxytryptamine at isolated mast cells. *Anal. Chem.* **67**: 4514–4521.

Pihel, K., Walker, Q. D. and Wightman, R. M. (1996) Overoxidized polypyrrole-coated carbon fiber microelectrodes for dopamine measurements with fast-scan cyclic voltammetry. *Anal. Chem.* **68**: 2084–2087.

Pishko, M. V., Michael, A. C. and Heller, A. (1991) Amperometric glucose microelectrodes prepared through immobilization of glucose oxidase in redox hydrogels. *Anal. Chem.* **63**: 2268–2272.

Piunno, P. A. E., Krull, U. J., Hudson, R. H. E., Damha, M. J. and Cohen, H. (1995) Fiber-optic DNA sensor for fluorometric nucleic acid determination. *Anal. Chem.* **67**; 2635–2643.

Plant, A. L. (1993) Self-assembled phospholipid/alkanethiol biomimetic bilayers and gold. *Langmuir* **9**: 2764–2767.

Plant, A. L., Brigham–Burke, M., Petrella, E. C. and O'Shannessy, D. J. (1995) Phospholipid/alkanethiol bilayers for cell-surface receptor studies by surface plasmon resonance. *Anal. Biochem.* **226**: 342–348.

Polzius, R., Schneider, T., Bier, F. F. and Bilitewski, U. (1996) Optimization of biosensing using grating couplers: immobilization on tantalum oxide waveguides. *Biosens. Bioelec.* **11**: 503–514.

Popp, J., Silber, A., Braeuchle, C. and Hampp, N. (1995) Sandwich enzyme membranes for amperometric multi-biosensor applications: improvement of linearity and reduction of chemical cross-talk. *Biosens. Bioelec.* **10**: 243–249.

Poscio, P., Emery, Y., Bauerfeind, P. and Depeursinge, C. (1994) *In vivo* measurements of dye concentration using an evanescent-wave optical sensor. *Med. Biol. Eng. Comput.* **32**: 362–366.

Postlethwaite, T. A., Hutchison, J. E., Murray, R., Fosset, B. and Amatore, C. (1996) Interdigitated array electrode as an alternative to the rotated ring-disk electrode for determination of the reaction products of dioxygen reduction. *Anal. Chem.* **68**: 2951–2958.

Pötter, W., Dumschat, C. and Cammann, K. (1995) Miniaturized reference electrode based on a perchlorate-selective field effect transistor. *Anal. Chem.* **67**: 4586–4588.

Pravda, M., Adeyoju, O., Iwuoha, E. I., Vos, J. G., Smyth, M. R. and Vytřas, K. (1995a) Amperometric glucose biosensors based on an osmium (2+/3+) redox polymer-mediated electron transfer at carbon paste electrodes. *Electroanalysis* **7**: 619–625.

Pravda, M., Jungar, C. M., Iwuoha, E. I., Smyth, M. R., Vytřas, K. and Ivaska, A. (1995b) Evaluation of amperometric glucose biosensors based on co-immobilization of glucose oxidase with an osmium redox polymer in electrochemically generated polyphenol films. *Anal. Chim. Acta* **304**: 127–138.

Preininger, C., Klimant, I. and Wolfbeis, O. S. (1994) Optical fiber sensor for biological oxygen demand. *Anal. Chem.* **66**: 1841–1846.

Preston, J. P. and Nieman, T. A. (1996) An electrogenerated chemiluminescence probe and its application utilizing Tris (2,2′-bipyridyl)ruthenium(II) and luminol chemiluminescence without a flowing stream. *Anal. Chem.* **68**: 966–970.

Preuss, M. and Hall, E. A. H. (1995) Mediated herbicide inhibition in a PET biosensor. *Anal. Chem.* **67**: 1940–1949.

Pritchard, D. J., Morgan, H. and Cooper, J. M. (1995a) Patterning and regeneration of surfaces with antibodies. *Anal. Chem.* **67**: 3605–3607.

Pritchard, D. J., Morgan, H. and Cooper, J. M. (1995b) Micron-scale patterning of biological molecules. *Angew. Chem.* (*Int. Ed. Eng.*) **34**: 91–93.

Qian, P., Matsuda, M. and Miyashita, T. (1993) Chiral molecular recognition in polymer Langmuir–Blodgett films containing axially chiral binaphthyl groups. *J. Amer. Chem. Soc.* **115**: 5624–5628.

Quinn, C. P., Pathak, C. P., Heller, A. and Hubbell, J. A. (1995a) Photo-crosslinked copolymers of 2-hydroxyethyl methacrylate poly(ethylene glycol) tetra-acrylate and ethylene dimethacrylate for improving biocompatibility of biosensors. *Biomaterials* **16**: 389–396.

Quinn, C. P., Pishko, M. V., Schmidtke, D. W., Ishikawa, M., Wagner, J. G., Raskin, P., Hubbell, J. A. and Heller, A. (1995b) Kinetics of glucose delivery to subcutaneous tissue in rats measured with 0.3-mm amperometric microsensors. *Amer. J. Physiol.* **269**: E155–161.

Rabbany, S. Y., Piervincenzi, R., Judd, L., Kusterbeck, A. W., Bredehorst, R., Hakansson, K. and Ligler, F. S. (1997) Assessment of heterogeneity in antibody–antigen displacement reactions. *Anal. Chem.* **69**: 175–182.

Raether, H. (1977) Surface plasma oscillations and their applications. In Hass, G., Francombe, M. and Hoffman, R. W. (eds) *Physics of Thin Films*, Vol. 9. Academic Press, New York.

Raether, H. (1980) *Excitation of Plasmons and Interband Transitions by Electrons*. Springer Tracts in Modern Physics, Vol. 88. Springer-Verlag, Berlin.

Ragsdale, S. R., Lee, J. and White, H. S. (1997) Analysis of the magnetic force generated at a hemispherical microelectrode. *Anal. Chem.* **69**: 2070–2076.

Ramanathan, K., Sundaresan, N. S. and Malhotra, B. D. (1995) Ion-exchanged polypyrrole-based glucose biosensor: enhanced loading and response. *Electroanalysis* **7**: 579–582.

Ramsden, J. J., Bachmanova, G. I. and Archakov, A. I. (1996) Immobilization of proteins to lipid bilayers. *Biosens. Bioelec.* **11**: 523–528.

Ramström, O., Ye, L. and Mosbach, K. (1996) Artificial antibodies to corticosteroids prepared by molecular imprinting. *Chem. Biol.* **3**: 471–477.

Rank, M., Gram, J. and Danielsson, B. (1993) Industrial on-line monitoring of penicillin V, glucose and ethanol using a split-flow modified thermal biosensor. *Anal. Chim. Acta* **281**: 521–526.

Rank, M., Gram, J., Nielsen, K. S. and Danielsson, B. (1995) On-line monitoring of ethanol, acetaldehyde and glycerol during industrial fermentations with *Saccharomyces cerevisiae*. *Appl. Microbiol. Biotechnol.* **42**: 813–817.

Ratcliff, B. B., Klancke, J. W., Koppang, M. D. and Engstrom, R. C. (1996) Microderivatization of anodized glassy carbon. *Anal. Chem.* **68**: 2010–2014.

Rawson, D. M., Willmer, A. J. and Turner, A. P. F. (1989) Whole cell biosensors for environmental monitoring. *Biosensors* **4**: 299–312.

Reach, G. and Wilson, G. S. (1992) Can continuous glucose monitoring be used for the treatment of diabetes? (Report). *Anal. Chem.* **64**: 381A–386A.

Redepenning, J., Miller, B. R. and Burnham, S. (1994) Reversible voltammetric response of electrodes coated with permselective redox films. *Anal. Chem.* **66**: 1560–1565.

Reiken, S. R., Van Wie, B. J., Sutisna, H., Moffett, D. F., Koch, A. R., Silber, M. and Davis, W. C. (1996) Bispecific antibody modification of nicotinic acetylcholine receptors for biosensing. *Biosens. Bioelec.* **11**: 91–102.

Reinhoudt, D. N. (1992) Molecular materials for the transduction of chemical information into electronic signals by chemical field-effect transistors. In Edelman, P. G. and Wang, J. (eds), *Biosensors and Chemical Sensors*. ACS Symposium Series #487, American Chemical Society, Washington, DC.

Reinhoudt, D. N., Engbersen, J. F. J., Brzozka, Z., van den Vlekkert, H. H., Honig, G. W. N., Holterman, H. A. J. and Verkerk, U. H. (1994) Development of durable K^+-selective chemically modified field effect transistors with functionalized polysiloxane membranes. *Anal. Chem.* **66**: 3618–3623.

Renken, J., Dahint, R., Grunze, M. and Jossé, F. (1996) Multifrequency evaluation of different immunosorbents on acoustic plate mode sensors. *Anal. Chem.* **68**: 176–182.

Rhodes, R. K., Shults, M. C. and Updike, S. J. (1994) Prediction of pocket-portable and implantable glucose enzyme electrode. *Anal. Chem.* **66**: 1520–1529.

Ricco, A. S., Martin, S. J., Niemcyzk, T. M. and Frye, G. C. (1989) Liquid-phase sensors based on acoustic plate mode devices. In Murray, R. W., Dessy, R. E., Heineman, W. R., Janata, J. and Seitz, W. R. (eds) *Chemical Sensors and Microinstrumentation*, ACS Symposium Series #403, American Chemical Society, Washington, DC.

Richter, P., Ruiz, B. L., Sánchez-Cabezudo, M. and Mottola, H. A. (1996) Immobilized enzyme reactors. Diffusion/convection, kinetics, and a comparison of packed-column and rotating bioreactors for use in a continuous flow system. *Anal. Chem.* **68**: 1701–1705.

Rickert, J., Göpel, W., Beck, W., Jung, G. and Heiduschka, P. (1996) A 'mixed' self-assembled monolayer for an impedimetric immunosensor. *Biosens. Bioelec.* **11**: 757–768.

Rickert, J., Brecht, A. and Göpel, W. (1997) QCM operation in liquids: constant sensitivity during formation of extended protein multilayers by affinity. *Anal. Chem.* **69**: 1441–1448.

Rieger, P. H. (1994) *Electrochemistry*, 2nd Edition. Chapman & Hall, New York.

Riklin, A., Katz, E., Willner, I., Stocker, A. and Buckmann, A. F. (1995a) Improving enzyme–electrode contacts by redox modification of cofactors. *Nature* **376**: 672–675.

Riklin, A. and Willner, I. (1995) Glucose and acetylcholine sensing multilayer enzyme electrodes of controlled enzyme layer thickness. *Anal. Chem.* **67**: 4118–4126.

Rivot, J. P., Cespuglio, R., Puig, S., Jouvet, M. and Besson, J. M. (1995) *In vivo* electrochemical monitoring of serotonin in spinal dorsal horn with Nafion-coated multi-carbon fiber electrodes. *J. Neurochem.* **65**: 1257–1263.

Roberts, G. G. (1990) *Langmuir–Blodgett Films*. Plenum Press, New York.

Roberts, M. A. and Durst, R. A. (1995) Investigation of liposome-based immunomigration sensors for the detection of polychlorinated biphenyls. *Anal. Chem.* **67**: 482–491.

Roberts, M. A., Rossier, J. S., Bercier, P. and Girault, H. (1997) UV laser machined polymer substrates for the development of microdiagnostic systems. *Anal. Chem.* **69**: 2035–2042.

Robinson, G., Leech, D. and Smyth, M. R. (1995) Electrically "wired" tyrosinase enzyme inhibition electrode for the detection of respiratory poisons. *Electroanalysis* **10**: 952–957.

Robinson, T. E. and Justice, J. B., Jr. (eds) (1991) *Microdialysis in Neuroscience*. Elsevier, Amsterdam.

Rodahl, M., Höök, F. and Kasemo, B. (1996) QCM operation in liquids: an explanation of measured variations in frequency and *Q* factor with liquid conductivity. *Anal. Chem.* **68**: 2219–2227.

Rogers, K. R., Fernando, J. C., Thompson, R. G., Valdes, J. J. and Eldefrawi, M. E. (1992) Detection of nicotinic receptor ligands with a light addressable potentiometric sensor. *Anal. Biochem.* **202**: 111–116.

Rogers, K. R., Mulchandani, A. and Zhou, W. (eds) (1995) *Biosensor and Chemical Sensor Technology. Process Monitoring and Control.* ACS Symposium Series #613, American Chemical Society, Washington, DC.

Rogers, K. R. (1995) Biosensors for environmental applications. *Biosens. Bioelec.* **10**: 533–541.

Rogers, K. R. and Williams, L. R. (1995) Biosensors for environmental monitoring: a regulatory perspective. *Trends Anal. Chem.* **14**: 289–294.

Rohm, I., Kuennecke, W. and Bilitewski, U. (1995) UV-polymerizable screen printed enzyme pastes. *Anal. Chem.* **67**: 2304–2307.

Rosenberg, A. and Yeung, E. S. (1994) Laser-based particle-counting microimmunoassay for the analysis of single human erythrocytes. *Anal. Chem.* **66**: 1771–1776.

Rosenberg, E., Krska, R. and Kellner, R. (1994) Theoretical and practical response evaluation of a fiber optic sensor for chlorinated hydrocarbons in water. *Fresenius' J. Anal. Chem.* **348**: 560–562.

Rosenzweig, Z. and Kopelman, R. (1995) Development of a submicrometer optical fiber oxygen sensor. *Anal. Chem.* **67**: 2650–2654.

Rosenzweig, Z. and Kopelman, R. (1996) Analytical properties and sensor size effects of a micrometer-sized optical fiber glucose biosensor. *Anal. Chem.* **68**: 1408–1413.

Ross, C. B., Sun, L. and Crooks, R. M. (1993) Scanning probe lithography. 1. Scanning tunneling microscope induced lithography of self-assembled n-alkanethiol monolayer resists. *Langmuir* **9**: 632–636.

Rouillon, R., Sole, M., Carpentier, R. and Marty, J. L. (1995) Immobilization of thyalkoids in poly(vinyl alcohol) for the detection of herbicides. *Sens. Actuators* (B) **27**: 477–479.

Ruegg, N., Williams, G., Birch, A., Robinson, J. A., Schlatter, D. and Huber, W. (1995) Mutagenesis of immunoglobulin-like domains from the extracellular human interferon-gamma receptor alpha chain and their recognition by neutralizing antibodies monitored by surface plasmon resonance technology. *J. Immunol. Methods* **183**: 95–101.

Russell, A. A., Doughty, D. H., Ballentine, D. S., Jr., and Hart, R. (1994) Frequency and attenuation response of acoustic plate mode devices coated with porous oxide films. *Anal. Chem.* **66**: 3108–3116.

Sabatani, E., Rubinstein, I., Maoz, R. and Sagiv, J. (1987) Organized self-assembling monolayers on electrodes. Part I. Octadecyl derivatives on gold. *J. Electroanal. Chem.* **219**: 365–371.

Sabatani, E. and Rubinstein, I. (1987) Organized self-assembling monolayers on electrodes. 2. Monolayer-based ultramicroelectrodes for the study of very rapid electrode kinetics. *J. Phys. Chem.* **91**: 6663-6666.

Saby, C., Mizutani, F. and Yabuki, S. (1995) Glucose sensor based on carbon-paste electrode incorporating poly(ethylene glycol)-modified glucose oxidase and various mediators. *Anal. Chim. Acta* **304**: 33–39.

Sackmann, E. (1996) Supported membranes: scientific and practical applications. *Science* **271**: 43–48.

Sadana, A. and Beelaram, A. M. (1995) Antigen-antibody diffusion-limited binding kinetics of biosensors: a fractal analysis. *Biosens. Bioelec.* **10**: 301–16.

Sadik, O. A. and Van Emon, J. M. (1996) Application of electrochemical immunosensors to environmental monitoring. *Biosens. Bioelec.* **11**: i-xi.

Sadowski, J. W., Lekkala, J. and Vikholm, I. (1991) Biosensors based on surface plasmons excited in non-noble metals. *Biosens. Bioelec.* **6**: 439–444.

Sampath, S. and Lev, O. (1996) Inert metal-modified, composite ceramic–carbon, amperometric biosensors: renewable, controlled reactive layer. *Anal. Chem.* **68**: 2015–2021.

Sanchez, E. and Kowalski, B. R. (1988a) Tensorial calibration: I. First-order calibration. *J. Chemomet.* **2**: 247–264.

Sanchez, E. and Kowalski, B. R. (1988b) Tensorial calibration: II. Second-order calibration. *J. Chemomet.* **2**: 265–280.

Sangodkar, H., Sukeerthi, S., Srinivasa, R. S., Lai, R. and Contractor, A. Q. (1996) A biosensor array based on polyaniline. *Anal. Chem.* **68**: 779–783.

Santandreu, M., Céspedes, F., Alegret, S. and Martinez-Fàbregas, E. (1997) Amperometric immunosensors based on rigid conducting immunocomposites. *Anal. Chem.* **69**: 2080–2085.

Satoh, I. and Iijima, Y. (1995) Multi-ion biosensor with use of a hybrid-enzyme membrane. *Sens. Actuators* (B) **24**: 103–106.

Sauerbrey, G. Z. (1959) Use of a quartz vibrator for weighing thin films on a microbalance. *Z. Physik* **155**: 206–210.

Schaller, U., Bakker, E., Spichiger, U. and Pretsch, E. (1994) Ionic additives for ion-selective electrodes based on electrically charged carriers. *Anal. Chem.* **66**: 391–398.

Schaller, U., Bakker, E. and Pretsch, E. (1995) Carrier mechanism of acidic ionophores in solvent polymeric membrane ion-selective electrodes. *Anal. Chem.* **67**: 3123–3132.

Scheller, F. W., Schubert, F., Neumann, B., Pfeiffer, D., Hintsche, R., Dransfeld, I., Wollenberger, U., Renneberg, R., Warsinke, A., Johansson, G., Skoog, M., Xiurong, Y., Bogdanovskaya, V., Bückmann, A. and Zaitsev, S. Y. (1991) Second generation biosensors. *Biosens. Bioelec.* **6**: 245–253.

Scheller, F. W. and Schubert, F. (1992) *Biosensors: Fundamentals, Technologies and Applications.* (revised translation) Elsevier, Amsterdam.

Scheller, F. W., Makower, A., Ghindilis, A. L., Bier, F. F., Förster, E., Wollenberger, U., Bauer, C., Michael, B., Pfeiffer, D., Szepanik, J., Michael, N. and Kaden, H. (1995) Enzyme sensors for subnanomolar concentrations. In Rogers, K. R., Mulchandani, A. and Zhou, W. (eds.) *Biosensor and Chemical Sensor Technology. Process Monitoring and Control.* ACS Symposium Series #613, American Chemical Society, Washington, DC.

Schierbaum, K. D., Weiss, T., Thoden van Velzen, E. U., Engbersen, J. F. J., Reinhoudt, D. N. and Göpel, W. (1994) Molecular recognition by self-assembled monolayers of cavitand receptors. *Science* **265**: 1413–1415.

Schipper, E. F., Kooyman, R. P. H., Heideman, R. G. and Greve, J. (1995) Feasibility of optical waveguide immunosensors for pesticide detection: physical aspects. *Sens. Actuators* (B) **24**: 90–93.

Schipper, E. F., Kooyman, R. P. H., Borreman, A. and Greve, J. (1996) The critical sensor: a new type of evanescent wave immunosensor. *Biosens. Bioelec.* **11**: 295–304.

Schleimer, M., Fluck, M. and Schurlg, V. (1994) Enantiomer separation by capillary SFC and GC on chirasil–nickel: observation of unusual peak broadening phenomena. *Anal. Chem.* **66**: 2893–2897.

Schmid, E. L., Keller, T. A., Dienes, Z. and Vogel, H. (1997) Reversible oriented surface immobilization of functional proteins on oxide surfaces. *Anal. Chem.* **69**: 1979–1985.

Schmid, R.D. and Scheller, F. (eds) (1989) *Biosensors: Applications in Medicine, Environmental Protection and Process Control.* VCH, Weinheim.

Schmid, R. D. and Scheller, F. (eds) (1992) *Biosensors: Fundamentals, Technologies and Applications.* GBF-Monographs, Vol. 17, VCH, Weinheim.

Schmidt, A., Rohm, I., Rueger, P., Weise, W. and Bilitewski, U. (1994) Application of screen printed electrodes in biochemical analysis. *Fresenius' J. Anal. Chem.* **349**: 607–612.

Schmidt, H-L. and Schuhmann, W. (1996) Reagentless oxidoreductase sensors. *Biosens. Bioelec.* **11**: 127–135.

Schmidtke, D. W., Pishko, M. V., Quinn, C. P. and Heller, A. (1996) Statistics for critical clinical decision making based on readings of pairs of implanted sensors. *Anal. Chem.* **68**: 2845–2849.

Schneider, B. H., Hill, M. R. S. and Prohaska, O. J. (1990) Microelectrode probes for biomedical application. *Amer. Biotech. Lab.*, February: 17–23.

Schneiderheinze, J. M. and Hogan, B. L. (1996) Selective *in vivo* and *in vitro* sampling of proteins using miniature ultrafiltration sampling probes. *Anal. Chem.* **68**: 3758–3762.

Schnur, J. M. and Peckerar, M. (1992) *Synthetic Microstructures in Biological Research.* Plenum Press, New York.

Schroeder, T. J., Jankowiski, J. A., Kawagoe, K. T., Wightman, R. M., Lefrou, C. and Amatore, C. (1992) Analysis of diffusional broadening of vesicular packets of catecholamines released from biological cells during exocytosis. *Anal. Chem.* **64**: 3077–3083.

Schuhmann, W., Ohara, T. J., Schmidt, H-L. and Heller, A. (1991) Electron transfer between glucose oxidase and electrodes via redox mediators bound with flexible chains to the enzyme surface. *J. Amer. Chem. Soc.* **113**: 1394–1397.

Schulte, A. and Chow, R. H. (1996) A simple method for insulating carbon-fiber microelectrodes using anodic electrophoretic deposition of paint. *Anal. Chem.* **68**: 3054–3058.

Schultz, N. M., Huang, L. and Kennedy, R. T. (1995) Capillary electrophoresis-based immunoassay to determine insulin content and insulin secretion from single islets of Langerhans. *Anal. Chem.* **67**: 924–929.

Schweitz, L., Andersson, L. I. and Nilsson, S. (1997) Capillary electrochromatography with predetermined selectivity obtained through molecular imprinting. *Anal. Chem.* **69**: 1179–1183.

Scott, D. L. and Bowden, E. F. (1994) Enzyme–substrate kinetics of adsorbed cytochrome *c* peroxidase on pyrolytic graphite electrodes. *Anal. Chem.* **66**: 1217–1223.

Scott, D. L., Ramanathan, S., Shi, W., Rosen, B. P. and Daunerts, S. (1997) Genetically engineered bacteria: electrochemical sensing systems for antimonite and arsenite. *Anal. Chem.* **69**: 16–20.

Scott, E. R., White, H. S. and Phipps, J. B. (1993) Ionophoretic transport through porous membranes using scanning electrochemical microscopy: application to *in vitro* studies of ion fluxes through skin. *Anal. Chem.* **65**: 1537–1545.

Seiler, K. (1991) *Ionenselektive Optodenmembrane.* Fluka Chemie, AG. Buchs, Switzerland.

Seiler, K., Harrison, D. J. and Manz, A. (1993) Planar glass chips for capillary electrophoresis: repetitive simple injection quantitation, and separation efficiency. *Anal. Chem.* **65**: 1481–1488.

Seitz, W. R. (1988) Chemical sensors based on immobilized indicators and fiber optics. *Crit. Revs Anal. Chem.* **19**: 135–173.

Sekulic, S., Seasholtz, M. B., Wang, Z., Kowalski, B. R., Lee, S. E. and Holt, B. R. (1993) Nonlinear multivariate calibration methods in analytical chemistry. *Anal. Chem.* **65**: 835A–845A.

Selinger, J. V. and Rabbany, S. Y. (1997) Theory of heterogeneity in displacement reactions. *Anal. Chem.* **69**: 170–174.

Seller, K., Fan, Z. H., Fluri, K. and Harrison, D. J. (1994) Electroosmotic pumping and valveless control of fluid flow within a manifold of capillaries on a glass chip. *Anal. Chem.* **66**: 3485–3491.

Sellergren, B. (1997) Imprinted polymers: stable, reusable antibody-mimics for highly selective separations. *Amer. Laboratory* **29** (12): 14–20.

Service, R. E. (1996) Atomic landscapes beckon chip makers and chemists. *Science* **274**: 723–724.

Sethi, R. S. (1994) Transducer aspects of biosensors. *Biosens. Bioelec.* **9**: 243–264.

Severinghaus, J. W. and Bradley, A. F. (1958) Electrodes for blood pO_2 and pCO_2 determinations. *J. Appl. Physiol.* **13**: 515–520.

Shaffer, R. E., Small, G. W. and Arnold, M. A. (1996) Genetic algorithm-based protocol for coupling digital filtering and partial least-squares regression: application to the near-infrared analysis of glucose in biological matrices. *Anal. Chem.* **68**: 2663–2675.

Shakhsher, Z., Seitz, W. R. and Legg, K. D. (1994) Single fiber-optic pH sensor based on changes accompanying polymer swelling. *Anal. Chem.* **66**: 1731–1735.

Shana, Z. A. and Jossé, F. (1994) Quartz crystal resonators as sensors in liquid using the acoustoelectric effect. *Anal. Chem.* **66**: 1955–1964.

Shao, Y., Mirkin, M. V., Fish, G., Kokotov, S., Palanker, D. and Lewis, A. (1997) Nanometer-sized electrochemical sensors. *Anal. Chem.* **69**: 1627–1634.

Sharaf, M. A., Illman, D. L. and Kowalski, B. R. (1986) *Chemometrics.* John Wiley and Sons, New York.

Shealy, D. B., Lipowska, M., Lipowski, J., Narayanan, N., Sutter, S., Strekowski, L. and Patonay, G. (1995) Synthesis, chromatographic separation, and characterization of near-infrared-labeled DNA oligomers for use in DNA sequencing. *Anal. Chem.* **67**: 247–251.

Shear, J. B., Fishman, H. A., Allbritton, N. L., Garigan, D., Zare, R. N. and Scheller, R. H. (1995) Single cells as biosensors for chemical separations. *Science* **267**: 74–77.

Sheppard Jr, N. F., Tucker, R. C. and Wu, C. (1993) Electrical conductivity measurements using microfabricated interdigitated electrodes. *Anal. Chem.* **65**: 1199–1202.

Sheppard Jr, N. F., Lesho, M. J., McNally, P. and Francomacaro, A. S. (1995) Microfabricated conductimetric pH sensor. *Sens. Actuators* (B) **28**: 95–102.

Shi-Hui, S., Yuan-Jin, X., Li-Hua, N. and Shou-Zhou, Y. (1995) Electropolymerized *m*-phenylenediamine as a means to immobilize active protein on thickness-shear-mode quartz crystal. *Talanta* **42**: 469–474.

Shiku, H., Takeda, T., Yamada, H., Matsue, T. and Uchida, I. (1995) Microfabrication and characterization of diaphorase-patterned surfaces by scanning electrochemical microscopy. *Anal. Chem.* **67**: 312–317.

Shiku, H., Matsue, T. and Uchida, I. (1996) Detection of microspotted carcinoembryonic antigen on a glass substrate by scanning electrochemical microscopy. *Anal. Chem.* **68**: 1276–1278.

Shimohigoshi, M., Yokoyama, K. and Karube, I. (1995) Development of a bio-thermochip and its application for the detection of glucose in urine. *Anal. Chim. Acta* **303**: 295–299.

Shimohigoshi, M. and Karube, I. (1996) Development of uric acid and oxalic acid sensors using a bio-thermochip. *Sens. Actuators* (B) **30**: 17–21.

Shimura, K. and Karger, B. L. (1994) Affinity probe capillary electrophoresis: analysis of recombinant human growth hormone with a fluorescent labeled antibody fragment. *Anal. Chem.* **66**: 9–15.

Shin, M. C. and Kim, H. S. (1995) Effects of enzyme concentration and film thickness on the analytical performance of a polypyrrole/glucose oxidase biosensor. *Anal. Lett.* **28**: 1017–1031.

Shortreed, M., Bakker, E. and Kopelman, R. (1996a) Minature sodium-selective ion-exchange optode with fluorescent pH chromoionophores and tunable dynamic range. *Anal. Chem.* **68**: 2656–2662.

Shortreed, M., Kopelman, R., Kuhn, M. and Hoyland, B. (1996b) Fluorescent fiber-optic calcium sensor for physiological measurements. *Anal. Chem.* **68**: 1414–1418.

Shortreed, M., Monson, E. and Kopelman, R. (1996c) Lifetime enhancement of ultrasmall fluorescent liquid polymeric film based optodes by diffusion-induced self-recovery after photobleaching. *Anal. Chem.* **68**: 4015–4019.

Shrive, J. D. A., Brennan, J. D., Brown, R. S. and Krull, U. J. (1995) Optimization of the self-quenching response of nitrobenzoxadiazole dipalmitoylphoshatidylethanolamine in phospholipid membranes for biosensor development. *Appl. Spectros.* **49**: 304–313.

Shriver-Lake, L. C., Anderson, G. P., Golden, J. P. and Ligler, F. S. (1992) Effect of tapering the optical fibre on evanescent wave measurements. *Anal. Lett.* **25**: 1183–1199.

Shriver-Lake, L. C., Ogert, R. A. and Ligler, F. S. (1993) Fibre-optic evanescent-wave immunosensor for large molecules. *Sens. Actuators* (B) **11**: 239–243.

Shriver-Lake, L. C., Breslin, K. A., Charles, P. T., Conrad, D. W., Golden, J. P. and Ligler, F. S. (1995) Detection of TNT in water using an evanescent wave fibre-optic biosensor. *Anal. Chem.* **67**: 2431–2435.

Shul'ga A. A., Koudelka-Hep, M., de Rooij, N. F. and Netchiporouk, L. I. (1994) Glucose-sensitive enzyme field effect transistor using potassium ferricyanide as an oxidizing substrate. *Anal. Chem.* **66**: 205–210.

Sierra, J. F., Galbán, J. and Castillo, J. R. (1997) Determination of glucose in blood based on the intrinsic fluorescence of glucose oxidase. *Anal. Chem.* **69**: 1471–1476.

Sigal, G. B., Bamdad, C., Barberis, A., Strominger, J. and Whitesides, G. M. (1996) A self-assembled monolayer for the binding and study of histidine tagged proteins by surface plasmon resonance. *Anal. Chem.* **68**: 490–497.

Silber, A., Bisenberger, M., Bräuchle, C. and Hampp, N. (1996) Thick-film multichannel biosensors for simultaneous amperometric and potentiometric measurements. *Sens. Actuators* (B) **30**: 127–132.

Singhal, P., Kawagoe, K. T., Christian, C. N. and Kuhr, W. G. (1997) Sinusoidal voltammetry for the analysis of carbohydrates at copper electrodes. *Anal. Chem.* **69**: 1662–1668.

Slama, M., Zaborosch, C., Wienke, D. and Spencer, F. (1996) Simultaneous mixture analysis using a dynamic microbial sensor combined with chemometrics. *Anal. Chem.* **68**: 3845–3850.

Slovacek, R. E., Furlong, S. C. and Love, W. F. (1993) Feasibility study of a plastic evanescent-wave sensor. *Sens. Actuators* (B) **11**: 307–311.

Small, G. W., Arnold, M. A. and Marquardt, L. A. (1993) Strategies for coupling digital filtering with partial least-squares regression: application to the determination of glucose in plasma by Fourier transform near-infrared spectroscopy. *Anal. Chem.* **65**: 3279–3289.

Smilde, A. K., Tauler, R., Henshaw, J. M., Burgess, L. W. and Kowalski, B. R. (1994) Multicomponent determination of chlorinated hydrocarbons using a reaction-based chemical sensor. 3. Medium-rank second-order calibration with restricted Tucker models. *Anal. Chem.* **66**: 3345–3351.

Smit, M. H. and Rechnitz, G. A. (1992) Reagentless enzyme electrode for the determination of manganese through biocatalytic enhancement. *Anal. Chem.* **64**: 245–249.

Smit, M. H. and Rechnitz, G. A. (1993) Toxin detection using a tyrosinase-coupled oxygen electrode. *Anal. Chem.* **65**: 380–385.

Smith, C. L., Kricka, L. and Krull, U. J. (1995) The development of single molecule environmental sensors. *Gen. Anal.* **12**: 33–37.

Smith, C. P. and White, H. S. (1992) Theory of the interfacial potential distribution and reversible voltammetric response of electrodes coated with electroactive molecular films. *Anal. Chem.* **64**: 2398–2405.

Smith, C. P. and White, H. S. (1993) Theory of the voltammetric response of electrodes of submicron dimensions. Violation of electroneutrality in the presence of excess supporting electrolyte. *Anal. Chem.* **65**: 3343–3353.

Smith, E. A., Lillenthal, R. P., Fonseca, R. J. and Smith, D. K. (1994) Electrochemical characterization of a viologen-based redox-dependent-receptor for neutral organic molecules. *Anal. Chem.* **66**: 3013–3020.

Smolander, M., Marko-Varga, G. and Gorton, L. (1995) Aldose dehydrogenase-modified carbon paste electrodes as amperometric aldose sensors. *Anal. Chim. Acta* **302**: 233–240.

Snyder, A. W. and Love, J. D. (1983) *Optical Waveguide Theory*. Chapman & Hall, New York.

Solomon, T. and Bard, A. J. (1995) Scanning electrochemical microscopy. 30. Application of glass micopipet tips and electron transfer at the interface between two immiscible electrolyte solutions for SECM imaging. *Anal. Chem.* **67**: 2787–2790.

Song, A., Parus, S. and Kopelman, R. (1997) High-performance fiber-optic pH microsensors for practical physiological measurements using a dual-emission sensitive dye. *Anal. Chem.* **69**: 863–867.

Song, M. I., Iwata, K., Yamada, M., Yokoyama, K. Takeuchi, T., Tamiya, E. and Karube, I. (1994) Multisample analysis using an array of microreactors for an alternating-current field-enhanced latex immunoassay. *Anal. Chem.* **66**: 778–781.

Speiser, B. (1996) Numerical simulation of electroanalytical experiments: recent advances in methodology. In Bard, A. J. and Rubinstein, I. (eds), *Electroanalytical Chemistry*, Vol. 19. Marcel Dekker, New York.

Sprules, S. D., Hart, J. P., Wring, S. A. and Pittson, R. (1994) Development of a disposable amperometric sensor for reduced nicotinamide adenine dinucleotide based on a chemically modified screen-printed carbon electrode. *Analyst* **119**: 253–257.

Sprules, S. D., Hart, J. P., Wring, S. A. and Pittson, R. (1995) A reagentless, disposable biosensor for lactic acid based on a screen-printed carbon electrode containing Meldola's blue and coated with lactate dehydrogenase, NAD^+ and cellulose acetate. *Anal. Chim. Acta* **304**: 17–24.

Sreenivas, G., Ang, S. S., Fritch, I., Brown, W. D., Gerhardt, G. A. and Woodward, D. J. (1996) Fabrication and characterization of sputtered-carbon microelectrode arrays. *Anal. Chem.* **68**: 1858–1864.

Stamford, J. A. and Justice Jr, J. B. (1996) Probing brain chemistry. *Anal. Chem.* **68**: 359A–363A.

Stancil, L., Macholan, L. and Scheller, F. (1995) Biosensing of tyrosinase inhibitors in non-aqueous solvents. *Electroanalysis* **7**: 649–651.

Sternesjo, A., Mellgren, C. and Bjorck, L. (1995) Determination of sulfamethazine residues in milk by a surface plasmon resonance-based biosensor assay. *Anal. Biochem.* **226**: 175–181.

Stimpson, D. I., Hoijer, J. V., Hsieh, W. T., Jou, C., Gordon, J., Theriault, T., Gamble, R. and Baldeschwieler, J. D. (1995) Real-time detection of DNA hybridization and melting on oligonucleotide arrays by using optical wave guides. *Proc. Natl Acad. Sci. USA* **92**: 6379–6383.

Strachan, N. J. and Gray, D. I. (1995) A rapid general method for the identification of PCR products using a fibre-optic biosensor and its application to the detection of *Listeria. Lett. Appl. Microbiol.* **21**: 5–9.

Straub, A. E. and Seitz, W. R. (1993) Fiber-optic temperature sensor based on the temperature dependence of the pK_a of Tris. *Anal. Chem.* **65**: 1491–1492.

Strehlitz, B., Gründig, B., Schumacher, W., Kroneck, P. M. H., Vorlop, K-D. and Kotte, H. (1996) A nitrite sensor based on a highly sensitive nitrite reductase mediator-coupled amperometric detection. *Anal. Chem.* **68**: 807–816.

Strein, T. G. and Ewing, A. G. (1992) Characterization of submicron-sized carbon electrodes insulated with a phenol–allylphenyl copolymer. *Anal. Chem.* **64**: 1368–1373.

Strein, T. G. and Ewing, A. G. (1993) Characterization of small noble metal microelectrodes by voltammetry and energy dispersive X-ray analysis. *Anal. Chem.* **65**: 1203–1209.

Strein, T. G. and Ewing, A. G. (1994) Laser activation of microdisk electrodes examined by fast-scan rate voltammetry and digital simulation. *Anal. Chem.* **66**: 3864–3872.

Strojek, J. W., Granger, M. C., Swain, G. M., Dallas, T. and Holtz, M. W. (1996) Enhanced signal-to-background ratios in voltammetric measurements made at diamond thin-film electrochemical interfaces. *Anal. Chem.* **68**: 2031–2037.

Su, H., Kallury, K. M. R., Thompson, M. and Roach, A. (1994) Interfacial nucleic acid hybridization studied by random primer ^{32}P labelling and liquid-phase acoustic network analysis. *Anal. Chem.* **66**: 769–777.

Su, H., Williams, P. and Thompson, M. (1995) Platinum anticancer drug binding to DNA detected by thickness-shear mode acoustic wave sensor. *Anal. Chem.* **67**: 1010–1013.

Su, Y. S. and Mascini, M. (1995) AP-GOD biosensor based on a modified poly(phenol) film electrode and its application in the determination of low levels of phosphate. *Anal. Lett.* **28**: 1359–1378.

Suelter, C. H. and Kricka, L. (eds) (1992) *Bioanalytical Applications of Enzymes.* John Wiley and Sons, New York.

Sugawara, K., Tanaka, S. and Nakamura, H. (1995) Electrochemical assay of avidin and biotin using a biotin derivative labeled with an electroactive compound. *Anal. Chem.* **67**: 299–302.

Sundaram, S., Padmanabhan, R., Gowrishankar, S. and Krithiga, S. (1995) Modeling and optimizing dextrose fermentation using a fluorosensor. *Biomed. Sci. Instrum.* **31**: 251–256.

Sundberg, S. A., Barrett, R. W., Pirrung, M., Lu, A. L., Kiangsoontra, B. and Holmes, C. P. (1995) Spatially-addressable immobilization of macromolecules on solid supports. *J. Amer. Chem. Soc.* **117**: 12050–12057.

Suri, C. R., Jain, P. K. and Mishra, G. C. (1995) Development of piezoelectric crystal based microgravimetric immunoassay for determination of insulin concentration. *J. Biotechnol.* **39**: 27–34.

Sutherland, R. M. and Dähne, C. (1989) IRS devices for optical immunoassays. In Turner, A. P. F., Karube, I. and Wilson, G. S. (eds) *Biosensors: Fundamentals and Applications.* Oxford University Press, New York.

Sutherland, R. M., Daehne, C. and Place, J. F. (1994a) Preliminary results obtained with a no-label, homogeneous, optical immunoassay for human immunoglobulin G. *Anal. Lett.* **17**: 43–53.

Sutherland, R. M., Daehne, C., Place, J. F. and Ringrose, A. R. (1994b) Immunoassays at a quartz–liquid interface: theory, instrumentation and preliminary application to the fluorescent immunoassay of human immunoglobulin G. *J. Immunol. Methods* **74**: 253–265.

Sutter, J. M. and Jurs, P. C. (1997) Neural network classification and quantification of organic vapors based on fluorescence data from a fiber-optic sensor array. *Anal. Chem.* **69**: 856–862.

Suzuki, H. (1994) Disposable Clark oxygen electrode using recycled materials and its applications. *Sens. Actuators* (B) **21**: 17-22.

Suzuki, K., Watanabe, K., Matsumoto, Y., Kobayashi, M., Sato, S., Siswanta, D. and Hisamoto, H. (1995) Design and synthesis of calcium and magnesium ionophores based on double-armed diazacrown ether compounds and their application to an ion-sensing component for an ion-selective electrode. *Anal. Chem.* **67**: 324–334.

Svendsen, C. N. (1989) Multi-electrode array detectors in high performance liquid chromatography: a new dimension in electrochemical analysis. *Analyst* **118**: 123–129.

Švorc, J., Miertuš, S., Katrlîk, J. and Stred'anský, M. (1997) Composite transducers for amperometric biosensors. The glucose sensor. *Anal. Chem.* **69**: 2086–2090.

Swalen, J. D., Allara, D. L., Andrade, J. D., Chandross, E. A., Garoff, S., Israelachivili, J., McCarthy, T. J., Murray, R., Pease, R. F., Rabolt, J. F., Wyn, K. J. and Yu, H. (1987) Molecular monolayers and films. *Langmuir* **3**: 932–950.

Swanek, F. D., Chen, G. and Ewing, A. G. (1996) Identification of multiple compartments of dopamine in a single cell by CE with scanning electrochemical detection. *Anal. Chem.* **68**: 3912–3916.

Swerdlow, H., Jones, B. J., and Wittwer, C. T. (1997) Fully automated DNA reaction and analysis in a fluidic capillary instrument. *Anal. Chem.* **69**: 848–855.

Synovec, R. E., Sulya, A. W., Burgess, L. W., Foster, M. D. and Bruckner, C. A. (1995) Fiber-optic-based mode-filtered light detection for small-volume chemical analysis. *Anal. Chem.* **67**: 473–481.

Szmacinski, H. and Lakowicz, J. R. (1993) Optical measurements of pH using fluorescence lifetimes and phase-modulation fluorometry. *Anal. Chem.* **65**: 1668–1674.

Tabei, H., Takahashi, M., Hoshino, S., Niwa, O. and Horiuchi, T. (1994) Subfemtomole detection of catecholamines with interdigitated array carbon microelectrodes in HPLC. *Anal. Chem.* **66**: 3500-3502.

Takada, K., Haseba, T., Tatsuma, T., Oyama, N., Li, Q. and White, H. S. (1995) *In situ* interferometric microscopy of temperature- and potential-dependent volume change of a redox gel. *Anal. Chem.* **67**: 4446–4451.

Takahashi, T., Forsythe, I. D., Tsujimoto, T., Barnes-Davies, M. and Onodera, K. (1996) Presynaptic calcium current modulation by a metabotropic glutamate receptor. *Science* **274**: 594–597.

Tan, H. and Yeung, E. S. (1997) Integrated on-line system for DNA sequencing by capillary electrophoresis: from template to called bases. *Anal. Chem.* **69**: 664–674.

Tan, W., Shi, Z-Y. and Kopelman, R. (1992a) Development of submicron chemical fiber-optic sensor. *Anal. Chem.* **64**: 2985–2990.

Tan, W., Shi, Z-Y., Smith, S., Birnbaum, D. and Kopelman, R. (1992b) Submicrometer intracellar chemical optical fiber sensors. *Science* **258**: 778–781.

Tan, W., Parpura, V., Haydon, P. G. and Yeung, E. S. (1995a) Neurotransmitter imaging in living cells based on native fluorescence detection. *Anal. Chem.* **67**: 2575–2579.

Tan, W., Shi, Z-Y. and Kopelman, R. (1995b) Miniaturized fiber-optic chemical sensors with fluorescent dye-doped polymers. *Sens. Actuators* (B) **28**: 157–163.

Tan, W. and Kopelman, R. (1996) Nanoscale imaging and sensing by near-field optics. In Wang, X. F. and Herman, B. (eds), *Fluorescence Imaging Spectroscopy and Microscopy*. John Wiley and Sons, New York.

Tao, L. and Kennedy, R. T. (1996) On-line competitive immunoassy for insulin based on capillary electrophoresis with laser-induced fluorescence detection. *Anal. Chem.* **68**: 3899–3906.

Tarnowski, D. J., Bekos, E. J. and Korzeniewski, C. (1995) Oxygen transport characteristics of refunctionalized fluoropolymeric membranes and their application in the design of biosensors based upon the Clark-type oxygen probe. *Anal. Chem.* **67**: 1546–1552.

Tatsuma, T. and Watanabe, T. (1992a) Peroxidase model electrodes: sensing of imidazole derivatives with heme peptide-modified electrodes. *Anal. Chem.* **64**: 142–147.

Tatsuma, T. and Watanabe, T. (1992b) Model analysis of enzyme monolayer- and bilayer-modified electrodes: the steady-state response. *Anal. Chem.* **64**: 625–630.

Tatsuma, T., Gondaira, M. and Watanabe, T. (1992a) Peroxidase-incorporated polypyrrole membrane electrodes. *Anal. Chem.* **64**: 1183–1187.

Tatsuma, T., Watanabe, T. and Okawa, Y. (1992b) Model analysis of enzyme monolayer- and bilayer-modified electrodes: the transient response. *Anal. Chem.* **64**: 631–635.

Tatsuma, T. and Watanabe, T. (1993) Theoretical evaluation of mediation efficiency in enzyme-incorporated electrodes. *Anal. Chem.* **65**: 3129–3133.

Tatsuma, T., Saito, K. and Oyama, N. (1994a) Enzyme electrodes mediated by a thermoshrinking redox polymer. *Anal. Chem.* **66**: 1002-1006.

Tatsuma, T., Watanabe, T., Tatsuma, S. and Watanabe, T. (1994b) Substrate-purging enzyme electrodes: peroxidase/catalase electrodes for H_2O_2 with an improved upper sensing limit. *Anal. Chem.* **66**: 290–294.

Tatsuma, T., Ariyama, K. and Oyama, N. (1995) Electron transfer from a polythiophene derivative to compounds I and II of peroxidases. *Anal. Chem.* **67**: 283–287.

Tatsuma, T. and Oyama, N. (1996) H_2O_2-generating peroxidase electrodes as reagentless cyanide sensors. *Anal. Chem.* **68**: 1612–1615.

Tatsuma, T., Tani, K., Oyama, N. and Yeoh, H-H. (1996) Linamarin sensors: interference-based sensing of linamarin using linamarase and peroxidase. *Anal. Chem.* **68**: 2946–2950.

Tatsuma, T. and Buttry, D. A. (1997) Electrochemical/piezoelectric dual-response biosensor for heme ligands. *Anal. Chem.* **69**: 887–893.

Tauler, R. and Kowalski, B. (1993) Multivariate curve resolution applied to spectral data from multiple runs of an industrial process. *Anal. Chem.* **65**: 2040-2047.

Tauler, R., Smilde, A. K., Henshaw, J. M., Burgess, L. W. and Kowalski B. R. (1994) Multicomponent determination of chlorinated hydrocarbons using a reaction-based chemical sensor. 2. Chemical speciation using multivariate curve resolution. *Anal. Chem.* **66**: 3337–3344.

Taylor, C. E., Garvey, S. D. and Pemberton, J. E. (1996) Carbon contamination at silver surfaces: surface preparation procedures evaluated by Raman spectroscopy and X-ray photoelectron spectroscopy. *Anal. Chem.* **68**: 2401–2408.

Taylor, R. F. (ed.) (1991) *Protein Immobilization: Fundamentals and Applications*. Marcel Dekker, New York.

Taylor, R. F. and Schultz, J. S. (eds) (1996) *Handbook of Chemical and Biological Sensors*. Institute of Physics, Bristol and Philadelphia.

Telting-Diaz, M., Collison, M. E. and Meyerhoff, M. E. (1994) Simplified dual-lumen catheter design for simultaneous potentiometric monitoring of carbon dioxide and pH. *Anal. Chem.* **66**: 576–583.

Tender, L., Carter, M. T. and Murray, R. W. (1994) Cyclic voltammetric analysis of ferrocene alkanethiol monolayer electrode kinetics based on Marcus theory. *Anal. Chem.* **66**: 3173–3181.

Tessier, L., Patat, F., Schmitt, N. and Feuillard, G. and Thompson, M. (1994) Effect of the generation of compressional waves on the response of the thickness-shear mode acoustic wave sensor in liquids. *Anal. Chem.* **66**: 3569–3574.

Teuscher, J. H. and Garrell, R. L. (1995) Stabilization of quartz oscillators by a conductive adhesive. *Anal. Chem.* **67**: 3372-3375.

Thevenot, D. R., Toth, K., Durst, R. A. and Wilson, G. S. (1996) Electrochemical biosensors: proposed definitions and classification. Synopsis of Report. *Sens. Actuators* (B) **30**: 81.

Thomé-Duret, V., Reach, G., Gangnerau, M. N., Lemonnier, F., Klein, J. C., Zhang, Y., Hu, Y. and Wilson, G. S. (1996) Use of a subcutaneous glucose sensor to detect decreases in glucose concentration prior to observation in blood. *Anal. Chem.* **68**: 3822–3826.

Thomas, E. V. (1994) A primer on multivariate calibration. *Anal. Chem.* **66**: 795A–804A.

Thompson, M. and Stone, D. C. (1997) *Surface-Launched Acoustic Wave Sensors: Chemical Sensing and Thin-Film Characterization*. John Wiley and Sons, New York.

Thompson, R. B. and Patchan, M. W. (1995) Lifetime-based fluorescence energy transfer biosensing of zinc. *Anal. Biochem.* **227**: 123–128.

Thundat, T., Chen, G. Y., Warmack, R. J., Allison, D. P. and Wachter, E. A. (1995) Vapor detection using resonating microcantilevers. *Anal. Chem.* **67**: 519–521.

Tierney, M. J. and Kim, H-O. (1993) Electrochemical gas sensor with extremely fast response. *Anal. Chem.* **65**: 3435–3440.

Tobias-Katona, E. and Pecs, M. (1995) Multienzyme-modified ion-sensitive field-effect transistor for sucrose measurement. *Sens. Actuators* (B) **28**: 17–20.

Tom-Moy, M., Baer, R. L., Spira-Solomon, D. and Doherty, T. P. (1995) Atrazine measurements using surface transverse wave devices. *Anal. Chem.* **67**: 1510–1516.

Towe, B. C. and Guilbeau, E. J. (1996) A vibrating probe thermal biochemical sensor. *Biosens. Bioelec.* **11**: 247–252.

Tran, C. D. and Gao, G-H. (1996) Characterization of an erbium-doped fiber amplifier as a light source and development of a near-infrared spectrophotometer based on the EDFA and an acoustooptic tunable filter. *Anal. Chem.* **68**: 2264–2269.

Treloar, P. H., Christie, I. M. and Vadgama, P. M. (1995) Engineering the right membranes for electrodes at the biological interface: solvent cast and electropolymerised. *Biosens. Bioelec.* **10**: 195–201.

Trojanowicz, M. (1996) Electrochemical detectors in automated analytical systems. In Vanýsek, P. (ed.), *Modern Techniques in Electroanalysis*. John Wiley and Sons, New York.

Trudeau, F., Daigle, F. and Leech, D. (1997) Reagentless mediated laccase electrode for the detection of enzyme modulators. *Anal. Chem.* **69**: 882–886.

Tsai, W. C. and Cass, A. E. G. (1995) Ferrocene-modified horseradish peroxidase enzyme electrodes. A kinetic study on reactions with hydrogen peroxide and linoleic hydroperoxide. *Analyst* **120**: 2249–2254.

Tschuncky, P. and Heinze, J. (1995) An improved method for the construction of ultramicro-electrodes. *Anal. Chem.* **67**: 4020–4023.

Tsien, R. Y. (1994) Fluorescence imaging creates a window on the cell. *C & E News* July 18: 34–44.

Tsionsky, M., Gun., G., Glezer, V. and Lev, O. (1994) Sol–gel-derived ceramic-carbon composite electrodes: introduction and scope of applications. *Anal. Chem.* **66**: 1747–1753.

Tsionsky, M. and Lev, O. (1995) Electrochemical composite carbon–ceramic gas sensor: introduction and oxygen sensing. *Anal. Chem.* **67**: 2409–2414.

Tsuchida, T. and Yoda, K. (1983) Multi-enzyme membrane electrodes for determination of creatinine and creatine in serum. *Clin. Chem.* **29**: 51–56.

Tsujimura, Y., Yokoyama, M., and Kimura, K. (1995) Practical applicability of silicone rubber membrane sodium-selective electrode based on oligosiloxane-modified calix{4}arene neutral carrier. *Anal. Chem.* **67**: 2401–2404.

Turner, A. P. F., Karube, I. and Wilson, G. S. (eds) (1989) *Biosensors: Fundamentals and Applications*. Oxford University Press, New York.

Turner, A. P. F. (1991-) *Advances in Biosensors*. Multivolume. JAI Press, Greenwich, CT.

Turner, A. P. F. (1995) Electrochemical sensors for continuous monitoring during surgery and intensive care. *Acta Anaesthesiol. Scand.* Suppl **104**: 15–19.

Turyan, I. and Mandler, D. (1994) Self-assembled monolayers in electroanalytical chemistry: application of (omega)-mercaptocarboxylic acid monolayers for electrochemical determination of ultralow levels of cadmium(II). *Anal. Chem.* **66**: 58–63.

Turyan, I. and Mandler, D. (1997) Selective determination of Cr(VI) by a self-assembled monolayer-based electrode. *Anal. Chem.* **69**: 894–897.

Uda, T., Hifumi, E., Kobayashi, T., Shimizu, K., Sata, T. and Ogino, K. (1995) An approach for an immunoaffinity AIDS sensor using the conservative region of the HIV envelope protein (gp41) and its monoclonal antibody. *Biosens. Bioelec.* **10**: 477–483.

Ueno, K. and Yeung, E. S. (1994) Simultaneous monitoring of DNA fragments separated by electrophoresis in a multiplexed array of 100 capillaries. *Anal. Chem.* **66**: 1424–1431.

Ugo, P., Moretto, L. M., Ballomi, S., Menon, V. P. and Martin, C. R. (1996) Ion-exchange voltammetry at polymer film-coated nanoelectrode ensembles. *Anal. Chem.* **68**: 4160–4165.

Uiga, E. (1995) *Optoelectronics*. Prentice-Hall, Englewood Cliffs, NJ.

Ulman, A. (1991) *An Introduction to Ultrathin Organic Films from Langmuir–Blodgett to Self Assembly*. Academic Press, San Diego, CA.

Umezawa, Y. (1990) *Handbook of Ion-Selective Electrodes: Selectivity Coefficients*. CRC Press, Boca Raton, FL.

Umezawa, Y., Umezawa, K. and Sato, H. (1995) Selectivity coefficients for ion-selective electrodes: recommended methods for reporting K_{AB}^{pot} values. *Pure Appl. Chem.* **67**: 507–518.

Underwood, A. L. and Burnett, R. W. (1973) Electrochemistry of biological compounds. In Bard, A. J. (ed.), *Electroanalytical Chemistry*, Vol. 6. Marcel Dekker, New York.

Updike, S. J. and Hicks, G. P. (1967) The enzyme electrode. *Nature* **214**: 986–988.

Urban, G., Kamper, H., Jachimowics, A., Kohl, F., Kuttner, H., Olcaytug, F., Goiser, P., Pittner, F., Schalkhammer, T. and Mann-Buxbaum, E. (1991) The construction of microcalorimetric biosensors by use of thin-film thermistors. *Biosens. Bioelec.* **6**: 275–280.

Vaezi, M. F. and Richter, J. E. (1995) Fiberoptic sensor for bilirubin. Letter. *Amer. J. Surg.* **170**: 310–312.

Vaillo, E., Walde, P. and Spichiger, U. E. (1995) Development of micellar bio-optode membranes. *Anal. Methods Instrum.* **2**: 145–153.

Valcárel, M. and Luque deCastro, M. D. (1994) *Flow-Through (Bio)Chemical Sensors*. Elsevier, Amsterdam.

Valencia-González, M. J. and Díaz-Garcia, M. E. (1994) Enzymatic reactor/room-temperature phosphorescence sensor system for cholesterol determination in organic solvents. *Anal. Chem.* **66**: 2726–2731.

Valkó, K., Bevan, C. and Reynolds, D. (1997) Chromatographic hydrophobicity index by fast-gradient RP-HPLC: a high throughput alternative to log *P*/log *D*. *Anal. Chem.* **69**: 2022–2029.

Van Emon, J. M. and Lopez-Avila, V. (1992) Immunochemical methods for environmental analysis. *Anal. Chem.* **64**: 79A–88A.

VanderNoot, V. A. and Lai, E. P. C. (1991) Detecting surface plasmon resonance using photothermal deflection spectroscopy: development of an optical immunosensor. Spectroscopy. **6**: 28–33.

Van der Schoot, B. H. and Bergveld, P. (1988) Coulometric sensors, the application of a sensor-actuator for long-term stability in chemical sensing. *Sens. Actuators* **13**: 251–262.

Van Kerkhof, J. C., Bergveld, P., and Schasfoort, R. B. M. (1995) The ISFET-based heparin sensor with a monolayer of protamine as affinity ligand. *Biosens. Bioelec.* **10**: 269–282.

Vansant, E. F., Van der Voort, P. and Vrancken, K. C. (1995) *Characterization and Chemical Modification of the Silica Surface*. Elsevier, Amsterdam.

Vellekoop, M. J., Lubking, G. W., Sarro, P. M. and Venema, A. (1994) Integrated-circuit-compatible design and technology of acoustic-wave-based microsensors. *Sens. Actuators* (A) **44**: 249–263.

Virta, M., Lampinen, J. and Karp, M. (1995) A luminescence-based mercury biosensor. *Anal. Chem.* **67**: 667–669.

Vo-Dinh, T., Houck, K. and Stokes, D. L. (1994a) Surface-enhanced Raman gene probes. *Anal. Chem.* **66**: 3379–3383.

Vo-Dinh, T., Viallet, P., Ramirez, L. and Pal, A. (1994b) Gel-based indo-1 probe for monitoring calcium(II) ions. *Anal. Chem.* **66**: 813–817.

Voegel, P. D., Zhou, W. and Baldwin, R. P. (1997) Integrated capillary electrophoresis/electrochemical detection with metal film electrodes directly deposited onto the capillary tip. *Anal. Chem.* **69**: 951–957.

Vreeke, M. S., Maidan, R. and Heller, A. (1992) Hydrogen peroxide and β-nicotinamide adenine dinucleotide sensing amperometric electrodes based on electrical connection of horseradish peroxidase redox centers to electrodes through a three dimensional electron relaying polymer network. *Anal. Chem.* **64**: 3084–3090.

Vreeke, M. S., Rocca, P. and Heller, A. (1995a) Direct electrical detection of dissolved biotinylated horseradish peroxidase, biotin, and avidin. *Anal. Chem.* **67**: 303–306.

Vreeke, M. S., Yong, K. T. and Heller, A. (1995b) A thermostable hydrogen peroxide sensor based on "wiring" of soybean peroxidase. *Anal. Chem.* **67**: 4247–4249.

Walczak, I. M., Love, W. F., Cook, T. A. and Slovacek, R. E. (1992) Application of evanescent wave sensing to a high-sensitivity fluoroimmunoassay. *Biosens. Bioelec.* **7**: 39–48.

Walmsley, A. D., Haswell, S. J. and Metcalfe, E. (1991) Chemometrics—the key to sensor array development. *Anal. Proc. (London)* **28**: 115–117.

Walt, D. R., Munkholm, C., Yuan, P., Luo, S. and Barnard, S. (1989) Design, preparation and applications of fiber-optic chemical sensors for continuous monitoring. In Murray, R. W. *et al.* (eds), *Chemical Sensors and Microinstrumentation*. American Chemical Society, Washington, DC.

Wang, A-J. and Rechnitz, G. A. (1993) Prototype transgenic biosensor based on genetically modified plant tissue. *Anal. Chem.* **65**: 3067–3070.

Wang, A. W., Costello, B. J. and White, R. M. (1993) An ultrasonic flexural plate-wave sensor for measurement of diffusion in gels. *Anal. Chem.* **65**: 1639–1642.

Wang, C-L. and Mulchandani, A. (1995) Ferrocene-conjugated polyaniline-modified enzyme electrodes for determination of peroxides in organic media. *Anal. Chem.* **67**: 1109–1114.

Wang, E., Meyerhoff, M. E. and Yang, V. C. (1995) Optical detection of macromolecular heparin via selective coextraction into thin polymeric films. *Anal. Chem.* **67**: 522–527.

Wang, H., Brennan, J. D., Gene, A. and Krull, U. J. (1995) Assembly of antibodies in lipid membranes for biosensor development. *Appl. Biochem. Biotech.* **53**: 1163–1181.

Wang, J., Wu, L-H, and Li, R. (1989) Scanning electrochemical microscopic monitoring of biological processes. *J. Electroanal. Chem.* **272**: 285–292.

Wang, J., Martinez, T., Yaniv, D. R. and McCormick, L. (1990) Characterization of the microdistribution of conductive and insulating regions of carbon paste electrodes with scanning tunneling microscopy. *J. Electroanal. Chem.* **286**: 265–272.

Wang, J., Wu, L-H. and Angnes, L. (1991) Organic-phase enzymatic assay with ultramicroelectrodes. *Anal. Chem.* **63**: 2993–2994.

Wang, J. and Angnes, L. (1992) Miniaturized glucose sensors based on electrochemical codeposition of rhodium and glucose oxidase onto carbon-fiber electrodes. *Anal. Chem.* **64**: 456–459.

Wang, J., Naser, N., Angnes, L., Wu, H. and Chen, L. (1992) Metal-dispersed carbon paste electrodes. *Anal. Chem.* **64**: 1285–1288.

Wang, J. and Tian, B. (1992) Screen-printed stripping voltammetric-potentiometric electrodes for decentralized testing of trace lead. *Anal. Chem.* **64**: 1706–1709.

Wang, J. and Chen, Q. (1993) *In situ* elimination of metal inhibitory effects using ligand-containing carbon paste electrodes. *Anal. Chem.* **65**: 2698–2700.

Wang, J. and Tian, B. M. (1993a) Mercury-free disposable lead sensors based on potentiometric stripping analysis at gold-coated screen-printed electrodes, *Anal. Chem.* **65**: 1529–1532.

Wang, J. and Tian, B. (1993b) Screen-printed electrodes for stripping measurements of trace mercury. *Anal. Chim. Acta* **274**: 1–6.

Wang, J., Wu, H. and Angnes, L. (1993) On-line monitoring of hydrophobic compounds at self-assembled monolayer modified amperometric flow detectors. *Anal. Chem.* **65**: 1893–1896.

Wang, J. (1994) Selectivity coefficients for amperometric sensors. *Talanta* **41**: 857–863.

Wang, J. and Chen, Q. (1994a) Enzyme microelectrode array strips for glucose and lactate. *Anal. Chem.* **66**: 1007–1011.

Wang, J. and Chen, Q. (1994b) Screen-printed glucose strip based on palladium-dispersed carbon ink. *Analyst* **119**: 1849–1851.

Wang, J. and Liu, J. (1994) Calixarene-coated amperometric detectors. *Anal. Chim. Acta* **294**: 201–206.

Wang, J., Liu, J., Chen, L. and Lu, F. (1994a) Highly selective membrane-free, mediator-free glucose biosensor. *Anal. Chem.* **66**: 3600–3603.

Wang, J., Baomin, T. and Setiadjii, R. (1994b) Disposable electrodes for field screening of trace uranium. *Electroanalysis* **6**: 317–320.

Wang, J., Chen, Q., Renschler, C. L. and White, C. (1994c) Ultrathin porous carbon films as amperometric transducers for biocatalytic sensors. *Anal. Chem.* **66**: 1988–1992.

Wang, J. and Chen, L. (1995a) Hydrazine detection using a tyrosinase-based inhibition biosensor. *Anal. Chem.* **67**: 3824–3827.

Wang, J. and Chen, Q. (1995b) Microfabricated phenol biosensors based on screen printings of tyrosinase containing carbon ink. *Anal. Lett.* **28**: 1131–1142.

Wang, J. and Pamidi, P. V. A. (1995) Disposable screen-printed electrodes for monitoring hydrazines. *Talanta* **42**: 463–467.

Wang, J., Cai, X., Wang, J., Jonsson, C. and Paleček, E. (1995a) Trace measurements of RNA by potentiometric stripping analysis at carbon paste electrodes. *Anal. Chem.* **67**: 4065–4070.

Wang, J., Chen, Q. and Pedrero, M. (1995b) Highly selective biosensing of lactate at lactate oxidase-containing rhodium-dispersed carbon-paste electrodes. *Anal. Chim. Acta* **304**: 41–46.

Wang, J., Chen, Q., Pedrero, M. and Pingarron, J. M. (1995c) Screen-printed amperometric biosensors for glucose and alcohols based on ruthenium-dispersed carbon inks. *Anal Chim. Acta* **300**: 111–116.

Wang, J., Lu, F., Angnes, L., Liu, J., Sakslund, H., Chen, Q., Pedrero, M., Chen, L. and Hammerich, O. (1995d) Remarkably selective metallized-carbon amperometric biosensors. *Anal. Chim. Acta* **305**: 3–7.

Wang, J., Pedrero, M. and Cai, X. (1995e) Palladium-doped screen-printed electrodes for monitoring formaldehyde. *Analyst* **120**: 1969–1972.

Wang, J. and Chen, L. (1996) Selectivity coefficients of class-selective enzyme electrodes. *Biosens. Bioelec* **11**: 751–756.

Wang, J., Cai, X., Rivas, G., Shiraishi, H., Farias, P. A. M. and Dontha, N. (1996a) DNA electrochemical biosensors for the detection of short DNA sequences related to the human immunodeficiency virus. *Anal. Chem.* **68**: 2629–2634.

Wang, J., Chicharro, M., Rivas, G., Cai, X., Dontha, N., Farias, P. A. M. and Shiraishi, H. (1996b) DNA biosensors for the detection of hydrazines. *Anal. Chem.* **68**: 2251–2254.

Wang, J., Pamidi, P. V. A. and Park, D. S. (1996c) Screen-printable–sol-gel enzyme-containing carbon inks. *Anal. Chem.* **68**: 2705–2708.

Wang, J., Rivas, G., Luo, D., Cai, X., Valera, F. S. and Dontha, N. (1996d) DNA-modified electrodes for the detection of aromatic amines. *Anal. Chem.* **68**: 4365–4369.

Wang, J., Rivas, G., Ozsoz, M., Grant, D. H., Cai, X. and Parrado, C. (1997) Microfabricated electrochemical sensors for the detection of radiation-induced DNA damage. *Anal. Chem.* **69**: 1457–1460.

Wang, Joseph (1994) *Analytical Electrochemistry*. VCH, New York.

Wang, Juan, Ward, M. D., Ebersole, R. C. and Foss, R. P. (1993) Piezoelectric pH sensors: AT-cut quartz resonators with amphoteric polymer films. *Anal. Chem.* **65**: 2553–2562.

Wang, R., Narang, U., Prasad, P. N. and Bright, F. V. (1993) Affinity of antifluorescein antibodies encapsulated within a transparent sol–gel glass. *Anal. Chem.* **65**: 2671–2675.

Wang, R., Sun, S. Y., Bekos, E. J. and Bright, F. V. (1995) Dynamics surrounding Cys-34 in native, chemically denatured and silica-adsorbed bovine serum albumin. *Anal. Chem.* **67**: 149–159.

Wang, S. X., Mure, M., Medzihradszky, K. L., Burlingame, A. L., Brown, D. E., Dooley, D. M., Smith, A. J., Kagan, H. M. and Klinman, J. P. (1996) A crosslinked cofactor in lysyl oxidase: redox function for amino acid side chains. *Science* **273**: 1078–1084.

Wang, Y. and Kowalski, B. R. (1993) Standardization of second-order instruments. *Anal. Chem.* **65**: 1174–1180.

Wang, Z., Hwang, J-N. and Kowalski, B. R. (1995) ChemNets: theory and application. *Anal. Chem.* **67**: 1497–1504.

Ward, L. D., Howlett, G. J., Hammacher, A., Weinstock, J., Yasukawa, K., Simpson, R. J. and Winzor, D. J. (1995) Use of a biosensor with surface plasmon resonance detection for the determination of binding constants: measurement of interleukin-6 binding to the soluble interleukin-6 receptor. *Biochemistry* **34**: 2901–2907.

Ward, W. K., Wilgus, E. S. and Troupe, J. E. (1994) Rapid detection of hyperglycaemia by a subcutaneously-implanted glucose sensor in the rat. *Biosens. Bioelec.* **9**: 423–428.

Watts, H. J., Lowe, C. R. and Pollard-Knight, D. V. (1994) Optical biosensor for monitoring microbial cells. *Anal. Chem.* **66**: 2465–2470.

Weber, K. and Creager, S. E. (1994) Voltammetry of redox-active groups irreversibly adsorbed onto electrodes. Treatment using the Marcus relation between rate and overpotential. *Anal. Chem.* **66**: 3164–3172.

Weber, S. G. (1992) Internal volume of competitive binding biosensors controls sensitivity: equilibrium theory. *Anal. Chem.* **64**: 330–332.

Weber, S. G. and Weber, A. (1993) Biosensor calibration. *In situ* recalibration of competitive binding sensors. *Anal. Chem.* **65**: 223-230.

Weber, S. G. (ed.) (1995) Special Issue: Single Cell Analysis. *Trends Anal. Chem.*

Weetall, H. H. (1976) Covalent coupling methods for inorganic support materials. *Methods Enzymology Immobilized Enzymes* **44**: 134–148.

Weetall, H. H. (1993) Preparation of immobilized proteins covalently coupled through silane coupling agents to inorganic supports. *Appl. Biochem. Biotech.* **41**: 157–188.

Weetall, H. H. (1996) Biosensor technology: what? where? when? and why? *Biosens. Bioelec.* **11**: i-iv.

Wei, A-P., Blumenthal, D. K. and Herron, J. N. (1994) Antibody-mediated fluorescence enhancement based on shifting the intramolecular dimer–monomer equilibrium of fluorescent dyes. *Anal. Chem.* **66**: 1500–1506.

Weigl, B. H. and Wolfbeis, O. S. (1994) Capillary optical sensors. *Anal. Chem.* **66**: 3323–3327.

Weisenhorn, A. L., Schmitt, F-J., Knoll, W. and Hansma, P. K. (1992) Streptavidin binding observed with an atomic force microscope. *Ultramicroscopy* **42–44**: 1125–1132.

Welsch, W., Klein, C., vonSchickfus, M. and Hunklinger, S. (1996) Development of a surface acoustic wave immunosensor. *Anal. Chem.* **68**: 2000-2004.

Westmark, P. R. and Smith, B. D. (1994) Boronic acids selectively facilitate glucose transport through a liquid bilayer. *J. Amer. Chem. Soc.* **116**: 9343–9344.

Whitcombe, M. J., Rodriquez, M. E., Villar, P. and Vulfson, E. N. (1995) A new method for the introduction of recognition site functionality into polymers prepared by molecular imprinting: synthesis and characterization of polymeric receptors for cholesterol. *J. Amer. Chem. Soc.* **117**: 7105–7111.

White, J., Kauer, J. S., Dickinson, T. A. and Walt, D. R. (1996) Rapid analyte recognition in a device based on optical sensors and the olfactory system. *Anal. Chem.* **68**: 2191-2202.

White, S. F., Turner, A. P., Bilitewski, U., Bradley, J. and Schmid, R. D. (1995) On line monitoring of glucose, glutamate and glutamine during mammalian cell cultivations. *Biosens. Bioelec.* **10**: 543–551.

Wienke, D., Slama, M. and Cammann, K. (1996a) Analytical 4D infrared tomography using an InSb focal plane array sensor. 1. 3D infrared tomography (single-wavelength approach). *Anal. Chem.* **68**: 3987–3993.

Wienke, D., Slama, M. and Cammann, K. (1996b) Analytical 4D infrared tomography using an InSb focal plane array sensor. 2. 4D infrared tomography (multiwavelength approach.) *Anal. Chem.* **68**: 3994-3999.

Wightman, R. M. and Wipf, D. O. (1989) Voltammetry at ultramicroelectrodes. In Bard, A. J. (ed.), *Electroanalytical Chemistry*, Vol. 15. Marcel Dekker, New York.

Wilding, P., Shoffner, M. A. and Kricka, L. J. (1994) PCR in a silicon microstructure. *Clin. Chem.* **40**: 1815–1818.

Wilkins, E., Atanasov, P. and Muggenburg, B. A. (1995) Integrated implantable device for longterm glucose monitoring. *Biosens. Bioelec.* **10**: 485–494.

Williams, R. A. and Blanch, H. W. (1994) Covalent immobilization of protein monolayers for biosensor applications. *Biosens. Bioelec.* **9**: 159–167.

Williams, R. J., Narayanan, N., Casay, G. A., Lipowska, M., Strekowski, L., Patonay, G., Pecalta, J. M. and Tsang, V. C. W. (1994) Instrument to detect near-infrared fluorescence in solid-phase immunoassay. *Anal. Chem.* **66**: 3102–3107.

Willner, I., Riklin, A., Shoham, B., Rivenzon, D. and Katz, E. (1993) Development of novel biosensor enzyme electrodes: glucose oxidase multilayer arrays immobilized onto self-assembled monolayers on electrodes. *Adv. Mater.* **5**: 912–915.

Willner, I. and Riklin, A. (1994) Electrical communication between electrodes and NAD(P)+-dependent enzymes using pyrroloquinolinequinone-enzyme electrodes in a self-assembled monolayer configuration: design of a new class of amperometric biosensors. *Anal. Chem.* **66**: 1535–1539.

Willner, I., Heleg-Shabtai, V., Blonder, R., Katz, E., Tao, G., Bückmann, A. F. and Heller, A. (1996) Electrical wiring of glucose oxidase by reconstitution of FAD-modified monolayers assembled onto Au-electrodes. *J. Amer. Chem. Soc.* **118**: 10321–10322.

Wilson, R. and Turner, A. P. F. (1992) Glucose oxidase: an ideal enzyme. *Biosens. Bioelec.* **7**: 165–185.

Wirth, M. J. and Fatunmbi, H. O. (1994) Spectroscopic and chromatographic characterization of a self-assembled monolayer as a stationary phase. In Pesek, J. J. and Leigh, I. E. (eds), *Chemically Modified Surfaces*. The Royal Society of Chemistry, Cambridge, UK.

Wise, D. L. (1989) *Applied Biosensors*. Butterworth, Stoneham, MA.

Wise, D. L. and Wingard, L. B., Jr. (eds) (1991) *Biosensors with Fiber Optics*. Humana Press, Clifton, NJ.

Witkowski, A., Daunert, S., Kindy, M. S. and Bachas, L. G. (1993) Enzyme-linked immunosorbent assay for an octapeptide based on a genetically engineered fusion protein. *Anal. Chem.* **65**: 1147–1151.

Witkowski, A., Kindy, M. S., Daunert, S. and Bachas, L. G. (1995) Preparation of biotinylated β-galactosidase conjugates for competitive binding assays by posttranslational modification of recombinant proteins. *Anal. Chem.* **67**: 1301–1306.

Wittstock, G., Yu, K., Halsall, H. B., Ridgway, T. H. and Heineman, W. R. (1995) Imaging of immobilized antibody layers with scanning electrochemical microscopy. *Anal. Chem.* **67**: 3578–3582.

Wohltjen, H., Ballentine Jr, D. S. and Jarvis, N. L. (1989) Vapor detection with surface acoustic wave microsensors. In Murray, R. W., Dessy, R. E., Heineman, W. R., Janata, J. and Seitz, W. R. (eds), *Chemical Sensors and Microinstrumentation*. ACS Symposium Series #403, American Chemical Society, Washington, DC.

Wolfbeis, O. S. (1990) Chemical sensors–survey and trends. *Fresenius' J. Anal. Chem.* **337**: 522–527.

Wolfbeis, O. S. (ed.) (1991) *Fiber Optic Chemical Sensors and Biosensors*, 2 Vols. CRC Press, Boca Raton, FL.

Wolowacz, S. E., Yon Hin, B. F. Y. and Lowe, C. R. (1992) Covalent electropolymerization of glucose oxidase in polypyrrole. *Anal. Chem.* **64**: 1541–1545.

Wong, J. Y., Kuhl, T. L., Israelachvili, J. W., Mullah, N. and Zalipsky, S. (1997) Direct measurements of a tethered ligand–receptor interaction potential. *Science* **275**: 820–822.

Wood, K. V. and Gruber, M. G. (1996) Transduction in microbial biosensors using multiplexed bioluminescence. *Biosens. Bioelec.* **11**: 207–214.

Woolley, A. T., Hadley, D., Landre, P., deMello, A. J., Mathies, R. A. and Northrup, M. A. (1996) Functional integration of PCR amplification and capillary electrophoresis in a microfabricated DNA analysis device. *Anal. Chem.* **68**: 4081–4086.

Woolley, A. T., Sensabaugh, G. F. and Mathies, R. A. (1997) High-speed DNA genotyping using microfabricated capillary array-electrophoresis chips. *Anal. Chem.* **69**: 2181–2186.

Wright, J. D., Rawson, K. M., Ho, W. O., Athey, D. and McNeil, C. J. (1995) Specific binding assay for biotin based on enzyme channeling with direct electron transfer electrochemical detection using horseradish peroxidase. *Biosens. Bioelec.* **10**: 495–500.

Wring, S. A., Hart, J. P., Bracey, L. and Birch, B. J. (1990) Development of screen-printed carbon electrodes, chemically modified with cobalt phthalocyanine, for electrochemical sensor applications. *Anal. Chim. Acta* **231**: 203–212.

Wring, S. A., Hart, J. P. and Birch, B. J. (1991) Voltammetric behaviour of screen-printed carbon electrodes, chemically modified with selected mediators, and their application as sensors for the determination of reduced glutathione. *Analyst* **116**: 123–129.

Wring, S. A., Hart, J. P. (1992) Chemically modified, screen-printed carbon electrodes. *Analyst* **117**: 1281–1286.

Wring, S. A., Hart, J. P. and Birch, B. J. (1992) Development of an amperometric assay for the determination of reduced glutathione, using glutathione peroxidase and screen-printed carbon electrodes chemically modified with cobalt phthalocyanine. *Electroanalysis* **4**: 299–309.

Wu, H. P. (1993) Fabrication and characterization of a new class of microelectrode arrays exhibiting steady-state current behavior. *Anal. Chem.* **65**: 1643–1646.

Wu, Z., Johnson, K. W., Choi, Y. and Ciardelli, T. L. (1995) Ligand binding analysis of soluble interleukin-2 receptor complexes by surface plasmon resonance. *J. Biol. Chem.* **270**: 16045–16051.

Wyman, J. and Gill, S. J. (1990) *Binding and Linkage: Functional Chemistry of Biological Macromolecules*. University Science Books, Mill Valley, CA.

Xia, Y., Kim, E., Zhoa, X-M., Rogers, J. A., Prentiss, M. and Whitesides, G. M. (1996) Complex optical surfaces formed by replica molding against elastomeric masters. *Science* **273**: 347–349.

Xiao, K. P., Bühlmann, P., Nishizawa, S., Amemiya, S. and Umezawa, Y. (1997) A chloride ion-selective solvent polymeric membrane electrode based on a hydrogen bond forming ionophore. *Anal. Chem.* **69**: 1038–1044.

Xie, B., Danielsson, B., Norberg, P., Winquist, F. and Lundstroem, I. (1992) Development of a thermal micro-biosensor fabricated on a silicon chip. *Sens. Actuators* (B) **10**: 127–130.

Xie, B., Khayyami, M., Nwosu, T., Larsson, P-O. and Danielsson, B. (1993) Ferrocene-mediated thermal biosensor. *Analyst* **118**: 845–848.

Xie, B., Harborn, U., Mecklenburg, M. and Danielsson, B. (1994a) Urea and lactate determined in 1-μl whole-blood samples with a miniaturized thermal biosensor. *Clin. Chem.* **40**: 2282–2287.

Xie, B., Mecklenburg, M., Danielsson, B., Ohman, O. and Winquist, F. (1994b) Microsensor based on an integrated thermopile. *Anal. Chim. Acta* **299**: 165–170.

Xie, B., Mecklenburg, M., Danielsson, B., Ohman, O., Norlin, P. and Winquist, F. (1995) Development of an integrated thermal biosensor for the simultaneous determination of multiple analytes. *Analyst* **120**: 155–160.

Xie, Y., Lui, T. Z. and Osteryoung, J. G. (1996) Steady-state limiting currents determined by coupled diffusion, migration, and chemical equilibrium. *Anal. Chem.* **68**: 4124–4129.

Xu, W., McDonough III, R. C., Langsdorf, B., Demas, J. N. and DeGraff, B. A. (1994) Oxygen sensors based on luminescence quenching: interactions of metal complexes with the polymer supports. *Anal. Chem.* **66**: 4133–4141.

Xu, W., Schmidt, R., Whaley, M., Demas, J. N., DeGraff, B. A., Karikari, E. K. and Farmer, B. L. (1995) Oxygen sensor based on luminescence quenching: interactions of pyrene with the polymer supports. *Anal. Chem.* **67**: 3172–3180.

Xu, X.-H. and Yeung, E. S. (1997) Direct measurement of single-molecule diffusion and photodecomposition in free solution. *Science* **275**: 1106–1109.

Yabuki, S. and Mizutani, F. (1995) Modifications to a carbon paste glucose-sensing enzyme electrode and a reduction in the electrochemical interference from L-ascorbate. *Biosens. Bioelec.* **10**: 353–358.

Yamaguchi, S., Shimomura, T., Tatsuma, T. and Oyama, N. (1993) Adsorption, immobilization, and hybridization of DNA studied by the use of quartz crystal oscillators. *Anal. Chem.* **65**: 1925–1927.

Yamato, H., Ohwa, M. and Wernet, W. (1995) A polypyrrole/three-enzymes electrode for creatinine detection. *Anal. Chem.* **67**: 2776–2780.

Yamazoe, K. (ed.) (1991) *Chemical Sensor Technology*, Vol. 3. Elsevier, New York.

Yang, H., Leland, J. K., Yost, D. and Massey, R. J. (1994) Electrochemiluminescence: a new diagnostic and research tool. *Bio/technology* **12**: 193–194.

Yang, Lin, Saavedra, S. S., Armstrong, N. and Hayes, J. (1994) Fabrication and characterization of low-loss, sol–gel planar waveguides. *Anal. Chem.* **66**: 1254–1263.

Yang, Lin and Saavedra, S. S. (1995) Chemical sensing using sol–gel derived planar waveguides and indicator phases. *Anal. Chem.* **67**: 1307–1314.

Yang, Lin, Saavedra, S. S. and Armstrong, N. R. (1996) Sol–gel-based planar waveguide sensor for gaseous iodine. *Anal. Chem.* **68**: 1834–1841.

Yang, Liu and Murray, R. W. (1994) Spectrophotometric and electrochemical kinetic studies of poly(ethylene glycol)-modified horseradish peroxidase reactions in organic solvents and aqueous buffers. *Anal. Chem.* **66**: 2710–2718.

Yang, Liu, Janle, E., Huang, T., Gitzen, J., Kissinger, P. T., Vreeke, M. and Heller, A. (1995) Applications of 'wired' peroxidase electrodes for peroxide determination in liquid chromatography coupled to oxidase immobilized enzyme reactors. *Anal. Chem.* **67**: 1326–1331.

Yang, M., Chung, F. L. and Thompson, M. (1993) Acoustic network analysis as a novel technique for studying protein adsorption and denaturation at surfaces. *Anal. Chem.* **65**: 3713–3716.

Yang, M. and Thompson, M. (1993) Multiple chemical information from the thickness shear mode acoustic wave sensor in the liquid phase. *Anal. Chem.* **65**: 1158–1168.

Yang, R., Naoi, K., Evans, D. F., Smyrl, W. H. and Hendrickson, W. A. (1991) Scanning tunneling microscope study of electropolymerized polypyrrole with polymeric anion. *Langmuir* **7**: 556–558.

Yang, R., Evans, D. F. and Hendrickson, W. A. (1995) Writing and reading at nanoscale with a scanning tunneling microscope. *Langmuir* **11**: 211–213.

Yaniv, D. R., McCormick, L., Wang, J. and Naser, N. (1991) Scanning tunneling microscopy of polypyrrole glucose oxidase electrodes. *J. Electroanal. Chem.* **314**: 353–361.

Ye, J. and Baldwin, R. P. (1994) Determination of amino acids and peptides by capillary electrophoresis and electrochemical detection at a copper electrode. *Anal. Chem.* **66**: 2669–2674.

Ye, L., Hammerle, M., Olsthoorn, A. J. J., Schuhmann, W., Schmidt, H.-L., Duine, J. A. and Heller, A. (1993) High current density "wired" quinoprotein glucose dehydrogenase electrode. *Anal. Chem.* **65**: 238–241.

Ye, L., Katakis, K., Schuhmann, W., Schmidt, H.-L., Duine, J. A. and Heller, A. (1994) Enhacement of the stability of wired quinoprotein glucose dehydrogenase electrode. In Usmani, A. M. and Akmal, N. (eds), *Diagnostic Biosensor Polymers*. ACS Symposium Series #556, American Chemical Society, Washington, DC.

Yeung, D., Gill, A., Maule, C. H. and Davies, R. J. (1995) Detection and quantification of biomolecular interactions with optical biosensors. *Trends Anal. Chem.* **14**: 49–56.

Yim, H.-S., Kibbey, C. E., Ma, S.-C., Kliza, D. M., Liu, D., Park, S.-B., Torre, C. E. and Meyerhoff, M. E. (1993) Polymer membrane-based ion-, gas-and bio-selective potentiometric sensors. *Biosens. Bioelec.* **8**: 1–38.

Yip, C. M., and Ward, M. D. (1994) Self-assembled monolayers with charge-transfer groups: n-mercaptoalkyl tetrathiafulvalenecarboxylate on gold. *Langmuir*, **10**: 549–556.

Yoshida, M., Shigemori, K., Sugimura, M. and Matano, M. (1993) Sensitivity enhancement of evanescent wave immunoassay. *Meas. Sci. Tech.* **4**: 1077–1079.

Yoshinobu, T., Oba, N., Tanaka, H. and Iwasaki, H. (1996) High-speed and high-precision chemical imaging sensors. *Sens. Actuators.* (A) **51**: 231–235.

Yun, S. Y., Hong, Y. K., Oh, B. K., Cha, G. S., Nam, H., Lee, S. B. and Jin, J.-I. (1997) Potentiometric properties of ion-selective electrode membranes based on segmented polyether urethane matrices. *Anal. Chem.* **69**: 868–873.

Zaks, A. and Klibanov, A. M. (1988) The effect of water on enzyme action in organic media. *J. Biol. Chem.* **263**: 8017–8021.

Zellers, E. T., Batterman, S. A., Han, M. and Patrash, S. J. (1995) Optimal coating selection for the analysis of organic vapor mixtures with polymer-coated surface acoustic wave sensor arrays. *Anal. Chem.* **67**: 1092–1096.

Zen, J.-M. and Lo, C.-W. (1996) A glucose sensor made of an enzymatic clay-modified electrode and methyl viologen mediator. *Anal. Chem.* **68**: 2635–2640.

Zhang, D., Crean, G. M., Flaherty, T. and Swallow, A. (1993) Development of interdigitated acoustic wave transducers for biosensor applications. *Analyst* **118**: 429–432.

Zhang, D., Wilson, G. S. and Niki, K. (1994) Electrochemistry of adsorbed cytochrome c_3 on mercury, glassy carbon, and gold electrodes. *Anal. Chem.* **66**: 3873–3881.

Zhang, H., Liu, H. Y., Liu, Y. C., Deng, J. Q. and Yu, T. Y. (1995) Structure of the blend membrane of regenerated silk fibroin and glucose oxidase and its application to glucose sensor. *Anal. Lett.* **28**: 1593–1609.

Zhang, S., Tanaka, S., Wickramasinghe, Y. E. and Rolfe, P. (1995) Fibre-optical sensor based on fluorescent indicator for monitoring physiological pH values. *Med. Bio. Eng. Comp.* **33**: 152–156.

Zhang, X., Zhang, W., Zhou, X. and Ogoreve, B. (1996) Fabrication, characterization, and potential application of carbon fiber cone nanometer-size electrodes. *Anal. Chem.* **68**: 3338–3343.

Zhang, Y., Hu, Y., Wilson, G. S., Moatti-Sirat, D., Poitout, V. and Reach, G. (1994) Elimination of the acetaminophen interference in an implantable glucose sensor. *Anal. Chem.* **66**: 1183–1188.

Zhao, G., Giolando, D. M. and Kirchoff, J. R. (1995a) Carbon ring-disk ultramicroelectrodes. *Anal. Chem.* **67**: 1491–1495.

Zhao, G., Giolando, D. M. and Kirchoff, J. R. (1995b) Chemical vapor deposition fabrication and characterization of silica-coated carbon fiber ultramicroelectrodes. *Anal. Chem.* **67**: 2592–2598.

Zhong, C. J. and Porter, M. D. (1995) Designing interfaces at the molecular level. *Anal. Chem.* **67**: 709A–715A.

Zhou, Q. and Swager, T. M. (1995) Fluorescent chemosensors based on energy migration in conjugated polymers: the molecular wire approach to increased sensitivity. *J. Amer. Chem. Soc.* **117**: 12593–12602.

Zhou, S. Y., Zuo, H., Stobaugh, J. F., Lunte, C. E. and Lunte, S. M. (1995) Continuous *in vivo* monitoring of amino acid neurotransmitters by microdialysis sampling with electrophoresis separation. *Anal. Chem.* **67**: 594–599.

Zhou, X. Z. and Arnold, M. A. (1995) Internal enzyme fibre-optic biosensors for hydrogen peroxide and glucose. *Anal. Chim. Acta* **304**: 147–156.

Zhou, X. Z. and Arnold, M. A. (1996) Response characteristics and mathematical modeling for a nitric oxide fiber optic chemical sensor. *Anal. Chem.* **68**: 1748–1754.

Zhou, Y., Laybourn, P. J. R., Magill, J. V. and de La Rue, R. M. (1991) An evanescent fluorescence biosensor using ion-exchanged buried waveguides and the enhancement of peak fluorescence. *Biosens. Bioelec.* **6**: 595–607.

Zhou, Y., Magill, J. V., de La Rue, R. M., Laybourn, P. J. R. and Cushley, W. (1993) Evanescent fluorescence immunoassays performed with a disposable ion-exchanged patterned waveguide. *Sens. Actuators* (B) **11**: 245–250.

Zhu, Z. Y. and Yappert, M. C. (1994) Sensitivity enhancement in capillary/fiber-optic fluorometric sensors. *Anal. Chem.* **66**: 761–764.

Zook, L. A. and Leddy, J. (1996) Density and solubility of Nafion: recast, annealed, and commercial films. *Anal. Chem.* **68**: 3793–3796.

Zoski, C. G. (1996) Steady-state voltammetry at microelectrodes. In Vanýsek, P. (ed.), *Modern Techniques in Electroanalysis*. John Wiley and Sons, New York.

Zupan, J. and Gasteiger, J. (1993) *Neural Networks for Chemists*. VCH, New York.

Index

In this index, *italic* page numbers designate figures; page numbers followed by "t" designate tables.